Decision science

Decision science

Theory and applications

edited by:
G.D.H. Claassen
Th.H.B. Hendriks
E.M.T. Hendrix

Mansholt publication series - Volume 2

ISBN 978-90-8686-001-2
ISSN 1871-9309

First published, 2007
Reprint, 2010
Second revised reprint, 2013
Third reprint, 2017

Wageningen Academic Publishers
The Netherlands, 2017

Mansholt Publication Series

The Mansholt Publication Series (MPS) contains peer-reviewed textbooks, conference proceedings and thematic publications focussing on social changes and control processes in rural areas and (agri)food chains as well as the institutional contexts in which these changes and processes take place. MPS provides a platform for researchers and educators who would like to increase the quality, status and international exposure of their teaching materials or of their research output.

The Series is named after Sicco Mansholt (1908-1995), who was Minister of Agriculture in The Netherlands from 1945 until 1958. From 1958 until 1972 he was Commissioner of Agriculture and Vice-President of the European Commission.

MPS is supported by the Mansholt Graduate School of Social Sciences (MG3S) and CERES Research School for Resource Studies for Development. The quality and contents of the Series is monitored by an interdisciplinary editorial board. MPS is published and marketed internationally by Wageningen Academic Publishers.

Table of contents

Preface

Decision making is part of daily life. Governments, organisations, workers, are faced with making decisions on a daily basis in order to solve different problems. Everyday, from the moment we wake up, we start making decisions and try to do things in the best way possible. Decisions can be automatic, i.e. programmed decisions already learned in such a way that they have been made many times before. Other decisions are non-programmed, that is, they are taken in unusual situations in which decisions are made on the basis of information or intuition. Decision Science, also called Operations Research (OR), is mainly concerned with the development and application of quantitative methods and techniques to support decision-making processes.

Courses on decision science are taught all over the world. Main topics are dealing with how to model decision situations effectively and how to find solutions to the models that can serve as aids to decision makers. The latter part is called mathematical programming and focuses on the mathematical analysis of models and the construction of solution techniques mainly based on so-called algorithms. At Wageningen University, decision science has been applied in research projects dealing with, for example, engineering, food science, agricultural planning, health care, environmental management, forestry, landscape ecology, water management and supply chain optimization.

Decision science has been taught at Wageningen University for more than thirty years already. Students at our university traditionally have a heterogeneous background coming from different countries all over the world with various study disciplines and with varying competence in mathematical abstraction. This diversity offers a continuous challenge with regard to lecturers as well as teaching material, both of which are the main sources of guidance for helping students to obtain knowledge, understanding, skills and insights. Throughout all these years, the lecturers in the Operations Research and Logistics Group have used their vast didactical experience to improve the teaching material and find a proper balance between mathematical exactness, knowledge and readability on the one hand and understanding, insights and applicability of the subjects on the other hand. Now that one of our colleagues, Theo Hendriks, is retiring from the university, the group thought it was time to gather all the material in a book.

The authors have tried to adapt the book to fit the diverse requirements without compromising the original purpose i.e. *to deliver an introduction to Operations Research; to explain how, why and when OR models and the related (basic) solution techniques can be expected to work; and to provide a basis for future study in OR topics.*

Virtually every topic in the text is illustrated by small numerical examples and graphical illustrations before going into mathematical abstraction. The result is a rich source of interesting examples, cases and explanations of the foundations of OR. It contains material for undergraduate and graduate courses. Basic knowledge of calculus, matrix algebra and probability theory is useful when studying the book. Furthermore, appendices on linear algebra and probability theory are provided.

Chapter 1 introduces the concept of modelling decision problems. Many cases are sketched and elaborated upon.
Chapter 2 to 4 constitute an interrelated text dealing with Linear Programming (LP), a frequently used technique. Chapter 2 introduces the concepts of variables, constraints, objective function and what can be concluded from solutions of a problem by performing a simple graphical analysis. Chapter 3 is devoted to the Simplex Method, an algorithm that is widely used for finding solutions of LP models. Chapter 4 deals with duality theory and its relationship with sensitivity analysis.
Chapter 5 introduces the problem of dealing with several, possibly conflicting objectives called multi-objective programming.
Chapter 6 focuses on a basic algorithm for solving LP problems in which variables are bound to take integer values.
Chapter 7 shows that many practical optimisation problems can be written as network problems. Several solution methods are outlined.
Chapter 8 shows that many optimisation problems consist of a sequence of linked or interrelated decisions which can be solved by breaking up (decomposing) the original problem into smaller, more tractable sub-problems. Problems related to a sequence of interrelated decisions are often associated with a mathematical technique called Dynamic Programming (DP). There is an illustration of which problems can be decomposed and how they can be solved using DP.
Chapter 9 deals with the interaction between modelling and solving models. Several examples are given of modelling techniques where the resulting mathematical structure can be used to find solutions in an efficient way.
Chapter 10 deals with popular types of algorithms which find good solutions for optimisation problems that are not necessarily optimal. In OR literature these methods are called heuristics.
Chapter 11 discusses several models that can be used for inventory management. Cases with a constant demand as well as cases with fluctuating demand, leading to stochastic models, are dealt with.
Chapter 12 sketches models where objective and constraint functions typically take a nonlinear shape. An outline is given of how they can be analysed and how the output of algorithms can be interpreted. Furthermore, a discussion takes place on how the effectiveness and efficiency of optimisation algorithms can be investigated using several examples of simple algorithms.
Chapter 13 shows how real-life processes can be modelled by discrete event simulation models. There is a discussion on how these models can be analysed and used for decision support.

Acknowledgments
We would like to thank all those people without whom this book would not have been possible. A crucial factor has been the patience of and stimulating collaboration with Ria Slootman-Vermeer, who has done the typing, and checked the lay-out and style of the manuscript. We thank Jaap Bijkerk for creating a homogeneous style for the figures that originate from various sources and Wageningen Academic Publishers for stimulating and facilitating the publication of the book.

The editors

List of notations

The definition of symbols below is followed generally through the book. Occasionally, there may be some deviations where appropriate. First the general notation is given and then chapter specific symbols are introduced.

- Italic, lower case letters are used to represent scalars
- Italic, lower case and underlined letters are used to represent vectors. Vectors are default column vectors, unless otherwise noted
- Italic, upper case letters are used to represent matrices
- Italic, letters with tilde denote random variables, e.g. $\tilde{x}$, $\tilde{r}$, $\tilde{d}$
- A prime or superscript letter T is used to denote the transpose of a matrix or a vector, e.g. A', $\underline{c}'$ or A^T, $\underline{c}^T$

$\mathbb{R}$	:	set of real numbers
$\mathbb{N}$	:	set of natural numbers
$\varnothing$	:	an empty set
$\mathbb{R}^n$	:	real space of dimension n
$\underline{c} \in \mathbb{R}^n$	:	vector $\underline{c}$ is a member of the n-dimensional real space
$\underline{x}$	:	vector of decision variables
x_i	:	the i-element of vector $\underline{x}$
A	:	coefficient matrix in a mathematical programming problem
$A^{m \times n}$	:	matrix A with m rows, n columns
A^{-1}	:	the inverse of matrix A
a_{ij}	:	the element in row i and column j of matrix A
α^n	:	a scalar α raised to the power n
E	:	the unit matrix
E_n	:	the unit matrix $E^{n \times n}$
$\underline{e}_1, \underline{e}_2$,	:	unit vectors
$\mid$	:	such that. Example: $\{\underline{x} \mid A\underline{x} \leq \underline{b}\}$ means the set of all $\underline{x}$ such that $A\underline{x} \leq \underline{b}$ holds
$f(x)$	:	a function (value)
$f'(x)$	:	the first derivative of function $f(x)$
$f''(x)$	:	second derivative of function $f(x)$
$\sum_{i=1}^{I} a_i$	:	$a_1 + a_2 + a_3 + \ldots + a_I$
$\sum_i a_i$	:	$a_1 + a_2 + a_3 + \ldots$
$\Phi(x)$	:	cumulative distribution function of standard normal $\tilde{x} \sim N(0,1)$
$E(\tilde{x})$	:	expected value of random variable $\tilde{x}$
$V(\tilde{x})$	:	variance of random variable $\tilde{x}$
$P(X)$	:	Probability $\tilde{x}$ is an element of set X

Chapter 3

$\underline{c}$	:	vector of coefficients in the objective function
A	:	coefficient matrix of a linear set of equations

B	:	basis matrix of an LP problem; it contains the basic columns of A
A_N	:	non-basic columns of A
$\underline{b}$	:	the right-hand-side vector of a mathematical programming problem
$\begin{pmatrix} \underline{x}_B \\ \underline{x}_N \end{pmatrix}$	:	basic solution corresponding to basis B
$\underline{x}_B$	:	the vector of basic variables (corresponding to basis B of an LP problem)
$\underline{x}_N$	:	the vector of non-basic variables (corresponding to basis B of an LP problem)

Chapter 6

$[x]$	:	round-down, largest integer with a value less than x
P_i	:	continuous sub-problem i
k	:	iteration counter, gives order in which sub-problems are evaluated
w_b	:	objective bound; objective function value of best feasible integer solution found thus far

Chapter 7

$G=(V,E)$	:	network G with set of nodes V and set of edges E
V	:	set of nodes
E	:	set of edges
A	:	set of arcs

Chapter 8

$[x]$	:	round-down, largest integer with a value less than x
$\underline{s}_k$	:	state vector of a system in stage k
X_k	:	set of all feasible decision $\underline{x}_k$ in stage k
T_k	:	transformation function in stage k, arguments: state and decision
G_k	:	cost function in stage k, arguments: state and decision
V_k	:	value function in stage k, argument: state

Chapter 9

λ_j	:	interpolation variable

Chapter 10

$N(\underline{x})$	:	neighbourhood of $\underline{x}$, set of solutions (neighbours) that can be reached from $\underline{x}$ in one move

Chapter 11

c	:	unit cost, the value of an item (€)
k	:	fixed ordering cost, every time an order is placed (€) or fixed set up cost, every time production is started (€)
h	:	holding cost per item per year (€/year)
r	:	carrying charge (interest rate) (€/€/year)
d	:	demand rate (units/year)
Q	:	order quantity (units)
$TC(Q)$	:	total costs per year (€) as a function of Q
$TRC(Q)$	:	total relevant costs per year (€) as a function of Q

N : number of periods
$TRCP(N)$: total relevant costs per period (€) as a function of N
T : time lapse between to consecutive orders (year)
t : time (a moment in time)
t_i : a particular moment in time $i = 1,...$
p : production rate (units/year)
T_p : time needed to make a replenishment of size Q (year)
M : maximum inventory level (units)
$I(t)$: inventory level as a function of time t (units)
$\bar{I}$: average inventory level (units)
I_S : safety stock (units)
t_l : lead time (year)
b_i : discount break (units)
s : order point (units)
S : order-up-to-level (units)
R : review time (year)
re : selling price (retail price) (€)
u : salvage value, rest value (€)
$\rho(Q)$: expected profit when ordering Q (€)
α : service level (%)
Basic relations:
h : $r \times c$ (€/year)
Q : $d \times T$ (units)

Chapter 12

$\nabla f(\underline{x})$: gradient, vector of partial derivatives $\nabla f(\underline{x}) = \left(\frac{\partial f}{\partial x_1}(\underline{x}), \frac{\partial f}{\partial x_2}(\underline{x}), \ldots, \frac{\partial f}{\partial x_n}(\underline{x}) \right)^T$
$H_f(\underline{x})$: Hessean, matrix with second order derivatives of f

$$H_f(\underline{x}) = \begin{pmatrix} \frac{\partial^2 f}{\partial x_1 \partial x_1}(\underline{x}) & . & . & . & . & \frac{\partial^2 f}{\partial x_1 \partial x_n}(\underline{x}) \\ . & . & . & . & . & . \\ . & . & . & . & . & . \\ . & . & . & . & . & . \\ \frac{\partial^2 f}{\partial x_n \partial x_1}(\underline{x}) & . & . & . & . & \frac{\partial^2 f}{\partial x_n \partial x_n}(\underline{x}) \end{pmatrix}$$

$g_i(\underline{x})$: constraint function
$\lfloor x \rfloor$: round-down, largest integer with a value less than or equal to x
$\lceil x \rceil$: round-up, smallest integer with a value bigger than or equal to x

Chapter 13

X_i : sample of generated random number or observation
$\overline{X}(n), \overline{X}$: mean (average) of data
$S^2(n)$: variance over data
$\hat{a}$: estimate of a based on data

ρ_j	:	correlation between data that are j points apart
X^2	:	test statistic of the chi-square test
χ^2	:	critical value of the chi-square test
$\chi^2_{p,\alpha}$	:	critical value of the chi-square test with p degrees of freedom and significance level α
R_i	:	sample of observations
N	:	number of observations
N_j	:	number of observations in replication j
$Y_{i,j}$	:	observation j in replication i
$W_{i,j}$	:	waiting time of customer i in replication j
α	:	significance level

1. General introduction to operations research

Th.H.B. (Theo) Hendriks

Operations Research (OR) can be described as 'a scientific way to support decision processes'. The aim of OR is to support problem solving. The application area is very wide and includes logistics, telecommunications, production planning, health care, environment, nature conservation and land use planning. Several application areas are discussed in the following sections.

1.1. History of operations research

Operations Research is a young scientific discipline. During the Second World War, OR groups were set up in the U.K. and U.S.A. to support military operations. Successes included the support of the co-ordinated use of radar systems, the warfare against submarines and the design of strategies for bombardments. The members of these first OR groups were established scientists from various areas such as physics, statistics and psychology. Specific supportive OR methods and techniques still had to be developed. In order to find solutions to their problems, scientists introduced ingenious ideas and quantitative and logic reasoning.
After the Second World War, the OR approach helped to reshape the society by solving complex problems. Large companies (oil, steel, automotive industry, telecommunications, air transport) established central OR groups to support strategic managerial problems.
During the sixties, several calculation methods were developed for solving specific problems. These methods were refined later and many new methods and application areas appeared. The developments of computers, information and communication technology had an enormous impact on OR. Large amounts of data became available for problem analysis and decision support. The availability of computers in daily life had a substantial impact on the practical progress of OR.

In the 21st century, many application areas and challenges remain for OR. The world has become more complex due to amongst other things globalisation and competition. For private companies as well as for governments the application of OR for supporting daily decisions, may be profitable.

1.2. Characterisation of operations research

One of the main characteristics of Operations Research (also called Management Science or Decision Science) is the attempt to quantify aspects of decision problems with abstract (mathematical) models.

The following stages can be identified (see Figure 1.1):

1. Problem analysis: analyse the situation with which the decision process is concerned.
2. Model building: construct a (mathematical) model that describes reality.
3. Model analysis: analyse the model and use calculation methods to generate optimal or high quality solutions for the problem.

4. Implementation: translate the solutions of the model into solutions of the actual decision process.

Feedback between the various stages is of course necessary.

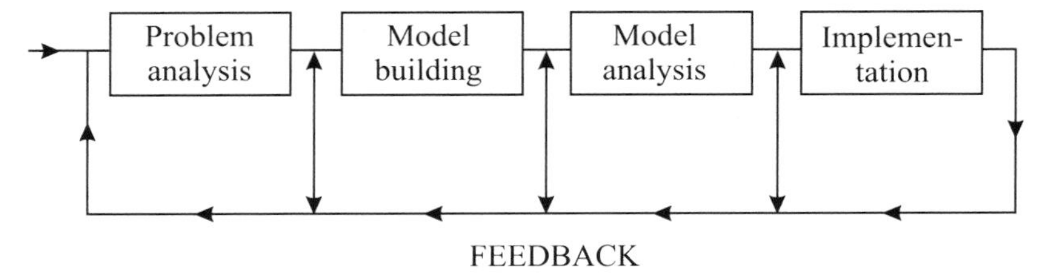

Figure 1.1. Four stages of applying models to support decision processes.

A model is an abstract description of a decision situation. In the model decisions are represented by variables. The interrelations between variables and the quality of decisions are represented by mathematical relations. We mention a number of motives for constructing models:

- Building a model requires profound insight into the decision situation. Therefore, the focus on the problem itself leads to better insight into the decision situation and part of the problem is already solved. On the other hand, the exercise of building a model often reveals relationships that are scarcely apparent to decision makers. As a result, there is an increase in the insight and understanding of the object being modelled.
- Models can be designed to evaluate decisions in a simple and cheap way. Experimentation is possible with a model whereas it is often undesirable to experiment with the object being modelled. Moreover, models can be used to answer 'what-if' questions rapidly at limited costs. Simulation, for instance, can be used to derive useful solutions on the basis of varying input.
- Models can also be designed to generate decision alternatives. Optimisation techniques (e.g. linear, integer and nonlinear programming) are usually applied.

A model is never a one-to-one image of reality. In order to handle the model and to use it to generate solutions, assumptions are necessary; e.g. nonlinear relations are approximated by linear relations. This implies that the solution of a model should be tested on practical applicability. Therefore, in practical decision problems from formulation to implementation phase, interaction is necessary between the modeller and customer (end user).
The success of most projects depends on multidisciplinary co-operation: managers, technicians, information scientists, biologists, and mathematicians all have a specific contribution. The final result of a model will never replace the decision maker. On the other hand, many (planning) problems cannot be solved without the use of quantitative techniques. Decision support systems or interactive planning tools should combine the strong points of human insight and experience on the one hand, with the efficiency and accuracy of quantitative optimisation techniques on the other hand.

The increasing success of applying quantitative techniques for practical decision problems and the interest from society in a quantitative approach is due to several factors:

- The progress of information technology implies the availability of enormous amounts of data on all kinds of aggregation levels. These data can be presented in an accessible (graphical) way.
- Better, faster, user-friendly and cheap software modules become available.
- Applying optimisation software no longer requires the use of expensive packages on mainframe computers; some software is easily accessible via internet.
- Available software on PC's and changing modes of communication facilitate the realisation of projects in co-operation with daily planners, who can take the initiative. Higher management can be confronted with working prototypes. In the past, the initiative for developing planning software was mainly due to higher management, while the planner (who has to work with the system) was confronted with a non-flexible system in the end.
- Due to practical OR successes, more students become convinced of the added value of quantitative techniques and choose to study OR subjects.
- Theory-oriented OR experts have their place in the 'ivory tower' of science and focus on the complex issue of designing optimal or approximation algorithms. The OR applicant can make use of such formal tools to solve practical problems. In the last few decades, there has been a shift from more theory-oriented research to research that has been inspired by practical applications.

1.3. Some application areas of operations research

In this section several application areas of OR are outlined.

Telecommunication

Telecommunications is a fast growing industry. In many countries, governments are aware that in the long run economic progress depends on the construction of a modern telecommunication network. As various types of data are to be transported through the network (data, voice, video, etc.), the design of such a network has many requirements. OR can play an important role in the design of an efficient and robust network.

Environmental management

For several years now, a worldwide discussion has been taking place on the greenhouse effect. The greenhouse effect is caused by so-called greenhouse gases, carbon dioxide (CO_2) being the most important among them. The reduction of CO_2 emissions can be achieved by many measures. Every method has its own costs and benefits. OR models can be used to choose the combination of measures such that the effect is maximised while the reduction costs are as low as possible. Another environmental problem in the Netherlands, the manure problem, requires choices with respect to the design of animal housing, means of transport, methods of storage, elaboration technology, etc. For this topic and other environmental areas, OR models have been and can be applied.

Aviation

Airlines have been among the first companies to apply OR techniques. Increasing competition requires very efficient planning. In aviation, many application areas of OR can be identified:

- Efficient use of the fleet is essential for every airline company. Given a list of flights, an aircraft should be allocated to every flight. Many aspects are important, e.g. the number of seats required for every flight, requirements with respect to the maintenance schedule, applicability for transatlantic flights, etc.
- Crew scheduling: every flight requires the allocation of a crew. The allocation has many requirements: qualification of pilots for different types of airplanes, number of hours a crew is allowed to work in a week, return of the crew to their home port, etc.
- Strategic decisions, like the acquisition of planes and licenses for airports.

Water management

Our earth seems to be unique among the other celestial bodies. It has water, which covers three-quarters of its surface. Water is one of the most precious commodities in our world. It is necessary for survival and is used in agriculture, in industry and for the generation of electricity. However, only 1 percent of the world's water is usable to us. In many parts of the world, the demand and availability are not balanced throughout the year. Water managers are facing floods and shortages.
OR can support strategic decisions with respect to water management. Governments are allocating scarce financial resources to long-term and costly projects aimed at regulating the water availability in several areas, e.g. constructing dams, canals, dykes, reservoirs, etc. Good planning is essential. OR techniques can be useful in developing plans that are economically solid and fulfilling physical, environmental and political requirements.
OR applications in water management are often related to the generation of energy. For instance, OR is used for the design of release schedules for a reservoir feeding an energy plant and irrigation areas. One has to take into account the capacity of canals, the jurisdictional limitations, the water level, and the energy requirement, etc.

Forestry management

OR methods and models are used for various planning problems in forest management and wood processing. A major topic in this area is land-use planning. The authority responsible first has to decide which part of the forest is to be used for which task, e.g. wood production, recreation, conservation of bio-diversity. OR models can help in the development of plans taking into account political, economical and ecological aspects.
OR is also of importance in wood processing. It can be used, for example, to decide on efficient saw patterns to convert a tree as efficiently as possible into timber of the required dimensions.

Landscape ecology

In landscape planning, spatial claims of nature conservation and of competing land use, such as agriculture, urbanisation, infrastructure, recreation, and so on may conflict.
The optimisation of land use allocation aims to enhance the biodiversity with minimum disadvantages for the competing land uses. OR can be helpful in investigating the hypothesis that agriculture has unused potential to supply 'green services' related to biodiversity and can combine these services with food production in an economically sustainable way.

Fishery

In many countries the amount of fish that can be caught is limited (quota). Fishing companies face the question of how to distribute the quota over the fleet with respect to timing, location and type of fish. OR models deepen the insight into the consequences of various decisions, so that the quality of decision making can be improved.

Health care

In health care, OR is applied at several places in organisations. It provides tools to generate working schedules for the medical staff in hospitals. Other applications are concerned with the treatment of patients. OR models are applied to predict the length of stay for individual patients in a certain department, which indeed depends on the type of illness. Knowledge is relevant for the optimisation of occupancy rates for hospitals.
OR techniques can be applied to determine the location of health centres such that the positive effect on national health is as big as possible.

Financial management

In financial management, OR is used to design portfolios with a trade-off between expected return and risk. OR can also be used in cost accounting at financial institutes to assign costs to the different services and to compare the performance of branch-offices of a bank.

OR in chains

Every product in a shop has its own history. For example, the chain of an apple in the grocer's shop starts at the grower who owns an orchard of apple trees, usually of several varieties. After the harvest the apples are sorted, chilled and stored until the moment of further trade. The apples are packed according to the requirements of the client. All such activities can take place at different locations and moments in time. Location and timing of activities influence costs in the chain and the quality of the product. Models of OR can be used to determine what, when and where actions should be taken in different links of the supply chain. This, not only applies to the chain of hard fruit, but also for chains of meat, dairy products, PCs, etc. Quite often problems in *supply chains* are related to so-called *location – allocation problems* (see Sections 1.4.6 and 1.4.7). In Section 1.4.6 we will present a typical application of location theory in agricultural chains.

1.4. Examples of OR models

In this section, several OR models are outlined. The examples are not all chosen from the application areas as mentioned in the preceding Section 1.3.

1.4.1. Crop farming

A farmer has to decide which crops to cultivate yearly on 6 hectares (ha) of land. There are two possible crops: corn and potatoes. The yearly return per hectare of corn is € 20,000 and the return per ha of potatoes is € 10,000. The labour requirements per ha of corn and potatoes are 20 and 40 hours respectively. The available number of working hours is 200. Environmental requirements imply that at maximum two-thirds of the available surface can be

used for growing corn and crop rotation requirements tell us that only half of the area can be cultivated for potatoes.
The question is how many hectares of corn and potatoes should be cultivated, taking all restrictions into account, such that the total return w is as high as possible.

The following variables are defined:
Let x_1 be the number of ha cultivated with corn and x_2 the number of ha with potatoes. In order to describe the decision problem mathematically, the following *model* can be formulated:

$$\text{maximise } \{w = 20{,}000x_1 + 10{,}000x_2\}$$

$$\text{subject to the constraints:} \qquad (1.1)$$

$$\begin{aligned} 20x_1 + 40x_2 &\leq 200 \text{ (labour)} \\ x_1 &\leq 4 \text{ (environment)} \\ x_2 &\leq 3 \text{ (crop rotation)} \\ x_1 + x_2 &\leq 6 \text{ (arable land)} \end{aligned}$$

Moreover x_1 and x_2 are non-negative: $x_1 \geq 0$ and $x_2 \geq 0$.

The function $20{,}000x_1 + 10{,}000x_2$ should be maximised. This function is also called the *objective function*.
Problem (1.1) is an example of a *linear programming problem* (LP-problem), as both the objective function and the left-hand sides of the constraints are *linear* functions of the decision variables x_1 and x_2. The problem can be solved graphically (Figure 1.2).

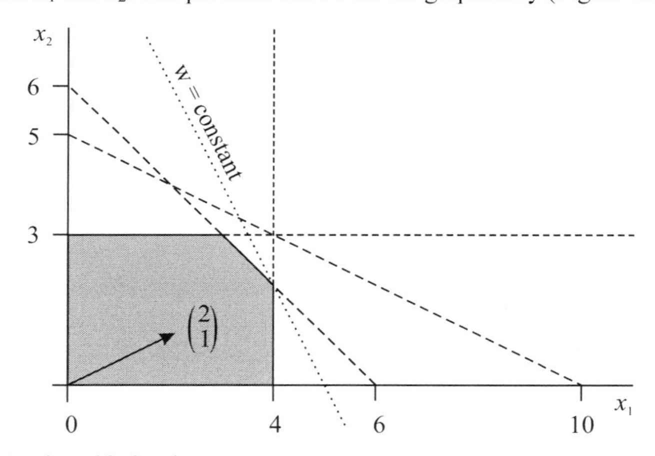

Figure 1.2. Optimal arable land use.

Figure 1.2 shows that for $x_1 = 4$ and $x_2 = 2$ the value of the objective function w is at its maximum. This maximum value $w^* = 20{,}000*4 + 10{,}000*2 =$ € 100,000.
Note that the labour constraint ($20x_1 + 40x_2 \leq 200$) is not binding. As a matter of fact, the constraint is even redundant as it does not affect the *feasible area* (the shaded area) at all. The

crop rotation constraint for potatoes ($x_2 \leq 3$) is a non-binding constraint too. However, it is not redundant!
As the optimal solution $(x_1, x_2) = (4, 2)$ is the intersection of the constraints ($x_1 + x_2 \leq 6$) and ($x_1 \leq 4$), these constraints are also called the binding constraints.
In practice the farmer can choose from a number of crops that can be sown on different days and treated in several ways, e.g. fertilising. The number of decision variables expands considerably with all these options. Taking into account all crop rotation requirements, fertilizer restrictions, weekly labour requirements, capacities of the machinery, etc. leads to large LP models (i.e. many variables and restrictions) that cannot be solved graphically. In Chapter 3, we deal with optimisation techniques for solving LP-problems.

1.4.2. Blending problems; composition of meals and diets

A dietician wants to compose a meal and can choose from n different ingredients; vegetables, rice, meat, etc. The unit costs of ingredient j is c_j Euro, ($j = 1, 2, ..., n$). The meal should be composed to meet the nutritional requirements. Assume that the requirements concern a minimum amount of $b_1, b_2, ..., b_m$ of certain vitamins necessary to compose a healthy meal.
Let a_{ij} be the amount of vitamin i in one unit of ingredient j. The meal should be composed, such that:

- the meal is as cheap as possible,
- the minimum requirements with respect to vitamins are met.

Besides vitamins, the fibre content, percentages of fat and protein, amount of minerals, etc. usually play a role. The meal composition problem or *blending problem* can be formulated mathematically as follows:

$$
\begin{array}{ll}
\text{minimise } \{w = c_1x_1 + ... c_nx_n\} & \\
\text{subject to:} & (1.2) \\
a_{11}x_1 + a_{12}x_2 + ... + a_{1n}x_n \geq b_1 & \\
a_{21}x_1 + a_{22}x_2 + ... + a_{2n}x_n \geq b_2 & \\
\quad . \quad\quad . \quad\quad . \quad\quad . \quad\quad . & \\
\quad . \quad\quad . \quad\quad . \quad\quad . \quad\quad . & \\
a_{m1}x_1 + a_{m2}x_2 + ... + a_{mn}x_n \geq b_m & \\
x_j \geq 0 \qquad\qquad \text{for all } j &
\end{array}
$$

Variable x_j represents the amount of ingredient j used in the meal. Of course the amounts x_j should be nonnegative.

Note that the objective function $w = c_1x_1 + ... + c_nx_n$ is minimised; it represents the (prime) costs. The objective function as well the left-hand sides of the constraints are linear functions of $x_1, x_2, ..., x_n$ such that problem (1.2) is an LP-problem.
For example, in the fodder industry, LP is used daily for decision-making related to the composition of concentrates for animal food. In practice, a fodder-producing company mainly produces several final products on a daily basis (concentrates for pigs, cattle, sheep, chickens, etc.). The inventory or stock levels of the available raw materials are mostly restricted to certain quantities.

1.4.3. Knapsack problems

Many OR problems can be classified as so-called *knapsack problems*. Literally this implies that a knapsack should be filled to its maximum capacity, i.e. a number of products should be selected such that the total priority (value) of the goods taken along with the knapsack is as high as possible, meanwhile not exceeding the maximum weight (or volume) of the knapsack. The volume of the knapsack might be an available budget or the length of a tree that must be cut to timber. The products in this context represent investment projects and timber assortments respectively. A small example is given with respect to sawing wood. For a more practical equivalent, see Section 1.4.9.

A straight tree, the same thickness all over, with a length of 7 meter, should be cut into assortments (see Table 1.1).

Given:

Table 1.1. Length and return of assortments.

	length	return
assortment 1	4 m	€ 20,=
assortment 2	3 m	€ 14,=
assortment 3	2 m	€ 8,=

How should the tree be cut into pieces, such that the total return w of the assortments is as high as possible?
We choose x_1 to represent the number of assortments 1; x_2 the number of assortments 2 and x_3 the number of assortments 3 cut from the tree. In this way the decision problem can be formulated as:

$$\begin{aligned}
&\text{maximise } \{w = 20x_1 + 14x_2 + 8x_3\} \\
&\text{subject to:} \\
&\quad 4x_1 + 3x_2 + 2x_3 \leq 7 \\
&\quad x_1 \geq 0,\ x_2 \geq 0,\ x_3 \geq 0 \\
&\quad x_1, x_2, x_3 \text{ integer valued}
\end{aligned} \qquad (1.3)$$

Formulation (1.3) is similar to an LP problem (linear objective function, linear constraints). The additional complication is that the values of the variables x_j should be integral. Integer LP programming is discussed in more detail in Chapter 6.

The knapsack problem can also be considered as a stepwise decision problem that can be solved by dynamic programming (Chapter 8). Therefore, the network of Figure 1.3 can be drawn. The nodes or states given at the vertical axis represent the remaining length of the tree. The arcs in the network represent the possible decisions. The numbers at the arcs represent the values of the decision variables. The numbers between brackets at the arcs represent the return values for every decision.

Optimal cutting of the tree can be determined by finding the longest path through the network (i.e. the path that corresponds to the highest total of return values). It can be easily seen that the optimal solution is $x_1 = 1$, $x_2 = 1$ and $x_3 = 0$ with a total return of $w = 34$.

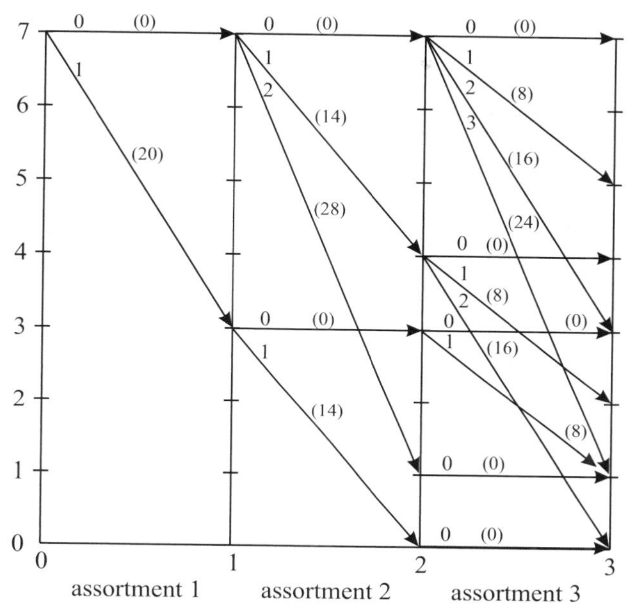

Figure 1.3. Network for determining the optimal cross-cutting of a tree.

- At time $t = 0$, the number of assortments 1 with a length of 4 m cut from the tree is decided. Possible decisions are $x_1 = 0$ and $x_1 = 1$ with values 0 and 20.
- At $t = 1$, two states (depending on the decision taken at $t = 0$) are possible: state 7 (for $x_1 = 0$) and state 3 (for $x_1 = 1$). In both states the number of assortments x_2 with a length of 3 m is determined.
- At $t = 2$, in every possible state the feasible decisions are depicted with respect to x_3 (length 2 m).

1.4.4. Optimal order quantity

Let us assume that the demand for a good is D units every week. The products are delivered directly from inventory. Every now and then the inventory is replenished with an order quantity of Q units. Assume that the order quantity is replenished directly, i.e. there is no delivery time. The course of the inventory level is given in Figure 1.4.

In this model, two types of costs appear:

a. *Holding cost.* These are € H per unit per week. Figure 1.4 shows that the average inventory level is $Q/2$. So, the average inventory cost per week is $HQ/2$.

b. *Order cost*. These are € K for every order quantity Q independent of the size of Q. Once every Q/D weeks an order is placed, so on average D/Q times a week. The average order costs per week are KD/Q.

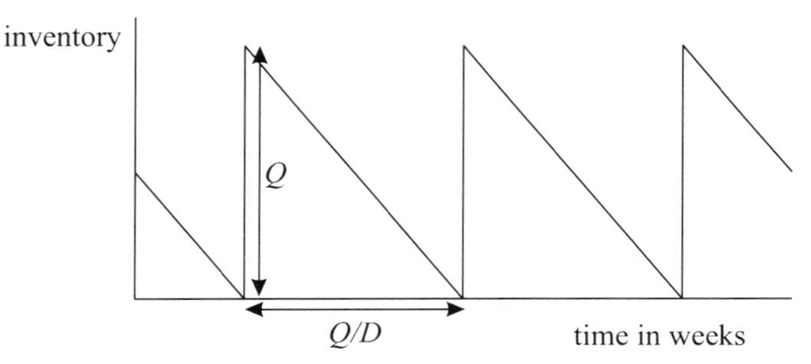

Figure 1.4. Inventory level at a constant demand per week.

It may be clear that ordering frequently (small Q), leads to high weekly order costs and low holding costs (see Figure 1.5). Ordering less frequently (large Q) gives low order costs and high holding costs. The question is: how to choose the size of Q, such that the average weekly costs of storage and ordering together are as low as possible?
The weekly average total costs C_{tot} are:

$$C_{tot} = \frac{HQ}{2} + \frac{KD}{Q} \qquad (1.4)$$

Let Q^* be the order quantity minimising the average total cost (1.4).
Q^* can be derived from the first order condition:

$$\frac{dC_{tot}}{dQ} = \frac{H}{2} - \frac{KD}{Q^2} = 0$$

This leads to:

$$Q^* = \sqrt{2KD/H} \qquad (1.5)$$

For the second order condition $\frac{d^2C_{tot}}{dQ^2} > 0$ holds, so Q^* is a minimum point.

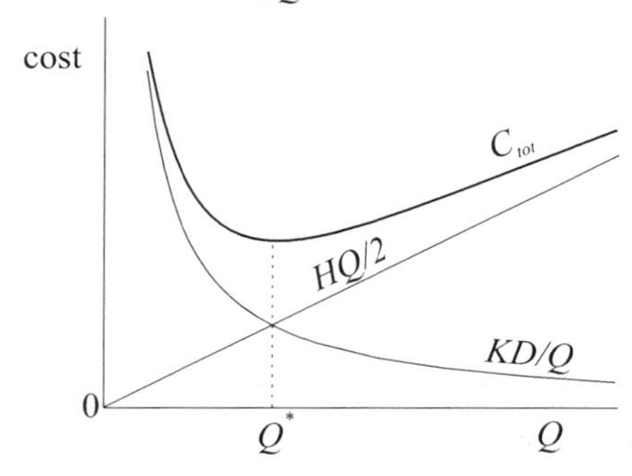

Figure 1.5. Optimal order quantity.

The model presented here, is usually too restricted for practical use. In practice the demand as well as the delivery time usually fluctuates. Moreover, quantity discounts may play a role. Equation (1.5) is widely known as '*Camps' formula*', 'EOQ' (*Economic Order Quantity*) or the '*formula of Wilson*' and applied as rule of thumb despite the unrealistic underlying assumptions (see also Chapter 11).

1.4.5. The travelling salesman problem

A salesperson wants to visit a number of cities starting from his home Wageningen (W) and finally returning to his departure site W. The question is how to determine the optimal *tour* i.e. find the sequence of the cities to visit, such that the total travel distance (or time) w is as small as possible. This well-known classical problem is called the *Travelling Salesman Problem* (TSP). The TSP has many practical extensions and occurs in various problem environments (e.g. scheduling different jobs on a production line such that the total set-up times between the orders are minimised).

Suppose:

- city 1 is the departure and return place W
- n-1 is the number of places to visit (city 2, 3, ..., n)
- d_{ij} is the distance between city i and city j $(i, j = 1 \ldots n\,;\ i \neq j)$.

To determine the optimal tour, the binary variables x_{ij} are defined: (a *binary variable* can only take the values 0 and 1)

$x_{ij} = 1$ if city j directly follows city i
$x_{ij} = 0$ if city j does not directly follow city i in the tour.

The total distance equals $\sum_{i=1}^{n} \sum_{j=1}^{n} d_{ij} x_{ij}$ and should be minimised.

Every city j is entered only once:

$$\sum_{i=1}^{n} x_{ij} = 1 \qquad \text{for all } j \text{ and } j \neq i \qquad (1.6)$$

Every city i is left exactly once:

$$\sum_{j=1}^{n} x_{ij} = 1 \qquad \text{for all } i \text{ and } i \neq j \qquad (1.7)$$

A feasible solution of (1.6) and (1.7) not always fulfils the 'tour-requirement'. Let for instance $n = 6$, then

$x_{12} = 1, x_{23} = 1, x_{31} = 1;\ x_{45} = 1, x_{56} = 1, x_{64} = 1$

is a feasible solution of (1.6) and (1.7). See Figure 1.6 for a graphical illustration of the solutuion.

Figure 1.6 shows that the given solution represents two sub tours and not one single tour. This implies that the constraints (1.6) – (1.7) are not sufficient to represent the complete problem.

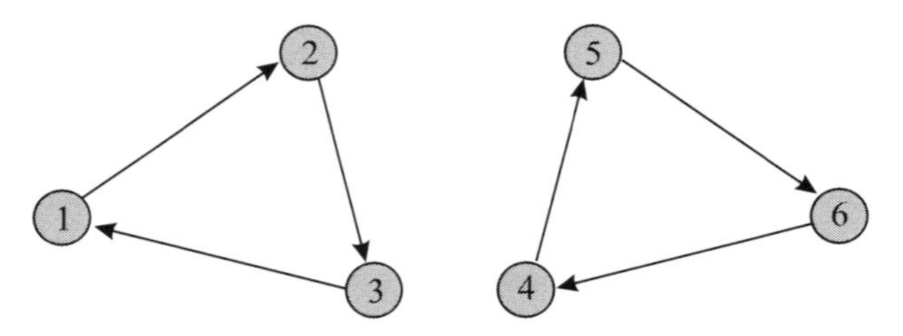

Figure 1.6. Two sub tours.

Therefore, an additional constraint (in words) should be added:

'no sub tours are allowed'

Now, the *travelling salesman problem* can be formulated as:

$$\text{minimise}\left\{ w = \sum_{i=1}^{n} \sum_{j=1}^{n} d_{ij} x_{ij} \right\}$$

subject to: (1.8)

$\sum_{i=1}^{n} x_{ij} = 1$ for all j and $j \neq i$

$\sum_{j=1}^{n} x_{ij} = 1$ for all i and $i \neq j$

'no sub tours are allowed'

$x_{ij} = 0$ or 1 for all i, j

The number of possible tours, starting from city 1 along $(n-1)$ cities equals $(n-1)!$. Namely, from city 1 the salesperson can leave to one of the other $(n-1)$ cities.
Then the next city is chosen from the $(n-2)$ remaining cities, etcetera. The total number of tours is $(n-1)$ $(n-2)$ ... (1) = $(n-1)!$. When n increases the number of possible tours will explode ($5! = 120$; $20! = 2.43 \times 10^{18}$; $69! = 1.71 \times 10^{98}$).

In practice, the TSP appears for example in vehicle routing problems (VRP) of transportation companies. Given a number of vehicles, the VRP consists of constructing delivery (and/or pick-up) routes for a fleet of vehicles at minimum costs. The capacity of each vehicle is fixed and may not be exceeded. Moreover, each vehicle must return to its departure site. Customers have a known demand that must be fully satisfied. Each customer is visited exactly once by a single vehicle. There may be constraints that limit the distance travelled by each vehicle or additional requirements regarding travel times or time windows at the customers. Another practical application of the TSP is that of production planning where the set-up cost (or time) depends on the production order.
The mathematical formulation of the travelling salesman problem is relatively simple. However, the development of fast algorithms to come to optimal or good solutions for big values of n, remains a challenge for scientific research (Lawler *et al.*, 1985).

1.4.6. Location problems

In this section the so-called *facility location problem* will be discussed. Usually the problem is to determine an optimal subset of sites for locating production sites. First we will formulate the problem mathematically. At the end of this section we will present a typical application of location theory in *supply chains*.

For the construction of one or more new distribution centres (facilities), the management of a company has $i = 1...I$ potential locations at its disposal. How many facilities should be constructed and how big should they be? Where should they be located?

In order to understand this, two types of costs are relevant:
- Yearly exploitation costs for each of the facilities (personnel, depreciation, etc.)
- Yearly costs for the transportation of goods from facility i to $j = 1...J$ clients.

In the following, the index i represents the location of a facility and the index j represents a specific client. The transportation costs from a facility to any client are assumed to be proportional to the transported quantity. A client is supplied by only one facility. Furthermore, it is assumed that the exploitation costs of a facility consist of a fixed amount plus an amount that is proportional to the flow of goods handled by the facility (the so-called throughput).

Some notations:

i : index for locations or facilities; $i = 1, 2, ..., I$
j : index for clients; $j = 1, 2, ..., J$
D_j : yearly demand of client j
α_i : yearly variable cost of facility i
c_{ij} : costs to transport the demand D_j from location i to client j
F_i : yearly fixed costs of facility i
T_i : yearly throughput (a decision variable) of facility i

The binary variable y_i indicates whether at location i a facility will be constructed or not. If $y_i = 0$ no facility will be built and $y_i = 1$ represents the decision to build a facility at location i. The variable x_{ij} represents whether client j is supplied by facility i. If $x_{ij} = 1$ client j obtains the complete demand D_j from i, if $x_{ij} = 0$ it does not. In this way, the total yearly throughput T_i of location i is determined by:

$$T_i = \sum_{j=1}^{J} D_j x_{ij} \tag{1.9}$$

The yearly transportation costs are: $\sum_{i=1}^{I} \sum_{j=1}^{J} c_{ij} x_{ij}$

The yearly exploitation costs are: $\sum_{i=1}^{I} (F_i y_i + \alpha_i T_i)$

Remarks
- If no facility is opened at location i, no clients can be supplied from i, so $x_{ij} \leq y_i$ for all i,j.

- Client j can only be supplied from a single facility:

$$\sum_{i=1}^{I} x_{ij} = 1 \qquad \text{for all } j$$

Values for the variables y_i and x_{ij} have to be determined, such that the total yearly costs are as low as possible.

The total costs can be written as:

$$\sum_{i=1}^{I}\sum_{j=1}^{J} c_{ij}x_{ij} + \sum_{i=1}^{I}(F_i y_i + \alpha_i T_i) \text{ using equation (1.9):}$$

$$\sum_{i=1}^{I}\sum_{j=1}^{J}(c_{ij} + \alpha_i D_j)x_{ij} + \sum_{i=1}^{I} F_i y_i$$

The complete problem can be formulated as:

$$\text{minimise} \left\{ \sum_{i=1}^{I}\sum_{j=1}^{J}(c_{ij} + \alpha_i D_j)x_{ij} + \sum_{i=1}^{I} F_i y_i \right\}$$

subject to: (1.10)

$$x_{ij} \le y_i \qquad \text{for all } i, j$$

$$\sum_{i=1}^{I} x_{ij} = 1 \qquad \text{for all } j$$

x_{ij}, y_i are binary variables for all i,j

Problem (1.10) is the so-called *uncapacitated facility location problem*. An efficient algorithm exists for solving this problem (Erlenkotter,1978).

A location problem for bio-ethanol

The first oil crisis in 1974 followed by the second ongoing crisis in 2000, made the world recognize the dangers of its high level of energy dependence and more in particular its addiction to oil.

Nowadays, the search for resources to substitute fossil fuels is becoming more and more important. A potential substitute for oil may be ethanol. There are many widely-available raw materials from which bio-ethanol can be produced such as grains, sugar beets, corn, sorghum and other crops. Moreover, there are numerous environmental advantages to using alcohol as a fuel, particularly with regard to the emission of lead, CO_2, SO_2, particulates, hydrocarbons and CO. Simple hydrocarbon fuels like ethanol or methanol burn clearly, forming mostly CO_2 and water which have less polluting emissions than those from fossil fuels.

As the prices for pure sugar are kept to the mark artificially within the European Union (EU), the use of sugar beets as a raw material for the production of bio-ethanol has been recognized in particular. Due to the new EU-policy, it is expected that the guaranteed price of white sugar will go down substantially in the near future. Given the alternative added value of sugar beets for the production of bio-ethanol as a substitute for common fuels, the question is: 'how to develop a logistical structure for the production of bio-ethanol and more in particular the sugar beet chain in the Netherlands'.

The existing situation for white sugar is that there are two big (white) sugar-producing factories in the North and one located in the South of the country. As sugar beets are only cultivated in a limited number of agricultural areas, large transportation costs are involved for transporting the voluminous raw materials to the sugar-producing companies. On the other hand, there are only three factories in the whole country, which implies that the fixed and variable production costs for expensive, large-scale production facilities are relatively low. A major problem in the white sugar chain is that the raw materials (beets) cannot be stored for a long time. Both the campaign for sugar beets and the production of white sugar are restricted to a few months of the year.

From a technical point of view, the processing technology is available to produce two intermediate products (i.e. crude sugar and crude ethanol) in single, small-scale pre-processing factories. The production of crude sugar or bio-ethanol from sugar beets near to the growing areas will reduce the transportation costs of voluminous raw materials. In contrast with beets, crude sugar and crude ethanol are less voluminous and can be stored, transported and finally processed into white sugar or pure ethanol at any time of the year in a final processing facility. Bio-ethanol can be produced in pre-processing factories throughout the year too, using all kinds of different raw materials. Moreover, the introduction of pre-processing factories as an additional link in the sugar beet chain will enable the choice of processing beets for either bio-ethanol or white sugar. The prices for both final products on the international market will be guiding.

Now, the questions are: 'is it wise to invest in pre-processing factories near to the growing areas of the raw materials, how many pre-processing and final processing companies should be constructed and at which capacity levels? Where should they be located and at which price levels of the final products is the production of white sugar or bio-ethanol worthwhile?' These decisions can be supported by developing location models.

1.4.7. Allocation problems

In this section we will discuss two examples of so-called allocation problems.

Allocation of dairy inspectors

A cooperative of dairy farmers in a province offers services to its members. One of the services is the inspection of milk. The members of the cooperative can have the milk inspected once every three or four weeks. At the inspection the milk production of every cow is measured and a sample is taken for investigation at the laboratory. The exact day of inspection is announced 24 hours in advance. The inspection takes place in the evening and following morning, which also implies that an inspector can only deal with one farmer at a time. The province is partitioned into several districts. In every district about 800 farmers (members of the cooperative) are dealt with by 50 inspectors. Farmers and inspectors live

scattered across the province, where direct connection roads are not always available due to the presence of canals and lakes.
Besides a fixed fee for every inspection, the inspector is paid travel expenses for the trips (four times) for every inspection between his home and the farm. The travel expenses make up a substantial part of the total inspection costs. The allocation of inspectors to farmers should be done in such a way that the total travel distance is as small as possible. A system to support decisions on the allocation should take into account many hard and soft requirements and circumstances:

- A fluctuating database of farmer members of the cooperative.
- Due to the voluntary character of the inspection, the farmer has limited options for shifting the inspection several days in case of his absence.
- The list of available inspectors also fluctuates. Many inspectors work part time and their skills vary; not all inspectors are able to use all equipment.
- Not every combination of a farmer and inspector is feasible due to family ties.
- An inspector should not visit the same farmer all the time.
- Some farmers have the appropriate measurement equipment available at their farm. For other farmers the inspector has to bring along the necessary equipment.

The aim of an OR investigation into this planning problem was to reduce the variable costs and to reduce the time needed for generating weekly and daily (allocation) plans. The planning requires many updates due to the absence of inspectors and farmers. Moreover, many soft constraints appear when stakeholders are confronted with the generated plans.

The main dialogue screen of the user interface in the resulting decision support system (DSS) resembles the former used planning board as much as possible. The underlying database, optimisation techniques and user interface were constructed such that the tool could be applied in nearly all districts of the province.

The complete system should not be considered as an optimizer but rather as a tool for generating high-quality plans to be used for further analyses. In this connection the various utilities of a user-friendly and fully interactive man-machine-interface are essential. Simplicity and speed of the generated plans resulted in savings of up to 20-30% on the variable costs.

Managing the waiting list for child care

Many types of child care exist for nursing the children of working or studying parents. The research reported here focuses on the day care for children under five in a nursery with several facilities. The organisation responsible aims at an occupancy rate of 90%.

The waiting list for child care is long. The lists contain (unborn) babies with a starting day for child care in the future and older children requiring day care in the short term. The day care is allocated per day period, what we will call a shift, i.e. a morning or afternoon. A week contains 10 shifts. The records of every child on the waiting list include information concerning the necessary (combination of) shifts, preferred shifts and undesirable shifts. The children that flow into the baby group obtain a guarantee to flow further into the toddler group at the same location around their second birthday maintaining the same shifts. This guarantee complicates the planning and affects the occupancy rate substantially. For example, a child may have to flow further into the toddler group, but there are not always sufficient shifts free in the age group.

The planning problem for the day care organisation is quite complicated: how to distribute the children on the waiting lists over the available facilities, groups and shifts, such that the occupancy rate is as high as possible? The final plan should comply with additional restrictions related to the maximum group size, the flow guarantee, the preference of parents for shifts, the location of facilities and the age group. Every group requires sufficient staff, where the competence of the staff has to be taken into account, the hours the employees work and their availability for certain shifts. Moreover, the people working in a particular group should not change continually.

1.4.8. Set-covering problems; designing farm maps for the reallocation of land

A design problem is outlined here for situations where it is necessary to have geographical information. The practical background is due to big land consolidation (reallocation) projects in the Netherlands during the seventies and eighties. At that time individual farms were redesigned by exchanging plots of land (initially owned by different farmers) in order to obtain a better manageable area on individual farms. In order to support decisions on a higher level, farm maps were necessary with a predefined size and scale that had to cover all the plots of one farm. Due to the scattering of the plots, it was not always possible to cover all plots on a single map, which meant that several maps were needed. The main question was how to design geographical maps so that as few maps as possible were necessary to cover all plots.

Problem in words:
Given the location of all the plots of a certain farm, determine the minimum number of maps with a fixed size such that all plots are covered by at least one map.

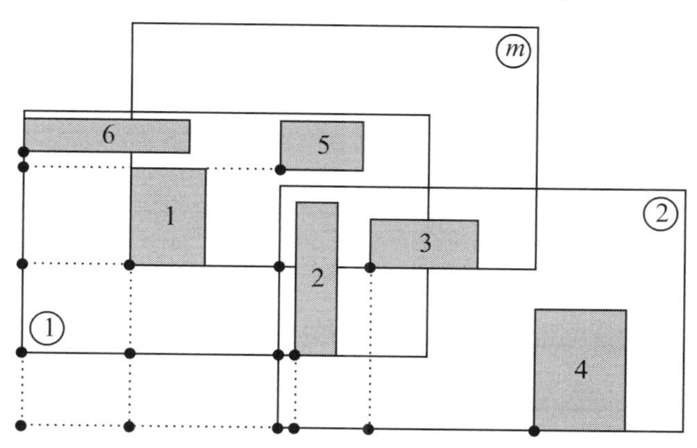

Figure 1.7. Farm maps for land consolidation.

In Figure 1.7 the plots 1, 2, 5 and 6 are covered completely by map 1; plot 3 is not completely covered by map 1. Map 2 covers plots 2, 3 and 4 completely. Map m covers plots 1, 3 and 5 completely.

The farm maps can be located everywhere in the plane. In other words: there are an infinite number of ways to locate a map. However, it can easily be seen that a logical choice for the corners of a map is to be on the interceptions of the dotted lines. The dotted lines extend the boundaries (left, right, upper and lower extremes) of the individual plots, such that the interceptions define a finite set of m possible locations (black dots in Figure 1.7).
In Figure 1.7 the 6 plots are depicted, but only three possible maps are outlined; 1, 2 and m. There are $m = 17$ interceptions (black dots) to place the lower left corner of the map.

Table 1.2 is a so-called *incidence matrix*, in which the number 1 indicates that a plot is covered by a map, e.g. plot 1 by map 1, and the value 0 means that a plot is not completely covered by a map, e.g. plot 3 and map 1.

Table 1.2. The incidence matrix.

plot	x_1 map 1	x_2 map 2	...	x_m map m
1	1	0	...	1
2	1	1	.	0
3	0	1	.	1
4	0	1	.	0
5	1	0	.	1
6	1	0	...	0

The binary variable x_i is now defined by: $x_i = 1$ means that map i is used and $x_i = 0$ means that the map is not included in the final set of maps. The total number w of required maps has to be minimised. So, $w = x_1 + x_2 + ... + x_m$ should be as small as possible. Every plot should appear on at least one map. For example, plot 1 can be represented by map 1, by map m or by both of them. Plot 1 does not appear on map 2.
Mathematically this can be written as $x_1 + 0x_2 + x_m \geq 1$. This constraint implies that at least one of the variables x_1 and x_m has a value of 1 in the final solution. Analogously, plot 2 leads to the constraint $x_1 + x_2 + 0x_m \geq 1$, and plot 6 gives $x_1 + 0x_2 + 0x_m \geq 1$.

Now, the complete model looks as follows:

$$\begin{aligned}
&\text{minimise } \{w = x_1 + x_2 + ... + x_m\} \\
&\text{subject to:} \qquad\qquad (1.11) \\
&\quad x_1 + 0x_2 + ... + x_m \geq 1 \\
&\quad x_1 + x_2 + ... + 0x_m \geq 1 \\
&\quad 0x_1 + x_2 + ... + x_m \geq 1 \\
&\quad 0x_1 + x_2 + ... + 0x_m \geq 1 \\
&\quad x_1 + 0x_2 + ... + x_m \geq 1 \\
&\quad x_1 + 0x_2 + ... + 0x_m \geq 1 \\
&\quad x_i = 0 \text{ or } x_i = 1 \qquad \text{for all } i
\end{aligned}$$

This problem is an example of a so-called *set-covering* problem and like the uncapacitated facility location problem in Section 1.4.6 it is a special case of integer linear programming.

1.4.9. Cutting problems: an example for wood processing

Wood processing is an industry converting trees into lumber and timber. We will sketch a *Decision Support System* (DSS) that has been developed specifically for this sector with respect to investment and inventory decisions.

The definition in literature of a DSS is not unique. As a working definition we state that it is a system consisting of three integrated modules (see Figure 1.8). Essential is the existence of an underlying mathematical model for decision support.

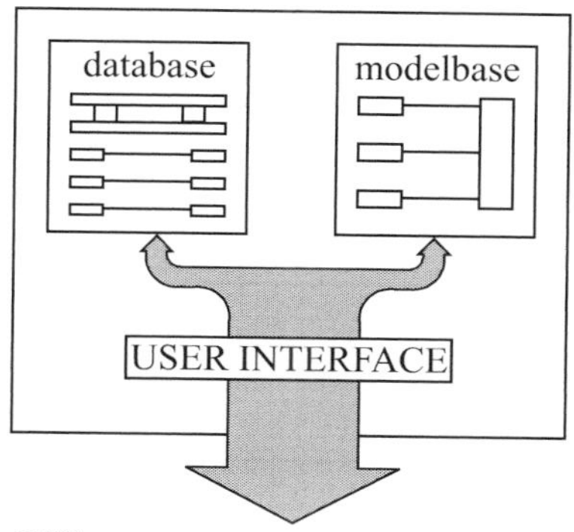

Figure 1.8. Components of a DSS.

The three components that can be found in the DSS for wood processing are:

- *Database.*
  A database contains data related to products (type, dimension, quality), raw materials (trees), semi-finished products (assortments), end products (timber), inventory and demand figures of the products, machinery specification, work force, available capacities, lay-out, etc.

- *Model base.*
  The model base is the actual core of the system. It contains OR models and solution techniques for generating high-quality plans to be used for further analyses in the user-interface.

- *User-interface.*
  The end-user of a DSS is confronted with the system via the user-interface. The user-friendliness of this component is of crucial importance for the acceptance of the DSS in practice. When designing this component of the DSS feedback is necessary from the users that have to work with the system.

The aim of the DSS is to support decisions on the strategic, tactical and/or the operational control level (Anthony, 1965):

- On the *strategic level*, the system can be used for long-term decision making related to investment decisions, extension of the workforce, product development, etc.

- On the *tactical level*, the emphasis is on effective and efficient use of all available resources. Production plans for several periods can be generated. Amongst other things, the production capacity, the demand and inventory figures of the end products (timber) and inventory figures of assortments and raw material. A complicating factor is the appearance of co-production; sawing an end product with a high value automatically generates other products that may be less valuable.
- On the *operational level*, the DSS deals with day-to-day operational decision making. For example, a tree must be sawn optimally into final end products.

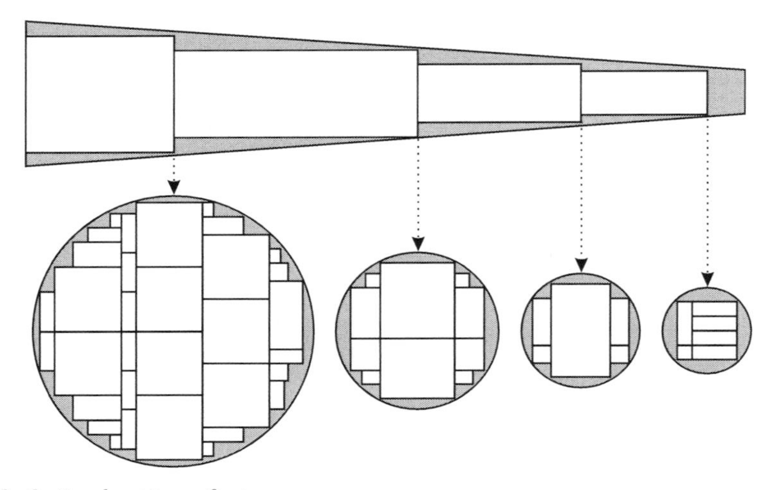

Figure 1.9. Optimal cutting of a tree.

The cutting of trees has two phases. In the first phase, a tree is crosscut into assortments (logs). In the second phase, these assortments are sawn into timber. The determination of an optimal saw pattern is a three-dimensional knapsack problem (see Section 1.4.3 and Chapter 8). Depending on the length of the logs generated in phase one, the value of the assortment is optimised by fitting rectangular end products into the circular form (see Figure 1.9). In phase one, a tree is converted to assortments. In phase two, the assortments are converted to end products. Optimal cutting of a tree implies integration of the two phases.

2. Introduction to linear programming: a graphical example

Th.H.B. (Theo) Hendriks

Ever since Linear Programming (LP) was introduced, it has been in wide use. From a certain point of view LP is one of the bedrocks of Operations Research (OR). Concepts of LP contributed to the development of other OR techniques, such as integer, stochastic and nonlinear programming. The Simplex Algorithm, developed by G.B. Dantzig in 1947 and the appearance of computers stimulated the application of LP on a large scale. Practical LP problems often contain tens of thousands of decision variables and thousands of constraints. In Chapter 3, we will deal with the background of iterative methods to find solutions for LP problems, i.e. the simplex method. In this chapter, basic concepts of LP are introduced using a simple example in only two dimensions. These kinds of LP problems can be solved graphically. The graphical solution procedure helps in understanding the simplex method as described in Chapter 3 and the interpretation of all results.

2.1. LP model formulation

Concepts and terminology for LP model formulations are introduced with a small example.

Example 2.1. The A-B production Example.

Company ABE can produce two products: product *A* and product *B*.

- The production capacity of product *A* is 4 tons and only 3 tons can be produced of product *B*.
- Product *A* and product *B* use the same raw material. Only 6 tons of raw materials are available. The production of 1 ton *A* requires 1 ton of raw materials. The same applies when producing *B*.
- The profit from product *A* is 2 units per ton; the profit from product *B* is 5 units per ton.

Problem formulation:

How many tons of products *A* and *B* should be produced such that the total profit for the company is maximised, taking all requirements into account?

This problem can be translated into a mathematical (decision) model, containing *decision variables*, an *objective function* and *constraints*:

Decision variables

The company should produce *A* and *B* in such amounts that the profit is maximised. It appears reasonable to define two decision variables denoted by x_1 and x_2 with

x_1 = number of tons of product *A* to be produced

x_2 = number of tons of product *B* to be produced

Objective function

Company ABE wants to maximise the total profit. Every ton of *A* and *B* produced results in a profit of 2 and 5 units respectively. Let w be the total profit, then the variables x_1 and x_2 should have a value such that $w = 2x_1 + 5x_2$ is maximised. The function $w = 2x_1 + 5x_2$ is called the objective function.

Constraints

The available raw materials and processing capacity limit the values of the variables x_1 and x_2.

Production capacity product *A*	:	$x_1 \le 4$
Production capacity product *B*	:	$x_2 \le 3$
Raw material	:	$x_1 + x_2 \le 6$

Moreover, the values of the variables should fulfil the *non-negativity constraints*: $x_1 \ge 0$ and $x_2 \ge 0$.

Summarised, the mathematical formulation of the *ABE problem* looks as follows:

maximise$\{w = 2x_1 + 5x_2\}$

subject to: (ABE)

$$\begin{aligned} x_1 \quad\; &\le 4 \\ x_2 &\le 3 \\ x_1 + x_2 &\le 6 \\ x_1 \ge 0,\ x_2 &\ge 0 \end{aligned}$$

Remarks

1 The ABE problem is used as a reference many times in this book. This problem is a tiny example of a *linear programming problem* (LP problem). An LP problem has an objective function and constraints that are linear functions of the variables.

2 Problem ABE is an LP problem which consists of only two decision variables (two products can be produced). In practical situations the number of decision variables is substantially larger.
- A company can produce e.g. $p = 100$ different products.
- Packing these products in 10 different boxes makes 1000 different articles.
- The company has to decide on which production line l an article is produced e.g. $l = 5$ implies the existence of 5 different production lines.
- One will also decide on which day $t = 1, 2, ..., 14$ to produce an article.

This 'practical' production planning problem can be formulated with decision variables x_{alt} = number of articles a to be produced on line l at day t ($a = 1, ... 1000$; $l = 1, ..., 5$; $t = 1, ..., 14$). The number of decision variables in this problem is $1000 \times 5 \times 14 = 70{,}000$.

2.2. Solving LP problems graphically

Problem ABE of Section 2.1 can be solved graphically, as it has only two decision variables x_1 and x_2. LP problems with 3 or more variables cannot be solved graphically. Solution methods for these problems are discussed in Chapter 3.

2.2.1. The feasible area

First the *feasible area* of the ABE example is defined by the constraints:

$$x_1 \leq 4 \quad (2.1)$$
$$x_2 \leq 3 \quad (2.2)$$
$$x_1 + x_2 \leq 6 \quad (2.3)$$
$$x_1 \geq 0, x_2 \geq 0 \quad (2.4)$$

The feasible area is situated in the positive quadrant due to the non-negativity constraints (2.4) and looks like Figure 2.1.

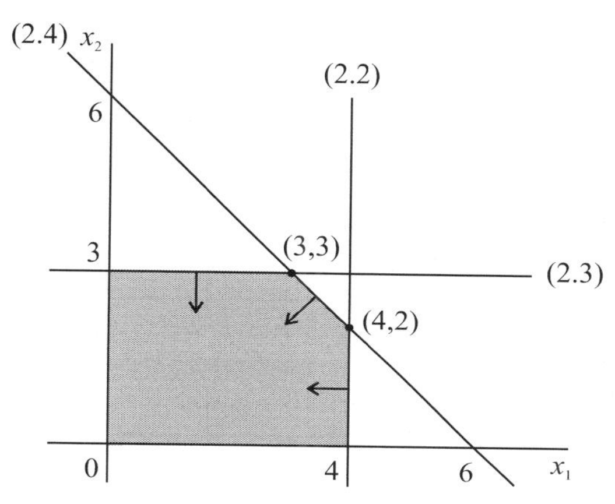

Figure 2.1. Feasible area of problem ABE.

The feasible area (shaded) in Figure 2.1 can be determined as follows.
For constraint (2.1): replace the ≤-sign by a = -sign and draw the line $x_1 = 4$. The area $x_1 \leq 4$ consists of the points at the left side of the line $x_1 = 4$ indicated by the arrow ←. The feasible area of problem ABE consists of the intersection of the areas at the left side of the line $x_1 = 4$; beneath the line $x_2 = 3$; under line $x_1 + x_2 = 6$ and the positive quadrant defined by $x_1 \geq 0$ and $x_2 \geq 0$.

2.2.2. The optimal solution

In the feasible area of Figure 2.1, all values for x_1 and x_2 should be found such that the objective function value $w = 2x_1 + 5x_2$ is at its maximum. For example:

In the origin $(x_1, x_2) = (0, 0)$ holds $w = 2x_1 + 5x_2 = 0$.
In $(x_1, x_2) = (0, 2)$ holds $w = 2x_1 + 5x_2 = 10$.
In $(x_1, x_2) = (3, 3)$ holds $w = 2x_1 + 5x_2 = 21$.

In Figure 2.2 three so-called *iso-profit lines* (i.e. $w = 2x_1 + 5x_2 =$ constant) are depicted. The corresponding constants are 0, 10 and 21.

$w = 2x_1 + 5x_2$ can be written as $w = (2, 5)\begin{pmatrix} x_1 \\ x_2 \end{pmatrix} = \underline{c}'\underline{x}$ with $\underline{c} = \begin{pmatrix} 2 \\ 5 \end{pmatrix}$ and $\underline{x} = \begin{pmatrix} x_1 \\ x_2 \end{pmatrix}$

The vector $\underline{c} = \begin{pmatrix} 2 \\ 5 \end{pmatrix}$ is called the objective vector. Vector $\underline{c}$ consists of the coefficients for the variables x_1 and x_2 in the objective function. From Figure 2.2 it follows that the iso-profit lines run parallel and are perpendicular to vector $\underline{c}$. Moving in the direction of $\underline{c}$, the objective function values increase. Finding the 'highest' iso-profit line that has at least one point in common with the feasible area, leads to the optimal solution.

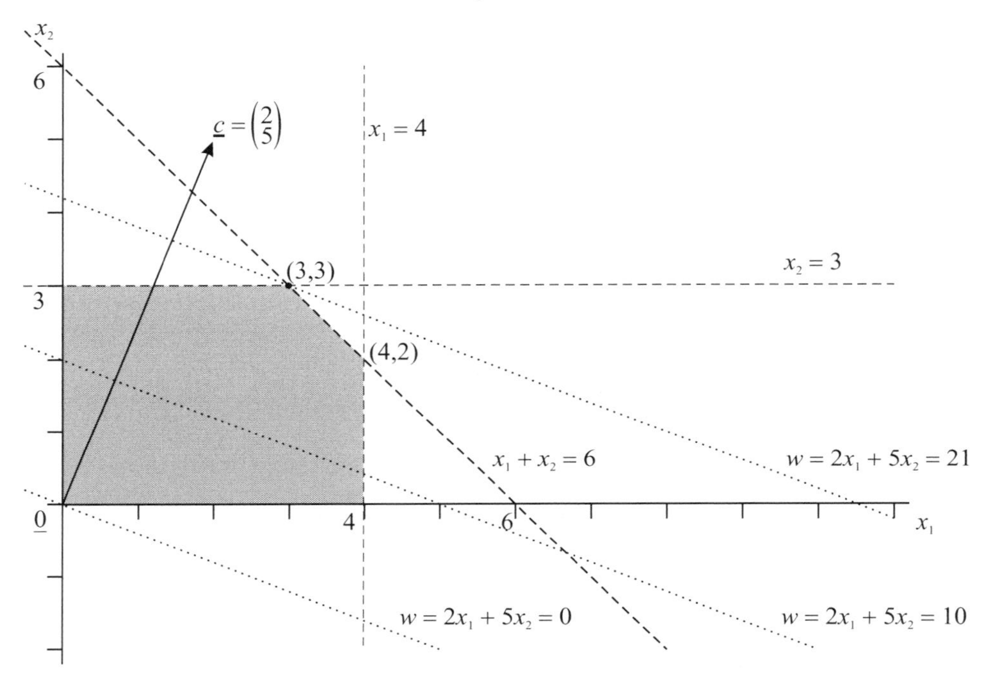

Figure 2.2. A single optimal solution.

In Figure 2.2 the optimal solution is x_1=3, x_2=3 with objective function value $w = 2x_1 + 5x_2 = 21$. One can shift the isoprofit lines perpendicular to the vector $\underline{c}$ in the direction of $\underline{c}$ up to the moment that line $w = \underline{c}'\underline{x}$ has just one point in common with the boundary of the feasible area. In vertex (3, 3) the objective value w reaches its optimal value of 21.

Remark

In Chapter 3 it is proven that if an optimal solution to an LP problem exists, then the solution is attained in a *vertex* (i.e. corner point).

The ABE example is now used to introduce some other basic concepts that are relevant for solving LP problems with more than two variables.

Concept 2.1.

A *hyperplane* V ($V \subset \mathbb{R}^n$) is the set of points with $\underline{a}'\underline{x} = \beta$ i.e. $a_1x_1 + a_2x_2 + ... + a_nx_n = \beta$.

Vector $\underline{a} \in \mathbb{R}^n$ is called the normal vector and $\beta \in \mathbb{R}$ is a scalar. The normal vector is orthogonal (perpendicular) to hyperplane V and gives the direction of increasing values of β.

In the ABE example, $w = 2x_1 + 5x_2 = (2, 5)\begin{pmatrix} x_1 \\ x_2 \end{pmatrix}$. Vector $\begin{pmatrix} 2 \\ 5 \end{pmatrix}$ is orthogonal (perpendicular) to line (hyperplane) $2x_1 + 5x_2 = \beta$ and gives the direction of increasing function values β. In Figure 2.2 the values are $\beta = 0$, $\beta = 10$ and $\beta = 21$.

Concept 2.2

The set V_1 defined by: $\underline{x} \in V_1$ satisfying $\underline{a}'\underline{x} \leq \beta$ and V_2 with $\underline{x} \in V_2$ satisfying $\underline{a}'\underline{x} \geq \beta$, are called *closed half-spaces*.

Concept 2.3

The intersection of a (finite) number of closed half-spaces is called a *polyhedral set* or *polyhedron*.

2.3. Possible solutions to LP problems

In the ABE example the feasible area is bounded and there is one optimal solution in the vertex (3, 3) with optimal objective value $w = 21$. An LP problem can have more than one optimal solution (alternative optimal solutions). Moreover, the feasible area can be unbounded; the feasible area can be empty (the LP problem has no feasible solution). This is illustrated with the following LP examples in $\mathbb{R}^2$.

2.3.1. Alternative optimal solutions

The next examples demonstrate the existence of alternative optimal solutions. We make a distinction between a line segment (Example 2.2) and a half-line (Example 2.3) of alternative optimal solutions.

Line segment of alternative optimal solutions

Example 2.2.

maximise $\{w = 2x_1 + 2x_2\}$

subject to:

$$\begin{aligned} x_1 \quad\quad &\leq 4 \\ x_2 &\leq 3 \\ x_1 + x_2 &\leq 6 \\ x_1 \geq 0, x_2 &\geq 0 \end{aligned}$$

The graphical solution is given in Figure 2.3. The iso-profit lines, perpendicular to $\underline{c} = \begin{pmatrix} 2 \\ 2 \end{pmatrix}$, are defined by the equation $2x_1 + 2x_2 = \gamma$. The value of γ corresponding to the iso-profit line that is as far as possible in the direction of $\underline{c}$ and still has points in common with the feasible area, is $\gamma = 12$. The set of optimal solutions in this case are all points located on the line segment between the vertices (3,3) and (4,2). The *line segment* of alternative optimal solutions can be described with the parameter equation (parameter λ):

$$\underline{x} = \begin{pmatrix} x_1 \\ x_2 \end{pmatrix} = \lambda \begin{pmatrix} 3 \\ 3 \end{pmatrix} + (1-\lambda) \begin{pmatrix} 4 \\ 2 \end{pmatrix}, \qquad 0 \le \lambda \le 1$$

All points on the line segment e.g. $\begin{pmatrix} 3 \\ 3 \end{pmatrix}$ for $\lambda = 1$; $\begin{pmatrix} 3\frac{1}{2} \\ 2\frac{1}{2} \end{pmatrix}$ for $\lambda = \frac{1}{2}$; or $\begin{pmatrix} 4 \\ 2 \end{pmatrix}$ for $\lambda = 0$ have the same optimal objective function value, i.e. $w = 12$.

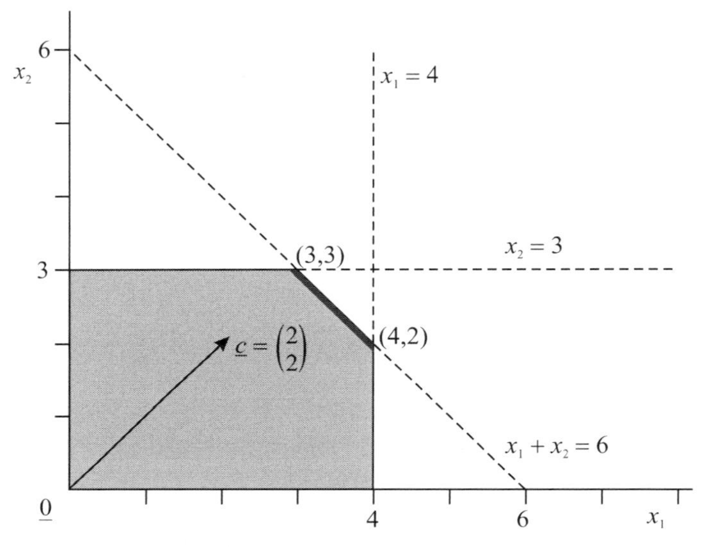

Figure 2.3. Line segment of alternative optimal solutions.

Half-line of alternative optimal solutions

Example 2.3.

maximise$\{w = -3x_1 + 3x_2\}$

subject to:

$$\begin{aligned} x_1 + \ x_2 &\ge 3 \\ -2x_1 + 2x_2 &\le 2 \\ x_1 \ge 0, x_2 &\ge 0 \end{aligned}$$

The solution set (feasible area) is an *unbounded* polyhedral set. The optimal iso-profit line is perpendicular to vector $\underline{c} = \begin{pmatrix} -3 \\ 3 \end{pmatrix}$ going through point $\underline{x} = \begin{pmatrix} 1 \\ 2 \end{pmatrix}$ such that the optimal objective value w equals 3. All points on the half-line $\underline{x} = \begin{pmatrix} 1 \\ 2 \end{pmatrix} + \lambda \begin{pmatrix} 1 \\ 1 \end{pmatrix}$, $\lambda \ge 0$ have value $w = 3$.

The general parameter equation (parameter λ) $\underline{x} = \underline{a} + \lambda \underline{b}$, $\lambda \geq 0$ is used to represent a half-line starting in $\underline{a}$ in the direction $\underline{b}$ (see Appendix A). So Example 2.3 has a *half-line* of alternative optimal solutions (see Figure 2.4).

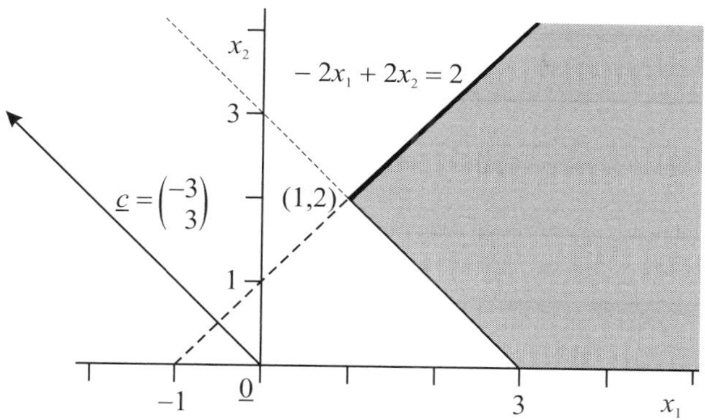

Figure 2.4. Half-line of alternative optimal solutions.

2.3.2. Unbounded solution

Next, we will give an example of an *unbounded solution.*

Example 2.4.

maximise $\{w = 4x_1 + x_2\}$

subject to:

$$\begin{aligned} x_1 + x_2 &\geq 3 \\ -2x_1 + 2x_2 &\leq 2 \\ x_1 \geq 0, x_2 &\geq 0 \end{aligned}$$

As already shown in Example 2.3 the feasible set is an unbounded polyhedron. In this example the iso-profit line, perpendicular to the vector $\underline{c}$, can be shifted infinitely in the direction of $\underline{c}$, always having points in common with the feasible area (see Figure 2.5).

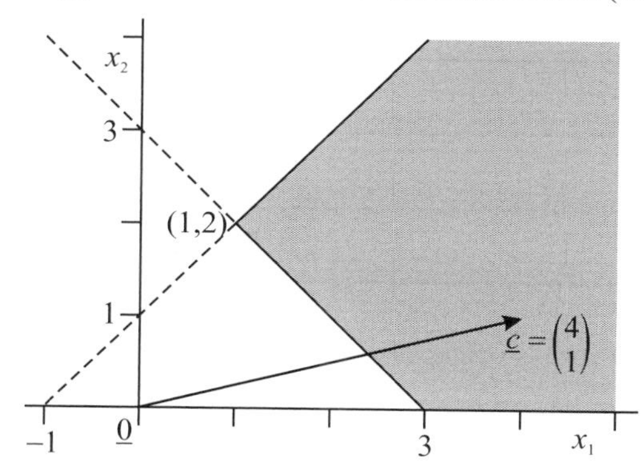

Figure 2.5. Unbounded solution.

Now, the objective function value can become arbitrarily large. For instance $\underline{x} = \begin{pmatrix} 100{,}000 \\ 100{,}000 \end{pmatrix}$ gives $w = 500{,}000$. Note that the set of optimal solutions of an LP problem with an unbounded feasible area depends on the objective function and can be either:

- one (finite) optimal solution;
- a line segment or a half line of alternative optimal solutions; or
- unbounded.

2.3.3. No feasible solution

In contrast with the Examples 2.1, 2.2, 2.3 and 2.4 LP problems can have no feasible solution at all. Such LP problems are called *infeasible*.

Example 2.5.

maximise $\{w = x_1 + 4x_2\}$

subject to:

$$\begin{aligned} x_1 \quad &\geq 4 \\ x_1 + x_2 &\leq 3 \\ x_1 \geq 0, x_2 &\geq 0 \end{aligned}$$

For Example 2.5 there are no (nonnegative) values of x_1 and x_2 that also fulfil the constraints $x_1 \geq 4$ and $x_1 + x_2 \leq 3$. The feasible area is empty and consequently no feasible solutions exist.

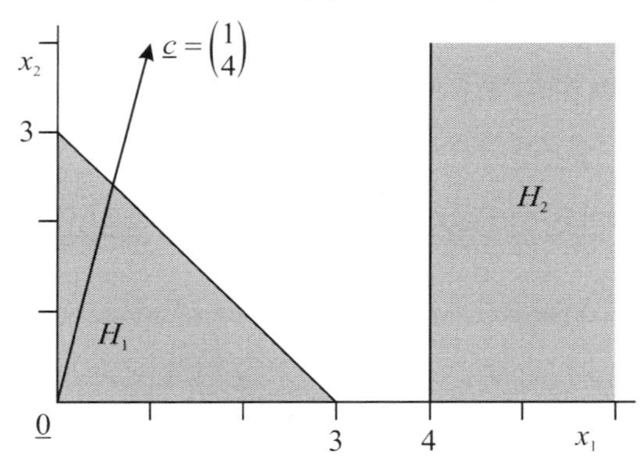

Figure 2.6. No feasible solution.

Summarising

The possible solutions of an LP problem are:

- An LP problem has one unique optimal solution — Figure 2.2
- An LP problem has alternative optimal solutions
  - Line segment — Figure 2.3
  - Half-line — Figure 2.4
- An LP problem has an unbounded solution — Figure 2.5
- An LP problem has no feasible solution — Figure 2.6

Chapter 3 discusses how the different types of solutions can be determined from the simplex tableau.

2.4. Binding and nonbinding constraints; slack variables

In Figure 2.2, the optimal solution $S = (x_1, x_2) = (3, 3)$ is the intersection of two lines, $x_2 = 3$ and $x_1 + x_2 = 6$. The raw material constraint $x_1 + x_2 \leq 6$ in the optimal solution S is exactly fulfilled: $x_1 + x_2 = 6$, i.e. all available raw material is used. There is no 'slack' (space or margin) left with regard to the available amount of raw material in the optimal vertex. A constraint with an equality sign in a point P is called *binding* in point P. Constraint $x_2 \leq 3$ is also binding in S; in S holds $x_2 = 3$.
Constraint $x_1 \leq 4$ is called *nonbinding* in $S = (3, 3)$. Slack is still present with respect to the capacity constraint of product A. The available production capacity is 4 units of A and in the optimal solution ($x_1 = 3$; $x_2 = 3$) only 3 units of A are produced i.e. there is still slack (space) for 1 unit more.

So called *slack variables* $y_i \geq 0$ can be used to transform inequalities into equality constraints.

Example 2.6.

Inequality constraint I: $x_1 + x_2 \leq 6$ is equivalent to

II: $$\begin{cases} x_1 + x_2 + y = 6 \quad \text{and} \\ y \geq 0 \end{cases}$$

- If point P e.g. $(x_1, x_2) = (3, 1)$ fulfils condition I, then P also fulfils II: $x_1 = 3$, $x_2 = 1$, $y = 2$ (so $y \geq 0$).
- If point Q does not fulfil condition I, e.g. $(x_1, x_2) = (3, 5)$, then Q neither fulfils II: $x_1 = 3$, $x_2 = 5$, $y = -2$ (so $y \geq 0$ does not hold).

The slack variable $y \geq 0$ represents the slack (surplus margin, excess), how much 'capacity' is left.

Note that:
- for any binding constraint the corresponding slack variable $y = 0$
- for any point in the interior of the feasible area, the slack variables $y_i \geq 0$ for all constraints i
- for any point outside the feasible area there will be negative slacks ($y_i < 0$) for at least one constraint i and/or at least one of the non-negativity constraints ($x_i \geq 0$) does not hold.

2.5. Sensitivity analysis

Sensitivity analysis is used to determine how sensitive the optimal solution is for changes in the original model. Usually, sensitivity analysis of LP problems refers to:

- consequences of changes in the objective vector $\underline{c}$. What happens to the optimal solution if prices and consequently objective coefficients change (Section 2.5.1)?
- consequences of changes in the constraints: the so-called *right hand side* (RHS) vector $\underline{b}$. For example: what happens to the optimal solution when more or less raw material is available (Section 2.5.2)?

When solving practical LP problems on a computer, valuable information regarding these questions becomes available. The graphical ABE example will be used to illustrate changes in objective vector $\underline{c}$ and the constraint (RHS) vector $\underline{b}$.

2.5.1. Changes in the objective vector

Consider again the ABE problem (Example 2.1 and Figure 2.2). Let the expected return of product A be 20 instead of 2. Now, the objective function to be maximised is:

$$\text{maximise}\{w = 20x_1 + 5x_2\}$$

It can be derived from Figure 2.7 that point (4, 2) is the optimal solution with $w = 20x_1 + 5x_2 = 20 \times 4 + 5 \times 2 = 90$.

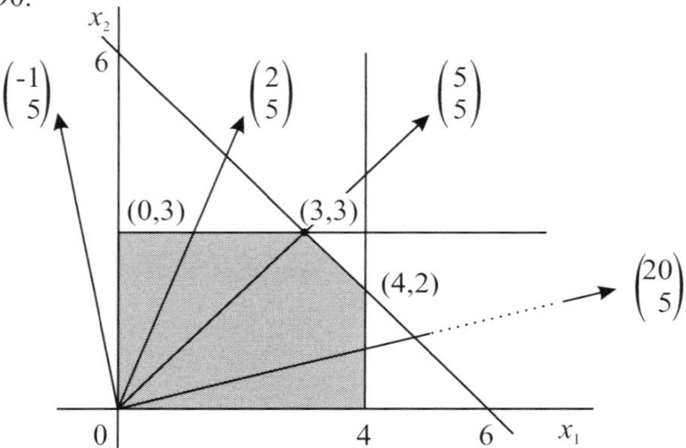

Figure 2.7. Optimal solution for different values of α.

The objective function $w = \alpha x_1 + 5x_2$ will be examined now for different values of α. This can be seen as turning the vector $\underline{c} = \begin{pmatrix}\alpha\\5\end{pmatrix}$ around the origin. The optimal value of w will be a function of the return value or parameter α of product A.

- For $\alpha < 0$, product A gives a loss. Point (0, 3) is optimal with $w = \alpha \times 0 + 5 \times 3 = 15$.
- For $\alpha = 0$, vertex (0, 3) as well as vertex (3, 3) are optimal with objective value $w = 15$. All points at the line segment between (0, 3) and (3, 3) are optimal:

  $$\underline{x} = \begin{pmatrix}x_1\\x_2\end{pmatrix} = \lambda\begin{pmatrix}0\\3\end{pmatrix} + (1-\lambda)\begin{pmatrix}3\\3\end{pmatrix},\ 0 \le \lambda \le 1 \quad \text{with} \quad w = 15.$$

  So, there are alternative optimal solutions for $\alpha = 0$.

- For $0 < \alpha < 5$ the point $(x_1, x_2) = (3, 3)$ is the optimal vertex with $w = 3\alpha + 15$.
- For $\alpha = 5$, there exist alternative optimal solutions too with $w = 30$. All points at the line segment between (3, 3) and (4, 2) are optimal:

$$\underline{x} = \begin{pmatrix} x_1 \\ x_2 \end{pmatrix} = \lambda \begin{pmatrix} 3 \\ 3 \end{pmatrix} + (1 - \lambda) \begin{pmatrix} 4 \\ 2 \end{pmatrix},\ 0 \leq \lambda \leq 1$$

- For $\alpha > 5$ the optimal solution is $(x_1, x_2) = (4, 2)$ with $w = 4\alpha + 10$.

Figure 2.8 illustrates the objective value w as function of α. These typical piece-wise linear functions are associated with the term *parametric linear programming* (Chapter 4).

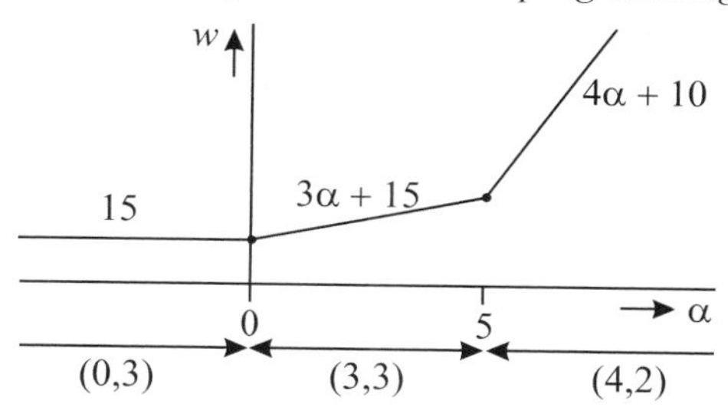

Figure 2.8. Optimal objective function value w as a function of α.

2.5.2. Changes in the constraints (RHS vector $\underline{b}$).

Together, the right hand sides in the constraints of problem ABE (capacity constraint $x_1 \leq 4$; capacity constraint $x_2 \leq 3$; raw material constraint $x_1 + x_2 \leq 6$) are called the $\underline{b}$-vector or (in literature) the RHS (Right-Hand-Side). In the ABE example $\underline{b}' = (4, 3, 6)$. The consequences of changes in the $\underline{b}$-vector will be analysed graphically.

The second constraint $x_2 \leq 3$ is taken as an example and is changed into: $x_2 \leq \beta$. The optimal objective value w is determined as a function of β with the help of Figure 2.9.

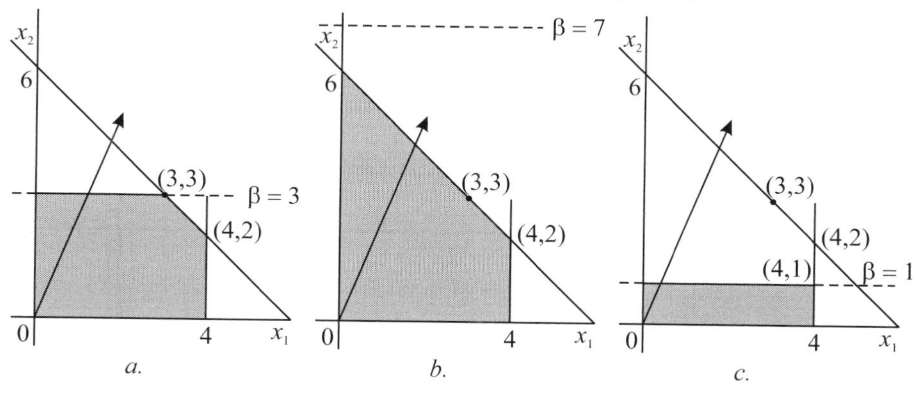

Figure 2.9. Optimal solution for varying values of β.

Problem ABE as a function of β looks as follows:

$$\text{maximise}\{w = 2x_1 + 5x_2\} \qquad (\text{ABE}_\beta)$$

$$x_1 \leq 4 \qquad (2.1)$$
$$x_2 \leq \beta \qquad (2.2')$$
$$x_1 + x_2 \leq 6 \qquad (2.3)$$
$$x_1 \geq 0, x_2 \geq 0 \qquad (2.4)$$

- For $\beta \geq 6$ the second constraint (2.2') is 'redundant'; the constraint does not influence the feasible area determined by (2.1), (2.3) and (2.4). In Figure 2.9b, $x_2 \leq 7$ is depicted. For $\beta \geq 6$ point $\underline{x}' = (0, 6)$ is the optimal vertex and $w = 2x_1 + 5x_2 = 30$.
- For $2 \leq \beta \leq 6$ the intersection of the lines $x_2 = \beta$ and $x_1 + x_2 = 6$ determines the optimal solution (constraints (2.2') and (2.3) are both binding constraints). The optimal vertex is $\underline{x}' = (6-\beta, \beta)$ with $w = 2x_1 + 5x_2 = 2(6-\beta) + 5\beta = 12 + 3\beta$. In Figure 2.8a, the optimal point is given for $\beta = 3$.
- For $0 \leq \beta \leq 2$ the vertex $(x_1, x_2) = (4, \beta)$ is optimal with $w = 2x_1 + 5x_2 = 8 + 5\beta$. Constraints (2.1) and (2.2') are binding constraints. In Figure 2.8c the optimal point is given for $\beta = 1$.
- For $\beta < 0$ no feasible solutions exist. The intersection of the half spaces defined by $x_2 \leq \beta < 0$ and $x_2 \geq 0$ is empty.

In Figure 2.10 the optimal objective function value w is given as a function of β.

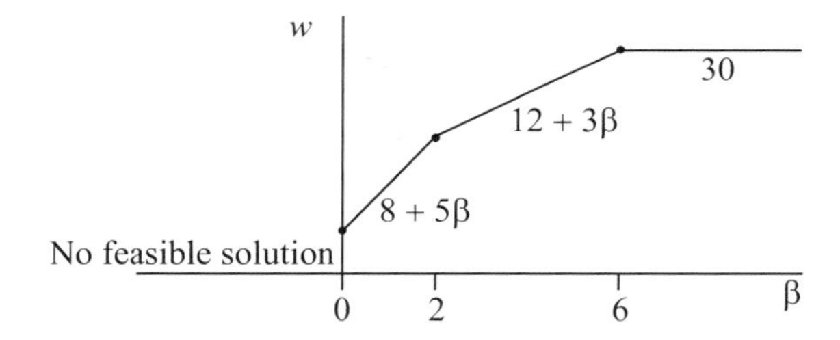

Figure 2.10. Optimal objective function value w as a function of β.

2.6. Shadow prices

The *shadow price* of a constraint in an LP problem is the change of the optimal objective value w when the RHS value of the constraint is increased by one unit (i.e. a unit change of the RHS value). At the end of this section, some comments will be made with respect to this definition. The easiest way to think about this concept is to have a maximisation problem with ≤ - constraints in mind.

The concept of shadow price is illustrated by means of the ABE example.

$$\text{maximise}\{w = 2x_1 + 5x_2\} \qquad \text{(ABE)}$$

$$x_1 \leq 4 \qquad (2.1)$$
$$x_2 \leq 3 \qquad (2.2)$$
$$x_1 + x_2 \leq 6 \qquad (2.3)$$
$$x_1 \geq 0, x_2 \geq 0 \qquad (2.4)$$

The optimal solution for problem (ABE) is $\underline{x}' = (3, 3)$ and $w = 21$.

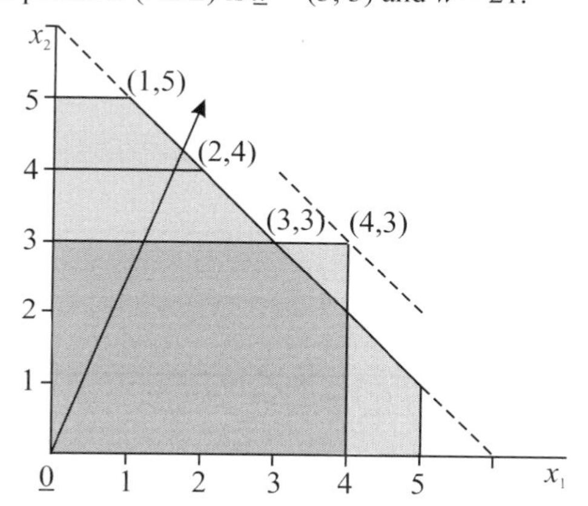

Figure 2.11. Relaxing constraints.

- Increasing the RHS of the second constraint $x_2 \leq 3$ by one unit (so the constraint becomes $x_2 \leq 4$) gives an optimal solution in $\underline{x}' = (2, 4)$ and $w = 2x_1 + 5x_2 = 24$ such that the optimal objective value increases with $\Delta w = 24 - 21 = 3$. The shadow price of the second constraint is 3. This shadow price can be derived from Figure 2.10 too; for $\beta = 3$ holds $w = 12 + 3\beta = 21$; for $\beta = 4$ holds $w = 12 + 3\beta = 24$ so $\Delta w = 3$. Increasing the RHS further towards $x_2 \leq 5$ gives the optimal solution $\underline{x}' = (1, 5)$, $w = 27$; again $\Delta w = 3$.

- Suppose we increase the RHS of the first constraint (2.1) $x_1 \leq 4$ with one unit. So it becomes $x_1 \leq 5$. Now, the vertex $\underline{x}' = (3, 3)$ remains the optimal solution with $w = 21$. The optimal objective value has not changed; the increase of the objective value is $\Delta w = 0$. The shadow price is 0 for constraint (2.1).

What determines the difference between both situations - a shadow price of 3 for constraint (2.2) and a shadow price of 0 for constraint (2.1)?

- In the first situation, constraint (2.2), i.e. $x_2 \leq 3$, is binding (slack variable $y_2 = 0$) in the optimal solution $\underline{x}' = (x_1, x_2) = (3, 3)$. This means that the available capacity for product *B* is fully used. Extension of the RHS, so enlarging the capacity for product *B* results into an improvement of $\Delta w = 3$ for the objective function value w.

- In the second situation, constraint (2.1), i.e. $x_1 \leq 4$, is nonbinding in the optimal solution $(x_1, x_2) = (3, 3)$. There is still slack with respect to the first constraint ($y_1 = 1$); the available capacity is not fully used. Extending the capacity towards $x_1 \leq 5$ is not useful and does not result in an improvement of the objective function value ($\Delta w = 0$). The corresponding shadow price is 0.

The shadow price of the third constraint (2.3); $x_1 + x_2 \leq 6$ equals 2. Relaxing the constraint towards $x_1 + x_2 \leq 7$ gives the optimal solution $x_1 = 4$, $x_2 = 3$ with objective value $w = 23$, so $\Delta w = 2$. In contrast to the second constraint, an additional relaxation (e.g. $x_1 + x_2 \leq 7\frac{1}{2}$) gives no further improvement of the objective function value. Constraint $x_1 + x_2 \leq 7\frac{1}{2}$ is no longer binding.

In general it can be stated that:

- The shadow price of a nonbinding constraint equals zero. The available scarce resources are not fully used (slack exists); extending the availability is useless.
- The shadow price of a binding constraint is non-negative in a maximisation problem with $\leq$ - constraints. The available scarce resource is fully used (zero slack). Increasing the RHS value of the corresponding constraint leads to higher production and an increase in the objective value.

Remarks

1. The shadow price represents the 'gross' increase in the objective value. Extending an RHS representing the capacity or available scarce resource will mostly involve additional costs. One can say that an extension of, for instance, the available area only makes sense if the shadow price is larger than the costs of the extension per hectare.

2. Given problem (ABE_1):

$$\begin{aligned} &\text{maximise}\{w = 2x_1 + 5x_2\} \qquad\qquad (ABE_1)\\ &x_1 \leq 3\\ &x_2 \leq 3\\ &x_1 + x_2 \leq 6\\ &x_1 \geq 0, x_2 \geq 0 \end{aligned}$$

   In the optimal solution $(x_1, x_2) = (3, 3)$ three (instead of two) constraints are binding. Extending the binding constraint $x_1 + x_2 \leq 6$ towards $x_1 + x_2 \leq 7$ does not result in an increase in the objective value w. The solution is called a *degenerate solution.* Degeneracy is discussed in Chapter 3.

3. Increasing the RHS of a constraint with one unit can be of no use due to some other reason than in remark 2. For instance, consider Figure 2.10 for $\beta = 5\frac{1}{2}$.

   Given problem (ABE_2):

$$\text{maximise}\{w = 2x_1 + 5x_2\} \qquad (\text{ABE}_2)$$

$$\begin{aligned} x_1 \qquad &\leq 3 \\ x_2 &\leq 5\tfrac{1}{2} \\ x_1 + x_2 &\leq 6 \\ x_1 \geq 0, x_2 &\geq 0 \end{aligned}$$

The optimal solution to problem (ABE_2) is $(x_1, x_2) = (\frac{1}{2}, 5\frac{1}{2})$; $w = 28\frac{1}{2}$. Increasing the RHS of the second constraint by one unit to $x_2 \leq 6\frac{1}{2}$ gives the optimal solution $(x_1, x_2) = (0, 6)$ with $w = 30$. So, $\Delta w = 30 - 28\frac{1}{2} = 1\frac{1}{2}$.

However, the shadow price in the solution $(\frac{1}{2}, 5\frac{1}{2})$ equals 3; the increase Δw equals only $1\frac{1}{2}$. The reason is that after an increase of half a unit of $x_2 \leq 5\frac{1}{2}$ the constraint $x_1 + x_2 \leq 6$ becomes binding. Shadow prices are always associated with a certain domain or range. The shadow price of the second constraint equals 3 but this value is only valid for $2 \leq \beta \leq 6$. The range [2, 6] is called the *RHS range* (see Chapter 3).

4. In Chapter 4, so-called dual variables are introduced. These variables are related to shadow prices and can often be interpreted as such.

3. The Simplex method

Th.H.B. (Theo) Hendriks

3.1. Introduction

In Chapter 1, many practical LP problems were discussed. In Chapter 2, the basic concepts of LP problems were illustrated graphically. In practice, LP problems are solved by mathematical programming software, usually based on the Simplex Method. This method, developed and introduced in 1947 by G.B. Dantzig (Dantzig, 1997), is discussed in this chapter. Basic knowledge of linear algebra as briefly reviewed in Appendix A, is necessary. Other techniques for solving LP problems will not be discussed in this textbook. We refer to Roos *et al.* (2006) for an overview on interior point methods.

3.2. LP problems in standard form

In this section we show how any LP problem can be transformed into the so-called standard form. An LP problem is a problem in which the maximum (or minimum) of a linear, so-called objective function, has to be found while a set of linear equality and inequality constraints is satisfied.

Example 3.1.

$$\text{maximise}\{w = 2x_1 - 1.3x_2 + x_3 + 5x_4 - 2.1x_5 + 0.3\}$$

subject to:

$$\begin{aligned} -0.3x_1 + 4x_2 - 2.1x_3 + x_4 + x_5 &\leq 7.5 \\ 2x_1 - 0.14x_2 + 4.5x_3 - 2x_4 &\geq -8 \\ 1.2x_1 - 2x_2 - 2.3x_3 + x_5 &= 9.3 \\ x_1 \geq 0,\ x_2 \leq 0,\ x_3 \geq 0,\ x_4 \leq 0,\ x_5 &\text{ free} \end{aligned}$$

It may be clear that general solution techniques for LP problems require uniformity in formulation. We will prove that any LP problem can be written in the so-called standard form. Methods used to solve LP problems are based on this standard form. An LP problem written in *standard form* looks like:

$$\text{maximise } \{w = c_1x_1 + c_2x_2 + \ldots + c_jx_j + \ldots + c_nx_n\}$$

subject to:

$$\begin{array}{ccccccccccc} a_{11}x_1 &+& a_{12}x_2 &+& \ldots &+& a_{1j}x_j &+& \ldots &+& a_{1n}x_n &=& b_1 \\ a_{21}x_1 &+& a_{22}x_2 &+& \ldots &+& a_{2j}x_j &+& \ldots &+& a_{2n}x_n &=& b_2 \\ \cdot && \cdot && \cdot && \cdot && \cdot && \cdot && \cdot \\ a_{i1}x_1 &+& a_{i2}x_2 &+& \ldots &+& a_{ij}x_j &+& \ldots &+& a_{in}x_n &=& b_i \\ \cdot && \cdot && \cdot && \cdot && \cdot && \cdot && \cdot \\ a_{m1}x_1 &+& a_{m2}x_2 &+& \ldots &+& a_{mj}x_j &+& \ldots &+& a_{mn}x_n &=& b_m \end{array} \qquad (3.1)$$

$$x_1 \geq 0, x_2 \geq 0, \ldots, x_n \geq 0.$$

The real numbers c_j, b_i and a_{ij} ($i = 1, \ldots, m; j = 1, \ldots, n$) are given data and the values for the variables x_j ($j = 1, \ldots, n$) should be determined. The term 'subject to' is often left out or replaced by 's.t.'. The terms 'maximise' and 'minimise' are mostly abbreviated by 'max' and 'min', respectively.

Example 3.2.

$$
\begin{aligned}
\max\{w = 3x_1 - 5x_2 + 7x_3\}& \\
-2x_1 + 3x_2 + 2.9x_3 &= 7.5 \\
4x_1 + x_2 + 5x_3 &= 9 \\
3x_1 \quad - x_3 &= 10 \\
x_1 \geq 0,\ x_2 \geq 0,\ x_3 &\geq 0
\end{aligned}
$$

Note that in the standard form all constraints are equalities and all variables should have a non-negative value. There are several ways to formulate an LP problem in standard form:

- Using summation signs:

$$
\begin{aligned}
&\max\left\{w = \sum_{j=1}^{n} c_j x_j\right\} \\
&\text{s.t.} && && (3.2) \\
&\sum_{j=1}^{n} a_{ij} x_j = b_i && \text{for all } i = 1, \dots, m \\
&x_j \geq 0 && \text{for all } j = 1, \dots, n
\end{aligned}
$$

- Using matrix notation the standard form (S) looks like:

$$
\begin{aligned}
&\max\ \{w = \underline{c}'\underline{x}\} \\
&\text{s.t.} && (S) && (3.3) \\
&\quad A\underline{x} = \underline{b} \\
&\quad \underline{x} \geq \underline{0} \quad \text{where}
\end{aligned}
$$

$$
\underline{c}' = (c_1, c_2, \dots, c_n); \quad A = \begin{pmatrix} a_{11} & a_{12} & \dots & a_{1n} \\ a_{21} & a_{22} & \dots & a_{2n} \\ . & . & \dots & . \\ . & . & \dots & . \\ a_{m1} & a_{m2} & \dots & a_{mn} \end{pmatrix}; \quad \underline{b}' = (b_1, b_2, \dots, b_m); \quad \underline{x}' = (x_1, x_2, \dots, x_n)
$$

If matrix A is partitioned into the columns $A = (\underline{a}_1, \underline{a}_2, \dots, \underline{a}_n)$ then (3.3) can be written as:

$$
\begin{aligned}
&\max\{w = c_1x_1 + \dots + c_jx_j + \dots + c_nx_n\} \\
&\quad x_1\underline{a}_1 + \dots + x_j\underline{a}_j + \dots + x_n\underline{a}_n = \underline{b} && (3.4) \\
&\quad x_1 \geq 0, \dots, x_j \geq 0, x_n \geq 0, \text{ where}
\end{aligned}
$$

$$
\underline{a}_j = \begin{pmatrix} a_{1j} \\ a_{2j} \\ . \\ . \\ a_{mj} \end{pmatrix} \quad \text{so } \underline{a}_j \in \mathbb{R}^m, \quad \text{for all } j = 1, \dots, n
$$

Depending on the context, one of the foregoing notations (3.2), (3.3) or (3.4) is used for LP problems in standard form.

Remarks

1. Unless stated otherwise, the following holds:
   $\underline{c} \in \mathbb{R}^n$, $\underline{x} \in \mathbb{R}^n$, $\underline{0} \in \mathbb{R}^n$; A is an $m \times n$-matrix and $\underline{b} \in \mathbb{R}^m$.
2. Depending on the context $\max\{w = \underline{c}'\underline{x}\}$ or $\max\{\underline{c}'\underline{x}\}$ is used.

3.3. Transforming LP problems

The LP solution technique that will be discussed uses the standard form. This section demonstrates procedures to transform any LP problem into an equivalent LP problem in standard form. The following operations or transformation rules transform any LP problem into an equivalent LP problem in the standard form:

1. Addition of a constant k to the objective function:
   $\max\{\underline{c}'\underline{x}\}$ is equivalent to $\max\{\underline{c}'\underline{x} + k\}$
2. Multiplication of an objective function $\underline{c}'\underline{x}$ with a constant k ($k > 0$):
   $\max\{\underline{c}'\underline{x}\}$ is equivalent to $\max\{k\,\underline{c}'\underline{x}\}$
3. An objective function $\min\{\underline{c}'\underline{x}\}$ can be replaced by the objective function $\max\{-\underline{c}'\underline{x}\}$ (and vice versa). Both functions have the same optimal vector $\underline{x}$; the function values have opposite signs.
4. The variable x_j with $x_j \leq 0$ can be replaced by a variable $x_j^* = -x_j$, with $x_j^* \geq 0$.
5. A free variable x_j can be replaced by $x_j = x_j^+ - x_j^-$, with $x_j^+ = \max\{x_j, 0\}$ and $x_j^- = \max\{-x_j, 0\}$ so that $x_j^+ \geq 0$ and $x_j^- \geq 0$.
6. The inequality $\underline{a}_i'\,\underline{x} \geq \beta$ can be replaced by $-\underline{a}_i'\,\underline{x} \leq -\beta$ (and vice versa).
7. The equality $\underline{a}_i'\,\underline{x} = \beta$ can be replaced by the two inequalities $\underline{a}_i'\,\underline{x} \leq \beta$ and $\underline{a}_i'\,\underline{x} \geq \beta$.
8. The inequality $\underline{a}_i'\,\underline{x} \leq \beta$ can be replaced by $\underline{a}_i'\,\underline{x} + y = \beta$, with $y \geq 0$. The variable y is called a *slack variable*.
9. Analogously $\underline{a}_i'\,\underline{x} \geq \beta$ can be replaced by $\underline{a}_i'\,\underline{x} - y = \beta$ with $y \geq 0$, (the variable y is also called a slack variable).

Example 3.3.

Given LP problem (P):

$$\begin{aligned} &\min\{20x_1 - 10x_2 + 30x_3 + 2002\} \\ &\quad 3x_1 - x_2 + 2x_3 \leq 9 \\ &\quad 2x_1 + x_2 - x_3 \geq 2 \\ &\quad -x_1 + 2x_2 + 3x_3 = 4 \\ &\quad x_1 \geq 0,\ x_2 \leq 0,\ x_3 \text{ free} \end{aligned} \qquad \text{(P)}$$

LP problem (P) can be transformed into the standard form by using the following steps:

1. Transform the objective function with the help of operations 1 and 2 into $\min\{2x_1 - x_2 + 3x_3\}$, and subsequently into $\max\{-2x_1 + x_2 - 3x_3\}$ (operation 3).
2. Add slack variable $y_1 \geq 0$ to the first restriction (operation 8): $3x_1 - x_2 + 2x_3 \leq 9$ is equivalent to $3x_1 - x_2 + 2x_3 + y_1 = 9$.
3. Add slack variable $y_2 \geq 0$ to the second restriction (operation 9): $2x_1 + x_2 - x_3 \geq 2$ is equivalent to $2x_1 + x_2 - x_3 - y_2 = 2$.
4. In the objective function as well as in the restrictions replace the variable x_2 by $x_2^* = -x_2 \geq 0$ (operation 4) and $x_3 = x_3^+ - x_3^-$ with $x_3^+ \geq 0$ and $x_3^- \geq 0$ (operation 5).

The preceding operations transform problem (P) into the equivalent problem $(\hat{\text{P}})$ in the standard form:

$$\max\{-2x_1 - x_2^* - 3x_3^+ + 3x_3^- + 0y_1 + 0y_2\} \qquad (\hat{\text{P}})$$

$$\begin{aligned} 3x_1 + x_2^* + 2x_3^+ - 2x_3^- + y_1 &= 9 \\ 2x_1 - x_2^* - x_3^+ + x_3^- - y_2 &= 2 \\ -x_1 - 2x_2^* + 3x_3^+ - 3x_3^- &= 4 \\ x_1 \geq 0,\ x_2^* \geq 0,\ x_3^+ \geq 0,\ x_3^- \geq 0,\ y_1 \geq 0,\ y_2 &\geq 0 \end{aligned}$$

3.4. Basic solutions

Given the LP problem in standard form:

$$\begin{aligned} &\max\{w = \underline{c}'\underline{x}\} \\ &A\underline{x} = \underline{b} \\ &\underline{x} \geq \underline{0} \end{aligned} \qquad (3.5)$$

We first consider the set of equalities $A\underline{x} = \underline{b}$ (see also Appendix A). The $m \times n$-matrix A is partitioned column-wise like in (3.4):

$$x_1\underline{a}_1 + x_2\underline{a}_2 + \ldots + x_j\underline{a}_j + \ldots x_n\underline{a}_n = \underline{b} \qquad (3.6)$$

where $\underline{a}_j$ is the jth column of matrix A. So $\underline{b} \in \mathbb{R}^m$ and $\underline{a}_j \in \mathbb{R}^m$, ($j = 1, \ldots, n$). From linear algebra (Appendix A) we know that:

- Space $\mathbb{R}^m$ is spanned by m linearly independent column vectors.
- A set of m linearly independent vectors is a basis of $\mathbb{R}^m$.
- Vector $\underline{b}$ and every vector $\underline{a}_j$ can be written uniquely as a linear combination of the m basic vectors.
- Rank r(A) of A equals the dimension of the space spanned by the column vectors of A. So r(A) equals the maximum number of linearly independent (column) vectors of A. Moreover, r(A) equals the dimension of the row space, such that r(A) $\leq m$ and r(A) $\leq n$. In

other words $r(A) \leq \min\{m, n\}$. The $m \times (n+1)$-matrix $(A, \underline{b})$ appears after adding vector $\underline{b}$ to matrix A. Of course $r(A, \underline{b}) \geq r(A)$.

- If $r(A, \underline{b}) = r(A)$ then system $A\underline{x} = \underline{b}$ has a solution. In this case, vector $\underline{b}$ is an element of the space that is spanned by the column vectors of A. If $r(A) = m$, then the column vectors span $\mathbb{R}^m$. Because $\underline{b} \in \mathbb{R}^m$, we have $r(A, \underline{b}) = r(A) = m$ and $A\underline{x} = \underline{b}$ is solvable.
- If $r(A, \underline{b}) > r(A)$ then system $A\underline{x} = \underline{b}$ is inconsistent (no feasible solution exists); $\underline{b}$ is not an element of the space that is spanned by the column vectors of A.

Example 3.4.

1. Suppose $A = (\underline{a}_1, \underline{a}_2, \underline{a}_3) = \begin{pmatrix} 1 & 2 & -1 \\ 2 & 4 & -2 \end{pmatrix}$; $\quad \underline{b}_1 = \begin{pmatrix} 3 \\ 4 \end{pmatrix}$; $\quad \underline{b}_2 = \begin{pmatrix} 5 \\ 10 \end{pmatrix}$

   $r(A) = r\begin{pmatrix} 1 & 2 & -1 \\ 2 & 4 & -2 \end{pmatrix} = 1, \ r(A, \underline{b}_1) = r\begin{pmatrix} 1 & 2 & -1 & 3 \\ 2 & 4 & -2 & 4 \end{pmatrix} = 2;$

   $r(A, \underline{b}_1) > r(A)$ so system $A\underline{x} = \underline{b}_1$ is inconsistent.

   $r(A, \underline{b}_2) = 1 = r(A)$ so $A\underline{x} = \underline{b}_2$ has a solution. Pivoting leads to the set of solutions.

2. Suppose $A = \begin{pmatrix} 1 & 2 \\ -1 & 3 \\ 1 & -2 \end{pmatrix}$; $\quad \underline{b}_1 = \begin{pmatrix} 3 \\ 2 \\ 7 \end{pmatrix}$; $\quad \underline{b}_2 = \begin{pmatrix} 3 \\ 2 \\ -1 \end{pmatrix}$

   $r(A) = 2$
   $r(A, \underline{b}_1) = 3 > r(A)$, so $A\underline{x} = \underline{b}_1$ has no solution (inconsistent).
   $r(A, \underline{b}_2) = 2 = r(A)$, so $A\underline{x} = \underline{b}_2$ has the unique solution $\underline{x}' = (x_1, x_2) = (1, 1)$.

3. Suppose $A = (\underline{a}_1, \underline{a}_2, \underline{a}_3, \underline{a}_4) = \begin{pmatrix} 1 & 1 & 3 & -2 \\ 2 & -1 & 4 & 6 \end{pmatrix}$; $\underline{b}$ arbitrary

   $r(A) = 2$
   $r(A, \underline{b}) = 2$, so for each $\underline{b}$ the system $A\underline{x} = \underline{b}$ is solvable.

Remark

In practise, the number of restrictions m is usually much smaller than the number of variables n. Note that slack variables are added to inequality constraints to obtain the standard form.

Assumption

In the following sections it is assumed that $n \geq m$ and $r(A) = m$. This assumption implies that:

If $r(A) = m$ then system $A\underline{x} = \underline{b}$ has a solution for all vectors $\underline{b}$ (as $r(A) = r(A, \underline{b}) = m$). (3.7)

If matrix A is partitioned into columns, a system of n column vectors $\underline{a}_j$ appears. Matrix A has rank $r(A) = m$, so there is at least one system of linearly independent column vectors of A that is a basis for $\mathbb{R}^m$. Suppose the first m column vectors $\underline{a}_j$ are linearly independent. If necessary renumber the columns.

$$A = \underbrace{(\underline{a}_1, \underline{a}_2, \quad \ldots, \quad \underline{a}_m,}_{\text{linearly independent basic vectors}} \quad \underbrace{\underline{a}_{m+1}, \quad \ldots, \quad \underline{a}_n)}_{\text{nonbasic vectors}} \tag{3.8}$$

$$\Downarrow \qquad\qquad\qquad\qquad \Downarrow$$

$$B \qquad\qquad\qquad\qquad A_N$$

In (3.8) the $m \times m$ basis matrix B is formed by the m linearly independent vectors $\underline{a}_1, \ldots, \underline{a}_m$; so B^{-1} exists. The $m\times(n-m)$ matrix A_N is formed by the remaining vectors $\underline{a}_{m+1}, \ldots, \underline{a}_n$.

Corresponding to the partitioning of A, vector $\underline{x}$ can be split:

$$\underline{x}' = \underbrace{(x_1, x_2, \quad \ldots, \quad x_m,}_{\text{basic variables}} \quad \underbrace{x_{m+1}, \quad \ldots, \quad x_n)}_{\text{nonbasic variables}} \tag{3.9}$$

$$\Downarrow \qquad\qquad\qquad\qquad \Downarrow$$

$$\underline{x}_B' \qquad\qquad\qquad\qquad \underline{x}_N'$$

The m elements $x_1, \ldots, x_m$ of vector $\underline{x}_B$ are called the *basic variables* corresponding to basis B. The $n-m$ elements $x_{m+1}, \ldots, x_n$ of vector $\underline{x}_N$ are called the *nonbasic variables*.

Theorem 3.1.

$\underline{x} = \begin{pmatrix} \underline{x}_B \\ \underline{x}_N \end{pmatrix} = \begin{pmatrix} B^{-1}\underline{b} \\ \underline{0} \end{pmatrix}$ is a solution of the system $A\underline{x} = \underline{b}$.

Proof

Partitioning according to $A = (B \mid A_N)$ and $\underline{x}' = (\underline{x}_B', \underline{x}_N')$ gives

$A\underline{x} = \underline{b} \Rightarrow (B \mid A_N)\begin{pmatrix} \underline{x}_B \\ \underline{x}_N \end{pmatrix} = \underline{b}$ or $B\underline{x}_B + A_N\underline{x}_N = \underline{b}$.

If $\underline{x}_N = \underline{0}$ then $B\underline{x}_B + A_N\,\underline{0} = \underline{b}$ or $B\underline{x}_B = \underline{b}$ with solution $\underline{x}_B = B^{-1}\underline{b}$.

So, $\underline{x} = \begin{pmatrix} \underline{x}_B \\ \underline{x}_N \end{pmatrix} = \begin{pmatrix} B^{-1}\underline{b} \\ \underline{0} \end{pmatrix}$ is a solution of $A\underline{x} = \underline{b}$.

Definition 3.1.

$\underline{x} = \begin{pmatrix} \underline{x}_B \\ \underline{x}_N \end{pmatrix} = \begin{pmatrix} B^{-1}\underline{b} \\ \underline{0} \end{pmatrix}$ is called the *basic solution* corresponding to basis B.

The basic solution corresponding to basis B is obtained by:
- giving the $n-m$ nonbasic variables the value $\underline{x}_N = \underline{0}$ and
- solving $B\underline{x}_B = \underline{b}$ or $\underline{x}_B = B^{-1}\underline{b}$ to obtain values for the basic variables.

Example 3.5.

Consider the system $A\underline{x} = \underline{b}$ with

$$A = (\underline{a}_1, \underline{a}_2, \underline{a}_3, \underline{a}_4) = \begin{pmatrix} 1 & 1 & 2 & 0 \\ 0 & 2 & 4 & 1 \end{pmatrix}; \quad \underline{b} = \begin{pmatrix} 1 \\ 3 \end{pmatrix}$$

B consists of two linearly independent vectors $\underline{a}_j$. Five different ways exist for choosing B:

1. $B = (\underline{a}_1, \underline{a}_2) = \begin{pmatrix} 1 & 1 \\ 0 & 2 \end{pmatrix}$ so $\underline{x}_B = \begin{pmatrix} x_1 \\ x_2 \end{pmatrix}$

   $A_N = (\underline{a}_3, \underline{a}_4) = \begin{pmatrix} 2 & 0 \\ 4 & 1 \end{pmatrix}$ so $\underline{x}_N = \begin{pmatrix} x_3 \\ x_4 \end{pmatrix} = \begin{pmatrix} 0 \\ 0 \end{pmatrix}$

   $\underline{x}_B = \begin{pmatrix} x_1 \\ x_2 \end{pmatrix}$ is determined by solving $B\underline{x}_B = \underline{b}$ or:

   $x_1 + \quad x_2 = 1$

   $\qquad 2x_2 = 3$ with the solution $x_1 = -1/2$; $x_2 = 3/2$.

   The basic solution is: $\underline{x}_1' = (x_1, x_2, x_3, x_4) = (-1/2, 3/2, 0, 0)$.

2. $B = (\underline{a}_1, \underline{a}_3) = \begin{pmatrix} 1 & 2 \\ 0 & 4 \end{pmatrix}$

   $\underline{x}_B = \begin{pmatrix} x_1 \\ x_3 \end{pmatrix}$ is determined by solving $B\underline{x}_B = \underline{b}$ or $\underline{x}_B = B^{-1}\underline{b}$

   $B^{-1} = \begin{pmatrix} 1 & -1/2 \\ 0 & 1/4 \end{pmatrix}$, so $\underline{x}_B = B^{-1}\underline{b} = \begin{pmatrix} 1 & -1/2 \\ 0 & 1/4 \end{pmatrix}\begin{pmatrix} 1 \\ 3 \end{pmatrix} = \begin{pmatrix} -1/2 \\ 3/4 \end{pmatrix}$

   The basic solution is $\underline{x}_2' = (-1/2, 0, 3/4, 0)$, as $\underline{x}_N' = (x_2, x_4) = (0, 0)$

3. $B = (\underline{a}_1, \underline{a}_4)$ corresponds to basic solution $\underline{x}_3' = (1, 0, 0, 3)$
4. $B = (\underline{a}_2, \underline{a}_4)$ corresponds to basic solution $\underline{x}_4' = (0, 1, 0, 1)$
5. $B = (\underline{a}_3, \underline{a}_4)$ corresponds to basic solution $\underline{x}_5' = (0, 0, 1/2, 1)$

Note that the two vectors $(\underline{a}_2, \underline{a}_3)$ do not form a basis, because $\underline{a}_2$ and $\underline{a}_3$ are linearly dependent. Besides the five basic solutions there are "infinitely" many (non-basic) solutions. For example, $\underline{x}_6' = (1/2, 1/2, 0, 2)$ is a *nonbasic solution*, because the vectors $\underline{a}_1$, $\underline{a}_2$ and $\underline{a}_4$ corresponding to the nonzero variables are linearly dependent, and do not constitute a basis.

3.4.1. Degenerate basic solution

Definition 3.2.

A basic solution is called *degenerate* if at least one of the elements of the m-dimensional vector $\underline{x}_B = B^{-1}\underline{b}$ has a value of zero.

Suppose in Example 3.5 we take the vector $\underline{b} = \begin{pmatrix} 3 \\ 6 \end{pmatrix}$, then $\underline{x}' = (0, 3, 0, 0)$ is a degenerate basic solution. If we choose x_1, x_2 to be basic variables, the basic variable x_1 has the value 0.

3.4.2. (In)feasible basic solution

Besides m equalities $A\underline{x} = \underline{b}$, LP problems in standard form (3.3) contain n non-negativity constraints $x_j \geq 0$ ($j = 1, 2, \ldots, n$). If all elements of a basic solution $\underline{x}$ of system $A\underline{x} = b$ are ≥ 0, then $\underline{x}$ is called a feasible basic solution. If at least one of the elements of basic solution $\underline{x}$ is negative, $\underline{x}$ is called an infeasible basic solution.

Definition 3.3.

The basic solution $\underline{x}$ of an LP problem in standard form should satisfy:

1. $A\underline{x} = \underline{b} \quad \rightarrow \quad (B|A_N)\begin{pmatrix} \underline{x}_B \\ \underline{x}_N \end{pmatrix} = \underline{b}$
2. $\underline{x} \geq \underline{0}$

If $\underline{x} = \begin{pmatrix} \underline{x}_B \\ \underline{0} \end{pmatrix}$ satisfies 1 and 2, then $\underline{x}$ is a *feasible basic solution.*

If $\underline{x} = \begin{pmatrix} \underline{x}_B \\ \underline{0} \end{pmatrix}$ satisfies 1 but does not satisfy 2, then $\underline{x}$ is an *infeasible basic solution.*

So, in Example 3.5:

$\underline{x}_3' = (1, 0, 0, 3)$	is called a feasible basic solution.
$\underline{x}_1' = (-1/2, 3/2, 0, 0)$	is called an infeasible basic solution

3.5. Main theorem of linear programming

Consider again an LP problem (P) in standard form.

$$\begin{aligned} \max\{w = \underline{c}'\underline{x}\} \\ A\underline{x} = \underline{b} \\ \underline{x} \geq \underline{0} \end{aligned} \qquad \text{(P)} \qquad (3.10)$$

Assume $r(A) = m$. If the vector $\underline{x}$ fulfils the constraints of (P) and for that $\underline{x}$ the objective function value w is at a maximum, then $\underline{x}$ is called an *optimal feasible solution.* Moreover, if $\underline{x}$ is a basic solution, then $\underline{x}$ is called an *optimal feasible basic solution.*

Theorem 3.2.

Main theorem of linear programming

a. If (P) has a feasible solution, then (P) also has a feasible basic solution.
b. If (P) has an optimal feasible solution with objective function value w^*, then (P) also has an optimal feasible basic solution with value w^*.

First the practical impact of Theorem 3.2 for solving LP problems is indicated. If we want to find an optimal solution $\underline{x}$ of (P), then we can limit the search to vectors $\underline{x}$ that are feasible basic solutions of (P). At first glance this appears very useful, as the number of basic solutions is finite. Every basic solution $\underline{x}$ corresponds to a selection B of m independent columns of A.

At most $\begin{bmatrix} n \\ m \end{bmatrix} = \dfrac{n!}{m!(n-m)!}$ basic solutions exist, as there are $\begin{bmatrix} n \\ m \end{bmatrix}$ ways to select m vectors out of n. Note that not every set of m vectors corresponds to a basis. Moreover, not every basic solution is necessarily feasible. In principle, one could try all (feasible) basic solutions and choose the one with the highest objective function value $w^* = \underline{c}'\underline{x}$. There are practical reasons why this is hardly an efficient procedure.
If, for instance, a (very) small LP problem consists of $n = 50$ variables and $m = 20$ constraints, then the total number of basic solutions is already $\begin{bmatrix} n \\ m \end{bmatrix} \approx 4{,}71 \times 10^{13}$. Simply trying all basic solutions implies that the system $B\underline{x}_B = \underline{b}$ should be solved for very many basic matrices B. In the following sections, the Simplex Algorithm is discussed. This is an algorithm that efficiently determines an optimal basic solution of (P) by moving along the boundaries of the feasible area and evaluating only a limited number of basic solutions.

A formal proof of the main theorem is omitted. We will do with illustrating the core of the proof from an example.

Example 3.6.
The core of the proof of the main theorem will be illustrated from an example. If $\underline{x}$ is a feasible solution of (3.10), then $A\underline{x} = \underline{b}$, $\underline{x} \geq \underline{0}$ or:

$$x_1\underline{a}_1 + \ldots + x_n\underline{a}_n = \underline{b} \text{ and } \underline{x} \geq \underline{0}$$

Let p be the number of positive elements of vector $\underline{x}$, such that n–p elements of $\underline{x}$ are zero. Given: $A = (\underline{a}_1, \underline{a}_2, \underline{a}_3, \underline{a}_4) = \begin{pmatrix} 4 & -2 & -3 & 2 \\ 2 & 4 & 1 & 6 \end{pmatrix}$. Note that $r(A) = m = 2$. Now, two cases can be distinguished:

Case 1; linear independence of $\underline{a}_1, \ldots, \underline{a}_p$.

- $(p = m = 2)$
  For $\underline{b} = \begin{pmatrix} 9 \\ 7 \end{pmatrix}$, $\underline{x}' = (3, 0, 1, 0)$ is a feasible basic solution; basic variables $x_1 = 3$ and $x_3 = 1$ and the nonbasic variables are $x_2 = 0$ and $x_4 = 0$

- $p < m \quad (p = 1\ ;\ m = 2)$
  For $\underline{b} = \begin{pmatrix} 1 \\ 3 \end{pmatrix}$, $\underline{x}' = (0, 0, 0, \frac{1}{2})$ is a feasible solution; $\underline{x}'$ contains only one positive element $x_4 = \frac{1}{2}$, so $p = 1$

Vector $\underline{a}_4$ does not constitute a basis of $\mathbb{R}^2$. Therefore, take $m - p = 2 - 1 = 1$ vector that, together with $\underline{a}_4$, constitutes a basis of $\mathbb{R}^2$, e.g. $\underline{a}_1$ and put $x_1 = 0$. Now $\underline{x}' = (0, 0, 0, \frac{1}{2})$ is a (degenerate) feasible basic solution with basic variables $x_1 = 0$ and $x_4 = \frac{1}{2}$. The non-basic variables are $x_2 = 0$ and $x_3 = 0$.

Case 2; linear dependence among $\underline{a}_1, ..., \underline{a}_p$

- Take $\underline{b} = \begin{pmatrix} 4 \\ 12 \end{pmatrix}$

$\underline{x}' = (3, 1, 2, 0)$ is a feasible solution with positive elements ($x_1 = 3, x_2 = 1, x_3 = 2$); $p = 3$. The vector $\underline{x}'$ is not a basic solution, because $\underline{a}_1$, $\underline{a}_2$ and $\underline{a}_3$ are linearly dependent. Starting from $\underline{x}' = (3, 1, 2, 0)$, a feasible basic solution can be constructed: $\underline{x}'$ is a feasible solution so:

$$3\begin{pmatrix} 4 \\ 2 \end{pmatrix} + 1\begin{pmatrix} -2 \\ 4 \end{pmatrix} + 2\begin{pmatrix} -3 \\ 1 \end{pmatrix} = \begin{pmatrix} 4 \\ 12 \end{pmatrix} \qquad \text{(A)}$$

As $\underline{a}_1$, $\underline{a}_2$ and $\underline{a}_3$ are linearly dependent vectors, there exist numbers α_j, not all equal to zero, such that $\alpha_1\underline{x}_1 + \alpha_2\underline{x}_2 + \alpha_3\underline{x}_3 = \underline{0}$

E.g.
$$1\begin{pmatrix} 4 \\ 2 \end{pmatrix} - 1\begin{pmatrix} -2 \\ 4 \end{pmatrix} + 2\begin{pmatrix} -3 \\ 1 \end{pmatrix} = \begin{pmatrix} 0 \\ 0 \end{pmatrix} \qquad \text{(B)}$$

Multiplication of (B) with ε gives: $\varepsilon\begin{pmatrix} 4 \\ 2 \end{pmatrix} - \varepsilon\begin{pmatrix} -2 \\ 4 \end{pmatrix} + 2\varepsilon\begin{pmatrix} -3 \\ 1 \end{pmatrix} = \begin{pmatrix} 0 \\ 0 \end{pmatrix}$ (C)

Subtraction of (C) from (A): $(3-\varepsilon)\begin{pmatrix} 4 \\ 2 \end{pmatrix} + (1+\varepsilon)\begin{pmatrix} -2 \\ 4 \end{pmatrix} + (2-2\varepsilon)\begin{pmatrix} -3 \\ 1 \end{pmatrix} = \begin{pmatrix} 4 \\ 12 \end{pmatrix}$

So, for any value of ε, $\underline{x}'_\varepsilon = (3-\varepsilon, 1+\varepsilon, 2-2\varepsilon, 0)$ is a solution of $A\underline{x} = \underline{b}$. If we require $\underline{x} \geq \underline{0}$, the solution $\underline{x}'_\varepsilon$ is feasible if $\underline{x}'_\varepsilon \geq \underline{0}'$ i.e. $-1 \leq \varepsilon \leq 1$. Taking for ε one of the boundary values: $\varepsilon = 1$ gives $\underline{x}' = (2, 2, 0, 0)$ and $\varepsilon = -1$ gives $\underline{x}' = (4, 0, 4, 0)$. By selecting for ε appropriate values the feasible basic solutions $\underline{x}' = (2, 2, 0, 0)$ and $\underline{x}' = (4, 0, 4, 0)$ have been constructed from the feasible nonbasic solution $\underline{x}' = (3, 1, 2, 0)$.

3.6. Feasible basic solutions and vertices (corner points)

From linear algebra (Appendix A) we know that if an LP problem has an optimal solution, the optimal value is (also) attained in a *vertex* (extreme point or corner point). The main theorem makes a similar statement about feasible basic solutions. Let us consider the relation between feasible basic solutions and vertices of polyhedra. Let the *polyhedron V* be determined by:

$$A\underline{x} = \underline{b} \qquad (3.11)$$
$$\underline{x} \geq \underline{0}$$
$$\text{So, } V = \{\underline{x} \mid A\underline{x} = \underline{b}, \underline{x} \geq \underline{0}\}$$

Theorem 3.3

Vector $\underline{x}$ is a vertex of V if, and only if, $\underline{x}$ is a feasible basic solution of V.

In other words: Vertex (extreme point or corner point) ⇔ feasible basic solution. We will omit the formal proof of Theorem 3.3.

Conclusion

Combining Theorem 3.2 and 3.3: if an LP problem has an optimal feasible solution, then there exists an optimal feasible basic solution corresponding to a vertex in which the object function value is optimal (see Figure 2.2).

Remark

Intuitively, the relation between feasible basic solutions and vertices can be seen as follows:

1. Let $\underline{x}$ be a feasible basic solution i.e. $\underline{x} = \begin{pmatrix} \underline{x}_B \\ \underline{x}_N \end{pmatrix}$, where $\underline{x}_B$ (m elements) is the unique solution of $B\underline{x}_B = \underline{b}$ and $\underline{x}_N$ (n-m elements) is a vector with all elements 0.
2. A vertex in $\mathbb{R}^n$ of $V = \{\underline{x} \mid A\underline{x} = \underline{b},\ \underline{x} \geq \underline{0}\}$ is a unique intersection of n independent hyperplanes. The system $A\underline{x} = \underline{b}$ provides m of the n hyperplanes. The other $n-m$ hyperplanes are given by the $n-m$ constraints $x_j = 0$ corresponding to the nonbasic variables.

3.7. Pivot operations and the basic form

The constraints $A\underline{x} = \underline{b}$ of an LP problem are a system of m equations with n variables x_j. In Section 3.4 has been shown that system $A\underline{x} = \underline{b}$ has a solution if $\text{r}(A) = \text{r}(A, \underline{b})$. The Simplex Method discussed in this chapter uses pivot operations (Section 3.5). In this section some formal definitions are given of pivot operations.

Definition 3.4.

Systems of equations are called *equivalent* if their solution sets coincide.

The following two operations transform system $A\underline{x} = \underline{b}$ into an equivalent system:
1. Multiplication of equation E_i with a scalar $k \neq 0$ ($i = 1, ..., m$).
2. Replacing equation E_i by the equation $E_i + kE_j$ ($i, j = 1, ..., m$) and $k \in \mathbb{R}$ (except for $i = j$ together with $k = -1$).

These two operations are called *elementary row operations* (*pivot operations*). They tell us that in a linear system every equation can be multiplied with a scalar k ($k \neq 0$) and added (subtracted) to another equation. The elementary row operations can be used to transform a linear system of equations into a system of equations that is easier to solve.

Example 3.7.

The system:

$$\begin{aligned} x_1 - 2x_2 + 3x_3 - \ \ x_4 &= 11 \\ 2x_1 - 2x_2 + 2x_3 + 4x_4 &= 12 \\ -x_1 - 2x_2 + \ \ x_3 - 7x_4 &= \ \ 5 \end{aligned}$$

can be written as (pivots are shaded):

x_1	x_2	x_3	x_4	
1	-2	3	-1	11
2	-2	2	4	12
-1	-2	1	-7	5

and is equivalent to

	x_1	x_2	x_3	x_4	
row 1:	1	-2	3	-1	11
row 2 - 2 row 1:	0	2	-4	6	-10
row 3 + row 1:	0	-4	4	-8	16

is equivalent to

	x_1	x_2	x_3	x_4	
row 1 - 1/2 row 3:	1	0	1	3	3
row 2 + 1/2 row 3:	0	0	-2	2	-2
- 1/4 row 3:	0	1	-1	2	-4

is equivalent to

	x_1	x_2	x_3	x_4	
row 1 + 1/2 row 2:	1	0	0	4	2
- 1/2 row 2:	0	0	1	-1	1
row 3 - 1/2 row 2:	0	1	0	1	-3

$(\hat{\text{P}})$

such that the solution of the system can be written as:

$$\underline{x} = \begin{pmatrix} x_1 \\ x_2 \\ x_3 \\ x_4 \end{pmatrix} = \begin{pmatrix} 2 \\ -3 \\ 1 \\ 0 \end{pmatrix} + x_4 \begin{pmatrix} -4 \\ -1 \\ 1 \\ 1 \end{pmatrix}$$

Elementary row operations have been used to transform system (P) into system $(\hat{\text{P}})$ which permits an easy representation of the solution set. The basic solution with x_4 as a nonbasic variable ($x_4 = 0$) is $x_1 = 2$, $x_2 = -3$, $x_3 = 1$, $x_4 = 0$. The following definition formalises what has been observed so far.

Definition 3.5.

A *pivot operation* with pivot $a_{rs} \neq 0$ consists of the following operations:

1. Replace equation E_i $(i \neq r)$ by $E_i - \frac{a_{is}}{a_{rs}} E_r$
2. Replace equation E_r by $\frac{E_r}{a_{rs}}$

In Example 3.7, the pivot in the first transformation is $a_{11} = 1$ ($r = 1$, $s = 1$), in the second transformation is $a_{32} = -4$ ($r = 3$, $s = 2$) and in the third transformation the pivot is $a_{23} = -2$.

Let $\hat{a}_{ij}$ and $\hat{b}_i$ denote the coefficients corresponding to a_{ij} and b_i in the transformed system, then:

$$\hat{a}_{ij} = a_{ij} - \frac{a_{is}}{a_{rs}} a_{rj} \qquad (i \neq r\,;\ j = 1, \ldots, n) \qquad (3.12)$$

$$\hat{a}_{rj} = \frac{a_{rj}}{a_{rs}} \qquad (j = 1, ..., n) \qquad (3.13)$$

$$\hat{b}_i = b_i - \frac{a_{is}}{a_{rs}} b_r \qquad (i \neq r) \qquad (3.14)$$

$$\hat{b}_r = \frac{b_r}{a_{rs}} \qquad (3.15)$$

By operations (3.12) - (3.15), column s is transformed into the rth unit vector:

$$\hat{a}_{is} = a_{is} - \frac{a_{is}}{a_{rs}} a_{rs} = 0 \qquad (i \neq r)$$

$$\hat{a}_{rs} = \frac{a_{rs}}{a_{rs}} = 1$$

In Example 3.7, column 1 has been transformed into $\underline{e}_1$, column 2 into $\underline{e}_3$ and column 3 into $\underline{e}_2$. During the pivoting operations, the pivot is used to convert column s into a unit vector. In Example 3.7 pivot $a_{11} = 1$ is used to convert the first column. In the second $a_{32} = -4$ is used as a pivot and finally a_{23} is used to convert the third column into a unit vector.

If $r(A) = m$, pivot operations can be used to transform system $A\underline{x} = \underline{b}$ in such a way that a unit matrix appears. In this case, matrix A can be partitioned as $A = (B \mid A_N)$ where basis matrix B can be inverted (Section 3.4).

So, $\quad A\underline{x} = \underline{b} \quad \rightarrow \quad (B \mid A_N)\begin{pmatrix} \underline{x}_B \\ \underline{x}_N \end{pmatrix} = \underline{b}$

Multiplication by B^{-1} gives:

$$B^{-1}A\underline{x} = B^{-1}\underline{b} \quad \rightarrow \quad B^{-1}(B \mid A_N)\begin{pmatrix} \underline{x}_B \\ \underline{x}_N \end{pmatrix} = B^{-1}\underline{b}$$

$$(E \mid B^{-1}A_N)\begin{pmatrix} \underline{x}_B \\ \underline{x}_N \end{pmatrix} = B^{-1}\underline{b}$$

Written in a different way:

$$\begin{pmatrix} 1 & 0 & \dots & 0 & \hat{a}_{1m+1} & \dots & \hat{a}_{1n} \\ 0 & 1 & \dots & 0 & \hat{a}_{2m+1} & \dots & \hat{a}_{2n} \\ . & . & & . & . & & . \\ . & . & & . & . & & . \\ . & . & & . & . & & . \\ 0 & 0 & \dots & 1 & \hat{a}_{mm+1} & \dots & \hat{a}_{mn} \end{pmatrix} \begin{pmatrix} x_1 \\ x_2 \\ . \\ . \\ . \\ x_m \\ x_{m+1} \\ . \\ . \\ x_n \end{pmatrix} = \begin{pmatrix} \hat{b}_1 \\ \hat{b}_2 \\ . \\ . \\ . \\ \hat{b}_m \end{pmatrix} \qquad (3.16)$$

or, $\quad (\underline{e}_1, ..., \underline{e}_m, \underline{\hat{a}}_{m+1}, ..., \underline{\hat{a}}_j, ..., \underline{\hat{a}}_n)\underline{x}_n = \underline{\hat{b}}$

Where $\hat{\underline{b}} = B^{-1}\underline{b}$ and $\hat{\underline{a}}_j = B^{-1}\underline{a}_j$ ($j = m+1, ..., n$). System (3.16) has a trivial basic solution $x_1 = \hat{b}_1$, $x_2 = \hat{b}_2$, ..., $x_m = \hat{b}_m$, $x_{m+1} = 0$, ..., $x_n = 0$.

Definition 3.6.

A system $A\underline{x} = \underline{b}$ in which matrix A contains m independent unit vectors is called a system in *basic form*.

Definition 3.7.

A system in basic form where $\underline{b} \geq \underline{0}$ holds, is called a system in *basic feasible form*.

A basic form (in which the order of the unit vectors is not relevant) can be obtained by
- either pivoting;
- or multiplication by B^{-1}.

Notice that the inverse B^{-1} can also be obtained by pivoting. A system in basic (feasible) form has a trivial basic (feasible) solution.

Example 3.8.

In Example 3.7 $A\underline{x} = \underline{b}$ is given by:

$$A = \begin{pmatrix} 1 & -2 & 3 & -1 \\ 2 & -2 & 2 & 4 \\ -1 & -2 & 1 & -7 \end{pmatrix} \text{ and } \underline{b} = \begin{pmatrix} 11 \\ 12 \\ 5 \end{pmatrix}$$

Three pivot operations transform system (P) into system $(\hat{\text{P}})$ in basic form. It is shown here that multiplication by B^{-1} gives the same result. In Example 3.7, the vectors $\underline{a}_1$, $\underline{a}_3$ and $\underline{a}_2$ have been transformed into $\underline{e}_1$, $\underline{e}_2$ and $\underline{e}_3$.

$$\text{So, } B = \begin{pmatrix} 1 & 3 & -2 \\ 2 & 2 & -2 \\ -1 & 1 & -2 \end{pmatrix} \text{ Mind the order!}$$

$$\text{It can be shown that } B^{-1} = \begin{pmatrix} -1/4 & 2/4 & -1/4 \\ 3/4 & -2/4 & -1/4 \\ 2/4 & -2/4 & -2/4 \end{pmatrix} = \frac{1}{4}\begin{pmatrix} -1 & 2 & -1 \\ 3 & -2 & -1 \\ 2 & -2 & -2 \end{pmatrix}$$

So, $B^{-1}A\underline{x} = B^{-1}\underline{b}$ gives:

$$\frac{1}{4}\begin{pmatrix} -1 & 2 & -1 \\ 3 & -2 & -1 \\ 2 & -2 & -2 \end{pmatrix}\begin{pmatrix} 1 & -2 & 3 & -1 \\ 2 & -2 & 2 & 4 \\ -1 & -2 & 1 & -7 \end{pmatrix}\begin{pmatrix} x_1 \\ x_2 \\ x_3 \\ x_4 \end{pmatrix} = \frac{1}{4}\begin{pmatrix} -1 & 2 & -1 \\ 3 & -2 & -1 \\ 2 & -2 & -2 \end{pmatrix}\begin{pmatrix} 11 \\ 12 \\ 5 \end{pmatrix}$$

$$\text{or } \begin{pmatrix} 1 & 0 & 0 & 4 \\ 0 & 0 & 1 & -1 \\ 0 & 1 & 0 & 1 \end{pmatrix}\begin{pmatrix} x_1 \\ x_2 \\ x_3 \\ x_4 \end{pmatrix} = \begin{pmatrix} 2 \\ 1 \\ -3 \end{pmatrix}$$

This coincides with the basic form $(\hat{\text{P}})$. Note that the basic form is infeasible: $x_2 = b_3 = -3 < 0$.

The corresponding infeasible basic solution $\underline{x}' = (x_1, x_2, x_3, x_4) = (2, -3, 1, 0)$.
Check now that $\underline{\hat{a}}_4 = B^{-1}\underline{a}_4$. So $\underline{a}_4' = (-1, 4, -7)$ is transformed by pivot operations as well as by multiplication by B^{-1} into $\underline{\hat{a}}_4' = (4, -1, 1)$.

3.8. Feasible basis transitions

A procedure is explained for moving from one feasible basic solution to another. This is called a *feasible basis transition*. The initial abstract notation will be illustrated by an example.

Consider the constraints of an LP problem $A\underline{x} = \underline{b}$; $\underline{x} \geq \underline{0}$ in feasible basic form, i.e. an $m \times m$-unit matrix exists and $b_i \geq 0$ $(i = 1, ..., m)$. Without loss of generality it can be assumed that the m unit vectors are the first vectors in the system:

x_1	...	x_k	...	x_m	x_{m+1}	...	x_p	...	x_n	
1	...	0	...	0	$a_{1(m+1)}$	...	a_{1p}	...	a_{1n}	b_1
⋮		⋮		⋮	⋮		⋮		⋮	⋮
0	...	1	...	0	$A_{k(m+1)}$	...	a_{kp}	...	a_{kn}	b_k
⋮		⋮		⋮	⋮		⋮		⋮	⋮
0	...	0	...	1	$a_{m(m+1)}$	...	a_{mp}	...	a_{mn}	b_m

Tableau 3.1. A feasible basic form, $b_i \geq 0$, $i = 1, ..., m$.

In Section 3.14, we will discuss the so-called two-phase method. This method is a systematic procedure to either

- convert an arbitrary system into a feasible basic form,
- or to detect that the system does not have feasible solutions.

An LP problem in feasible basic form permits trivial determination of a basic feasible solution. Tableau 3.1 corresponds to the following feasible basic solution:

$$\begin{aligned}\underline{x}' &= (x_1, \ ..., \ x_k, \ ..., \ x_m, \ x_{m+1}, \ ..., \ x_p, \ ..., \ x_n) \\ &= (b_1, \ ..., \ b_k, \ ..., \ b_m, \ \ 0, \ ..., \ \ 0, \ ..., \ 0)\end{aligned} \tag{3.17}$$

Feasible basis transition

Solution $\underline{x}$ in (3.17) is a feasible basic solution. The system of Tableau 3.1 probably has several feasible basic solutions. How do we find them? Suppose we want to include nonbasic variable x_p in the basis. As we will see in Section 3.9, this may be profitable from the objective point of view. Including x_p in the basis means we want x_p to have a value as big and positive as possible, while keeping all current basic variables non-negative $x_1, \ ..., \ x_m \geq 0$ and the other nonbasic variables $(x_{m+1}, \ ..., \ x_{p-1}, \ x_{p+1}, \ ..., \ x_n)$ at zero. As the number of basic variables remains m and x_p becomes basic too, one of the current basic variables $x_1, \ ..., \ x_m$ should make place and disappear from the basis. It becomes nonbasic. Including x_p in Tableau 3.1 implies $(x_{m+1} = 0, ..., \ x_{p-1} = 0, \ x_{p+1} = 0, \ ..., \ x_n = 0)$:

$$\begin{array}{llllll} x_1 & & + a_{1p}x_p = b_1 & \rightarrow & x_1 = b_1 - a_{1p}x_p \\ & x_2 & + a_{2p}x_p = b_2 & \rightarrow & x_2 = b_2 - a_{2p}x_p \\ & \vdots & & & \vdots \\ & x_k & + a_{kp}x_p = b_k & \rightarrow & x_k = b_k - a_{kp}x_p \\ & \vdots & & & \vdots \\ & x_m & + a_{mp}x_p = b_m & \rightarrow & x_m = b_m - a_{mp}x_p \end{array}$$

So $\underline{x}' = (b_1 - a_{1p}x_p, \ldots, b_k - a_{kp}x_p, \ldots, b_m - a_{mp}x_p, 0, \ldots, x_p, \ldots, 0)$ (3.18)

Notation (3.18) shows the influence of nonbasic variable x_p on basic variables $x_1, \ldots, x_m$. Note that if $x_p = 0$ then (3.17) and (3.18) are equal.

What happens to the basic variable $x_i = b_i - a_{ip}x_p$ if a nonbasic variable x_p becomes positive? We distinguish two cases:

1. If $a_{ip} \leq 0$ then $x_i \geq 0$ holds for all values $x_p \geq 0$
2. If $a_{ip} > 0$ then $x_i = b_i - a_{ip}x_p \geq 0$ holds as long as $x_p \leq \frac{b_i}{a_{ip}}$

We now look systematically for the basic variable(s), which is going to reach a value of zero first and therefore be removed from the basis.

Define $\varepsilon_i := \frac{b_i}{a_{ip}}$ and let $\varepsilon_k = \min_i \varepsilon_i$.

As long as $x_p \leq \varepsilon_k$ all basic variables $x_i \geq 0$. If exactly $x_p = \varepsilon_k$, then

$$x_k = b_k - a_{kp}x_p = b_k - a_{kp}\varepsilon_k = b_k - a_{kp}\frac{b_k}{a_{kp}} = 0.$$

So, filling in $x_p = \min_i \left\{ \frac{b_i}{a_{ip}} \mid a_{ip} > 0 \right\} = \frac{b_k}{a_{kp}} = \varepsilon_k$ into (3.18) leads to the new basic solution:

$$\underline{x}' = (\underbrace{b_1 - a_{1p}\varepsilon_k}_{\geq 0}, \underbrace{b_2 - a_{2p}\varepsilon_k}_{\geq 0}, \ldots, \underset{\substack{\uparrow \\ \text{element } k}}{0}, \ldots, \underbrace{b_m - a_{mp}\varepsilon_k}_{\geq 0}, \ldots, \underset{\substack{\uparrow \\ \text{element } p}}{\varepsilon_k}, \ldots, 0) \qquad (3.19)$$

In (3.19) $x_p = \varepsilon_k$ has come into the basis by removing the variable x_k.

The *feasible basis transition* is the same as pivoting Tableau 3.1 with pivot a_{kp}

For the pivot should hold:

Pivot a_{kp}

$$a_{kp} > 0 \text{ and } \frac{b_k}{a_{kp}} = \min_i \left\{ \frac{b_i}{a_{ip}} \mid a_{ip} > 0 \right\}$$

Pivoting Tableau 3.1 with a_{kp} gives:

x_1	...	x_k	...	x_m	x_{m+1}	...	x_p	...	x_n	
1	...	$\hat{a}_{1k}$	...	0	$\hat{a}_{1(m+1)}$	...	0	...	$\hat{a}_{1n}$	$\hat{b}_1$
.		.		.	.		.		.	.
.		.		.	.		.		.	.
0	...	$\hat{a}_{kk}$	...	0	$\hat{a}_{k(m+1)}$	...	1	...	$\hat{a}_{kn}$	$\hat{b}_k$
.				.	.		.		.	.
.				.	.		.		.	.
0	...	$\hat{a}_{mk}$	...	1	$\hat{a}_{m(m+1)}$	...	0	...	$\hat{a}_{mn}$	$\hat{b}_m$

Tableau 3.2. Feasible basic form ($\underline{\hat{b}} \geq \underline{0}$) after the basis transition.

Comparing Tableau 3.1 and 3.2 shows that x_k and x_p have been exchanged. The nonbasic variable x_p has become basic ($x_p = \hat{b}_k$) and has pushed the variable x_k out of the basis ($x_k = 0$). Tableau 3.2 corresponds to the basic solution:

$$\begin{aligned} \underline{x}' &= (x_1, \ ..., \ x_k, \ ..., \ x_m, \ x_{m+1}, \ ..., \ x_p, \ ..., \ x_n) \\ &= (\hat{b}_1, \ ..., \ 0, \ ..., \ \hat{b}_m, \ 0, \ ..., \ \hat{b}_k, \ ..., \ 0) \end{aligned} \tag{3.20}$$

The pivot operations have lead to:

$$\hat{b}_k = \frac{b_k}{a_{kp}} = \varepsilon_k$$

$$\hat{b}_1 = b_1 - \frac{a_{1p}}{a_{kp}} b_k = b_1 - a_{1p}\varepsilon_k$$

$$\hat{b}_m = b_m - \frac{a_{mp}}{a_{kp}} b_k = b_m - a_{mp}\varepsilon_k$$

So (3.20) corresponds to (3.19).

Remarks

1. Pivoting with a_{kp} corresponds to the multiplication of Tableau 3.1 with matrix B^{-1}, where $B = (\underline{e}_1, \underline{e}_2, ..., \underline{e}_{k-1}, \underline{a}_p, \underline{e}_{k+1}, ..., \underline{e}_m)$. The inverse B^{-1} of the basic matrix B can be found in Tableau 3.2 where in Tableau 3.1 the unit matrix can be found. So in this case, B^{-1} consists of the first m columns.
2. If $a_{ip} \leq 0$ for all $i = 1, ..., m$, then in (3.18) all elements $b_i - a_{ip}x_p \geq 0$ for all $x_p \geq 0$. No variable is removed from the basis, so no new basis appears. In this situation, the set $V = \{\underline{x} \mid A\underline{x} = \underline{b}, \underline{x} \geq \underline{0}\}$ is unbounded (see Figure 2.5).

Example 3.9.

Consider system:

$$\begin{aligned} x_1 \quad\quad &+ 2x_4 + 4x_5 - 6x_6 = 2 \\ x_2 \quad &+ \ x_4 - 2x_5 - 3x_6 = 3 \\ x_3 &- \ x_4 + 2x_5 + \ x_6 = 1 \end{aligned}$$

$x_1,\ x_2,\ x_3,\ x_4,\ x_5,\ x_6 \geq 0$

Written in tableau form:

x_1	x_2	x_3	x_4	x_5	x_6	
1	0	0	2	4	-6	2
0	1	0	1	-2	-3	3
0	0	1	-1	2	1	1

Tableau 3.3.

Tableau 3.3 corresponds to the (trivial) feasible basic solution $\underline{x}' = (x_1,\ x_2,\ x_3,\ x_4,\ x_5,\ x_6) =$ (2, 3, 1, 0, 0, 0). Including x_4 in the basis (so $p = 4$) and ($x_5 = x_6 = 0$) leads to:

$x_1 + 2x_4 = 2$	$\rightarrow$	$x_1 = 2 - 2x_4 \geq 0$	if	$x_4 \leq 2/2 = 1$
$x_2 + \ x_4 = 3$	$\rightarrow$	$x_2 = 3 - \ x_4 \geq 0$	if	$x_4 \leq 3/1 = 3$
$x_3 - \ x_4 = 1$	$\rightarrow$	$x_3 = 1 + \ x_4 \geq 0$	for any	$x_4 \geq \ 0$

The maximum value of x_4 such that all variables remain non-negative equals 1, because

$$x_4 = \min_i \left\{ \frac{b_i}{a_{i4}} \mid a_{i4} > 0 \right\} = \min \left\{ \frac{b_1}{a_{14}}\ ,\ \frac{b_2}{a_{24}} \right\} = \min \left\{ \frac{2}{2}\ ,\ \frac{3}{1} \right\} = 1$$

This corresponds to pivoting column 4 with $a_{14} = 2$. Pivot $a_{14} = 2$ has been chosen such that $a_{14} > 0$ and at the same time:

$$\frac{b_1}{a_{14}} = \min_i \left\{ \frac{b_i}{a_{i4}} \mid a_{i4} > 0 \right\} = \min \left\{ \frac{2}{2}\ ,\ \frac{3}{1} \right\}$$

Pivoting with $a_{14} = 2$ leads to:

x_1	x_2	x_3	x_4	x_5	x_6	
1/2	0	0	1	2	-3	1
-1/2	1	0	0	-4	0	2
1/2	0	1	0	4	-2	2

Tableau 3.4.

Tableau 3.4 corresponds to the feasible basic solution $\underline{x}' = (0, 2, 2, 1, 0, 0)$ such that x_1 has been pushed out of the basis and x_4 has been included. It can be shown that:

- $B = (\underline{a}_4, \underline{e}_2, \underline{e}_3) = \begin{pmatrix} 2 & 0 & 0 \\ 1 & 1 & 0 \\ -1 & 0 & 1 \end{pmatrix}; \quad B^{-1} = \begin{pmatrix} 1/2 & 0 & 0 \\ -1/2 & 1 & 0 \\ 1/2 & 0 & 1 \end{pmatrix}$

- pivoting in Tableau 3.3 with $a_{24} = 1$ does not lead to a feasible basic solution as $\hat{b}_1 < 0$ after pivoting. Neither pivot $a_{34} = -1$ leads to a feasible solution.

Making variable x_6 positive in Tableau 3.4 does not end in a feasible basic solution, as column $\underline{a}_6$ contains no positive elements. Variable x_6 cannot be included in the basis. Making x_6 positive and keeping $x_1 = x_5 = 0$ leads to:

$$\begin{aligned} x_4 - 3x_6 &= 1 \quad \rightarrow \quad x_4 = 1 + 3x_6 \geq 0 \quad \text{for all } x_6 \geq 0 \\ x_2 &= 2 \quad \rightarrow \quad x_2 = 2 \\ x_3 - 2x_6 &= 2 \quad \rightarrow \quad x_3 = 2 + 2x_6 \geq 0 \quad \text{for all } x_6 \geq 0 \end{aligned}$$

So, $\underline{x} = \begin{pmatrix} x_1 \\ x_2 \\ x_3 \\ x_4 \\ x_5 \\ x_6 \end{pmatrix} = \begin{pmatrix} 0 \\ 2 \\ 2 \\ 1 \\ 0 \\ 0 \end{pmatrix} + x_6 \begin{pmatrix} 0 \\ 0 \\ 2 \\ 3 \\ 0 \\ 1 \end{pmatrix}$ is a feasible solution for all $x_6 \geq 0$.

The solution set is unbounded (Figure 2.5).

3.9. Determination of the optimal basic feasible solution

After we have learned to move from one feasible basic solution (vertex) to the other, we want to find the optimal basis with respect to the objective function $w = \underline{c}'\underline{x}$. Consider Tableau 3.5:

x_1	x_2	x_3	x_4	x_5	
1	0	0	2	2	4
0	1	0	-1	0	6
0	0	1	1	1	3

Tableau 3.5.

The corresponding feasible basic solution is $\underline{x}' = (4, 6, 3, 0, 0)$. We don't know whether this basic solution has the highest objective function value. It might be possible to reach a better solution by making x_4 positive (it becomes a basic variable and pushes out x_1). In order to find this out, the objective function $w = \underline{c}'\underline{x}$ should be considered.

Suppose we want to find the maximum of $w = x_1 + 2x_2 + 3x_3 + 10x_4 + 7x_5$. The current basic solution yields $w = 1 \times 4 + 2 \times 6 + 3 \times 3 + 10 \times 0 + 7 \times 0 = 25$. In the objective function, x_4 has a high weight of $c_4 = 10$, so it looks reasonable to include x_4 in the current plan and bring x_4 into the basis.

Pivoting with $a_{14} = 2$ gives:

x_1	x_2	x_3	x_4	x_5	
1/2	0	0	1	1	2
1/2	1	0	0	1	8
-1/2	0	1	0	0	1

Tableau 3.6.

The corresponding feasible basic solution is $\underline{x}' = (0, 8, 1, 2, 0)$ with objective function value $w = 16 + 3 + 20 = 39$. This value is higher than the value $w = 25$ we found earlier. Next, we might consider including x_5 in the basis, etc.

Determination of the optimal basic solution for big problems by simply trying all feasible basic solutions is impracticable. The efficient method, which we will discuss next, makes use of the objective function in the tableau by means of the so-called r-row.

Definition 3.8.

The *r-row* is the objective function $w = \underline{c}'\underline{x}$ expressed in the nonbasic variables.

The r-row can be derived mathematically. Consider the LP problem (P) in standard form:

$$\begin{aligned} \max\{w = \underline{c}'\underline{x}\} & \\ A\underline{x} = \underline{b} & \qquad \text{(P)} \\ \underline{x} \geq \underline{0} & \end{aligned}$$

Let $\underline{x}_B = B^{-1}\underline{b}$ be a feasible basic solution. Matrix A can be partitioned into $A = (B \mid A_N)$.

$A\underline{x} = \underline{b} \rightarrow (B \mid A_N)\begin{pmatrix} \underline{x}_B \\ \underline{x}_N \end{pmatrix} = \underline{b}$, where $\underline{x}_B' = (x_1, \ldots, x_m)$ and $\underline{x}_N' = (x_{m+1}, \ldots, x_n)$

So, $B^{-1}(B \mid A_N)\begin{pmatrix} \underline{x}_B \\ \underline{x}_N \end{pmatrix} = B^{-1}\underline{b}$ or $(E \mid B^{-1}A_N)\begin{pmatrix} \underline{x}_B \\ \underline{x}_N \end{pmatrix} = B^{-1}\underline{b}$ such that $\underline{x}_B + B^{-1}A_N\,\underline{x}_N = B^{-1}\underline{b}$

or

$$\underline{x}_B = B^{-1}\underline{b} - B^{-1}A_N\underline{x}_N \tag{3.21}$$

Expression (3.21) gives the basic variables $\underline{x}_B$ as a function of the nonbasic variables $\underline{x}_N$. Note that if $\underline{x}_N = \underline{0}$ then $\underline{x}_B = B^{-1}\underline{b}$, which we have seen before. We started with $\underline{x}_B = B^{-1}\underline{b} \geq \underline{0}$. Consider the objective function $w = \underline{c}'\underline{x}$.

Partitioning $\underline{x} = \begin{pmatrix} \underline{x}_B \\ \underline{x}_N \end{pmatrix}$ means we can also partition $\underline{c}$ as $\underline{c} = \begin{pmatrix} \underline{c}_B \\ \underline{c}_N \end{pmatrix}$. These partitions are often used in this chapter. This means $w = \underline{c}'\underline{x} = (\underline{c}_B', \underline{c}_N')\begin{pmatrix} \underline{x}_B \\ \underline{x}_N \end{pmatrix} = \underline{c}_B'\,\underline{x}_B + \underline{c}_N'\,\underline{x}_N$

Substitution of $\underline{x}_B$ via (3.21) gives $w = \underline{c}_B'\,\underline{x}_B + \underline{c}_N'\,\underline{x}_N = \underline{c}_B'\,(B^{-1}\underline{b} - B^{-1}A_N\underline{x}_N) + \underline{c}_N'\,\underline{x}_N$, i.e.

$$w + (\underline{c}_B'\,B^{-1}A_N - \underline{c}_N')\underline{x}_N = \underline{c}_B'\,B^{-1}\underline{b} \tag{3.22}$$

In (3.22) the objective function value w is given as a function of the nonbasic variables $\underline{x}_N$. Define $w_0 := \underline{c}_B'\,\underline{x}_B = \underline{c}_B'\,B^{-1}\underline{b}$ the objective function value of the current basic solution and $\underline{r}_N' = \underline{c}_B'\,B^{-1}A_N - \underline{c}_N'$ then the objective function can be written as:

$$w + \underline{r}_N'\,\underline{x}_N = w_0 \tag{3.23}$$

Expressed in the nonbasic variables $\underline{x}'_N = (x_{m+1}, ..., x_n)$:

$$w + r_{m+1}x_{m+1} + ... + r_px_p + ... + r_nx_n = w_0 \tag{3.24}$$

$$\text{where } r_p = \underline{c}'_B B^{-1}\underline{a}_p - c_p \qquad p = m+1, ..., n \tag{3.25}$$

In (3.24) the objective function w is given as a function of the nonbasic variables. The objective function value corresponding to basis B is $w_0 = \underline{c}'_B B^{-1}\underline{b}$, because the nonbasic variables are all zero $x_{m+1} = ... = x_n = 0$. Expression (3.24) offers information regarding the usefulness of including a nonbasic variable x_p in the current plan (enter the basis):

- If $r_p \geq 0$ then it is not profitable to have x_p enter the basis. As $w = w_0 - r_px_p \leq w_0$, the objective function value w will decrease or does not change in case x_p becomes positive.
- If $r_p < 0$ then it is profitable to have x_p enter the basis. The relation $w = w_0 - r_px_p > w_0$ shows that the objective function value increases if x_p becomes positive.

In Section 3.8 we have seen that:

1. If $a_{ip} > 0$ for some value of i, then x_p can enter the basis via pivot a_{kp}, with:

   $$\frac{b_k}{a_{kp}} = \min\left\{\frac{b_i}{a_{ip}} \mid a_{ip} > 0\right\}$$

   If $b_k > 0$ then $x_p = \frac{b_k}{a_{kp}} > 0$, such that the objective function value increases with r_px_p.

   If $b_k = 0$ (degenerate basic solution) then $x_p = \frac{b_k}{a_{kp}} = 0$, such that $r_px_p = 0$. The objective function value w remains the same.

2. If $a_{ip} \leq 0$ for all $i = 1, ..., m$, then x_p can increase unlimited. Set $V = \{A\underline{x} = \underline{b}, \underline{x} \geq \underline{0}\}$ is unbounded and the solution remains feasible with an increasing value of x_p. The objective function value w can also increase unlimited with r_px_p. This is called an unbounded solution as we have seen in Figure 2.5.

Summary

Given the feasible basic solution $(\underline{x}'_B, \underline{x}'_N) = (\underline{x}'_B, \underline{0}')$; with objective function value $w_0 = \underline{c}'_B \underline{x}_B$.

Relation (3.24): $w + r_{m+1}x_{m+1} + ... + r_jx_j + ... + r_nx_n = w_0$, where $r_j = \underline{x}'_B B^{-1}\underline{a}_j - c_j$, offers a tool to check the optimality of the current feasible basic solution:

- If $r_j \geq 0$ for all $j = m+1, ..., n$, the objective function value $w = w_0$ cannot be improved.
- If for some $j = m+1, ..., n$ the value $r_j < 0$ holds, a feasible solution $\underline{x}$ exists with $w = \underline{c}'\underline{x} \geq w_0$.

Remarks

1. In a minimisation problem, x_p should enter the basis if $r_j > 0$.
2. In a minimisation problem a basic solution is optimal if all $r_j \leq 0$.

3. In a maximisation problem a basic solution is optimal if all $r_j \geq 0$.
Note that any minimisation problem $\min\{\underline{c}'\underline{x}\}$ can be transformed into a maximisation problem: $\max\{-\underline{c}'\underline{x}\}$.

3.10. Reduced costs

The elements in the r-row as introduced in Section 3.9 are called *reduced costs*. In this section we focus on the interpretation of this concept.

Consider again an LP problem in standard form:

$$\begin{aligned} &\max\{w = \underline{c}'\underline{x}\} \\ &\quad A\underline{x} = \underline{b} \\ &\quad \underline{x} \geq \underline{0} \end{aligned}$$

Let us interpret the problem as a production planning problem in which n products can be produced. The profit is c_j per unit produced product j. The constraint set $A\underline{x} = \underline{b}$ represents constraints regarding the available raw material, labour, capacity restrictions, etc.

A current production plan tells us that products $x_1 = \hat{b}_1, \ldots, x_m = \hat{b}_m$ are manufactured and others $x_{m+1}, \ldots, x_p, \ldots, x_n$ are not produced. This concept corresponds to a feasible basic solution as presented in Tableau 3.7.

c_1	...	c_m	c_{m+1}	...	c_p	...	c_n	
x_1	...	x_m	x_{m+1}	...	x_p	...	x_n	
1	...	0	$\hat{a}_{1m+1}$	...	$\hat{a}_{1p}$	...	$\hat{a}_{1n}$	$\hat{b}_1$
.		.	.		.		.	.
.		.	.		.		.	.
.		.	.		.		.	.
0	...	1	$\hat{a}_{mn+1}$	...	$\hat{a}_{mp}$	...	$\hat{a}_{mn}$	$\hat{b}_m$

Tableau 3.7.

The profit of the current plan equals:

$$w = c_1x_1 + \ldots + c_mx_m = c_1\hat{b}_1 + c_m\hat{b}_m = \underline{c}_B'\,\hat{\underline{b}} \tag{3.26}$$

Is the current production plan optimal? Is it useful for any product x_p to be included in the production plan ($p = m+1, \ldots, n$)? Including x_p in the production plan implies making variable x_p positive. As all requirements still have to hold, so $A\underline{x} = \underline{b}$, producing x_p influences the produced quantities of $x_1, \ldots, x_m$.

Making x_p positive implies values to change according to (Tableau 3.7):

$x_1 = \hat{b}_1 - \hat{a}_{1p}x_p, \ldots, x_m = \hat{b}_m - \hat{a}_{mp}x_p$. In vector notation:

$$\hat{\underline{x}}_B = \hat{\underline{b}} - \hat{\underline{a}}_p x_p$$

The profit of the production plan including x_p is:

$w = c_1x_1 + ... + c_mx_m + c_px_p = c_1(\hat{b}_1 - \hat{a}_{1p}x_p) + ... + c_m(\hat{b}_m - \hat{a}_{mp}x_p) + c_px_p$

In vector notation:

$$w = \underline{c}_B'\underline{\hat{b}} - \underline{c}_B'\underline{\hat{a}}_p x_p + c_p x_p \qquad (3.27)$$

The increase of the profit by including x_p equals (3.27) – (3.26):

The increase in profit $= -\underline{c}_B'\underline{\hat{a}}_p x_p + c_p x_p = (-\underline{c}_B'\underline{\hat{a}}_p + c_p)x_p = -r_p x_p$

in which $r_p = \underline{c}_B'\underline{\hat{a}}_p - c_p$

In other words, it is useful to include x_p in the production plan if $r_p < 0$, so $\underline{c}_B'\underline{\hat{a}}_p - c_p < 0$, in other words the direct profit c_p should be bigger than the loss of profit $\underline{c}_B'\underline{\hat{a}}_p$ caused by any change in the current plan. If $r_p \geq 0$, the profit will decrease when producing x_p. If $r_j \geq 0$ for all $j = m+1, ..., n$, the current plan is optimal. In this case the reduced costs r_j represent the loss of profit per unit of x_j if we decide to produce product j ($x_j > 0$).

A diet problem

An economic interpretation of the reduced costs will be given in an example, i.e. the so-called diet problem.

$$\min\{w = \underline{c}'\underline{x}\}$$
$$A\underline{x} = \underline{b}$$
$$\underline{x} \geq \underline{0}$$

In this problem, the elements a_{ij} of matrix A represent the amount of nutrient i (protein, vitamins, iron, carbohydrate) per unit of component j (e.g. meat, potatoes, salad, etc.). Column j of matrix A represents the 'nutrient value' of food component j. Coefficient b_i gives the requirement with respect to the amount of nutrient i that should be in the 'meal'. In the final optimal solution (meal) the tableau shows how the nutrient value of food component j can be composed from the food components that are in the basis (i.e. included in the current meal).

Suppose the variables $x_1 \dots x_m$, representing the first m components, are basic variables and $\underline{\hat{a}}_j$ is the column vector belonging to component j in the final tableau. So, $\underline{\hat{a}}_j$ consists of the elements $\hat{a}_{1j}, \hat{a}_{2j}, ..., \hat{a}_{mj}$ (compare with Tableau 3.7).

The price of the 'natural' component j (e.g. salad) is c_j. If component j is made 'artificially' from the products in the basis, the price per unit is $c_1\hat{a}_{1j} + c_2\hat{a}_{2j} + ... + c_m\hat{a}_{mj} = \underline{c}_B'\underline{\hat{a}}_j$.

If $\underline{c}_B'\underline{\hat{a}}_j - c_j > 0$, then the 'natural' food component j is apparently cheaper than composing it (artificially) from a combination of other components. In that case it is profitable to include food component j in the basis. Remember that in a minimisation problem, it is useful to include x_j in the plan, if $r_j > 0$!

3.11. The simplex algorithm

In the preceding sections, the most important aspects of the Simplex Algorithm have been discussed. We know how to move from one feasible basic solution to the other and the reduced costs r_j offer the information regarding the effectiveness of moving from one solution to the other. In other words, the reduced costs answer the question: how attractive is it to include a nonbasic variable (i.e. not in the current plan) in the basis.

To make life easy, let the LP problem be $\max\{w = \underline{c}'\underline{x} \mid A\underline{x} = \underline{b}, \underline{x} \geq \underline{0}\}$ and the current feasible basic solution $\underline{x}' = (x_1, ..., x_m, x_{m+1}, ..., x_n) = (\hat{b}_1, ..., \hat{b}_m, 0, ..., 0)$ (Tableau 3.8).
Tableau 3.8 appears after adding the objective function vector $\underline{c}$ and the r-row to Tableau 3.1. The r-row is $w + 0x_1 + ... 0x_m + r_{m+1}x_{m+1} + ... + r_px_p + ... r_nx_n = w_0$.

	$\underline{c}'$	c_1	...	c_k	...	c_m	c_{m+1}	...	c_p	...	c_n	
	w	x_1	...	x_k	...	x_m	x_{m+1}	...	x_p	...	x_n	
	0	1	...	0	...	0	$\hat{a}_{1m+1}$	...	$\hat{a}_{1p}$	...	$\hat{a}_{1n}$	$\hat{b}_1$
	.	.		.		.	.		.		.	.
	.	.		1		.	.		$\hat{a}_{kp}$		.	$\hat{b}_k$
	.	.		.		.	.		.		.	.
	0	0	...	0	...	1	$\hat{a}_{m\,m+1}$	...	$\hat{a}_{mp}$	...	$\hat{a}_{mn}$	$\hat{b}_m$
r-row $\rightarrow$	1	0	...	0	...	0	r_{m+1}	...	r_p	...	r_n	w_0

Tableau 3.8. The simplex tableau.

Given that the problem is a maximisation problem:
- If $r_j \geq 0$ for all $j = m+1, ..., n$, the current basic solution $\underline{x}' = (\underline{x}_B', \underline{x}_N') = (\hat{b}_1, ..., \hat{b}_m, 0, ..., 0)$ is optimal. The optimal objective function value $w = w_0$ can be found in the bottom right-hand corner of the tableau.
- If a negative r_j exists, it is profitable to include x_j in the basis.
- If more than one $r_j < 0$, selecting one of the x_j with $r_j < 0$ will probably improve the objective function value. Usually one selects x_p to enter the basis with:

$$r_p = \min_j\{r_j \mid r_j < 0\} \tag{3.28}$$

So, a ('promising') variable x_p with the most negative r_j will enter the basis (see remarks). An arbitrary choice can be made when more variables fulfil (3.28).

Remarks

1 Criterion (3.28) doesn't necessarily lead to the largest improvement of the objective function value w.

If x_p enters the basis with $r_p < 0$ and other nonbasic variables are kept at zero:

$$\begin{array}{rcccl} x_1 & = & \hat{b}_1 & - & \hat{a}_{1p}x_p \\ \vdots & & & & \\ x_k & = & \hat{b}_k & - & \hat{a}_{kp}x_p \\ \vdots & & & & \\ x_m & = & \hat{b}_m & - & \hat{a}_{mp}x_p \end{array}$$

and $$w = w_0 - r_px_p$$

The improvement of the objective function value equals $-r_px_p$. As with $x_1 \geq 0, ..., x_m \geq 0$, the maximum value of x_p is determined by (Section 3.8):

$$x_p = \min_i \left\{ \frac{\hat{b}_i}{\hat{a}_{ip}} \mid \hat{a}_{ip} > 0 \right\} = \frac{\hat{b}_k}{\hat{a}_{kp}}$$

So, the improvement $-r_px_p$ of the objective function value depends on r_p as well as on $x_p = \frac{\hat{b}_k}{\hat{a}_{kp}}$. Note that in case of a degenerate solution such as $\hat{b}_k = 0$, the new basic solution provides no improvement to the objective function value.

2 If the improvement $-r_px_p$ is as large as possible in successive iterations, it does not necessarily imply that the number of iterations for reaching the optimum is as small as possible. This is illustrated by Example 3.10. As the number of iterations for reaching the optimum can not be predicted in advance, a simple criterion like (3.28) is taken.

Example 3.10.

In Figure 3.1, vertex (4) is the optimal solution.

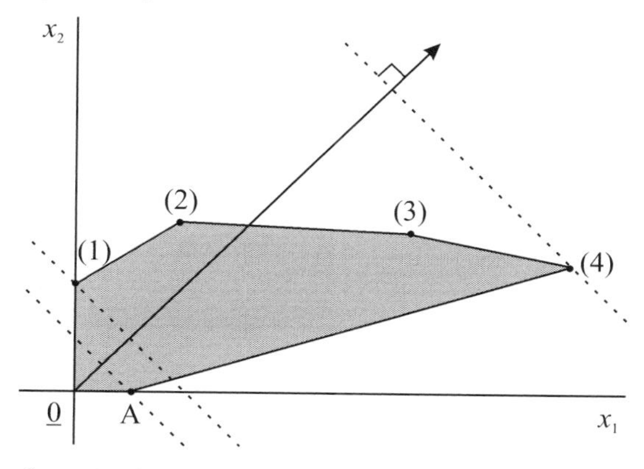

Figure 3.1. Number of required iterations.

- Moving towards vertex (1) from the origin 0, initially yields the largest improvement to the objective function value. So the optimal vertex (4) is reached from (1) by moving over vertices (2) and (3). So this path requires four iterations in total.

- Starting in the origin $\underline{0}$, we may decide not to apply criterion (3.28) and move to vertex A. Now, the improvement of the objective function value is less than moving to vertex (1). However, a first move to vertex A requires only two iterations in order to reach the optimal solution in vertex (4) (see Figure 3.1).

After selecting x_p to enter the basis, a pivoting operation is done with pivot a_{kp}. Now the r-row can be transformed comparable to the other rows i ($i \neq k$); i.e. if x_p enters the basis we have to make $r_p = 0$. This implies adding $-r_p / a_{kp}$ times the pivoting row to the r-row. The new objective function value appears in the lower right corner of the tableau.

Example 3.11.

$$\max\{w = 6x_1 - 15x_2 + 8x_3\}$$
$$-x_1 + x_2 - x_3 \leq 1$$
$$x_1 - 2x_2 + x_3 \leq 7$$
$$-x_1 + 2x_2 + x_3 \leq 4$$
$$x_1 \geq 0, x_2 \geq 0, x_3 \geq 0$$

Three slack variables y_1, y_2 and y_3 are added to the problem to convert the problem into standard form. The corresponding cost coefficients of the slack variables are of course zero. The objective function is:

$$\max\{w = 6x_1 - 15x_2 + 8x_3 + 0y_1 + 0y_2 + 0y_3\}$$

The starting tableau is:

	$\underline{c}'$	6	-15	8	0	0	0	
	w	x_1	x_2	x_3	y_1	y_2	y_3	
	0	-1	1	-1	1	0	0	1
$\underline{c}_B' = (0, 0, 0) \Leftrightarrow$	0	1	-2	1	0	1	0	7
	0	-1	2	**1**	0	0	1	4
	1	-6	15	-8	0	0	0	0

Tableau 3.9. Starting tableau.

$\underline{x}_B' = (y_1, y_2, y_3) = (1, 7, 4)$; $\underline{x}_N' = (x_1, x_2, x_3) = (0, 0, 0)$; objective function value $w = w_0 = 0$; feasible basic solution $\underline{x}' = (x_1, x_2, x_3, y_1, y_2, y_3) = (0, 0, 0, 1, 7, 4)$.

There are two ways to add the r-row to the tableau:

1. express w as a function of the nonbasic variables. In the tableau the nonbasic variables are x_1, x_2 and x_3, so the *r-row* in this example can be written as:
   $w - 6x_1 + 15x_2 - 8x_3 = 0$

2. using the formula $r_j = \underline{c}_B' B^{-1}\underline{a}_j - c_j = \underline{c}_B' \hat{\underline{a}}_j - c_j$; $\underline{c}_B' = (0, 0, 0)$

$$r_1 = (0, 0, 0)\begin{pmatrix}-1\\1\\-1\end{pmatrix} - 6 = -6; \quad r_2 = (0,0,0)\begin{pmatrix}1\\-2\\2\end{pmatrix} - (-15) = 15; \quad r_3 = -8$$

For the basic variables $r_4 = (0,\ 0,\ 0)\begin{pmatrix}1\\0\\0\end{pmatrix} - 0 = 0$, $r_5 = r_6 = 0$

The objective function value is $w = \underline{c}_B'\,\underline{x}_B = (0,\ 0,\ 0)\begin{pmatrix}1\\7\\4\end{pmatrix} = 0$

The corresponding solution is not optimal as r_1 and r_3 are negative. The variables x_1 or x_3 can be exchanged with a basic variable. As $r_3 = \min\{r_1, r_3\} = \min\{-6, -8\} = -8$, first x_3 will enter the basis. From $\min_i\left\{\frac{b_i}{a_{i3}} \mid a_{i3} > 0\right\} = \min\left\{\frac{7}{1}, \frac{4}{1}\right\} = 4$ the pivot $a_{33} = 1$ is derived, leading to:

	$\underline{c}'$	6	-15	8	0	0	0	
	w	x_1	x_2	x_3	y_1	y_2	y_3	
	0	-2	3	0	1	0	1	5
$\underline{c}_B' = (0, 0, 8) \Leftrightarrow$	0	2	-4	0	0	1	-1	3
	0	-1	2	1	0	0	1	4
	1	-14	31	0	0	0	8	32

Tableau 3.10.

Variable x_3 has entered the basis with a value of 4 pushing out the variable y_3. The value w of the objective function has increased by $-r_p x_p = r_3 x_3 = 8 \times 4 = 32$. Analogously, we can derive that if x_1 enters the basis in Tableau 3.9, an increase in one iteration of $w = r_1 x_1 = 6 \times 7 = 42$ will follow. Tableau 3.10 corresponds to: $\underline{x}_B' = (y_1, y_2, x_3) = (5, 3, 4)$; $\underline{c}_B' = (0, 0, 8)$.

Note that the order of the elements in $\underline{c}_B$ matters; the order corresponds to the order of the elements of $\underline{x}_B$. This is the order of the unit vectors in the tableau:

$$\underline{x}_B = \begin{pmatrix}y_1\\y_2\\x_3\end{pmatrix};\ \underline{c}_B = \begin{pmatrix}c_{y_1}\\c_{y_2}\\c_{x_3}\end{pmatrix} = \begin{pmatrix}0\\0\\8\end{pmatrix} \begin{matrix}\rightarrow & \underline{e}_1 \text{ corresponding to } y_1\\ \rightarrow & \underline{e}_2 \text{ corresponding to } y_2\\ \rightarrow & \underline{e}_3 \text{ corresponding to } x_3\end{matrix}$$

The objective function coefficient c_{y_1} belongs to variable y_1; c_{y_2} corresponds to y_2, etc.

Remarks

The r-row of Tableau 3.10 can be calculated in three ways.

1. The easiest way is to pivot the r-row. In the example of Tableau 3.9, row 3 should be added 8 times to the *r-row*.

2. Calculating r_j via $r_j = \underline{c}_B'\,\underline{\hat{a}}_j - c_j = \underline{c}_B'\,B^{-1}\underline{a}_j - c_j$.

   In this formula $\underline{a}_j$ is the column vector j in starting Tableau 3.9 and $\underline{\hat{a}}_j = B^{-1}\underline{a}_j$ is column vector j in Tableau 3.10 that appears after pivoting Tableau 3.9.

$$B^{-1} = \begin{pmatrix} 1 & 0 & 1 \\ 0 & 1 & -1 \\ 0 & 0 & 1 \end{pmatrix}; \quad \hat{\underline{a}}_1 = B^{-1}\underline{a}_1 = \begin{pmatrix} 1 & 0 & 1 \\ 0 & 1 & -1 \\ 0 & 0 & 1 \end{pmatrix}\begin{pmatrix} -1 \\ 1 \\ -1 \end{pmatrix} = \begin{pmatrix} -2 \\ 2 \\ -1 \end{pmatrix}; \quad \hat{\underline{a}}_2 = \begin{pmatrix} 3 \\ -4 \\ 2 \end{pmatrix} \text{ etc.}$$

$$r_1 = \underline{c}_B'\, \hat{\underline{a}}_1 - c_1 = (0,\ 0,\ 8)\begin{pmatrix} -2 \\ 2 \\ -1 \end{pmatrix} - 6 = -14; \quad r_2 = (0,\ 0,\ 8)\begin{pmatrix} 3 \\ -4 \\ 2 \end{pmatrix} + 15 = 31$$

$$r_3 = (0,\ 0,\ 8)\begin{pmatrix} 0 \\ 0 \\ 1 \end{pmatrix} - 8 = 0 \text{ etc. }; \quad w_0 = \underline{c}_B'\, \hat{\underline{b}} = (0,\ 0,\ 8)\begin{pmatrix} 5 \\ 3 \\ 4 \end{pmatrix} = 32$$

3. By expressing the objective function, $w = 6x_1 - 15x_2 + 8x_3$ in the nonbasic variables. In Tableau 3.10 the nonbasic variables are x_1, x_2 and y_3. Variable x_3 is basic, so it should be eliminated from the function w. This can be done with the help of the third equation in Tableau 3.10: $-x_1 + 2x_2 + x_3 + y_3 = 4$, or $x_3 = 4 + x_1 - 2x_2 - y_3$. So x_3 is expressed as a function of the nonbasic variables. Substitution in $w = 6x_1 - 15x_2 + 8x_3$ gives:

   $w = 6x_1 - 15x_2 + 8(4 + x_1 - 2x_2 - y_3)$, or
   $w - 14x_1 + 31x_2 + 8y_3 = 32$.

   The last equation corresponds to the r-row in Tableau 3.10.

Tableau 3.10 is not optimal as $r_1 = -14$. The nonbasic variable x_1 should enter the basis via the pivot $a_{21} = 2$; this leads to the final Tableau 3.11.

	$\underline{c}'$	6	-15	8	0	0	0	
	w	x_1	x_2	x_3	y_1	y_2	y_3	
	0	0	-1	0	1	1	0	8
$\underline{c}_B' = (0, 6\ 8) \Leftrightarrow$	0	1	-2	0	0	1/2	-1/2	3/2
	0	0	0	1	0	1/2	1/2	11/2
	1	0	3	0	0	7	1	53

Tableau 3.11. Final tableau.

As $r_j \geq 0$ ($j = 1, ..., 6$), $\underline{x}' = (3/2, 0, 11/2, 8, 0, 0)$ is the optimal, non-degenerate, basic solution. The optimal objective function value $w = \underline{c}'\underline{x} = 6\times3/2 - 15\times0 + 8\times11/2 + 0\times8 + 0\times0 + 0\times0 = 53$.

Remarks

1. The w-column does not change during the iterations.

2. The optimal basis B consists of the column vectors in starting Tableau 3.9 corresponding to the variables (y_1, x_1, x_3); in that order! So,

$$B = \begin{pmatrix} 1 & -1 & -1 \\ 0 & 1 & 1 \\ 0 & -1 & 1 \end{pmatrix}$$

3. B^{-1} can be found in final Tableau 3.11 where E_3 was situated in the starting tableau:

$$B^{-1} = \begin{pmatrix} 1 & 1 & 0 \\ 0 & 1/2 & -1/2 \\ 0 & 1/2 & 1/2 \end{pmatrix}$$

4. E_3 is located in final Tableau 3.11 where B was in starting Tableau 3.9; so corresponding to the columns of y_1, x_1, x_3.

5. In final Tableau 3.11, $r_1 = \underline{c}_B' B^{-1} \underline{a}_1 - c_1$ with $\underline{c}_B' = (0, 6, 8)$ and $\underline{a}_1 = \begin{pmatrix} -1 \\ 1 \\ -1 \end{pmatrix}$ the first column in starting Tableau 3.9. The first column in the final tableau is

$$\hat{\underline{a}}_1 = B^{-1}\underline{a}_1 = \begin{pmatrix} 1 & 1 & 0 \\ 0 & 1/2 & -1/2 \\ 0 & 1/2 & 1/2 \end{pmatrix} \begin{pmatrix} -1 \\ 1 \\ -1 \end{pmatrix} = \begin{pmatrix} 0 \\ 1 \\ 0 \end{pmatrix} \text{ such that } r_1 = (0, 6, 8)\begin{pmatrix} 0 \\ 1 \\ 0 \end{pmatrix} - 6 = 0$$

Analogously, $\hat{\underline{a}}_2 = B^{-1}\underline{a}_2$; $r_2 = 3$; $\hat{\underline{b}} = B^{-1}\underline{b}$ and $w = \underline{c}_B' \hat{\underline{b}} = 53$

Example 3.12.

$$\max\{w = 2x_1 + 5x_2\}$$

$$\begin{aligned} x_1 \qquad &\leq 4 \\ x_2 &\leq 3 \\ x_1 + x_2 &\leq 6 \\ x_1 \geq 0,\ x_2 &\geq 0 \end{aligned}$$

Example 3.12 is the ABE example of Chapter 2. The solution process of the Simplex Method can be compared to the graphical solution presented in Chapter 2. In particular, the resemblance of vertices to feasible basic solutions can be illustrated.

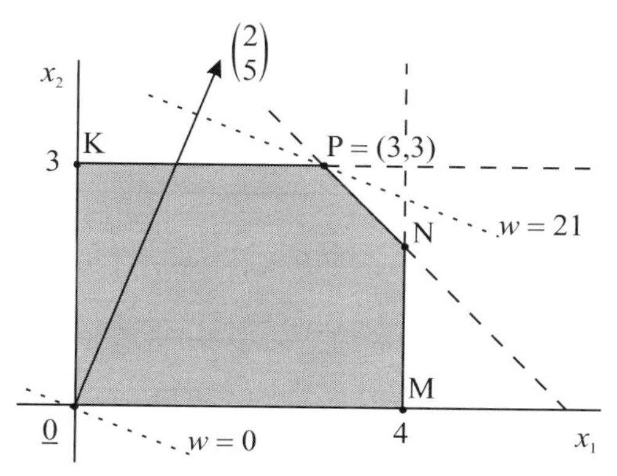

Figure 3.2. Graphical solution of the ABE example.

As we have seen in Chapter 2, the optimal vertex is point P = (3, 3); with optimal objective function value $w = 2\times3 + 5\times3 = 21$. Next, we follow the Simplex Algorithm.

	$\underline{c}'$	2	5	0	0	0		
	w	x_1	x_2	y_1	y_2	y_3		
	0	1	0	1	0	0	4	
$\underline{c}_B' = (0, 0, 0)$	0	0	**1**	0	1	0	3	(1)
	0	1	1	0	0	1	6	
	1	-2	-5	0	0	0	0	
	0	1	0	1	0	0	4	
$\underline{c}_B' = (0, 5, 0)$	0	0	1	0	1	0	3	(2)
	0	**1**	0	0	-1	1	3	
	1	-2	0	0	5	0	15	
	0	0	0	1	1	-1	1	
$\underline{c}_B' = (0, 5, 2)$	0	0	1	0	1	0	3	(3)
	0	1	0	0	-1	1	3	
	1	0	0	0	3	2	21	

Tableaus 3.12. (1) to (3).

Tableau 3.12 (3) is optimal because $\underline{r} \geq \underline{0}$. The optimal basic solution is $\underline{x}' = (x_1, x_2, y_1, y_2, y_3) = (3, 3, 1, 0, 0)$. The corresponding optimal objective function value $w = 21$.

Remarks

1. The w-column does not change during the pivoting. Therefore, in the sequel this column will be omitted.

2. Consider the correspondence of the vertices in Figure 3.2 to the feasible basic solutions in the tableaus of 3.12. Tableau (1) shows a feasible basic solution $\underline{x}' = (x_1, x_2, y_1, y_2, y_3) = (0, 0, 4, 3, 6)$, so $x_1 = 0$ and $x_2 = 0$.
   Tableau (1): $x_1 = 0, x_2 = 0 \rightarrow$ corresponds to vertex $\underline{0}$ in Figure 3.2,
   Tableau (2): $x_1 = 0, x_2 = 3 \rightarrow$ corresponds to vertex K = (0, 3) in Figure 3.2,
   Tableau (3): $x_1 = 3, x_2 = 3 \rightarrow$ corresponds to the optimal vertex P = (3, 3) in Figure 3.2.

3. When variable x_1 enters the basis in Tableau (1) via pivot $a_{11} = 1$, vertex P = (3, 3) is reached via tableaus that correspond to the vertices M = (4, 0), N = (4, 2) and P = (3, 3).

4. Consider the objective function $w = 2x_1 + 2x_2$. Graphically it can easily be seen that the vertices P = (3, 3) and N = (4, 2) are both optimal. In that case the problem has alternative optimal solutions (Section 3.13). All points on the line segment between P and N, described by the convex combination $\lambda\begin{pmatrix}3\\3\end{pmatrix} + (1-\lambda)\begin{pmatrix}4\\2\end{pmatrix}$, $0 \leq \lambda \leq 1$, are optimal.

   $\lambda = 0$ gives the optimal feasible basic solution $\underline{x}' = (4, 2, 0, 1, 0)$ and
   $\lambda = 1/2$ gives the optimal feasible nonbasic solution $\underline{x}' = (3\frac{1}{2}, 2\frac{1}{2}, \frac{1}{2}, \frac{1}{2}, 0)$.

   Solving with $w = 2x_1 + 2x_2$ also gives, besides the usual $r_j = 0$ for all basic variables, an element $r_j = 0$ for a nonbasic variable. Calculate the r-row in Tableau 3.12 (3) with $\underline{c}_B' = (0, 2, 2)$ and notice that $r_{y_2} = 0$.

Theorem 3.4.

If in every iteration the basic solution $\underline{x}_B$ is non-degenerate, the optimal basic solution is reached in a finite number of iterations.

Proof

As the number of feasible basic solutions is finite (at most $\begin{bmatrix} n \\ m \end{bmatrix}$), not reaching the optimal solution in a finite number of steps implies that a feasible basic solution is found repeatedly during the iterations of the Simplex Algorithm.
This also implies that a particular value of the objective function is found repeatedly. However, given the non-degeneracy of the bases, the value of the objective function is in every iteration strictly higher than the value of the objective function in the former iteration. Therefore, the value of the objective function cannot be the same during successive iterations. In other words: the same basis cannot appear twice.

3.12. Degeneracy, cycling, lexicographical method

The proof of Theorem 3.4 requires the assumption that basic solution $\underline{x}_B$ is not degenerate, i.e. $\underline{x}_B$ has m positive elements. Now we consider what happens if *degeneracy* appears during the iterations. Consider a basic variable x_k with value 0. When x_p enters the basis pushing out x_k with the pivot $a_{kp} > 0$ the entering variable x_p will have the value 0. This implies that the new solution is also degenerate and the objective function value does not change as the increase equals $-r_p x_p = 0$.

A theoretical problem that seldom appears in practice is called *cycling*; i.e. the same basic solution appears after a number of degenerate basic transitions. In this case the procedure does not converge, but repeats itself infinitely. Although cycling is very rare in practice, it is a fascinating phenomenon from a mathematical point of view. Several approaches have been introduced to avoid degeneracy and thus to prevent the possibility of cycling. One of the methods to prevent cycling is the so-called *lexicographical method*. By means of the lexicographical method it is possible to prove the convergence of the Simplex Algorithm. For details on resolving degeneracy and cycling we refer to more advanced textbooks on linear programming, e.g. (Dantzig and Thapa, 2003).

3.13. Alternative optimal solutions and unbounded solutions

This section discusses how the final simplex tableau indicates that the LP problem has alternative optimal solutions (3.13.1) or an unbounded solution (3.13.2). We refer to Chapter 2 (Section 2.3) for the graphical illustration of alternative optimal and unbounded solutions.

3.13.1. Alternative optimal solutions

Consider the final simplex tableau with $\underline{r} \geq \underline{0}$. If :

a. all nonbasic variables x_j have $r_j > 0$, the optimal solution is unique.
b. a nonbasic variable x_p exists with $r_p = 0$ and there is a $a_{kp} > 0$, then the variable x_p can enter the basis though the objective function value does not increase; $-r_p x_p = 0$. This

shows the existence of an alternative optimal basic solution and a *line segment of alternative optimal solutions* (Figure 2.3).

c. there exists a nonbasic variable x_p with $r_p = 0$ and all $a_{ip} \leq 0$ $(i = 1, ..., m)$, then x_p can be made positive unboundedly: this implies the problem has a *half-line of alternative optimal solutions* (Figure 2.4).

The complete set of optimal solutions of an LP problem can be described by combining the alternative optimal solutions in b. and c. This is illustrated in Example 3.13.

Example 3.13.

Simplex Tableau 3.13 is optimal as $\underline{r} \geq \underline{0}$ (maximisation problem).

	1	1	3	-4	
	x_1	x_2	x_3	x_4	
	1	0	2	-3	2
$\underline{c}_B' = (1, 1)$	0	1	1	-1	4
	0	0	0	0	6

Tableau 3.13.

The corresponding optimal basic solution is $\underline{x}_1' = (x_1, x_2, x_3, x_4) = (2, 4, 0, 0)$; and the objective function value $w = 6$. The fact that both $r_3 = r_4 = 0$ for the nonbasic variables x_3 and x_4 respectively, shows the existence of alternative optimal solutions. First we let nonbasic variable x_3 enter the basis with pivot $a_{13} = 2$ such that Tableau 3.14 appears:

	1	1	3	-4	
	x_1	x_2	x_3	x_4	
	1/2	0	1	-3/2	1
$\underline{c}_B' = (3, 1)$	-1/2	1	0	1/2	3
	0	0	0	0	6

Tableau 3.14.

Now, the optimal basic solution $\underline{x}_2' = (0, 3, 1, 0)$; and the objective function value $w = 6$ remains the same. The line segment of alternative optimal solutions can be described as a convex combination of the optimal basic solutions $\underline{x}_1$ and $\underline{x}_2$: $\underline{x} = \lambda_1\underline{x}_1 + \lambda_2\underline{x}_2$, $(\lambda_1 + \lambda_2 = 1, \lambda_1, \lambda_2 \geq 0)$.

Tableau 3.14 shows two variables that can enter the basis:

- x_1, because $r_1 = 0$; pivot $a_{11} = 1/2$ leads to Tableau 3.13.
- x_4, because $r_4 = 0$; pivot $a_{24} = 1/2$ leads to Tableau 3.15:

	1	1	3	-4	
	x_1	x_2	x_3	x_4	
	-1	3	1	0	10
$\underline{c}_B' = (3, -4)$	-1	2	0	1	6
	0	0	0	0	6

Tableau 3.15.

with the corresponding optimal basic solution $\underline{x}_3' = (0, 0, 10, 6)$ and the objective function value $w = 6$. In Tableau 3.13 $r_4 = 0$. However, x_4 can not enter the basis as column 4 does not contain any positive element. Making $x_4 = \mu_1 \geq 0$ leads to:

$$\begin{aligned} x_1 &= 2 + 3\mu_1 \\ x_2 &= 4 + \mu_1 \\ x_3 &= 0 \\ x_4 &= \mu_1 \end{aligned} \qquad \text{so} \qquad \underline{x}_4 = \begin{pmatrix} x_1 \\ x_2 \\ x_3 \\ x_4 \end{pmatrix} = \begin{pmatrix} 2 \\ 4 \\ 0 \\ 0 \end{pmatrix} + \mu_1 \begin{pmatrix} 3 \\ 1 \\ 0 \\ 1 \end{pmatrix} \text{ with } \mu_1 \geq 0$$

Note that x_3 is a nonbasic variable in Tableau 3.13 and remains nonbasic in case the variable x_4 enters the basis with value $x_4 = \mu_1$. Above, $\underline{x}_4$ represents a parameter equation describing a half-line of alternative feasible optimal solutions.

Tableau 3.15 also shows another half-line $\underline{x}_5' = (0, 0, 10, 6) + \mu_2(1, 0, 1, 1)$ with $\mu_2 \geq 0$.

Writing the convex combination of the alternative basic solutions ($\underline{x}_1$, $\underline{x}_2$ and $\underline{x}_3$) and the non-negative combination of the half-lines $\underline{x}_4$ and $\underline{x}_5$ gives the convex set C of all alternative optimal solutions:

$$C = \left\{ \underline{x} \;\middle|\; \underline{x} = \lambda_1 \begin{pmatrix} 2 \\ 4 \\ 0 \\ 0 \end{pmatrix} + \lambda_2 \begin{pmatrix} 0 \\ 3 \\ 1 \\ 0 \end{pmatrix} + \lambda_3 \begin{pmatrix} 0 \\ 0 \\ 10 \\ 6 \end{pmatrix} + \mu_1 \begin{pmatrix} 3 \\ 1 \\ 0 \\ 1 \end{pmatrix} + \mu_2 \begin{pmatrix} 1 \\ 0 \\ 1 \\ 1 \end{pmatrix}; \quad \begin{matrix} \lambda_1, \lambda_2, \lambda_3, \mu_1, \mu_2 \geq 0 \\ \text{and} \\ \lambda_1 + \lambda_2 + \lambda_3 = 1 \end{matrix} \right\}$$

For instance $\underline{x}_6' = (x_1, x_2, x_3, x_4) = (131, 12, 105, 113)$ fulfils the constraints and has objective function value $w = 6$. Vector $\underline{x}_6$ appears when choosing $\lambda_1 = 1/2$, $\lambda_2 = 0$, $\lambda_3 = 1/2$, $\mu_1 = 10$ and $\mu_2 = 100$.

3.13.2. Unbounded solutions

If in the simplex tableau (maximisation problem):

- $r_p < 0$, the tableau is not optimal yet; i.e. it is profitable to make variable x_p positive,
- column $\underline{a}_p$ corresponding to x_p contains no positive elements (all $a_{ip} \leq 0$), then the LP problem has an *unbounded solution* (Figure 2.5).

Example 3.14.

The next tableau (Tableau 3.16) is not optimal because $r_3 = -25 < 0$.
The objective function value $w - 25x_3 = 19$ so $w = 19 + 25x_3$.

	$\underline{c}'$:	4	5	2	
		x_1	x_2	x_3	
$\underline{c}_B' = (4, 5)$		1	0	-2	1
		0	1	-3	3
		0	0	-25	19

Tableau 3.16.

It is useful to make x_3 positive:

$$x_1 - 2x_3 = 1 \quad \rightarrow \quad x_1 = 1 + 2x_3 \geq 0 \quad \text{for all } x_3 \geq 0$$
$$x_2 - 3x_3 = 3 \quad \rightarrow \quad x_2 = 3 + 3x_3 \geq 0 \quad \text{for all } x_3 \geq 0 \quad \text{such that}$$

$$\underline{x} = \begin{pmatrix} x_1 \\ x_2 \\ x_3 \end{pmatrix} = \begin{pmatrix} 1 \\ 3 \\ 0 \end{pmatrix} + x_3 \begin{pmatrix} 2 \\ 3 \\ 1 \end{pmatrix} \text{ is feasible for all values } x_3 \geq 0$$

Conclusion
The objective function value $w = 19 + 25x_3$ is unbounded.

3.14. The two-phase method

In Section 3.8 we discussed a method to convert a basic feasible solution into another basic feasible solution (basic transition). The Simplex Method requires a starting basic feasible solution. In this section the so-called *two-phase method* is presented. This method is a systematic procedure with which we can either convert an LP problem into a basic feasible form or come to the conclusion that the LP problem has no feasible solutions at all.

For application of the Simplex Method the following conditions should be satisfied:

1. all variables ≥ 0;
2. $\underline{b} \geq \underline{0}$;
3. only equality restrictions.

These three conditions can be satisfied with the transformation rules in paragraph 3.3.

4. The problem has to be in basic feasible form; i.e. a set of m (linearly independent) unit vectors should exist.

In the examples discussed so far, condition 4 was always fulfilled. If $\underline{b} \geq \underline{0}$ and all constraints have the sign $\leq$, the addition of slack variables $y_i \geq 0$ automatically leads to a unit matrix. If the LP problem also contains greater than or equal ($\geq$) constraints and/or equality ($=$) constraints, the unit matrix does not necessarily appear in the starting tableau.

Example 3.15.
Given system (A) of an LP problem

$$\begin{aligned} -x_1 - 2x_2 - 3x_3 &= -7 \\ 4x_1 - 3x_2 + 2x_3 &\leq 6 \\ 4x_1 + 5x_2 - 6x_3 &\geq 5 \\ x_1 + x_2 + x_3 &= 4 \\ x_1 \geq 0, x_2 \geq 0, x_3 &\geq 0 \end{aligned} \qquad \text{(A)}$$

Adding the slack variables $y_2 \geq 0$, $y_3 \geq 0$ and transforming constraint 1 leads to Tableau (B):

x_1	x_2	x_3	y_2	y_3	
1	2	3	0	0	7
4	-3	2	1	0	6
4	5	-6	0	-1	5
1	1	1	0	0	4

(B)

Tableau 3.17.

Note that we index the slack variables corresponding to the sequence number of the constraints. This is not necessary, but clarifies the unit vectors they relate to in later iterations. Tableau 3.17 does not fulfil the fourth condition (a set of m unit vectors), because three unit vectors $\underline{e}_1$, $\underline{e}_3$ and $\underline{e}_4$ are absent. In the two-phase method the missing unit vectors $\underline{e}_i$ and corresponding so-called *artificial variables* $z_i \geq 0$ are added.

Example 3.16.

In Tableau 3.17 the unit vectors $\underline{e}_1$, $\underline{e}_3$ and $\underline{e}_4$ are missing. Adding these missing vectors and the corresponding artificial variables $z_1 \geq 0$, $z_3 \geq 0$ and $z_4 \geq 0$, yields Tableau 3.18 (C).

x_1	x_2	x_3	y_2	y_3	z_1	z_3	z_4	
1	2	3	0	0	1	0	0	7
4	-3	2	1	0	0	0	0	6
4	5	-6	0	-1	0	1	0	5
1	1	1	0	0	0	0	1	4

(C)

Tableau 3.18.

It can easily be seen that the basic solution corresponding to Tableau 3.18

$$\underline{x}' = (x_1, x_2, x_3, y_2, y_3, z_1, z_3, z_4) = (0, 0, 0, 6, 0, 7, 5, 4)$$

does not satisfy system (B), so neither does it satisfy (A). The values $x_1 = x_2 = x_3 = 0$ do not satisfy the first, third and fourth restriction of (A). Only if $z_1 = z_3 = z_4 = 0$, system (C) is equivalent with system (B) and system (A). To achieve this, the so-called *phase-1-objective function* $\max\{w_I = -z_1 - z_3 - z_4\}$ is added to Tableau 3.18. Now, the Simplex Method is applied with this phase-1 objective function.

For every solution $w_I \leq 0$. Two situations are possible:

1. If the optimal value of $w_I = 0$, then all artificial variables z_i have value $z_i = 0$. So all z_i are nonbasic variables (for degeneracy, see Example 3.19). So the original variables x_j and/or y_i have replaced the artificial variables as the basic variables. In other words: if $w_I = 0$, then after the optimization Tableau 3.18 contains a system of unit vectors associated with variables x_j en y_i. Of course, this also holds for Tableau 3.17.

2. If the optimal value of $w_I < 0$, then it is apparently impossible to determine a basic feasible solution for system (C) with all artificial variables $z_i = 0$. In this case system (B) does not have a feasible solution. If (B) has a feasible solution ($\underline{x}'$, $\underline{y}'$), then ($\underline{x}'$, $\underline{y}'$) is a feasible solution for (C) too (with all $z_i = 0$) and yields $w_I = 0$.

Conclusion phase-1

The optimization in *phase-1*:

- either yields a basic feasible solution (if $w_I = 0$).
- or concludes that the system has no feasible solution at all (if $w_I < 0$).

After determination of a basic feasible solution with $w_I = 0$ in phase-1, the original objective function $w = \underline{c}'\underline{x}$ can be optimised. This optimization is called *phase-2*.

Summarizing the two-phase method

Make sure (if necessary, use transformations) that the LP problem is written in standard form:

$$\begin{aligned} &\max\{w = \underline{c}'\underline{x}\} \\ &\qquad A\underline{x} = \underline{b} \quad \text{with } \underline{b} \geq \underline{0} \\ &\qquad \underline{x} \geq \underline{0} \end{aligned} \qquad \text{(P)}$$

Phase-1. Find a basic feasible solution of (P):

a. Add the missing unit vectors and the corresponding artificial variables $z_i \geq 0$.

b. Add the phase-1 objective function $\max\{w_I = -\Sigma z_i\}$ and find the optimal tableau.
   - If $w_I < 0$ then (P) has no feasible solutions.
   - If $w_I = 0$ go to phase-2.

Phase-2 Determine the optimal solution of (P)
Add the original objective function $\max\{w = \underline{c}'\underline{x}\}$ to the final tableau of phase 1 and optimise w over the alternative optimal solutions of the phase-1 problem . (Example 3.17).

Example 3.17.

This example illustrates the two-phase method. A graphical interpretation is given too. Given LP problem (P)

$$\begin{aligned} &\max\{w = x_1 + 4x_2\} \\ &\qquad -x_1 \qquad\quad \leq -2 \\ &\qquad\qquad\quad x_2 \geq \ \ 1 \\ &\qquad x_1 + \ \ x_2 \leq \ \ 6 \\ &\qquad x_1 \geq 0,\ x_2 \geq 0 \end{aligned} \qquad \text{(P)}$$

Problem (P) is transformed into standard form:

$$\begin{aligned} &\max\{w = \underline{c}'\underline{x}\}\ ; \\ &\quad A\underline{x} = \underline{b}\ ;\ \ \underline{b} \geq \underline{0} \\ &\quad \underline{x} \geq \underline{0}\,, \end{aligned}$$

yielding problem $(\hat{\text{P}})$:

$$\begin{aligned} &\max\{w = x_1 + 4x_2\} \\ &\qquad x_1 \quad - y_1 \qquad\qquad\quad = 2 \\ &\qquad\quad x_2 \qquad - y_2 \qquad = 1 \\ &\qquad x_1 + x_2 \qquad\qquad + y_3 = 6 \\ &\qquad x_1,\ x_2,\ y_1,\ y_2,\ y_3 \geq 0 \end{aligned} \qquad (\hat{\text{P}})$$

In problem $(\hat{P})$ the two unit vectors $\underline{e}_1$ and $\underline{e}_2$ are missing. So the two-phase method is used to determine a basic feasible solution of $(\hat{P})$.

Phase-1

Add the missing unit vectors $\underline{e}_1$ and $\underline{e}_2$ and the corresponding artificial variables $z_1 \geq 0$ and $z_2 \geq 0$. The phase-1 objective is $\max\{w_I = -z_1 - z_2\}$ i.e. $\max\{w_I = 0x_1 + 0x_2 + ... + 0y_3 - z_1 - z_2\}$ so $\max\{w_I = \underline{c}_I'\underline{x}\}$ with $\underline{c}_I'$=(0, 0, 0, 0, 0, −1, −1). The corresponding Tableau 3.19 is:

$\underline{c}_I'$:	0	0	0	0	0	−1	−1	
	x_1	x_2	y_1	y_2	y_3	z_1	z_2	
	1	0	−1	0	0	1	0	2
$\underline{c}_B'$=(−1, −1, 0)	0	1	0	−1	0	0	1	1
	1	1	0	0	1	0	0	6
	−1	−1	1	1	0	0	0	−3

Tableau 3.19. Initial tableau phase-1.

Tableau 3.19 is not optimal because $r_1 < 0$ and $r_2 < 0$. The values $x_1 = 0$ and $x_2 = 0$ (x_1 and x_2 are nonbasic) do not satisfy the constraints of problem (P). Choosing x_1 as entering basic variable with pivot $a_{11} = 1$ gives Tableau 3.20:

$\underline{c}_I'$:	0	0	0	0	0	−1	−1	
	x_1	x_2	y_1	y_2	y_3	z_1	z_2	
	1	0	−1	0	0	1	0	2
$\underline{c}_B'$=(0, −1, 0)	0	**1**	0	−1	0	0	1	1
	0	1	1	0	1	−1	0	4
	0	−1	0	1	0	1	0	−1

Tableau 3.20.

Tableau 3.20 is not optimal; $x_1 = 2$ and $x_2 = 0$ does not satisfy the constraints of (P). Variable x_2 enters the basis via pivot $a_{22} = 1$:

$\underline{c}_I'$:	0	0	0	0	0	−1	−1	
	x_1	x_2	y_1	y_2	y_3	z_1	z_2	
	1	0	−1	0	0	1	0	2
$\underline{c}_B'$=(0, 0, 0)	0	1	0	−1	0	0	1	1
	0	0	1	1	1	−1	−1	3
	0	0	0	0	0	1	1	0

Tableau 3.21.a Final tableau phase-1.

Tableau 3.21a is optimal because $\underline{r} \geq \underline{0}$.

- Phase-1 objective function value $w_I = 0$. The artificial variables z_1 and z_2 left the basis.
- There is a system of unit vectors corresponding to the original variables.
- $x_1 = 2$, $x_2 = 1$ satisfies the constraints of (P).

We can conclude that by phase-1 a basic feasible solution for problem $(\hat{P})$ has been determined.

Phase-2

The original objective function $w = x_1 + 4x_2$ is added to the final tableau of phase-1 and optimised over the alternative optimal solutions of the phase-1 problem. This implies that the variables that correspond to the positive elements in the r-row of the final tableau of phase-1, may not enter the basis in phase 2; this is called *blocking*. In this particular case the artificial variables z_1 and z_2 are blocked (denoted by an X under the r-value) in order to prevent them entering the basis again. A positive value of z_1 and/or z_2 implies that the conditions of (P) are no longer satisfied!

$\underline{c}'$:	1	4	0	0	0	0	0	
	x_1	x_2	y_1	y_2	y_3	z_1	z_2	
	1	0	−1	0	0	1	0	2
$\underline{c}_B'=(1, 4, 0)$	0	1	0	−1	0	0	1	1
	0	0	1	**1**	1	−1	−1	3
	0	0	−1	−4	0	1	4	6
						X	X	

Tableau 3.21.b. Initial tableau phase-2.

Tableau 3.21a is converted into the initial tableau of phase-2 (Tableau 3.21b) as follows:

1. the phase-1 objective vector $\underline{c}_1'\underline{x}$ is replaced by the phase-2 original objective vector $\underline{c}'$,
2. the r-row is calculated with the phase-2 objective vector $\underline{c}'$.

Tableau 3.21b is not optimal; y_2 enters the basis via pivot $a_{34} = 1$ yielding Tableau 3.22:

$\underline{c}'$:	1	4	0	0	0	0	0	
	x_1	x_2	y_1	y_2	y_3	z_1	z_2	
	1	0	−1	0	0	1	0	2
$\underline{c}_B'=(1, 4, 0)$	0	1	1	0	1	−1	0	4
	0	0	1	1	1	−1	−1	3
	0	0	3	0	4	−3	0	18
						X	X	

Tableau 3.22. Final tableau phase-2.

Tableau 3.22 is optimal because $\underline{r} \geq \underline{0}$ for the nonblocked variables. The artificial variable z_1 is blocked, so in spite of $r_6 = -3 \leq 0$, it may not enter the basis. For the optimal solution to problem (P) we found $x_1 = 2$, $x_2 = 4$ and $w = 18$.

Problem (P) in Example 3.17 has two variables which implies that the optimal solution can be determined graphically too (see Figure 3.3). The shaded area denotes the set of feasible solutions of (P).

Note that, contrary to prior examples, the origin is not a feasible vertex. The Tableaus 3.19, 3.20 and 3.21 in phase 1 correspond to the points A, B and C in Figure 3.3. Phase-1 ends after a basic feasible solution of $(\hat{P})$ is determined (vertex C in Figure 3.3). Phase-2 starts in C and finds the optimal vertex D.

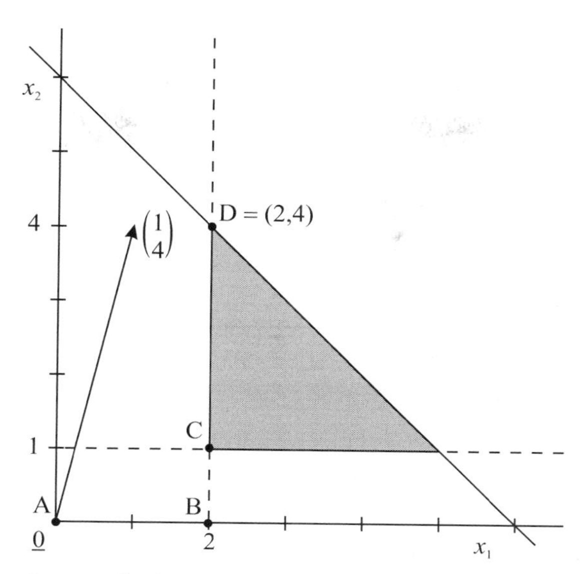

Figure 3.3. The two-phase method.

The next examples (3.18 and 3.19) illustrate the way in which the two-phase method detects:

- *inconsistent restrictions* i.e. there is no feasible solution (Example 3.18)
- *superfluous constraints* (Example 3.19).

Example 3.18.

$$\max\{w = 4x_1 + 5x_2\}$$
$$x_1 + x_2 \leq 3$$
$$x_1 + x_2 \geq 6$$
$$x_1 \geq 0,\ x_2 \geq 0$$

It may be obvious that the constraints $x_1 + x_2 \leq 3$ and $x_1 + x_2 \geq 6$ are inconsistent. Adding one artificial variable z_2 leads to Tableau 4.23 (1).

(1)

$\underline{c}_I'$:	0	0	0	0	−1	
	x_1	x_2	y_1	y_2	z_2	
$\underline{c}_B' = (0, -1)$	1	1	1	0	0	3
	1	1	0	−1	1	6
	−1	−1	0	1	0	−6

(2)

$\underline{c}_B' = (0, -1)$	1	1	1	0	0	3
	0	0	−1	−1	1	3
	0	0	1	1	0	−3

Tableau 4.23. (1) and (2) The initial and final tableau phase-1.

The final Tableau 4.23 (2) of phase 1 is optimal because $\underline{r} \geq \underline{0}$. However, $w_I = -3$ and z_2 is a basic variable with $z_2 = 3$. So it is not possible to remove z_2 from the basis. This implies that the restrictions are inconsistent; the system has no feasible solution.

Example 3.19.

Either one of the restrictions:

$$x_1 + x_2 = 2$$
$$2x_1 + 2x_2 = 4$$

is superfluous; the system is dependent. Dependent restrictions are detected in phase-1. Application of the two-phase method yields:

	$\underline{c}_1'$:	0	0	−1	−1		
		x_1	x_2	z_1	z_2		
$\underline{c}_B' = (-1, -1)$		1	1	1	0	2	(1)
		2	2	0	1	4	
		−3	−3	0	0	−6	
$\underline{c}_B' = (0, -1)$		1	1	1	0	2	(2)
		0	0	−2	1	0	
		0	0	3	0	0	

Tableaus 3.24. (1) and (2).

Tableau 3.24 (2) is an optimal final tableau. The artificial variable z_2 is a basic variable with value $z_2 = 0$ (degeneracy). As $\hat{a}_{21} = \hat{a}_{22} = 0$, z_2 can not be replaced by x_1 or x_2. The second restriction is superfluous (dependent) and can be omitted.

If phase-1 ends with $w_I = 0$ all artificial variables z_i will usually have left the basis. The optimal solution of phase-2 over the alternative optimal solutions of phase-1 is determined by blocking the variables with positive r-values in the final tableau of phase-1.

The artificial variables have usually become nonbasic variables. However, in case of degeneracy, it is possible that positive r-values occur at non-artificial variables in the final tableau of phase-1; $w_I = 0$ and at least one z_i is a basic variable with value 0. In these cases the columns with positive r-elements are blocked, and phase-2 is executed in the same way.

Example 3.20.

$$\max\{w = x_1 + 2x_2 + 3x_3 + 4x_4\}$$

$$x_1 + x_2 + 2x_4 = 4$$
$$-2x_2 + x_3 = 0$$
$$x_2 - 3x_3 = 0$$
$$x_1 \geq 0,\ x_2 \geq 0,\ x_3 \geq 0,\ x_4 \geq 0$$

Phase-1

	$\underline{c}_1'$:	0	0	0	0	−1	−1	
		x_1	x_2	x_3	x_4	z_2	z_3	
$\underline{c}_B' = (0, -1, -1)$		1	1	0	2	0	0	4
		0	−2	1	0	1	0	0
		0	1	−3	0	0	1	0
		0	1	2	0	0	0	0
			X	X				

Tableau 3.25. Initial and final tableau of phase-1.

Tableau 3.25 is optimal because $\underline{r} \geq \underline{0}$ although z_2 and z_3 are still basic variables (degeneracy). Phase-2 is optimised over the alternative optimal solutions of phase-1. The variables x_2 and x_3 are blocked to prevent them from entering the basis, as r_2 and r_3 are positive. The phase-1 objective function expressed in the nonbasic variables is:

$$w_I = -z_2 - z_3 = -(2x_2 - x_3) - (-x_2 + 3x_3) = -x_2 - 2x_3.$$

Restriction $w_I = -z_2 - z_3 = 0$ is equivalent to $w_I = -x_2 - 2x_3 = 0$. Therefore x_2 and x_3 are blocked. (Of course it is also possible that based on Tableau 3.25, x_2 and x_3 replace z_3 and z_2 as basic variables by pivoting. Due to degeneracy, it is even permissible to choose a negative pivot.)

Phase-2

	x_1	x_2	x_3	x_4	z_2	z_3		
$\underline{c}'$:	1	2	3	4	0	0		
$\underline{c}_B' = (1, 0, 0)$	1	1	0	**2**	0	0	4	
	0	–2	1	0	1	0	0	(1)
	0	1	–3	0	0	1	0	
	0	–1	–3	–2	0	0	4	
		X	X					
$\underline{c}_B' = (4, 0, 0)$	1/2	1/2	0	1	0	0	2	
	0	–2	1	0	1	0	0	(2)
	0	1	–3	0	0	1	0	
	1	0	–3	0	0	1	8	
		X	X					

Tableau 3.26. (1 and 2) Initial tableau and final tableau of phase-2.

Tableau 3.26 (2) is optimal because $\underline{r} \geq \underline{0}$ and variable x_3 is blocked.

Conclusions

At the end of phase-1 of the two-phase method, one of the following situations occurs:

1. $\max w_I < 0$

   The original set of restrictions is inconsistent and phase 2 is not started. No feasible solution exists.

2. $\max w_I = 0$

   - all artificial variables are nonbasic variables. In this case a basic feasible solution to the original problem is found.
   - there is at least one artificial variable z_k that is basic with value zero. Suppose it corresponds to unit vector $\underline{e}_i$. In this case block all variables with positive r.
     - if $\hat{a}_{ij} \neq 0$ then z_k can also be replaced by x_j by means of a pivot operation
     - if $\hat{a}_{ij} = 0$ for all original variables x and y then the set of restrictions is dependent; omit the superfluous restrictions.
   - Start phase-2.

Remark

If an artificial variable leaves the basis, then this variable and the corresponding column vector can be omitted from the tableau. However, including this column in further iterations allows for the possibility of deriving the matrix B^{-1} in a simple way from the final tableau (see Section 3.16.2). Also for the determination of the values of the dual variables (to be discussed in Chapter 4), it is useful only to block the 'artificial' columns and transform them according to the pivot operations.

3.15. The big *M*-method

In order to solve an arbitrary LP problem, the Simplex Algorithm actually has to be applied twice. In phase-1 a basic feasible solution is determined and then in phase-2 the optimum of the original objective function is calculated, using the final tableau of phase-1. The *big M-method* combines these two successive phases.

Given the LP problem (P)

$$\max\{w = \underline{c}'\underline{x}\} \qquad (P)$$
$$A\underline{x} = \underline{b} \quad \text{with } \underline{b} \geq \underline{0}$$
$$\underline{x} \geq \underline{0}$$

Analogous to the two-phase method, the missing unit vectors and corresponding artificial variables $z_i \geq 0$ are added (assume $i = 1, ..., m$).

In contrast to the two-phase method, the original variables x_j and the artificial variables z_i are both included in a single objective function. All artificial variables get coefficient $-M$. Problem $(\hat{P})$ follows from (P) by adding artificial variables:

$$\max(w_M = \underline{c}'\underline{x} - Mz_1 - Mz_2 - ... - Mz_m) \qquad (\hat{P})$$

$$A\underline{x} + \begin{pmatrix} z_1 \\ z_2 \\ \cdot \\ \cdot \\ z_m \end{pmatrix} = \begin{pmatrix} b_1 \\ b_2 \\ \cdot \\ \cdot \\ b_m \end{pmatrix}$$

$\underline{x} \geq \underline{0}$; $z_i \geq 0$ $(i = 1, ..., m)$; M represents a 'big' positive number.

The term $-Mz_1 - Mz_2 - ... - Mz_m$ in the objective function acts as a so-called 'penalty term'. For large values of M (big M), the penalty term in the optimal solution will become zero if possible, so that all artificial variables have a value $z_i = 0$.

The two-phase method uses the phase-1 objective function to give z_i the value $z_i = 0$; the big M-method uses the penalty term $-M\sum_i z_i$. The objective function w_M of the big M-method includes the original objective function $w = \underline{c}'\underline{x}$. In this way phase-2 is not necessary.

In theory the big M-method looks faster, but in practice it turns out to be scarcely more efficient. The big M-method and the two-phase method make the same sequence of pivots. In most computer programs the two-phase method is implemented because M (a large positive number) may cause round-off errors and other computational difficulties.

3.16. Concluding remarks

We will finalise Chapter 3 by a step-by-step summary of the complete *Simplex Algorithm* and an *algebraic formulation of the simplex tableau.*

Simplex algorithm:A summary

1. Convert the LP problem into a maximization problem (Sections 3.3 and 3.9).
2. Make sure that all variables $x_j \geq 0$ (Section 3.3).
3. Make sure that $\underline{b} \geq \underline{0}$ (Section 3.3).
4. Convert the problem into standard form by adding slack variables $y_i \geq 0$ (Section 3.3)
5. Make sure that there is a unit matrix E in the simplex tableau. Add missing unit vectors and the corresponding artificial variables $z_i \geq 0$ (Section 3.14).
6. After adding artificial variables z_i, the two-phase method is applied; in phase-1, the phase-1 objective function $\max\{w_{\text{I}} = -\sum_i z_i\}$ should be used (Section 3.14).
7. Calculate the r-row from $r_j = \underline{c}_B' B^{-1}\underline{a}_j - c_j = \underline{c}_B' \hat{\underline{a}}_j - c_j$ (Sections 3.9, 3.11).
8. If all $r_j \geq 0$ then the simplex tableau is optimal:
   a. If at the end of phase-1 $w_{\text{I}} < 0$, the set of restrictions is inconsistent; no feasible solution exists. (Section 3.14).
   b. If at the end of phase-1 $w_{\text{I}} = 0$, block the variables with positive r-values; start phase-2 by adding the original objective function $w = \underline{c}'\underline{x}$ to the final tableau of phase-1 (Section 3.14). Go to step 7.
9. a. If $r_p < 0$ and no pivot can be chosen (all $a_{ip} \leq 0$ for $i = 1, ..., m$), then the problem has an unbounded solution (Section 3.13).
   b. Determine the most negative r_p.

   Choose in column $\underline{a}_p$ the pivot a_{kp} such that $\dfrac{b_k}{a_{kp}} = \left\{\min_i \dfrac{b_i}{a_{ip}} \middle| a_{ip} > 0\right\}$.

   Perform a pivot operation including $\underline{b}$-column and r-row (Sections 3.8, 3.11). In case of degeneracy, the lexicographic method can be applied (Section 3.12).
10. If in the optimal final tableau $r_p = 0$ for a nonbasic variable x_p, then the problem has alternative optimal solutions (Section 3.13.1):
    a. If a pivot $a_{ip} > 0$ exists, the problem has a line segment of alternative solutions.
    b. If all $a_{ip} \leq 0$ $(i = 1, ..., m)$, there is a half-line of alternative optimal solutions.

In conclusion we present the algebraic formulation of the simplex tableau.

Given LP problem (P):

$$\max\{w = \underline{c}'\underline{x}\} \quad \underline{A}\underline{x} \le \underline{b} \quad \underline{x} \ge \underline{0} \qquad \text{or} \qquad \max\{ w = \underline{c}'\underline{x} + \underline{0}'\underline{y} \} \quad A\underline{x} + E\underline{y} = \underline{b} \quad \underline{x} \ge \underline{0},\ \underline{y} \ge \underline{0} \qquad (P)$$

Let $b \ge 0$ and the optimal basis B be constituted by the first m columns of A. The initial and final tableau of the Simplex Algorithm can be written as follows:

Algebraic formulation of the simplex tableau

$\underline{c}'$:	$c_1 \dots c_m$ $x_1 \dots x_m$	$c_{m+1} \dots c_n$ $x_{m+1} \dots x_n$	$0 \dots 0$ $y_1 \dots y_m$	
$\underline{c}_B' = (0, \dots, 0)$	B	A_N	E	$\underline{b}$
r-row:	$-c_1 \dots -c_m$	$-c_{m+1} \dots -c_n$	$0 \dots 0$	0

Tableau 3.28. Initial tableau.

$\underline{c}'$:	$c_1 \dots c_m$ $x_1 \dots x_m$	$c_{m+1} \dots c_n$ $x_{m+1} \dots x_n$	$0 \dots 0$ $y_1 \dots y_m$	
$\underline{c}_B' = (c_1, \dots, c_m)$	E	$B^{-1}A_N$	B^{-1}	$B^{-1}\underline{b}$
r-row:	$\underline{0}'$	$\underline{c}_B' B^{-1} A_N - \underline{c}_N'$	$\underline{c}_B' B^{-1}$	$\underline{c}_B' B^{-1} \underline{b} = w_0$

Tableau 3.29. Final tableau= B^{-1} × Initial tableau.

The following emerges from the final tableau:

- The basic variables $\underline{x}_B$:

  $\underline{x}_B = (x_1, x_2, \dots, x_m)' = B^{-1}\underline{b}$

- Non basic variables $\underline{x}_N$:

  $\underline{x}_N = (x_{m+1}, \dots, y_m)' = \underline{0}'$

- The optimal value of the objective function w_0 :

  $w_0 = \underline{c}_B' B^{-1} \underline{b}$

4. Duality and sensitivity analysis

Th.H.B. (Theo) Hendriks

In this chapter we will introduce the concept of duality and define the dual of any linear programming problem. Duality plays a major role in the economic interpretation of the optimal solution. It will be shown how the sensitivity of the optimal solution, due to changes in the problem parameters, can be derived from the optimal simplex tableau. The relation with the graphical sensitivity analysis of Chapter 2 is given too. An important role is played by the so-called dual problem, which will be explained in the next section.

4.1. The dual problem

Every LP problem may be called a *primal problem* and can be associated with a so-called *dual problem*. Sometimes it is easier to solve an LP problem via its dual. Moreover, *duality* plays a major role in the economic interpretation with respect to shadow prices of the optimal solution. The concept is first introduced in an intuitive way.

Example 4.1.

Given LP problem (P):

$$\max\{w_p = 5x_1 + 4x_2 + 6x_3\} \quad \text{(P)}$$

s.t.

$$7x_1 + x_2 + 2x_3 \leq 200 \quad (4.1)$$

$$x_1 + 4x_2 + 2x_3 \leq 100 \quad (4.2)$$

$$x_1 \geq 0, x_2 \geq 0, x_3 \geq 0$$

Without solving (P), something can be said with respect to the optimal objective function value w_p^* of problem (P).

a. The objective function $w_p = 5x_1 + 4x_2 + 6x_3$ can be compared with the first constraint (4.1): $7x_1 + x_2 + 2x_3 \leq 200$. If the first inequality (4.1) is multiplied by four then: $4(7x_1 + x_2 + 2x_3) \leq 4 \times 200$ such that $28x_1 + 4x_2 + 8x_3 \leq 800$. The value of the multiplier (four) is chosen such that each coefficient of the individual terms in the objective function is lower than or equal to the corresponding term in the constraint. So:

$$\begin{array}{l} 5x_1 \leq 28x_1 \\ 4x_2 \leq 4x_2 \\ \underline{6x_3 \leq 8x_3} \quad + \\ w_p = 5x_1 + 4x_2 + 6x_3 \leq 28x_1 + 4x_2 + 8x_3 \text{ (as all } x_i \geq 0) \end{array}$$

As $x_1, x_2, x_3 \geq 0$, it is clear that the value of the objective function $w_p = 5x_1 + 4x_2 + 6x3 \leq 28x_1 + 4x_2 + 8x_3 \leq 800$. So an upper bound on the optimal objective function value w_p^* is 800.

b. Similarly, inequality (4.2) can be multiplied by five: $5x_1 + 20x_2 + 10x_3 \leq 500$.
So $w_p = 5x_1 + 4x_2 + 6x_3 \leq 5x_1 + 20x_2 + 10x_3 \leq 500$ leads to a smaller upper bound on w_p^*, namely $5 \times 100 = 500$.

c. $1 \times$ constraint (4.1) + $2 \times$ constraint (4.2) gives $9x_1 + 9x_2 + 6x_3 \leq 400$, such that $5x_1 + 4x_2 + 6x_3 \leq 9x_1 + 9x_2 + 6x_3 \leq 400$ leads to the lower upper bound $w_p^* = 1 \times 200 + 2 \times 100 \leq 400$.

The question is whether a value of 400 is the lowest possible upper bound for w_p^* that can be found. Determination of the lowest possible upper bound w_p^* in this way, leads to the question: Find nonnegative multipliers u_1 and u_2 for the constraints of (P) such that:

- the upper bound of w_p^* is as low as possible,
- the coefficient of x_1 is greater than or equal to 5; of x_2 greater than or equal to 4 and of x_3 greater than or equal to 6.

Determination of the lowest possible upper bound apparently leads to LP problem (D):

$$
\begin{array}{rcl}
\min\{w_d = 200u_1 + 100u_2\} & \rightarrow & \text{determine the lowest possible upper bound} \\
7u_1 + u_2 \geq 5 & \rightarrow & \text{coefficient of } x_1 \text{ should be } \geq 5 \\
u_1 + 4u_2 \geq 4 & \rightarrow & \text{coefficient of } x_2 \text{ should be } \geq 4 \\
2u_1 + 2u_2 \geq 6 & \rightarrow & \text{coefficient of } x_3 \text{ should be } \geq 6 \\
u_1 \geq 0,\ u_2 \geq 0 & &
\end{array}
$$

Case a. represents $\underline{u} = \begin{pmatrix} u_1 \\ u_2 \end{pmatrix} = \begin{pmatrix} 4 \\ 0 \end{pmatrix}$ with $w_d = 800$.

Case b. implies $\underline{u} = \begin{pmatrix} u_1 \\ u_2 \end{pmatrix} = \begin{pmatrix} 0 \\ 5 \end{pmatrix}$ with $w_d = 500$ and

Case c. means $\underline{u} = \begin{pmatrix} u_1 \\ u_2 \end{pmatrix} = \begin{pmatrix} 1 \\ 2 \end{pmatrix}$ with $w_d = 400$.

The optimal solution for problem (D) is $(u_1, u_2) = \left(\frac{1}{3}, \frac{8}{3}\right)$ with objective value $w_d = 333\frac{1}{3}$. So the lowest possible upper bound of w_p^* is $333\frac{1}{3}$.

A feasible solution of (P) is $\underline{x}' = (x_1, x_2, x_3) = \left(\frac{50}{3}, 0, \frac{125}{3}\right)$ with objective function value $w_p = 333\frac{1}{3}$. The objective function value $w_p = 333\frac{1}{3}$ corresponding to this feasible solution $\underline{x}$ is equal to the lowest possible upper bound $w_d = 333\frac{1}{3}$. This implies that $w_p = 333\frac{1}{3}$ is the optimal objective function value of (P) (see Theorem 4.2).

Conclusion

A *dual LP* problem (D) can be associated with its primal problem (P):

$$\begin{array}{lll} \max\{w_p = 5x_1 + 4x_2 + 6x_3\} & & \min\{w_d = 200u_1 + 100u_2\} \\ \text{s.t.} \quad (P) & & \text{s.t.} \quad (D) \\ & & 7u_1 + u_2 \geq 5 \\ 7x_1 + x_2 + 2x_3 \leq 200 & \Rightarrow & u_1 + 4u_2 \geq 4 \\ x_1 + 4x_2 + 2x_3 \leq 100 & & 2u_1 + 2u_2 \geq 6 \\ x_1 \geq 0,\ x_2 \geq 0,\ x_3 \geq 0 & & u_1 \geq 0,\ u_2 \geq 0 \end{array}$$

Generalisation of Example 4.1 leads to the following definition.

Definition 4.1.

For each LP problem (P) an LP problem (D) can be defined:

$$\begin{array}{lll} \max\{w_p = \underline{c}'\underline{x}\} \quad (P) & & \min\{w_d = \underline{b}'\underline{u}\} \quad (D) \\ A\underline{x} \leq \underline{b} & \Rightarrow & A'\underline{u} \geq \underline{c} \\ \underline{x} \geq \underline{0} & & \underline{u} \geq \underline{0} \end{array}$$

Problem (P) is called the *primal problem*; problem (D) the *dual problem*. Problem (P) is formulated in so-called *canonical form*, i.e. the constraints are $A\underline{x} \leq \underline{b}$, $\underline{x} \geq \underline{0}$.

Note the 'symmetry' of both problems:

The objective function vector $\underline{c}$ of the primal problem is the right-hand side in the dual problem. The right-hand side vector $\underline{b}$ of the primal problem is the objective function vector of the dual problem. A is an $m\times n$-matrix; A' is an $n\times m$-matrix; $\underline{x} \in \mathbb{R}^n$; $\underline{c} \in \mathbb{R}^n$; $\underline{b} \in \mathbb{R}^m$; $\underline{u} \in \mathbb{R}^m$.

Example 4.2.

$$\begin{array}{lll} \max\{w_p = 2x_1 - 3x_2\} & & \min\{w_d = 3u_1 - 6u_2 + 9u_3\} \\ \text{s.t.} \quad (P) & & \text{s.t.} \quad (D) \\ x_1 + 2x_2 \leq 3 & & \\ 4x_1 - 5x_2 \leq -6 & \Rightarrow & u_1 + 4u_2 - 7u_3 \geq 2 \\ -7x_1 + 8x_2 \leq 9 & & 2u_1 - 5u_2 + 8u_3 \geq -3 \\ x_1 \geq 0,\ x_2 \geq 0 & & u_1 \geq 0,\ u_2 \geq 0,\ u_3 \geq 0 \end{array}$$

4.2. The dual of an arbitrary LP problem

An example is used to illustrate how Definition 4.1 and the transformation rules of Section 3.3 can be used to derive the dual.

Example 4.3.

Given primal problem (P):

$$\begin{array}{ll} \max\{x_1 - 3x_2 + 2x_3\} & \\ x_1 + x_2 - x_3 \leq 3 & (P) \\ 2x_1 - 3x_2 + x_3 = -5 & \\ -x_1 + 2x_2 - x_3 \geq 7 & \\ x_1 \geq 0,\ x_2 \leq 0,\ x_3 \text{ free} & \end{array}$$

First problem (P) is written in canonical form max$\{\underline{c}'\underline{x}\}$, $A\underline{x} \le \underline{b}$, $\underline{x} \ge \underline{0}$ using the transformation rules of Section 3.3:

- replace the equality by two inequalities,
- multiply the inequality with the $\ge$-sign by -1,
- replace variable x_2 by $-x_2^*$ with $x_2^* \ge 0$,
- replace free variable x_3 by $x_3^+ - x_3^-$ with $x_3^+ \ge 0$ and $x_3^- \ge 0$.

In this way (P) has been transformed into $(\hat{\text{P}})$:

$$\begin{array}{ll} \max\{x_1 + 3x_2^* + 2x_3^+ - 2x_3^-\} & \\ \text{s.t.} & (\hat{\text{P}}) \\ \quad x_1 - x_2^* - x_3^+ + x_3^- \le 3 & \\ \quad 2x_1 + 3x_2^* + x_3^+ - x_3^- \le -5 & \\ \quad -2x_1 - 3x_2^* - x_3^+ + x_3^- \le 5 & \\ \quad x_1 + 2x_2^* + x_3^+ - x_3^- \le -7 & \\ \quad x_1,\ x_2^*,\ x_3^+,\ x_3^- \ge 0 & \end{array}$$

As $(\hat{\text{P}})$ is in canonical form, Definition 4.1 can be used to derive the dual problem $(\hat{\text{D}})$:

$$\begin{array}{ll} \min\{3U_1 - 5U_2 + 5U_3 - 7U_4\} & \\ \text{s.t.} & (\hat{\text{D}}) \\ \quad U_1 + 2U_2 - 2U_3 + U_4 \ge 1 & \\ \quad -U_1 + 3U_2 - 3U_3 + 2U_4 \ge 3 & \\ \quad -U_1 + U_2 - U_3 + U_4 \ge 2 & \\ \quad U_1 - U_2 + U_3 - U_4 \ge -2 & \\ \quad U_1, U_2, U_3, U_4 \ge 0 & \end{array}$$

Problem $(\hat{\text{D}})$ is the dual of $(\hat{\text{P}})$ which in turn is equivalent to problem (P). However, $(\hat{\text{D}})$ doesn't have much in common with (P) at first sight. The vectors $\underline{c}' = (1, -3, 2)$, $\underline{b}' = (3, -5, 7)$ and the transpose A' do not appear directly in $(\hat{\text{D}})$. Therefore it is necessary to transform $(\hat{\text{D}})$ 'backwards'. Multiply the second inequality by -1 and combine the last two relations:

$$\begin{array}{l} \min\{3U_1 - 5(U_2 - U_3) - 7U_4\} \\ \quad U_1 + 2(U_2 - U_3) + U_4 \ge 1 \\ \quad U_1 - 3(U_2 - U_3) - 2U_4 \le -3 \\ \quad -U_1 + (U_2 - U_3) + U_4 = 2 \\ \quad U_1, U_2, U_3, U_4 \ge 0 \end{array}$$

Introduction of the variables $u_1 = U_1$, $u_2 = U_2 - U_3$ (so u_2 is free) and $u_3 = -U_4$ (so $u_3 \le 0$) leads to the dual (D):

$$\min\{3u_1 - 5u_2 + 7u_3\}$$
s.t. (D)
$$\begin{aligned} u_1 + 2u_2 - u_3 &\geq 1 \\ u_1 - 3u_2 + 2u_3 &\leq -3 \\ -u_1 + u_2 - u_3 &= 2 \\ u_1 \geq 0,\ u_2 \text{ free},\ u_3 &\leq 0 \end{aligned}$$

Objective vector $\underline{c}$, the transpose A' and right-hand side $\underline{b}$ of problem (P) are directly visible in (D). The signs of the constraints and variables are determined by the transformations.

Example 4.3 can be generalised into Scheme 4.1 to find the dual in an easy way without the transformations for an arbitrary LP problem. For instance, the equality constraint of (P) gave rise to two $\leq$-constraints in $(\hat{\text{P}})$; this resulted in two columns in $(\hat{\text{D}})$ leading to a free variable.

Scheme 4.1. Deriving the dual.

max	$\leftrightarrow$	min
matrix A	$\leftrightarrow$	matrix A'
i^{th} constraint $\leq$	$\leftrightarrow$	i^{th} variable ≥ 0
j^{th} constraint $\geq$	$\leftrightarrow$	j^{th} variable ≤ 0
k^{th} constraint $=$	$\leftrightarrow$	k^{th} variable free
i^{th} variable ≥ 0	$\leftrightarrow$	i^{th} constraint $\geq$
j^{th} variable ≤ 0	$\leftrightarrow$	j^{th} constraint $\leq$
k^{th} variable free	$\leftrightarrow$	k^{th} constraint $=$

If (P) max $\rightarrow$ read scheme from left to right
If (P) min $\leftarrow$ read scheme from right to left

If the primal is a maximisation problem, Scheme 4.1 should be read from left to right **($\rightarrow$)**.
If the primal is a minimisation problem, Schema 4.1 is interpreted from right to left **($\leftarrow$)**.

Scheme 4.1 can be used to derive the dual of any primal problem (P) and turns out to be much faster. We will demonstrate the use of Scheme 4.1 from Example 4.3.

Example 4.4.

a. The primal problem (P) in Example 4.3 is a maximisation problem so Scheme 4.1 should be read from left to right (max $\rightarrow$ min) :

Primal (P)				Dual (D)
$\max\{x_1 - 3x_2 + 2x_3\}$				min
$x_1 + x_2 - x_3 \leq 3$	:	1st constraint $\leq$	$\rightarrow$	1st variable $u_1 \geq 0$
$2x_1 - 3x_2 + x_3 = -5$	:	2nd constraint $=$	$\rightarrow$	2nd variable u_2 free
$-x_1 + 2x_2 - x_3 \geq 7$	:	3rd constraint $\geq$	$\rightarrow$	3rd variable $u_3 \leq 0$

$x_1 \geq 0$	:	1st variable ≥ 0	$\rightarrow$	1st constraint $\geq$
$x_2 \leq 0$	:	2nd variable ≤ 0	$\rightarrow$	2nd constraint $\leq$
x_3 free	:	3rd variable free	$\rightarrow$	3rd constraint $=$

Transposing A and switching $\underline{c}$ and $\underline{b}$ leads to dual problem (D):

$$
\begin{array}{ll}
\min\{3u_1 - 5u_2 + 7u_3\} & \\
\text{s.t.} & \text{(D)} \\
\quad u_1 + 2u_2 - u_3 \geq 1 & \\
\quad u_1 - 3u_2 + 2u_3 \leq -3 & \\
\quad -u_1 + u_2 - u_3 = 2 & \\
\quad u_1 \geq 0,\ u_2 \text{ free},\ u_3 \leq 0 &
\end{array}
$$

b. The dual of (D) appears to be (P). Considering (D) as the primal (minimisation problem) let us read Scheme 4.1 from right to left: max $\leftarrow$ min.

constraint 1 $\geq$ $\rightarrow$ 1st variable $x_1 \geq 0$		variable $u_1 \geq 0 \rightarrow 1^{st}$ constraint $\leq$
constraint 2 $\leq$ $\rightarrow$ 2nd variable $x_2 \leq 0$		variable u_2 free $\rightarrow 2^{nd}$ constraint $=$
constraint 3 $=$ $\rightarrow$ 3rd variable x_3 free		variable $u_3 \leq 0 \rightarrow 3^{rd}$ constraint $\geq$

$$
\begin{array}{llcll}
\min\{3u_1 - 5u_2 + 7u_3\} & & & \max\{x_1 - 3x_2 + 2x_3\} & \\
\text{s.t.} & \text{(D)} & & \text{s.t.} & \text{(P)} \\
\quad u_1 + 2u_2 - u_3 \geq 1 & & & \quad x_1 + x_2 - x_3 \leq 3 & \\
\quad u_1 - 3u_2 + 2u_3 \leq -3 & & \Rightarrow & \quad 2x_1 - 3x_2 + x_3 = -5 & \\
\quad -u_1 + u_2 - u_3 = 2 & & & \quad -x_1 + 2x_2 - x_3 \geq 7 & \\
\quad u_1 \geq 0,\ u_2 \text{ free},\ u_3 \leq 0 & & & \quad x_1 \geq 0,\ x_2 \leq 0,\ x_3 \text{ free} &
\end{array}
$$

In a. (P) $\rightarrow$ (D) ; in b. (D) $\rightarrow$ (P). This explains the notation $\leftrightarrow$ in Scheme 4.1.

4.3. Duality properties

In this section, some important properties on duality are discussed. Consider the primal problem (P): $\max\{w_p = \underline{c}'\underline{x}\}$ and the associated dual problem (D): $\min\{w_d = \underline{b}'\underline{u}\}$

$$
\begin{array}{lcl}
A\underline{x} \leq \underline{b} & \qquad\qquad & A'\underline{u} \geq \underline{c} \\
\underline{x} \geq \underline{0} & & \underline{u} \geq \underline{0}
\end{array}
$$

Theorem 4.1.

Let $\underline{x}$ be a feasible solution of (P) and $\underline{u}$ a feasible solution of (D) then $\underline{c}'\underline{x} \leq \underline{b}'\underline{u}$.

Proof

$\underline{x}$ is primal feasible, so: $\quad A\underline{x} \leq \underline{b} \quad (1)$

$\quad \underline{x} \geq \underline{0} \quad (2)$

$\underline{u}$ is dual feasible, so: $\quad A'\underline{u} \geq \underline{c}$ or $\underline{c} \leq A'\underline{u}$ so $\underline{c}' \leq \underline{u}'A \quad (3)$

$\quad \underline{u} \geq \underline{0} \quad (4)$

$$
\begin{array}{lcl}
(3) & \rightarrow & \underline{c}' \leq \underline{u}'A \qquad (2) \rightarrow \underline{x}' \geq \underline{0} \\
\text{combination } (3)+(2) & \rightarrow & \underline{c}'\underline{x} \leq \underline{u}'A\underline{x} \underset{\substack{\uparrow \\ (1)+(4)}}{\leq} \underline{u}'\underline{b} = \underline{b}'\underline{u}
\end{array}
$$

Example 4.5.

In Example 4.1, feasible dual solutions (4, 0), (0, 5), (1, 2) and $(\frac{1}{3}, \frac{8}{3})$ lead to upper bounds 800, 500, 400 and $333\frac{1}{3}$ respectively for the optimal objective function value $w_p^* = 333\frac{1}{3}$.

Some consequences of Theorem 4.1:

1. The objective function value w_d of a feasible solution of (D) is an upper bound for the maximum of the primal objective function.
2. If the primal problem (P) has feasible solutions and the primal objective function is unbounded, the dual problem (D) has no feasible solutions.
3. Objective function value w_p of a feasible solution for problem (P) is a lower bound for the minimum of the dual objective function.
4. If (D) has feasible solutions and the dual objective function is unbounded, the primal problem has no feasible solutions.

Remark

It is possible that both problems, the primal and the dual have no feasible solutions.

Example 4.6.

$$\begin{array}{lcl} \max\{w_p = x_1 + 2x_2\} \quad (P) & & \min\{w_d = -u_1 - u_2\} \quad (D) \\ x_1 - x_2 \leq -1 & & u_1 - u_2 \geq 1 \\ -x_1 + x_2 \leq -1 & \Leftrightarrow & -u_1 + u_2 \geq 2 \\ x_1 \geq 0, x_2 \geq 0 & & u_1 \geq 0, u_2 \geq 0 \end{array}$$

Adding both constraints in (P) as well as in (D) shows that both problems (P) and (D) have no feasible solutions.

Theorem 4.2.

Let $\underline{x}_0$ be a feasible solution of (P) and $\underline{u}_0$ a feasible solution of (D). If $\underline{c}'\underline{x}_0 = \underline{b}'\underline{u}_0$ then both $\underline{x}_0$ and $\underline{u}_0$ are optimal solutions of (P) and (D) respectively.

Proof

a. $\underline{u}_0$ is a feasible solution of (D). From Theorem 4.1 it follows that for every feasible solution $\underline{x}$ of (P) $\underline{c}'\underline{x} \leq \underline{b}'\underline{u}_0$. Given $\underline{c}'\underline{x}_0 = \underline{b}'\underline{u}_0$ so $\underline{c}'\underline{x} \leq \underline{c}'\underline{x}_0$ implies $\underline{x}_0$ is an optimal solution of (P).
b. Similarly for every feasible solution $\underline{u}$ of (D) $\underline{b}'\underline{u} \geq \underline{c}'\underline{x}_0 = \underline{b}'\underline{u}_0$, so $\underline{u}_0$ is optimal in (D).

Example 4.7.

In Example 4.1 we found the feasible solutions, $\underline{x}_0' = (\frac{50}{3}, 0, \frac{125}{3})$ and $\underline{u}_0' = (\frac{1}{3}, \frac{8}{3})$ for (P) and (D) respectively. The primal objective value $w_p = 333\frac{1}{3}$ and the dual objective value $w_d = 333\frac{1}{3}$. So $\underline{x}_0$ and $\underline{u}_0$ are optimal solutions.

Remark

Dual problems of equivalent LP problems are equivalent. Suppose (D) is the dual of (P), $(\hat{P})$ is an equivalent LP problem for problem (P), $(\hat{D})$ is the dual of $(\hat{P})$, then $(\hat{D})$ is equivalent to (D):

$$\begin{array}{ccccccc} & & & \text{dual} & & & \\ & (P) & & \Rightarrow & (D) & & \\ \text{equivalent} & \Downarrow & & & \Uparrow & \text{equivalent} & \\ & & & \text{dual} & & & \\ & (\hat{P}) & & \Rightarrow & (\hat{D}) & & \end{array} \tag{4.3}$$

Example 4.8.

$$\begin{array}{lcl} \max\{6x_1 + 7x_2\} & & \min\{3u_1 + 5u_2\} \\ x_1 + 2x_2 \le 3 & (P) \Rightarrow & u_1 + 3u_2 \ge 6 \quad (D) \\ 3x_1 + 4x_2 \le 5 & & 2u_1 + 4u_2 \ge 7 \\ x_1 \ge 0, x_2 \ge 0 & & u_1 \ge 0, u_2 \ge 0 \\ \text{equivalent} \Downarrow & & \Uparrow \text{equivalent} \\ \max\{6x_1 + 7x_2 + 0y_1 + 0y_2\} & & \min\{3u_1 + 5u_2\} \\ & & u_1 + 3u_2 \ge 6 \\ x_1 + 2x_2 + y_1 \qquad = 3 & (\hat{P}) \Rightarrow & 2u_1 + 4u_2 \ge 7 \quad (\hat{D}) \\ 3x_1 + 4x_2 \qquad + y_2 = 5 & & u_1 \qquad \ge 0 \\ & & u_2 \ge 0 \\ x_1 \ge 0, x_2 \ge 0, y_1 \ge 0, y_2 \ge 0 & & u_1 \text{ free}, u_2 \text{ free} \end{array}$$

According to (4.3), it does not matter in which form problem (P) has been formulated. Problem (P) can always be transformed into a desired form $(\hat{P})$. After all, the dual problems (D) and $(\hat{D})$ are equivalent. In the derivation of the theorems, we will use the canonical form of (P).

Theorem 4.3.

If (P) has an optimal solution $\underline{x}_0$ then (D) has an optimal solution $\underline{u}_0$ with $w_p = \underline{c}'\underline{x}_0 = \underline{b}'\underline{u}_0 = w_d$.

Proof

Let $\underline{x}' = (\underline{x}_B', \underline{x}_N') = (\underline{x}_B', \underline{0}')$ be an optimal solution of (P): $\max\{w_p = \underline{c}'\underline{x} \mid A\underline{x} \le \underline{b},\ \underline{x} \ge \underline{0}\}$.
The initial tableau of (P) looks as follows (Section 3.16):

$\underline{c}'$ $\underline{x}'$	$\underline{0}'$ $\underline{y}'$	
A	E	$\underline{b}$
$-\underline{c}'$	$\underline{0}'$	0

(1)

The final tableau with optimal basis B can be written as:

$\underline{c}'$ $\underline{x}'$	$\underline{0}'$ $\underline{y}'$		
$B^{-1}A$	B^{-1}	$B^{-1}\underline{b}$	(2)
$\underline{c}_B' B^{-1}A-\underline{c}'$	$\underline{c}_B' B^{-1}$	$\underline{c}_B' B^{-1}\underline{b}$	

Assume: $\underline{c}_B' B^{-1} = \underline{u}'$ (4.4)

Using (4.4), the final tableau of (P) can than be rewritten as:

$\underline{c}'$ $\underline{x}'$	$\underline{0}'$ $\underline{y}'$		
$B^{-1}A$	B^{-1}	$B^{-1}\underline{b}$	(3)
$\underline{u}'A-\underline{c}'$	$\underline{u}'$	$\underline{u}'\underline{b}$	

The tableaus (2) and (3) are final tableaus of (P). In other words: the $\underline{r}$-row is $\geq \underline{0}$. So, in tableau (3) $\underline{u}'A - \underline{c}' \geq \underline{0}'$ and $\underline{u}' \geq \underline{0}'$. This implies that for $\underline{u}$ in (4.4) holds: $A'\underline{u} \geq \underline{c}$ and $\underline{u} \geq \underline{0}$. Consequently, $\underline{u}' = \underline{c}_B' B^{-1}$ is a feasible solution of the dual problem (D):

$$\min\{w_d = \underline{b}'\underline{u} \mid A'\underline{u} \geq \underline{b},\ \underline{u} \geq \underline{0}\}$$

For the objective function value of (D) holds: $w_d = \underline{b}'\underline{u} = \underline{u}'\underline{b} = \underline{c}_B' B^{-1}\underline{b} = \underline{c}_B'\underline{x}_B = \underline{c}'\underline{x} = w_p$, so according to Theorem 4.2 we conclude that $\underline{u}' = \underline{c}_B' B^{-1}$ is an optimal solution of the dual problem (D).

The proof of Theorem 4.3 shows how an optimal solution of (D) can be derived from the final tableau of (P):

a. Optimal values for the *dual main variables* $\underline{u}$:
   - In tableau (1) in the proof of Theorem 4.2, the unit matrix E appears at the slack variables y. In that case, in the final tableau: $\underline{r}' = \underline{c}_B' B^{-1}E - \underline{0}' = \underline{u}' - \underline{0}'$, or $\underline{u} = \underline{r} + \underline{0}$
   - In an arbitrary problem, the unit matrix E appears spread over several main variables $\underline{x}$, slack variables $\underline{y}$ and artificial variables $\underline{z}$. For the corresponding r-values of those variables where the unit matrix E appears, holds:
     $\underline{r}' = \underline{c}_B' B^{-1}E - \underline{c}' = \underline{u}' - \underline{c}'$, or
     $\underline{u} = \underline{r} + \underline{c}$ (mind the order of the unit vectors!) (4.5)

b. The optimal values of the *dual slack variables* $\underline{v}$:
- Transform dual problem (D) into standard form:

$$\begin{array}{llll} \min\{w_d = \underline{b}'\underline{u}\} & \text{(D)} & & \min\{w_d = \underline{b}'\underline{u}\} \\ A'\underline{u} \geq \underline{c} & & \Rightarrow & A'\underline{u} - \underline{v} = \underline{c} \\ \underline{u} \geq \underline{0} & & & \underline{u}, \underline{v} \geq \underline{0} \end{array}$$

in which the variables $\underline{v}$ represent the dual slack variables. Now $\underline{v} = A'\underline{u} - \underline{c}$, such that $\underline{v}' = \underline{u}'A - \underline{c}'$.
- So apparently the optimal values of the dual slack variables $\underline{v}$ equal the r-elements in final tableau (3) of (P) corresponding to the main variables $\underline{x}$.

c. The optimal objective function value w_d = optimal value $w_p = \underline{c}_B' B^{-1}\underline{b} = \underline{u}'\underline{b}$.

Summary

The optimal solution of (P) as well as the optimal solution of (D) follow from the final tableau

	$\underline{c}'$ $\underline{x}'$	$\underline{0}'$ $\underline{y}'$	
	$B^{-1}A$	B^{-1}	$B^{-1}\underline{b}$
	$\underline{c}_B' B^{-1}A - \underline{c}'$	$\underline{c}_B' B^{-1}$	$\underline{c}_B' B^{-1}\underline{b} = w_p$
	$\Vert$	$\Vert$	$\Vert$
$\underline{u}' = \underline{c}_B' B^{-1} \Rightarrow$	$\underline{u}'A - \underline{c}'$	$\underline{u}'$	$\underline{u}'\underline{b} = \underline{b}'\underline{u} = w_d$
	$\Vert$		
	$\underline{v}'$		

Tableau 4.1. Optimal solution of (D) in the final tableau of (P) for (P) in canonical form.

Example 4.9.

$$\begin{array}{lll} \max\{w_p = 2x_1 + 5x_2\} & & \min\{w_d = 4u_1 + 3u_2 + 6u_3\} \\ \text{s.t.} \quad\quad\quad\quad\quad\quad\quad \text{(P)} & & \text{s.t.} \quad\quad\quad\quad\quad\quad\quad \text{(D)} \\ x_1 \leq 4 & & u_1 + u_3 \geq 2 \\ x_2 \leq 3 & \Rightarrow & u_2 + u_3 \geq 5 \\ x_1 + x_2 \leq 6 & & \\ x_1, \; x_2 \geq 0 & & u_1, \; u_2, \; u_3 \geq 0 \end{array}$$

Next, we will solve problem (P) and (D) by the simplex method. Problem (D) is solved as a minimisation problem (i.e. optimal if $\underline{r} \leq \underline{0}$).

Solving both problems is not necessary as the optimal solution of the dual problem follows directly from the final tableau of its primal (see Tableau 4.2).

(P)

$\underline{c}'$	2	5	0	0	0	
	x_1	x_2	y_1	y_2	y_3	
	1	0	1	0	0	4
	0	1	0	1	0	3
	1	1	0	0	1	6
	−2	−5	0	0	0	0
	1	0	1	0	0	4
	0	1	0	1	0	3
	1	0	0	−1	1	3
	−2	0	0	5	0	15
	0	0	1	1	−1	1
	0	1	0	1	0	3
	1	0	0	−1	1	3
	0	0	0	3	2	21
	↓	↓	↓	↓	↓	
	v_1	v_2	u_1	u_2	u_3	

(D)

$\underline{b}'$	4	3	6	0	0	
	u_1	u_2	u_3	v_1	v_2	
	1	0	1	-1	0	2
	0	1	1	0	-1	5
	0	0	1	-4	-3	23
	1	0	1	-1	0	2
	-1	1	0	1	-1	3
	-1	0	0	-3	-3	21
	↓	↓	↓	↓	↓	
	$-y_1$	$-y_2$	$-y_3$	$-x_1$	$-x_2$	

Tableau 4.2.

The optimal solution of (P) is $(x_1, x_2, y_1, y_2, y_3) = (3, 3, 1, 0, 0)$. The optimal objective function value $w_p = 21$. The optimal solution of (D) is $(u_1, u_2, u_3, v_1, v_2) = (0, 3, 2, 0, 0)$ with objective function value $w_d = 21$. Notice that the dual problem is a minimisation problem, so the tableau is optimal if $\underline{r} \leq \underline{0}$.

Remark

Note that the optimal final tableau of (D) also offers the optimal solution for (P). The variables x_i and y_i have the opposite signs. Alertness is required in general.

Example 4.10.

Given LP problem (P):

$$\min\{x_1 + 2x_2 + 3x_3\}$$

s.t.

$$\left.\begin{aligned} x_1 + 5x_2 + 7x_3 &\geq 1 \\ -2x_1 + x_2 + 8x_3 &\geq 2 \\ 6x_1 + 2x_2 - x_3 &\geq 3 \\ 4x_1 + 9x_2 - 2x_3 &\geq 4 \\ &\;\;. \\ &\;\;. \\ -3x_1 + 8x_2 + 4x_3 &\geq 10 \end{aligned}\right\} \quad (\text{ten } \geq -\text{constraints}) \qquad \text{(P)}$$

$$x_1,\; x_2,\; x_3 \geq 0$$

Applying the Simplex Method straightaway to problem (P) leads to a simplex tableau with three x-variables, ten slack variables y and ten artificial variables z (dimension of the tableau: 10 rows × 23 columns). At least ten iterations are required in phase 1 and some in phase 2.

Solving the dual problem (D), implies that phase 1 can be omitted and only three slack variables v have to be added to the ten main variables u. The dimension of the tableau is 3 rows × 13 columns. From a computational point of view it is more efficient to solve the dual and derive the optimal solution for problem (P) from the final tableau of (D).

Remark

The required number of iterations to solve the primal (P) or the dual (D) is different. Obviously, in the examples 4.9 and 4.10 it is more efficient to solve the dual (D). In general, we may state that the calculation time for solving LP problems depends on the number of constraints m, assuming that $m < n$ (number of columns).

Given (P) and (D):

$$\max\{\underline{c}'\underline{x}\} \quad A\underline{x} \le \underline{b} \quad \underline{x} \ge \underline{0} \qquad \text{(P)} \qquad \text{or} \qquad \max\{\underline{c}'\underline{x}\} \quad A\underline{x} + \underline{y} = \underline{b} \quad \underline{x} \ge \underline{0},\ \underline{y} \ge \underline{0} \qquad (\text{so } \underline{y} = \underline{b} - A\underline{x} \ge \underline{0})$$

$$\min\{\underline{b}'\underline{u}\} \quad A'\underline{u} \ge \underline{c} \quad \underline{u} \ge \underline{0} \qquad \text{(D)} \qquad \text{or} \qquad \min\{\underline{b}'\underline{u}\} \quad A'\underline{u} - \underline{v} = \underline{c} \quad \underline{u} \ge \underline{0},\ \underline{v} \ge \underline{0} \qquad (\text{so } \underline{v} = A'\underline{u} - \underline{c} \ge \underline{0})$$

Theorem 4.4.

Let $\underline{x}$ be a feasible solution of (P) and $\underline{u}$ a feasible solution of (D).

- If $\underline{x}$ and $\underline{u}$ are optimal solutions, then:
  $\underline{x}'(A'\underline{u} - \underline{c}) = 0$ or $\underline{x}'\underline{v} = 0$
  $\underline{u}'(\underline{b} - A\underline{x}) = 0$ or $\underline{u}'\underline{y} = 0$
- If $\underline{x}'(A'\underline{u} - \underline{c}) = 0$ and $\underline{u}'(\underline{b} - A\underline{x}) = 0$ (or $\underline{x}'\underline{v} = 0$ and $\underline{u}'\underline{y} = 0$), then $\underline{x}$ and $\underline{u}$ are optimal.

Proof (see also the remarks in Example 4.11):
The feasible solutions $\underline{x}$ of (P) and $\underline{u}$ of (D) are represented here by $\underline{x} \in$ P and $\underline{u} \in$ D.

a. Given $\underline{x}$ and $\underline{u}$ are optimal solutions, so:

$$\left.\begin{array}{lll} \underline{x} \in \mathrm{P} & \rightarrow & \underline{x} \ge \underline{0} \\ \underline{u} \in \mathrm{D} & \rightarrow & A'\underline{u} \ge \underline{c} \text{ or } A'\underline{u} - \underline{c} \ge \underline{0} \end{array}\right\} \rightarrow \alpha := \underline{x}'(A'\underline{u} - \underline{c}) \ge 0 \qquad (1)$$

$$\left.\begin{array}{lll} \underline{x} \in \mathrm{P} & \rightarrow & A\underline{x} \le \underline{b} \text{ or } \underline{b} - A\underline{x} \ge \underline{0} \\ \underline{u} \in \mathrm{D} & \rightarrow & \underline{u} \ge \underline{0} \end{array}\right\} \rightarrow \beta := \underline{u}'(\underline{b} - A\underline{x}) \ge 0 \qquad (2)$$

In this way $\alpha + \beta = \underline{x}'A'\underline{u} - \underline{x}'\underline{c} + \underline{u}'\underline{b} - \underline{u}'A\underline{x} \ge 0$
As $\underline{x}'A'\underline{u} = \underline{u}'A\underline{x}$, applies $\alpha + \beta = -\underline{x}'\underline{c} + \underline{u}'\underline{b} \ge 0$ (3)

$\underline{x}$ and $\underline{u}$ are optimal solutions so according to Theorem 4.2, $\underline{x}'\underline{c} = \underline{u}'\underline{b}$ such that

$$\left.\begin{array}{l}(3) \text{ implies } \alpha+\beta=0 \\ (1) \text{ and } (2) \text{ lead to } \alpha \geq 0,\ \beta \geq 0\end{array}\right\} \rightarrow \quad \alpha = 0 \text{ and } \beta = 0$$

b. Given $\underline{x}'(A'\underline{u}-\underline{c}) = 0$ and $\underline{u}'(\underline{b}-A\underline{x}) = 0$. The definition of α and β in a. tells us: $\alpha = 0$ and $\beta = 0$.
Relation (3) shows: $\alpha + \beta = \underline{x}'A'\underline{u} - \underline{x}'\underline{c} + \underline{u}'\underline{b} - \underline{u}'A\underline{x} = -\underline{x}'\underline{c} + \underline{u}'\underline{b} = 0$ such that $\underline{x}'\underline{c} = \underline{u}'\underline{b}$.
Theorem 4.1 tells us that $\underline{x}$ and $\underline{u}$ are optimal solutions of (P) and (D) respectively.

Definition 4.2.

The relations $\underline{x}'\underline{v} = 0$ and $\underline{u}'\underline{y} = 0$ in Theorem 4.4 are called the *complementary slackness relations* being memorised easily by:

Main variables (P) $\times$ Slack variables (D) = 0 $\quad (\underline{x}'\underline{v} = 0)$
Main variables (D) $\times$ Slack variables (P) = 0 $\quad (\underline{u}'\underline{y} = 0)$

Example 4.11. (follows on Example 4.9)
The optimal solution of (P) is: $x_1 = 3$, $x_2 = 3$, $y_1 = 1$, $y_2 = 0$, $y_3 = 0$
The optimal solution of (D) is: $u_1 = 0$, $u_2 = 3$, $u_3 = 2$, $v_1 = 0$, $v_2 = 0$

So $\underline{x}'\underline{v} = x_1v_1 + x_2v_2 = 3\times0 + 3\times0 = 0$
and $\underline{u}'\underline{y} = u_1y_1 + u_2y_2 + u_3y_3 = 0\times1 + 3\times0 + 2\times0 = 0$.

Although Theorem 4.4 has been proven formally, the complementary slackness relations also follow from the final tableau of an LP problem (see Tableau 4.1):
- if x_j is a basic variable, then $r_j = 0$ such that $v_j = 0$ or $x_jv_j = 0$
- if x_j is a nonbasic variable is then $x_j = 0$ such that $x_jv_j = 0$
- if y_i is a basic variable then $r_{y_i} = 0$ such that $u_i = 0$ and $y_iu_i = 0$
- if y_i is a nonbasic variable then $y_i = 0$ and $y_iu_i = 0$

Concluding remarks

1. If the primal problem has an optimal solution, the dual also has an optimal solution and vice versa. The optimal objective function value of both problems is equal. Let B be the optimal basis of the primal problem, then $\underline{u}' = \underline{c}_B' B^{-1}$ is the optimal dual solution.
2. The optimal solution of the dual problem can be derived from the simplex tableau corresponding to the optimal solution of the primal problem.

4.4. Economic interpretation of the dual problem; shadow price

In this section an economic interpretation of the dual problem and shadow prices are introduced. The concept of shadow prices will be illustrated both graphically (see also Chapter 2) and mathematically.

Example 4.12.

We will demonstrate from a simple production planning problem how the *dual problem* can be formulated from an economical point of view.
Given LP problem (P):

$$\begin{aligned} &\max\{5x_1 + 4x_2 + 6x_3\} \\ &\text{s.t.} \qquad\qquad\qquad\qquad\qquad \text{(P)} \\ &\quad 4x_1 + 3x_2 + 6x_3 \leq 200 \qquad \text{(A)} \\ &\quad 2x_1 + 5x_2 + 4x_3 \leq 100 \qquad \text{(B)} \\ &\quad x_1,\ x_2,\ x_3 \geq 0 \end{aligned}$$

Problem (P) represents a production planning problem for company Primo with the following characteristics.

- Primo can produce three products with a respective margin of € 5, € 4 and € 6 per unit. Primo uses raw materials A and B with an availability of 200 and 100 kg respectively. Producing one unit of product 1 requires 4 kg of A and 2 kg of B. Requirements for producing product 2 and 3 follow from formulation (P).
- Company Duo wants to buy the raw materials A and B from Primo. Let u_1 and u_2 be the prices that Duo wants to pay for A and B respectively. The commercial objective of Duo is: $\min\{200u_1 + 100u_2\}$.
- The margin of Primo on product 1 is € 5 per unit which requires 4 kg of A and 2 kg of B. Primo is only willing to sell the raw materials to Duo if the prices u_1 and u_2 are such that $4u_1 + 2u_2 \geq 5$; otherwise producing product 1 is more profitable. Analogously, the relations $3u_1 + 5u_2 \geq 4$ and $6u_1 + 4u_2 \geq 6$ should hold for the products 2 and 3 respectively.

Summarising

The prices u_1 and u_2 that Duo must pay for raw materials A and B, are determined by the objective of Duo and the willingness of Primo to sell. This can be formulated by the dual LP problem (D).

$$\begin{aligned} &\min\{200u_1 + 100u_2\} \\ &\text{s.t.} \qquad\qquad\qquad\qquad \text{(D)} \\ &\quad 4u_1 + 2u_2 \geq 5 \\ &\quad 3u_1 + 5u_2 \geq 4 \\ &\quad 6u_1 + 4u_2 \geq 6 \\ &\quad u_1 \geq 0,\ u_2 \geq 0 \end{aligned}$$

As we have seen in the other sections, the optimal objective function value of (P) and (D) are equal. The optimal values for the dual variables u_i represent the so-called shadow prices for the available raw material i.

The concept of shadow price

The concept of shadow prices is based on the value for additional capacity or available raw material. In terms of LP, we are considering the change in the optimal objective value, if capacity constraint i is relaxed (extended) by one unit.

Given LP problem (P) with optimal (feasible) basis B, so $B^{-1}\underline{b} \geq \underline{0}$:

$$\max\{w = \underline{c}'\underline{x}\}$$
$$A\underline{x} \leq \underline{b} \qquad \text{assuming } \underline{b} \geq \underline{0} \qquad \text{(P)}$$
$$\underline{x} \geq \underline{0}$$

Let's consider the profitability of increasing b_i, representing the available capacity i:

$$\max\{w = \underline{c}'\underline{x}\}$$
$$A\underline{x} \leq \underline{b} + \lambda\underline{e}_i \qquad \text{capacity } i \text{ is increased by } \lambda \geq 0$$
$$\underline{x} \geq \underline{0}$$

The r-row does not depend on $\underline{b}$ as $r_j = \underline{c}_B' B^{-1}\underline{a}_j - c_j$ and therefore basis B is optimal and feasible as long as $B^{-1}(\underline{b} + \lambda\underline{e}_i) \geq \underline{0}$. Usually this is the case for small values of λ (see Section 4.5). The optimal objective function value $w^* = \underline{c}_B' B^{-1}\underline{b}$ of (P) with increasing value of λ equals:

$$\hat{w} = \underline{c}_B' B^{-1}(\underline{b} + \lambda e_i) = \underline{c}_B' B^{-1}\underline{b} + \lambda\, \underline{c}_B' B^{-1} e_i$$

The optimal dual values due to (4.4) are $\underline{u}' = \underline{c}_B' B^{-1}$ i.e. $\underline{u}_i = \underline{c}_B' B^{-1}\underline{e}_i$ such that:

$$\hat{w} = \underline{c}_B' B^{-1}\underline{b} + \lambda\, \underline{c}_B' B^{-1}\underline{e}_i = \underline{c}_B' B^{-1}\underline{b} + \lambda u_i = w^* + \lambda u_i$$

So, the optimal value of dual variable u_i gives the increase in the optimal objective function value per unit ($\lambda = 1$) increase of capacity i.

If the costs for increasing b_i per unit are smaller than u_i, it is profitable to increase capacity i. Value u_i is called the *shadow price*.

- If in the optimal solution $\underline{x}$ constraint i is not binding, $a_{i1}x_1 + \ldots + a_{in}x_n < b_i$, the capacity is not fully used and therefore increasing b_i doesn't generate additional profit. In that case $u_i = 0$, which also follows from the complementary slackness relations, i.e. if the value of slack variable $y_i > 0$, then from $u_i y_i = 0$ it follows that $u_i = 0$.
- If in the optimal solution $\underline{x}$, the capacity is fully used, $a_{i1}x_1 + \ldots + a_{in}x_n = b_i$, increasing b_i is only useful if the shadow price u_i is bigger than the cost per unit of extending b_i . In Section 4.5 the maximum possible increase is discussed.

Considering the dual objective function gives the following point of view on the shadow price: In the optimal solution $w = \underline{c}'\underline{x} = \underline{b}'\underline{u} = b_1u_1 + \ldots + b_iu_i + \ldots + b_mu_m$.

So, $\dfrac{\partial w}{\partial b_i} = u_i$, approximately $\dfrac{\Delta w}{\Delta b_i} \approx u_i$, or $\Delta w \approx u_i \Delta b_i$

Changing b_i into $b_i + \Delta b_i$ changes the optimal objective value with $u_i\Delta b_i$ (Example 4.13).

The approximation $\Delta w = u_i\Delta b_i$ can be derived in another way too.
Extending b_i to $b_i + \Delta b_i$ ($\Delta b_i > 0$) leads to $\hat{w} = b_1u_1 + \ldots + (b_i + \Delta b_i)u_i + \ldots + b_mu_m$. So the objective function value changes with $\Delta w = \hat{w} - w = u_i\Delta b_i$. The value of dual variable u_i (shadow price) gives the change in optimal objective function value when relaxing constraint i with one unit, $\Delta b_i = 1$. Important: as will be discussed in Section 4.5, the basic solution should remain feasible so, $B^{-1}\underline{b} \geq \underline{0}$.

Example 4.13.

The shadow price for the ABE example is now derived from the dual solution.

$$\max\{w = 2x_1 + 5x_2\}$$
$$\begin{aligned} x_1 \quad &\leq 4 \\ x_2 &\leq 3 \\ x_1 + x_2 &\leq 6 \qquad (P) \\ x_1 \geq 0, x_2 &\geq 0 \end{aligned}$$

x_1	x_2	y_1	y_2	y_3	
0	0	1	1	−1	1
0	1	0	1	0	3
1	0	0	−1	1	3
0	0	0	3	2	21
		↓	↓	↓	
		u_1	u_2	u_3	

Final tableau from tableau 4.2.

The first constraint of (P), $x_1 \leq 4$ or $x_1 + y_1 = 4$ is not binding in the optimal solution $x_1 = 3$, $x_2 = 3$. The optimal value of the slack variable is $y_1 = 1$. Relaxing the first constraint to $x_1 \leq 5$ does not affect the optimal solution; it remains (3,3) with $w^* = 21$. The increase in the optimal objective function value equals zero; so, $u_1 = 0$ (see Figure 4.1).

The second constraint $x_2 \leq 3$ or $x_2 + y_2 = 3$ is binding; $y_2 = 0$ in $\underline{x}' = (3, 3)$. Relaxing the second constraint to $x_2 \leq 4$ will affect the optimal solution, which becomes $x_1 = 2$ and $x_2 = 4$ and objective function value $\hat{w} = 24$ (Figure 4.1). The increase of the optimal objective function values is $\Delta w = \hat{w} - w^* = 24 - 21 = 3$ which equals $u_2 = 3$.
Relaxing the second constraint with two units to $x_2 \leq 5$ leads to an increase of $\Delta w = 2 \times u_2 = 6$. The maximum possible increase for the second constraint is three units $x_2 \leq 3 + 3$ with maximum increase $\Delta w = 3 \times u_2 = 9$.

The third constraint $x_1 + x_2 \leq 6$ or $x_1 + x_2 + y_3 = 6$ is also binding; $y_3 = 0$ in $\underline{x}' = (3, 3)$. A relaxation to $x_1 + x_2 \leq 7$ changes the optimal solution. The solution becomes $x_1 = 4$ and $x_2 = 3$ with objective function value $\hat{w} = 23$. The increase $\Delta w = 23 - 21 = 2$ corresponds to $u_3 = 2$.
A further relaxation of the third constraint e.g $x_1 + x_2 = 7.1$ will not increase the optimal objective function value. The third constraint is not binding anymore.

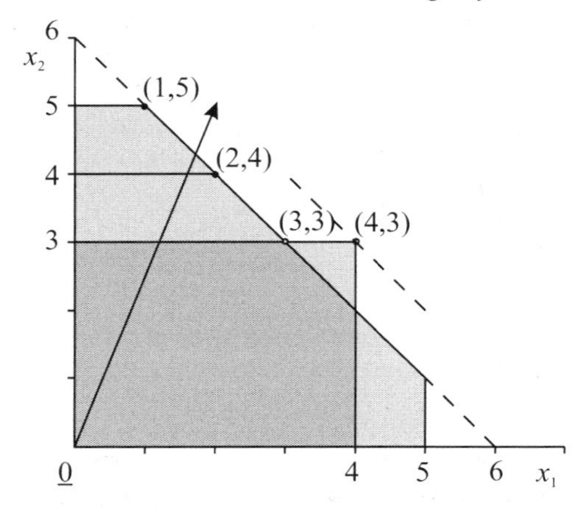

Figure 4.1. Relaxing constraints (see also Figure 2.11).

Remark

Alertness is required in the economic interpretation of the shadow price, i.e. the optimal value of u_i. In particular, we should take into account up to what level a constraint can be relaxed. This aspect will be studied in Section 4.5.

4.5. Sensitivity analysis and parametric linear programming

In practice data are mostly based on estimations and predictions. Uncertainty in the exact values can have a substantial impact on the optimal solution. This explains why it is important to know how changes in coefficient values of, for example, the $\underline{b}$ and $\underline{c}$-vector may influence the optimal solution. Moreover, management questions related to the data, the so-called 'what-if' questions, leads to questions on the behaviour of the optimal outcome with respect to the amount of scarce resources, possible price settings, etc. If we are interested in studying the impact of a continuous varying value of a parameter on the optimal solution, the term *parametric linear programming* is used. Usually the term *sensitivity analysis*, sometimes called *post-optimality analysis*, is used to analyse the effect of changes in the data on the optimal solution. This analysis can be more important in practice than finding the optimal solution. Information regarding the stability of the optimal solution under changes of parameter is usually desirable.

In Section 2.5 some exercises have been done based on a graphical analysis. In this section we will see how the optimal solution of a 'changed' LP problem $(\widetilde{\text{P}})$ can be determined efficiently from the optimal solution of the original LP problem (P). The tableau of the optimal solution gives useful information regarding these changes.

Given LP problem (P):

$$\begin{aligned} &\max\{w = \underline{c}'\underline{x}\} \qquad\qquad \text{(P)} \\ &A\underline{x} = \underline{b} \\ &\underline{x} \geq \underline{0} \end{aligned}$$

Let B be an optimal basis with corresponding basic values $\underline{x}_B = B^{-1}\underline{b} = \hat{\underline{b}}$; the r-row is known too.

We study the changes of the following coefficients:

- changing the elements of the objective function vector $\underline{c}$.
- changing the right-hand side vector $\underline{b}$.
- adding extra variables to the LP problem.

Changes of the matrix elements a_{ij} or adding an additional constraint are not discussed in this book.

4.5.1. Changing the objective function vector $\underline{c}$

- The vector $\underline{x}_B = B^{-1}\underline{b}$ does not depend on $\underline{c}$. So, $\underline{x}_B$ does not change and remains feasible ($\underline{x}_B \geq \underline{0}$).
- The values in the r-row depend on $\underline{c}$. So they change: $\underline{r}' = \underline{c}_B' B^{-1} A - \underline{c}'$. Two possibilities are conceivable:

1. All r_j remain ≥ 0.
   In this case $\underline{x}_B \geq \underline{0}$ and $\underline{r} \geq \underline{0}$ such that $\underline{x}_B$ is also optimal for $(\widetilde{\mathrm{P}})$. Note that the objective function value $w = \underline{c}_B^{\,\prime}\,\underline{x}_B$ may change.
2. At least one $r_j < 0$.
   In this case the tableau is no longer optimal and the Simplex Method can be applied to determine the optimal solution; $\underline{b} \geq \underline{0}$ and at least one $r_j < 0$.

Remark

If only one element c_k corresponding to a non-basic variable x_k changes, then only r_k has to be recalculated. However, if x_k is a basic variable, then the change in c_k also affects $\underline{c}_B$ such that the complete r-row should be recalculated.

4.5.2. Changing the right-hand side vector $\underline{b}$

- The r-row $\underline{r}' = \underline{c}_B^{\,\prime} B^{-1} A - \underline{c}'$ does not depend on $\underline{b}$. So in $(\widetilde{\mathrm{P}})$ all r_j remain ≥ 0.
- $\underline{x}_B = B^{-1}\underline{b}$ changes. Two possibilities are imaginable:
  1. The values in vector $\underline{x}_B \geq 0$. As $\underline{r} \geq \underline{0}$ and $\underline{x}_B \geq \underline{0}$, $\underline{x}_B = B^{-1}\underline{b}$ is also optimal for $(\widetilde{\mathrm{P}})$.
  2. At least one element of $\underline{x}_B$ is negative. Vector $\underline{x}_B$ is not feasible for $(\widetilde{\mathrm{P}})$. The *Dual Simplex Method* (Dantzig and Thapa, 2003 or Hillier and Lieberman, 2005) can be applied in order to find the optimal solution.

4.5.3. Adding one or several variables $x_{n+1}, \ldots, x_{n+k}$ $(k \geq 1)$

- The vector $\underline{x}_B = B^{-1}\underline{b}$ is not affected by the introduction of new variables.
- The r-values of the added variables $x_{n+1}, \ldots, x_{n+k}$ have to be determined. Two possibilities are imaginable:
  1. If all $r_j \geq 0$ $(j = n+1, \ldots, n+k)$, the tableau with $\underline{x}_B \geq \underline{0}$ and $\underline{r} \geq \underline{0}$ is optimal for $(\widetilde{\mathrm{P}})$.
  2. If at least one $r_j < 0$ $(j = n+1, \ldots, n+k)$, the Simplex Method (also called the primal Simplex Method) can be applied in order to find the new optimal solution.

Scheme 4.2. Overview sensitivity analysis.

		$\underline{x}_B$	$\underline{R}$
4.5.1	change in $\underline{c}$	$\underline{x}_B \geq \underline{0}$	1. if all $r_j \geq 0 \rightarrow$ tableau optimal 2. if one $r_j < 0 \rightarrow$ primal method
4.5.2	change in $\underline{b}$	1. if $\underline{x}_B \geq \underline{0}$ 2. if $\underline{x}_B$ contains a negative element	all $r_j \geq 0 \rightarrow$ tableau optimal all $r_j \geq 0 \rightarrow$ dual method
4.5.3	adding extra variables	$\underline{x}_B \geq \underline{0}$	1. if all $r_j \geq 0 \rightarrow$ tableau optimal 2. if one $r_j < 0 \rightarrow$ primal method

Example 4.15.

The sensitivity analysis of Section 2.5 for the ABE example is elaborated in this example from the optimal tableau.

$$\max\{w = 2x_1 + 5x_2\}$$
$$x_1 \leq 4$$
$$x_2 \leq 3$$
$$x_1 + x_2 \leq 6$$
$$x_1 \geq 0,\ x_2 \geq 0$$

$\underline{c}'$:	2	5	0	0	0	
	x_1	x_2	y_1	y_2	y_3	
	0	0	1	1	−1	1
$\underline{c}_B' = (0,5,2)$	0	1	0	1	0	3
	1	0	0	−1	1	3
$\underline{r}'$:	0	0	0	3	2	21

Tableau 4.3. Final tableau of ABE.

Changing the objective vector $\underline{c}$

In Section 2.5, we changed the objective function $\max\{2x_1 + 5x_2\}$ into $\max\{w = \alpha x_1 + 5x_2\}$. For which values of α is Tableau 4.3 optimal? This question is called *ranging*, i.e. there is a range (interval) of values for α such that the corresponding solution is optimal. Note again:

- The right-hand side vector $\underline{b}$ is not affected by $\underline{c}$ and does not change.
- The r-row changes; it is a function of $\underline{c}$ and therefore also of α.

$\underline{c}'$:	α	5	0	0	0	
	x_1	x_2	y_1	y_2	y_3	
	0	0	1	1	−1	1
$\underline{c}_B' = (0,5,\alpha)$	0	1	0	1	0	3
	1	0	0	−1	1	3
	0	0	0	$5-\alpha$	α	$15+3\alpha$

Tableau 4.3.a. (Optimal for $0 \leq \alpha \leq 5$).

Tableau 4.3.a is optimal if $\underline{r} \geq \underline{0}$, so for $0 \leq \alpha \leq 5$; this is the so-called corresponding *cost-coefficient range***.** The optimal objective function value is $w = 15 + 3\alpha$ and $\underline{x}' = (3, 3)$. Note that tableau 4.3 and tableau 4.3.a are the same for $\alpha = 2$.

- If $\alpha < 0$ then $r_{y_3} < 0$ and y_3 enters the basis. The optimal solution is:

$\underline{c}'$:	α	5	0	0	0	
	x_1	x_2	y_1	y_2	y_3	
	1	0	1	0	0	4
$\underline{c}_B' = (0,5,0)$	0	1	0	1	0	3
	1	0	0	−1	1	3
	$-\alpha$	0	0	5	0	15

Tableau 4.3.b. (Optimal for $\alpha \leq 0$).

Tableau 4.3.b is optimal for $\alpha \leq 0$; the optimal solution is $(x_1, x_2) = (0, 3)$ and $w = 15$.

- If $\alpha > 5$ in tableau 4.3a, then $r_{y_2} < 0$ and variable y_2 enters the basis:

$\underline{c}'$:	α	5	0	0	0	
	x_1	x_2	y_1	y_2	y_3	
	0	0	1	1	−1	1
$\underline{c}_B' = (0, 5, \alpha)$	0	1	−1	0	1	2
	1	0	1	0	0	4
	0	0	$\alpha-5$	0	5	$10+4\alpha$

Tableau 4.3.c. (Optimal for $\alpha \geq 5$).

Tableau 4.3c is optimal for the range $\alpha \geq 5$; the optimal solution is $(x_1, x_2) = (4, 2)$ and $w = 10+4\alpha$.

The optimal objective function value w as a function of α (see Figure 4.2) has been derived graphically in Section 2.5 (Figure 2.8) and in this example via optimal simplex tableaus. Derivation of the ranges and consequences of a continuously changing parameter value is called *parametric linear programming*.

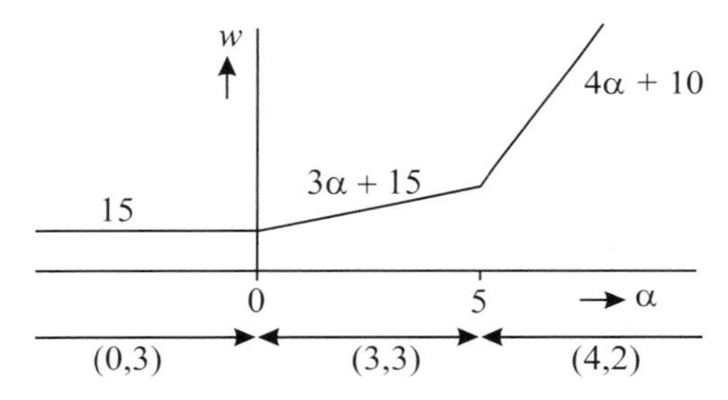

Figure 4.2. Optimal objective function value w as a function of α (see also Figure 2.8).

Changing the right-hand side vector $\underline{b}$

Consider the right-hand-side $\underline{b} = \begin{pmatrix} 4 \\ \beta \\ 6 \end{pmatrix}$ of the ABE problem, i.e. the second constraint has become $x_2 \leq \beta$.

- The r-row does not depend on $\underline{b}$ and does not change.
- The values of the basic variables become $\underline{x}_B = B^{-1}\underline{b}$.

$$\underline{x}_B = B^{-1}\underline{b} = \begin{pmatrix} 1 & 1 & -1 \\ 0 & 1 & 0 \\ 0 & -1 & 1 \end{pmatrix} \begin{pmatrix} 4 \\ \beta \\ 6 \end{pmatrix} = \begin{pmatrix} \beta-2 \\ \beta \\ -\beta+6 \end{pmatrix}$$

Next we determine the so-called *right-hand side range*, i.e. the values of β for which tableau 4.3 is feasible.

The basic solution corresponding to tableau 4.3 is feasible as long as $\underline{x}_B \geq \underline{0}$ such that the range is given by $2 \leq \beta \leq 6$. If $\beta = 3$ the values in tableau 4.3 appear, i.e. $\underline{x}_B' = (1, 3, 3)$. As the solution is also optimal ($\underline{r} \geq \underline{0}$), the optimal objective function value is:

$$w = \underline{c}_B' B^{-1} \underline{b} = (0,5,2) \begin{pmatrix} \beta - 2 \\ \beta \\ -\beta + 6 \end{pmatrix} = 12 + 3\beta$$

The range $2 \leq \beta \leq 6$ has also been found graphically in Section 2.5. From Figure 2.9 can be observed that the binding constraints are $x_2 = \beta$ (i.e. $y_2 = 0$) and $x_1 + x_2 = 6$ (i.e. $y_3 = 0$).
In the range $0 \leq \beta \leq 2$, the constraints $x_1 = 4$ and $x_2 = \beta$ are binding and $w = 8 + 5\beta$. This follows from Figure 2.9, but also when the Dual Simplex Method (Dantzig and Thapa, 2003) or (Hillier and Lieberman, 2005) would be applied.
For $\beta \geq 6$, the optimal vertex (0,6) is determined by the binding constraints $x_1 + x_2 = 6$ and $x_1 = 0$ and the optimal objective function value is $w = 30$.
The outcome of this *parametric linear program* in which the consequences of a continuously varying value for β is studied, is given in Figure 4.3.

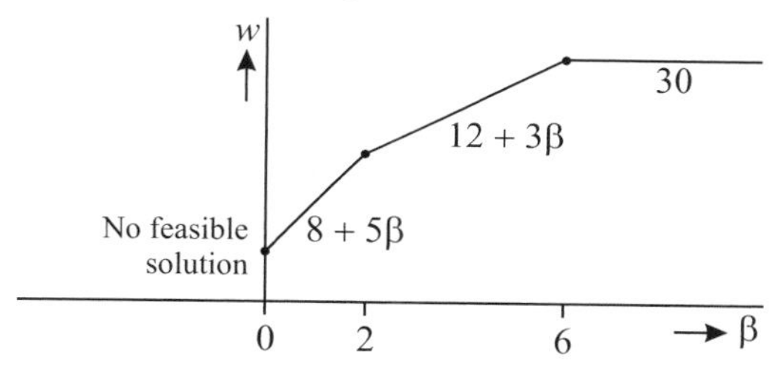

Figure 4.3. Optimal objective function value w as a function of β (*see also Figure 2.10*).

Adding an extra variable

Variable $x_3 \geq 0$ with column vector $\underline{a}_{x_3} = \begin{pmatrix} 3 \\ 2 \\ 1 \end{pmatrix}$ and objective coefficient c_{x_3} is added to problem ABE. After adding vector $\underline{a}_{x_3}$ to the initial tableau, the corresponding column $\underline{\hat{a}}_{x_3}$ in the final tableau can be calculated easily: $\underline{\hat{a}}_{x_3} = B^{-1}\underline{a}_{x_3}$.

$$\underline{\hat{a}}_{x_3} = B^{-1}\underline{a}_{x_3} = \begin{pmatrix} 1 & 1 & -1 \\ 0 & 1 & 0 \\ 0 & -1 & 1 \end{pmatrix} \begin{pmatrix} 3 \\ 2 \\ 1 \end{pmatrix} = \begin{pmatrix} 4 \\ 2 \\ -1 \end{pmatrix}$$

The corresponding r-value for x_3 is $r_{x_3} = \underline{c}_B' \underline{\hat{a}}_{x_3} - c_{x_3} = (0, 5, 2)\begin{pmatrix} 4 \\ 2 \\ -1 \end{pmatrix} - c_{x_3} = 8 - c_{x_3}$

The extra variable x_3 is a non-basic variable as long as $r_{x_3} \geq 0$, so $c_{x_3} \leq 8$; the current solution and corresponding basis B are optimal, $\underline{x}' = (x_1, x_2, x_3) = (3, 3, 0)$ is the optimal solution. However, if $c_{x_3} > 8$ then $r_{x_3} < 0$ which implies that the current basis is not optimal. The new variable x_3 must enter the basis.

5. Multi objective problems

Th.H.B. (Theo) Hendriks

Throughout the preceding chapters we assumed that the objectives of the organisation conducting the linear programming study can be encompassed within a single overriding objective, such as maximising total profit or minimising total cost. However, this assumption is not always realistic. In practice the management of corporations frequently focuses on a variety of other objectives, e.g., to maintain stable profits, increase (or maintain) market share, diversify products, maintain stable prices to restrict pollution, improve employment and worker morale, maintain family control of the business, and increase company prestige.

5.1. Mathematical formulation of the multi-objective problem

In this section, a mathematical formulation of the multi-criteria problem is presented, and some important topics are introduced.

In general a multi-criteria problem has n decision variables represented by the vector $\underline{x} = (x_1, x_2, ..., x_n)'$. Values of $\underline{x}$ have to be found that maximise p ($p \geq 2$) different *objective functions* $\underline{w}(\underline{x}) = (w_1(\underline{x}),\ w_2(\underline{x}),\ ...,\ w_p(\underline{x}))'$. The values of the decision variables $\underline{x}$ are subject to m restrictions $g_i(\underline{x}) \leq 0$, $i = 1, ..., m$. The general multi-criteria problem can be formulated as:

$$\begin{aligned} \max\ & \underline{w}(\underline{x}) = (w_1(\underline{x}),\ w_2(\underline{x}), ...,\ w_p(\underline{x}))' \\ & \underline{x} \in X \end{aligned} \qquad (5.1)$$

In problem (5.1) X is the set of all *feasible solutions*, i.e. the set of solutions that satisfy the restrictions $g_i(\underline{x}) \leq 0$ ($i = 1, ..., m$):

$$X = \{\underline{x} \mid g_i(\underline{x}) \leq 0,\ i = 1, ...,\ m\} \qquad (5.2)$$

In (5.1) vector $\underline{w}(\underline{x})$ has to be maximised. This is called a *Vector Maximization Problem* (VMP). The set of all vectors $\underline{x} = (x_1, x_2, ..., x_n)'$ constitutes the so-called *decision space* of the VMP. In general, not all vectors in the decision space satisfy the restrictions (5.2). The set X of vectors that satisfy (5.2) constitute the feasible region in the decision space. For a situation with two decision variables, the decision space and the feasible region can be represented graphically (see Figure 5.1).

Figure 5.1.a. The feasible area X in decision space $\mathbb{R}^2$.
Figure 5.1.b. The set W of feasible solutions in the objective space $\mathbb{R}^2$.

The set of all vectors $\underline{w}(\underline{x}) = (w_1(\underline{x}), w_2(\underline{x}), ..., w_p(\underline{x}))'$ constitutes the so-called *objective space.* For every feasible solution $\underline{x}$, the vector $\underline{w}(\underline{x})$ with the corresponding values of the objective functions can be calculated. The set of all these vectors is the set W of feasible function values: $W = \{\underline{w}(\underline{x}) \mid \underline{x} \in X\}$. So W is a subset of the objective space. In a VMP with two objective functions, the objective space can be represented graphically (see Figure 5.1b).

Figure 5.1 shows a decision problem with two decision variables x_1 and x_2 and two objective functions w_1 and w_2. We take a closer look at point $P \in X$, which has the values x_1^P and x_2^P for the decision variables x_1 and x_2, yielding the objective function values w_1^P and w_2^P. The point (w_1^P, w_2^P) in the objective space W is denoted by $\hat{P}$. 'Translating' all points of X into the objective space yields the area W with feasible objective function values.

Now the VMP can be formulated as:

$$\begin{aligned} &\max \underline{w}(\underline{x}) = (w_1(\underline{x}), w_2(\underline{x}), ..., w_p(\underline{x}))' \\ &\text{s.t. } \underline{w}(\underline{x}) \in W \end{aligned} \tag{5.3}$$

In Figure 5.1, the number of decision variables n equals the number of objective functions p ($n = p = 2$). This choice allows a graphical representation of the problem both in the solution and in the objective space. In general, the number of decision variables will not equal the number of objective functions.

Consider again point $\hat{P} = (w_1^P, w_2^P)$ in the objective space W. In Figure 5.1b it can be seen that w_1^P can be improved without decreasing w_2^P. This continues until the point $\hat{P}_1$ is reached. It is also possible, starting from $\hat{P}$, to improve w_2^P without decreasing w_1^P. Improving w_2^P without decreasing w_1^P is possible until the point $\hat{P}_2$ is reached. However, starting from $\hat{Q}$ it is not possible to improve w_1^Q without decreasing w_2^Q. Moreover, it is not possible to improve w_2^Q without decreasing w_1^Q.

The corresponding point Q in the solution space is called an *efficient solution* of the VMP. In general: for an efficient solution it is not possible to improve one of the objective functions without decreasing (at least) one of the other objective functions. Efficient solutions are also called *Pareto optimal* solutions.

Now we will consider the set of efficient solutions. It has been shown that in point $\hat{P}$ one objective function value can be improved without decreasing the other. So P is not an efficient solution. For points on the curve $\hat{A}\hat{B}$ it is not possible to improve one objective function value without decreasing the other. The solutions corresponding to this curve (vectors in the decision space) are the efficient solutions of the VMP. They are represented by the curve AB in the decision space.

A formal definition of an efficient solution is:

Definition 5.1.

A feasible solution $\underline{x}^Q \in X$ of a VMP is an *efficient solution* or *Pareto optimal solution* if there is no other feasible solution $\underline{x} \in X$ such that $w_k(\underline{x}^Q) \leq w_k(\underline{x})$ for all $k = 1, ..., p$ and $w_k(\underline{x}^Q) < w_k(\underline{x})$ for at least one $k = 1, ..., p$.

In words: a feasible solution is an efficient solution, if no other solution exists that is at least as good at every objective, and (at least) better at one objective.

Example 5.1.

Consider the following problem:

$$
\begin{array}{l}
\max \{w_1 = 3x_1 + x_2\} \\
\max \{w_2 = -x_1 + 4x_2\} \\
\text{s.t.} \\
\quad x_1 \leq 4 \\
\quad x_2 \leq 3 \\
\quad x_1 + x_2 \leq 6 \\
\quad x_1 \geq 0, \; x_2 \geq 0
\end{array}
\tag{5.4}
$$

The feasible decision space X and the corresponding set W of feasible objective function values are shown in Figure 5.2. The calculations are shown in Table 5.1.

Table 5.1. Calculations corresponding to Figure 5.2.

(x_1, x_2)	$W_1(\underline{x}) = 3x_1 + x_2$	$w_2(\underline{x}) = -x_1 + 4x_2$	(w_1, w_2)
$O = (0, 0)$	0	0	(0, 0)
$A = (4, 0)$	12	−4	(12,−4)
$B = (4, 2)$	14	4	(14, 4)
$C = (3, 3)$	12	9	(12, 9)
$D = (0, 3)$	3	12	(3, 12)
$E = (2, 2)$	8	6	(8, 6)

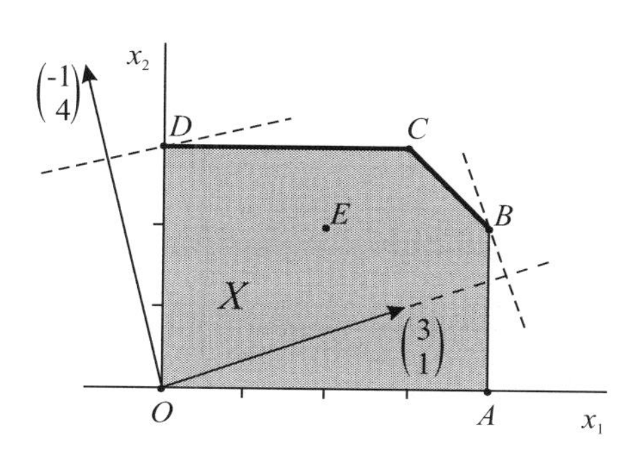

Figure 5.2.a. Decision space X.

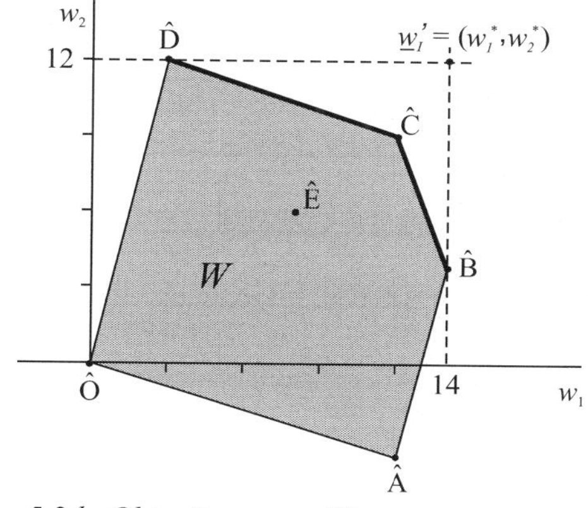

5.2.b. Objective space W.

In Figure 5.2.a and 5.2.b we can see the efficient solutions for problem (5.4). For the points on the line segments $\hat{B}\hat{C}$ and $\hat{C}\hat{D}$ it is not possible to improve w_1 without decreasing w_2 and vice versa.
The vectors $\underline{x}$ corresponding to these line segments are the efficient solutions. In the decision space X the efficient solutions are represented by the line segments BC and CD. So the Pareto set consists of the line segment BC (which can be written as $\lambda_1(4,2) + \lambda_2(3,3)$ with $\lambda_1 + \lambda_2 = 1$, $\lambda_1, \lambda_2 \geq 0$ and line segment CD (which can be written as $\mu_1(3,3) + \mu_2(0,3)$ with $\mu_1 + \mu_2 = 1$ and $\mu_1 \geq 0, \mu_2 \geq 0$.

Until now the two objective functions of (5.4) were studied simultaneously. Next we will study them separately. Problem (5.4) is split into two separate problems:

$$\begin{aligned} &\max \{w_1 = 3x_1 + x_2\} \\ &\text{s.t. } \underline{x} \in X \end{aligned} \tag{5.4a}$$

and

$$\begin{aligned} &\max \{w_2 = -x_1 + 4x_2\} \\ &\text{s.t. } \underline{x} \in X \end{aligned} \tag{5.4b}$$

The optimal solution of (5.4a) is $(x_1, x_2) = (4, 2)$ with corresponding objective function value $w_1^* = 14$.
For (5.4b) the optimum solution is $(x_1, x_2) = (0, 3)$, with optimal objective value $w_2^* = 12$.
The point in the objective space that corresponds to these two optimal objective function values is the so-called *ideal point* or $\underline{w}^I$. So, $\underline{w}^I = (w_1^*, w_2^*) = (14, 12)$.

Formally speaking: the ideal point $\underline{w}^I = (w_1^*, \ldots, w_p^*)$ is a point in the objective space such that w_k^*, $k = 1, \ldots, p$ is the optimal value of the objective function for problem:

$$\begin{aligned} &\max w_k(\underline{x}) \\ &\text{s.t. } \underline{x} \in X \end{aligned}$$

In Figure 5.2b it can be seen that 'ideal point' is a very suitable name for this point: like many ideals, the ideal point can not be reached (it does not belong to W). As the objective functions in a VMP are usually conflicting, the improvement of one objective function can only be accomplished at the cost of increasing distance to the optimal value for another objective function. The ideal point is important as a reference in determining targets for the objective function values and in methods for determining the Pareto set.

Only in exceptional cases the ideal point can be reached. If we define, for example, a new multi-objective problem with the same feasible area as in Figure 5.2a and the objective functions:

$$\begin{aligned} &\max \{ w_1 = 3x_1 + 2x_2\} \\ &\max \{w_2 = 7x_1 + x_2\} \end{aligned} \tag{5.5}$$

Splitting (5.5) into two sub-problems yields:

$$\max \{ w_1 = 3x_1 + 2x_2\} \quad \text{s.t. } \underline{x} \in X \tag{5.5a}$$
$$\max \{ w_2 = 7x_1 + x_2\} \quad \text{s.t. } \underline{x} \in X \tag{5.5b}$$

From Figure 5.2a it follows that the optimal solution $(x_1, x_2) = (4, 2)$ for (5.5a) coincides with the optimal solution $(x_1, x_2) = (4, 2)$ of (5.5b). Optimal values $w_1^* = 16$ and $w_2^* = 30$.
The corresponding ideal point is $\underline{w}^I = (w_1^*, w_2^*) = (16, 30)$. The corresponding point $\underline{x} = (4, 2)$ in the solution space is called the *superior solution* to problem (5.5), which indicates that this solution gives the optimal solution for both objective functions:

$$\underline{x}^S \text{ is a superior solution} \quad \Leftrightarrow \quad \underline{x}^S \in X \text{ and } w_k(\underline{x}^S) \geq w_k(\underline{x}) \;\; k = 1, ..., p \text{ for all } \underline{x} \in X.$$

A superior solution is also called the optimal solution. If a superior solution exists, then the ideal point can be reached. In this case the Pareto set consists of a single point, the superior point. Most VMPs do not have a superior solution. Therefore, one is usually more interested in the efficient solutions of a VMP. If we assume that a decision maker always prefers an efficient solution to an inefficient solution, then the decision-making process can concentrate on the solutions in the Pareto set. Many methods have been developed for choosing a solution of the Pareto set. The solution that will be chosen is called the *preferred solution.*

We discuss only two methods for 'solving' multi-objective problems (i.e. the determination of a preferred solution).

- In Section 5.2 the method of goal programming is discussed. In goal programming the decision maker has to establish specific numerical goals (targets) for each objective function. A linear programming model can be formulated to minimise the sum of deviations of these objective function values from their target values (i.e. goals).
- In Section 5.3 we discuss a method of interactive multi-criteria decision making. In this method there is an interaction between the decision maker and a quantitative model. The interaction is a dialogue between the model, which responds with outcomes on a set of preferences (targets, goals) given by the decision maker. Depending on the outcome, the decision maker offers a new set of preferences to the model and so on. The process proceeds in an interactive and iterative way until the decision maker has found a satisfactory (preferred) solution.

5.2. Linear goal programming

The basic approach of *goal programming* is to establish a specific numeric goal or target for each objective function, and then find a solution that minimises the (weighted) sum of deviations of the objective function values from their respective goals. Three possible types of goals exist:

1. A *lower, one-sided goal* sets a lower limit we do not want to fall under (but exceeding the limit is fine), e.g. $2x_1 + 3x_2 \geq 7$.
2. An *upper, one-sided goal* sets an *upper limit* we do not want to exceed (but falling under the limit is fine), e.g. $4x_1 + 5x_2 \leq 10$.
3. A *two-sided goal* sets a specific target that we prefer to reach exactly, e.g. $6x_1 + 7x_2 = 15$.

Goal-programming problems can be classified according to the type of mathematical programming model that it fits (linear programming, integer programming, non-linear programming, etc.), except for having several goals instead of a single objective. We only consider linear goal programming.

We first introduce so-called *auxiliary variables* $g^+ \geq 0$ and $g^- \geq 0$. In practice, it is not always possible to attain all goals simultaneously. With the help of auxiliary variables we can mathematically express the situation in which it might not be possible to achieve certain goals.

Example 5.2.
Suppose a lower, one-sided goal is set on the profit, e.g. $2x_1 + 3x_2 \geq 7$. We will penalise situations in which the profit is less than 7. Therefore, the auxiliary variables $g^+ \geq 0$ and $g^- \geq 0$ are introduced:

$$2x_1 + 3x_2 \geq 7 \quad \Rightarrow \quad 2x_1 + 3x_2 - g^+ + g^- = 7, \qquad g^+ \geq 0, g^- \geq 0 \tag{5.7}$$

We state that in an optimal situation if $g^+ > 0$ then $g^- = 0$, and if $g^- > 0$ then $g^+ = 0$. In other words it is not possible that both $g^+ > 0$ and $g^- > 0$ simultaneously (see also Section 9.3.1).
If the profit is 10 (so $2x_1 + 3x_2 = 10$) then from (5.7) follows $g^+ = 3$ and $g^- = 0$, and the goal is attained. If, on the other hand, the profit is 5 then $g^- = 2$ and $g^+ = 0$ and apparently the target value of the goal has not been reached. In other words: if $g^- > 0$ then the lower one-sided goal has not been attained. A suitable objective function to reach the goal (profit at least 7) is $\min\{g^-\}$.

Example 5.3.
Suppose the nitrogen emission is defined as $4x_1 + 5x_2$ and it must be less than or equal to 10 (an upper, one-sided goal). So, the goal is $4x_1 + 5x_2 \leq 10$. Adding auxiliary variables $g^+ \geq 0$ and $g^- \geq 0$ gives:

$$4x_1 + 5x_2 \leq 10 \quad \Rightarrow \quad 4x_1 + 5x_2 - g^+ + g^- = 10, \qquad g^+ \geq 0, g^- \geq 0$$

If the nitrogen emission turns out to be 8, e.g. for $(x_1, x_2) = (2, 0)$ then $g^+ = 0$ and $g^- = 2$ (the goal is attained). However, if the nitrogen emission turns out to be 13 then $g^+ = 3$ and $g^- = 0$ and the goal is not attained. A suitable objective function for an upper, one-sided goal is $\min\{g^+\}$.

Example 5.4.
In a similar way, for a two-sided employment goal of $6x_1 + 7x_2 = 15$, the auxiliary variables $g^+ \geq 0$ and $g^- \geq 0$ are introduced. If the employment level is 17 (so $g^+ = 2$) or the employment level is 12 (so $g^- = 3$), the desired goal is not attained. So, a suitable objective function for a two-sided goal is $\min\{g^+ + g^-\}$.
As $g^+ \geq 0$ and $g^- \geq 0$, the objective function is always ≥ 0. The minimum value is zero in case $g^+ = 0$ and $g^- = 0$. If $g^+ = 0$ and $g^- = 0$, we have attained our goal.

W_0 is the complete (light and dark) shaded area, $W_1 = W_0 \cap \{w_1 \geq 4\}$, $W_2 = W_1 \cap \{w_1 \geq 8\}$, $W_3 = W_2 \cap \{w_2 \geq 3\}$.

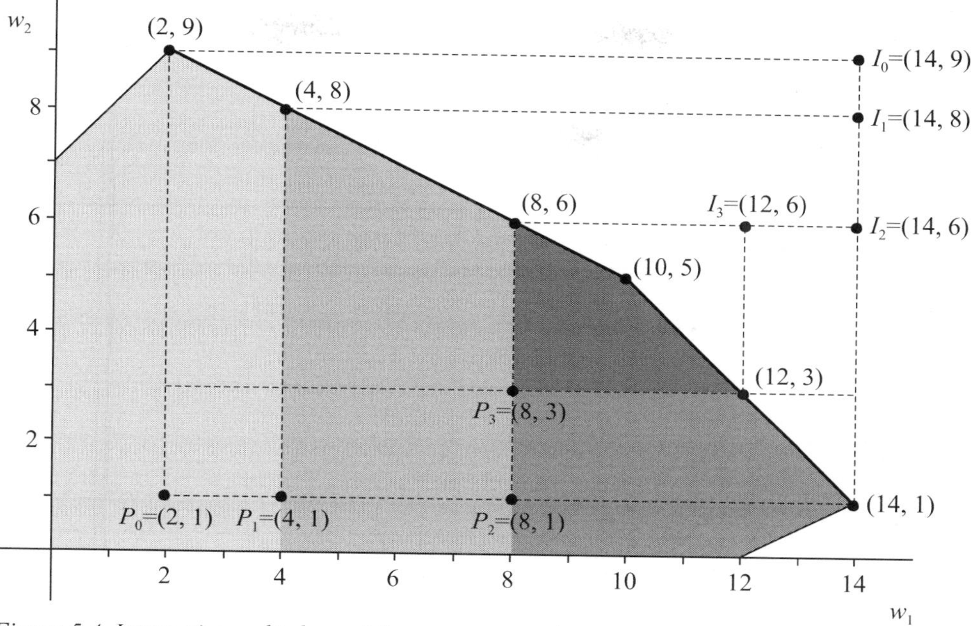

Figure 5.4. Interactive multiple goal linear programming.

Iteration 0

In the zero round (iteration 0) each objective function is maximised separately, yielding $\max\{w_1\} = 14$ and $\max\{w_2\} = 9$. So, the ideal point in iteration 0 is $I_0 = (14, 9)$. The worst values that are found for w_1 and w_2 are summarised in the pessimistic point $P_0 = (2, 1)$, which guarantees that w_1 is at least 2, and w_2 should be at least 1. Note that $w_2 = 1$ is obtained in case the first objective w_1 is maximised; $w_1 = 2$ in case w_2 is maximised (see also Table 5.2).

Iteration 1

Now the decision maker is asked whether any of the values in the pessimistic point $P_0 = (2, 1)$ are unacceptable. Suppose the decision maker does not accept the value $w_1 = 2$, and claims a lower bound of $w_1 \geq 4$. This additional restriction will be added to the model, which reduces the feasible area to $W_1 = W_0 \cap \{w_1 \geq 4\}$. The maximum value of w_2 on W_1 is $w_2 = 8$. So the lower bound $w_1 \geq 4$ implies that $w_2 = 9$ is no longer attainable. The new ideal point in the first iteration is $I_1 = (14, 8)$, and the new pessimistic point is $P_1 = (4, 1)$. The results of the first iteration are summarised in Table 5.2.

Iteration 2

Suppose the decision maker is still not satisfied with the lowest value for w_1 and requires $w_1 \geq 8$. This implies that the feasible area will be reduced to $W_2 = W_1 \cap \{w_1 \geq 8\}$. The maximum value of w_2 is lowered to $w_2 = 6$. The ideal point becomes $I_2 = (14, 6)$ and the pessimistic point becomes $P_2 = (8, 1)$.

Iteration 3

Now the decision maker wants $w_2 \geq 3$, which yields $I_3 = (12, 6)$ and $P_3 = (8, 3)$. In subsequent iterations the feasible area can be reduced even further.

Table 5.2. Summary of an IMGLP procedure.

Iteration	max w_1 yields		max w_2 yields		Ideal point	Pessimistic point	Area
	$\max\{w_1\}$	(w_1, w_2)	$\max\{w_2\}$	(w_1, w_2)			
It. 0	$w_1 = 14$	(14, 1)	$w_2 = 9$	(2, 9)	$I_0 = (14, 9)$	$P_0 = (2, 1)$	W_0
It. 1 $w_1 \geq 4$	$w_1 = 14$	(14, 1)	$w_2 = 8$	(4, 8)	$I_1 = (14, 8)$	$P_1 = (4, 1)$	W_1
It. 2 $w_1 \geq 8$	$w_1 = 14$	(14, 1)	$w_2 = 6$	(8, 6)	$I_2 = (14, 6)$	$P_2 = (8, 1)$	W_2
It. 3 $w_2 \geq 3$	$w_1 = 12$	(12, 3)	$w_2 = 6$	(8, 6)	$I_3 = (12, 6)$	$P_3 = (8, 3)$	W_3

In this example the generated points in successive iterations $(w_1, w_2) = (14, 1), (2, 9), (4, 8), (12, 3), (8, 6), \ldots$ are efficient (Pareto optimal) points. If the decision maker is satisfied, the procedure stops. In this example the solution $(w_1, w_2) = (12, 3)$ or $(8, 6)$ would be accepted as the preferred solution. In IMGLP the user becomes aware of the trade-off between the various objectives. Of course, decision makers with different interests can end in different (efficient) points.

Remark

1. For this small example, all efficient intermediate points in the objective space can be found as a convex combination of the vertices $\begin{pmatrix}2\\9\end{pmatrix}$ and $\begin{pmatrix}10\\5\end{pmatrix}$; respectively $\begin{pmatrix}10\\5\end{pmatrix}$ and $\begin{pmatrix}14\\1\end{pmatrix}$:

   a. $\lambda_1\begin{pmatrix}2\\9\end{pmatrix}+\lambda_2\begin{pmatrix}10\\5\end{pmatrix}$ with $\lambda_1 + \lambda_2 = 1$ and $\lambda_i \geq 0$. For example: $\begin{pmatrix}8\\6\end{pmatrix}=\frac{1}{4}\begin{pmatrix}2\\9\end{pmatrix}+\frac{3}{4}\begin{pmatrix}10\\5\end{pmatrix}$

   b. $\lambda_1\begin{pmatrix}10\\5\end{pmatrix}+\lambda_2\begin{pmatrix}14\\1\end{pmatrix}$ with $\lambda_1 + \lambda_2 = 1$ and $\lambda_i \geq 0$. For example: $\begin{pmatrix}12\\3\end{pmatrix}=\frac{1}{2}\begin{pmatrix}10\\5\end{pmatrix}+\frac{1}{2}\begin{pmatrix}14\\1\end{pmatrix}$

   In general it is not easy to determine all efficient (Pareto optimal) points.

2. If alternative optimal solutions exist for one (or more) objective function(s), the generated solutions may not be Pareto optimal.

6. Integer linear programming

Th.H.B. (Theo) Hendriks

6.1. Introduction

A wide class of practical problems can be modelled using integer variables. If such a model consists solely of integer variables, the problem is called a *pure integer programming model*. More commonly a model consists of both integer variables and continuous variables. Models of the latter class are called *mixed integer programming models*. The Simplex Method can solve LP problems, but there is no guarantee that the solution has an integral value. Many methods for solving LP problems with integrality constraints have been developed. The one that is most often used is the so-called *Branch-and-Bound method*.

In spite of the enormous development in calculation speeds of computers, there can be no guarantee that the optimal solution to an integer LP problem can be found within a reasonable time. Note that an integer LP problem with binary (zero-one) variables is just a special case of integrality constraints in linear programming. Although the Branch-and-Bound method presented in this chapter assumes integer values for the variables, the same concept is applicable to LP problems with binary variables.

An integer LP problem can be formulated as follows:

$$\begin{aligned} &\max\{w = \underline{c}'\underline{x}\} \\ &\text{s.t.} \\ &\quad A\underline{x} = \underline{b} \\ &\quad \underline{x} \geq \underline{0} \\ &\quad x_j \text{ integer, } j \in I \subset \{1, 2, 3, ..., n\} \end{aligned} \tag{6.1}$$

In problem (6.1) the following holds: $\underline{x} \in \mathbb{R}^n$, $\underline{c} \in \mathbb{R}^n$, $\underline{b} \in \mathbb{R}^m$, A is an $m \times n$-matrix, I is the index set of the integer variables. If $I = \{1, 2, ..., n\}$ then all x_j have to be integer. This problem is called a *pure integer LP problem*. Otherwise (6.1) is called a *mixed integer LP problem*. Without the integrality constraint, problem (6.1) is a continuous LP problem.

Two obvious methods to solve integer LP problems are:

1. For all combinations of values for the variables, check if they are feasible and optimal. In practice this method, i.e. complete enumeration, requires too much computational effort.
2. Solve the integer LP problem without integrality constraints as a continuous LP problem with the Simplex Method and round the variables to integer values. In general, rounding e.g. 999.2 to 999 will hardly cause any problems. However, rounding variables with smaller values (in particular 0-1 variables) has serious drawbacks:
   - A solution that is found by rounding might not be feasible. In Example 6.1 the optimal solution to the continuous LP problem is $(x_1, x_2) = (\frac{31}{4}, 0)$. The rounded solution (8, 0) is not feasible.
   - A 'neighbour' integral solution need not be optimal. In Example 6.1 the optimal integer solution $\underline{x}' = (3, 2)$ does not look like the optimal continuous solution $(\frac{31}{4}, 0)$.

Example 6.1.
Given integer LP problem (P):

$$
\begin{aligned}
&\max\{w = 12x_1 + 26x_2\} \\
&\quad 4x_1 + 9x_2 \leq 31 \\
&\qquad\quad 4x_2 \leq 10 \qquad\qquad\qquad \text{(P)} \\
&\quad x_1 \geq 0,\ x_2 \geq 0 \\
&\quad x_j \text{ integer, } j \in I = \{1, 2\} \qquad\qquad\qquad\qquad\qquad\qquad (6.2)
\end{aligned}
$$

Figure 6.1 gives a graphical representation of problem (P). Without the integrality constraints (6.2), the feasible area is *OPQR*. The optimal solution to the so-called *LP relaxation*, i.e. problem (P) without constraints (6.2), is vertex $R = (\frac{31}{4}, 0)$. The grid points indicated by '×' are the feasible solutions of the integer problem. The optimal integer solution to problem (P) is the point $V = (3, 2)$. Note that the optimal solution to problem (P) is not a vertex of the LP relaxation. Apparently, if an optimal solution exists for an integer LP problem, it may lie in the interior of the decision space of the LP relaxation. From a computational point of view, this observation has an enormous impact on solving (mixed) integer linear programming problems.

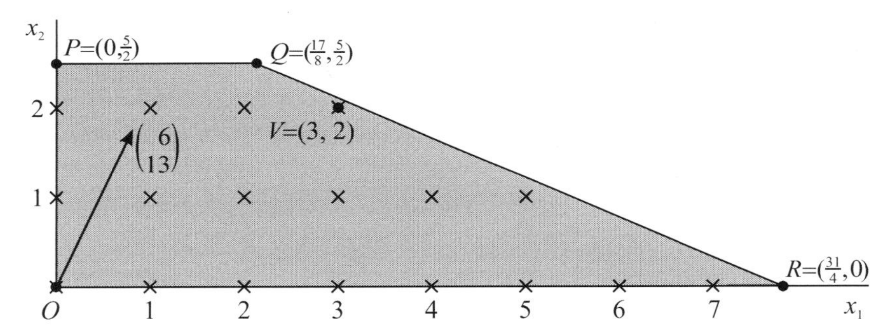

Figure 6.1. Integer solutions to problem (P) in Example 6.1.

6.2. The branch-and-bound method

We discuss the *Branch-and-Bound method* (B&B) as described by R. Dakin (1965). The algorithm was introduced by Land and Doig (1960). The method can be used both for pure and mixed integer problems. We will assume maximisation problems. The concept of the (B&B) method is based on the following steps:

- A continuous LP problem, e.g. the LP relaxation of the (mixed) integer linear programming problem, is branched into two continuous LP problems.
  Suppose the optimal solution $\underline{x}$ of continuous LP problem (P_k) has a non-integral value β_j for the integer variable x_j. No integer values for x_j exist in the interval $[\beta_j] < x_j < [\beta_j] + 1$, in which $[\beta_j]$ is the largest integer $\leq \beta_j$, e.g. $[3.6] = 3$, $[0.7] = 0$ and $[-\frac{3}{2}] = -2$. Now, LP problem (P_k) can be replaced by two new continuous LP problems:

1. Adding constraint $x_j \leq [\beta_j]$ to (P_k) yields problem (P_{k+1}).
2. Adding constraint $x_j \geq [\beta_j] + 1$ to (P_k) yields problem (P_{k+2}).

Variable x_j is called the *branching variable*.

- Bounds are calculated to determine whether it is useful to branch the generated (sub-) problems. The bound w_b is the objective function value of the best integer solution that has been found so far. If a continuous LP problem (P_k) has an optimal objective function value w smaller (in a maximisation problem) than the bound, $w < w_b$, then it is useless to continue branching problem (P_k). Note that adding an extra constraint will never yield a higher objective function value. Figure 6.2 is a schematic representation of a B&B iteration step.

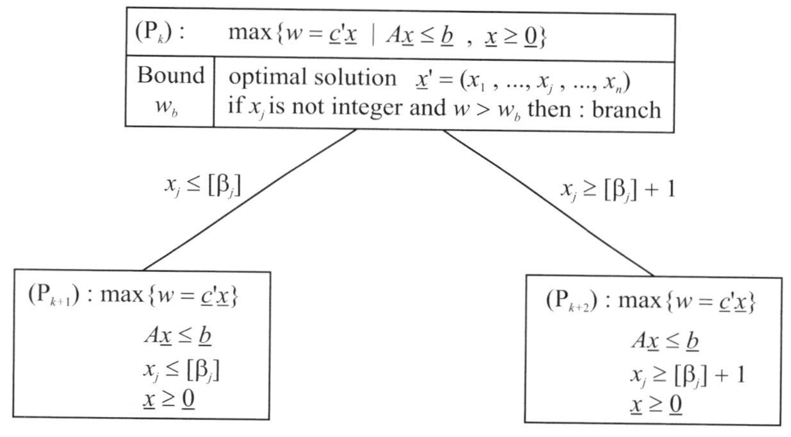

Figure 6.2. A Branch-and-Bound iteration step.

The B&B algorithm is an enumeration method that explicitly or implicitly enumerates all integer solutions. The efficiency of this (implicit) enumeration process is increased by (strong) bounds in an early phase of the search tree.

The B&B algorithm can be represented graphically in a tree structure (see also Figure 6.2). A node in the search tree corresponds to LP problem P_k (k = 1, 2, ...). If the bound and the solution to (P_k) are such that it is useful to branch (P_k), then node (P_k) has two leaving arcs. These arcs correspond to the extra constraints leading to the LP problems (P_{k+1}) and (P_{k+2}). Problem (P_{k+1}) is (P_k) in which $x_j \leq [\beta_j]$ is added as a restriction. Problem (P_{k+2}) is problem (P_k) in which $x_j \geq [\beta_j] + 1$ has been added.

In the following Example 6.2 is $\underline{x} \in \mathbb{R}^2$, so all sub-problems can be solved graphically.

Example 6.2.

$$\begin{aligned} &\max\{w = 5x_1 + 2x_2\} \\ &\quad 6x_1 + 4x_2 \leq 25 \\ &\quad x_1 \qquad\quad \leq 3 \qquad\qquad\qquad\qquad (P) \\ &\quad x_1, x_2 \geq 0 \text{ and integer} \end{aligned}$$

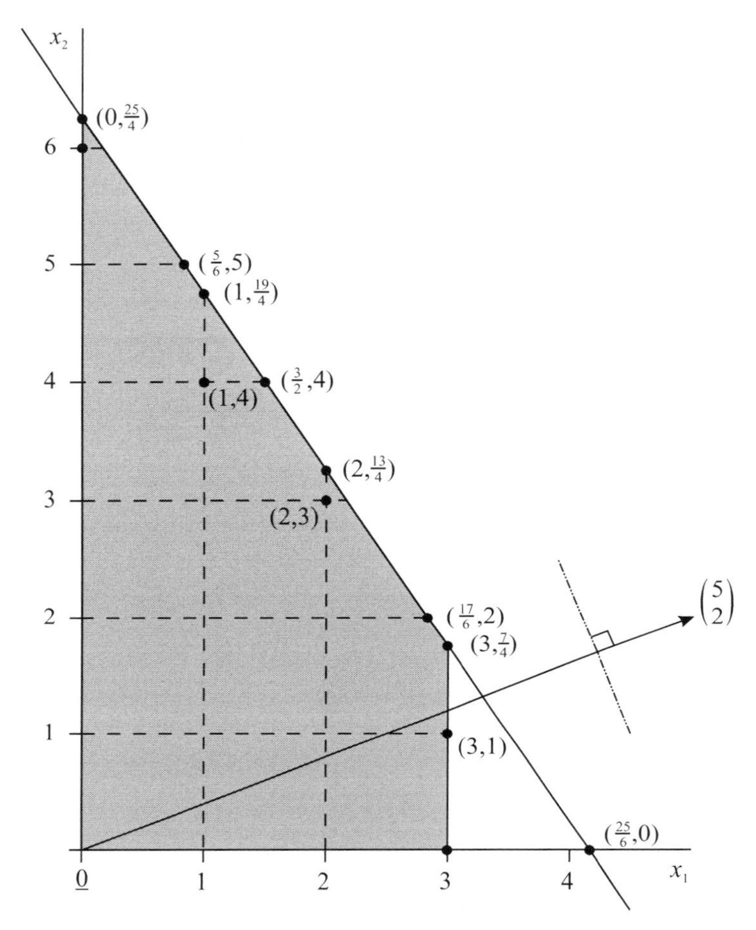

Figure 6.3. The Branch-and-Bound method.

Problem (P) is a pure integer LP problem. Omitting the integrality constraints yields the LP relaxation problem P_1, which can be solved graphically (see Figure 6.3). The optimal solution is $\underline{x}_1' = (x_1, x_2) = (3, \frac{7}{4})$. The corresponding objective function value is $w_1 = 18\frac{1}{2}$. The solution $\underline{x}_1$ is clearly not integral. At this stage no integral solution has yet been found. So, no bound w_b is available. We choose $w_b = 0$, as it can be seen easily that (0, 0) is a feasible integral solution to (P). Of course we can also use $w_b = -\infty$. Because of the fractional value $x_2 = \frac{7}{4}$, the variable x_2 is used as the branching variable.

Problem P_2 appears after the constraint $x_2 \leq [\frac{7}{4}] = 1$ is added to problem P_1. Problem P_3 consists of problem P_1 and the additional constraint $x_2 \geq [\frac{7}{4}] + 1 = 2$. Note that P_2 and P_3 together contain all feasible solutions of the original problem (P). Now the LP problems P_2 or P_3 are solved.

We discuss the search tree in Figure 6.4 that is obtained by systematically solving the left-hand side problem first. The index k indicates the order in which the problems are solved.

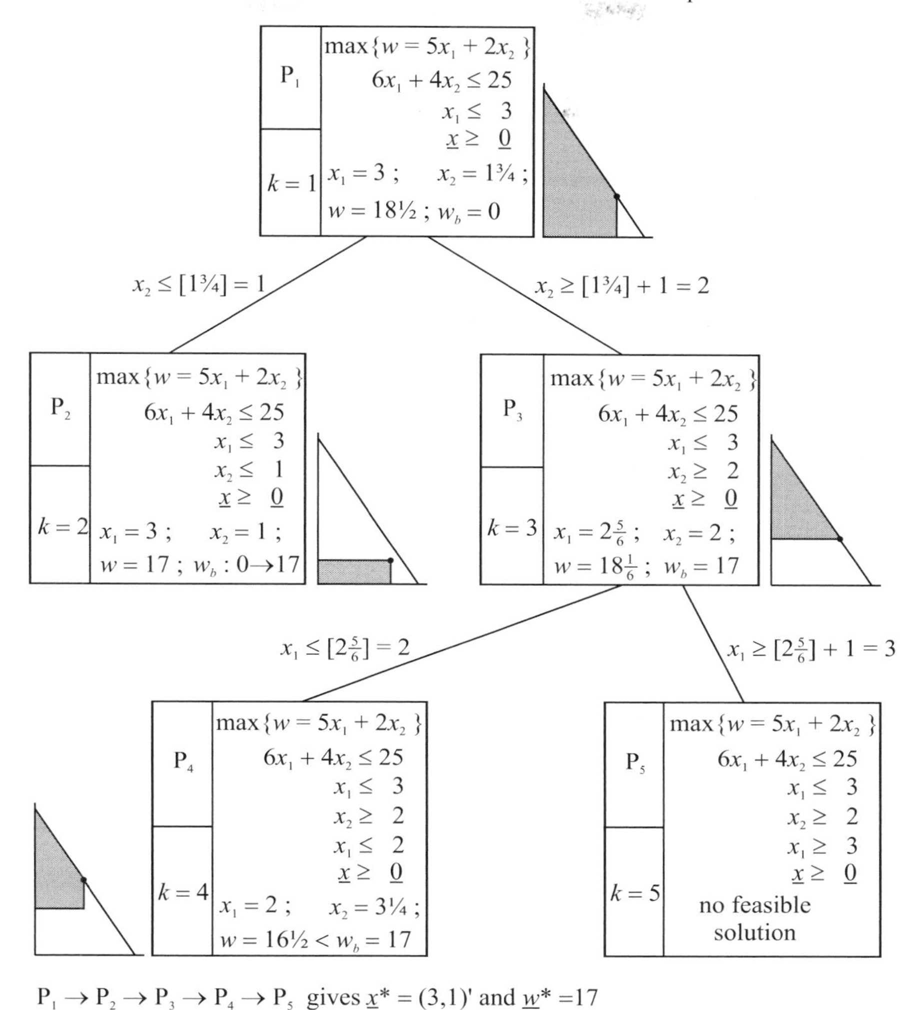

Figure 6.4. The B&B tree after the 'left-hand side problems' ($x_i \leq$) are solved first.

The optimal solution to P_2 is $\underline{x}_2' = (3, 1)$ with objective function value $w_2 = 17$. The solution $\underline{x}_2' = (3, 1)$ is a feasible integer solution. It is not known yet whether $\underline{x}_2$ is the optimal integer solution. However, what is certain is that the optimal solution $\underline{x}^*$ has optimal objective function value $w^* \geq 17$. So, the bound w_b becomes $w_b = 17$.

The optimal solution to P_3 is $\underline{x}_3' = (\frac{17}{6}, 2)$ with objective function value $w_3 = 18\frac{1}{6}$. As $x_1 = \frac{17}{6}$ and $w_2 = 18\frac{1}{6} > 17$, the branching variable is $x_1 = \frac{17}{6}$.

Problem P_4 arises after the restriction $x_1 \leq [\frac{17}{6}] = 2$ is added to P_3. The optimal solution to P_4 is $\underline{x}_4' = (2, \frac{13}{4})$ with $w_4 = 16\frac{1}{2}$. As the value of the objective function $w_4 = 16\frac{1}{2}$ is smaller than the bound $w_b = 17$, it is useless to branch problem P_4. Adding restrictions to P_4 will never lead to better objective function values. The value w will decrease or (at most) remain equal in any node beneath problem P_4. In other words: no better feasible integer solution can be generated from P_4.

Problem P_5 arises by adding the restriction $x_1 \geq [\frac{17}{6}] + 1 = 3$ to P_3. Problem P_5 has no feasible solutions.

All generated LP problems are solved.

Conclusion

The optimal integer solution $\underline{x}^* = (3, 1)'$ and $w^* = 17$ has been found in problem P_2. The required number of iterations to find this solution depends on the search strategy. In Figure 6.4, the optimal integer solution is found after solving 'only' 5 continuous LP problems. On the other hand, if the right-hand side problems, i.e. $x_j \geq [\beta_j] + 1$, are systematically examined first, then the search tree as illustrated in Figure 6.5 is obtained. The index k indicates the order in which the problems are solved.

After problem P_1 ($k = 1$), problem P_3 ($k = 2$) is solved first. The branching variable is x_1. Problem P_5 ($k = 3$) has no feasible solution. After problem P_4 ($k = 4$), the variable x_2 is the branching variable, yielding the LP problems P_6 and P_7, and so on (see Figure 6.5).
In problem P_{14} ($k = 12$), the first integer solution $\underline{x}_{14}' = (0, 6)$ is found, with $w_{14} = 12$. So, the lower bound becomes $w_b = 12$. Subsequently, the bound becomes $w_b = 13$ at P_{10} ($k = 13$) and $w_b = 16$ at P_6 ($k = 14$). In total, 15 continuous LP problems must be solved in order to identify and verify the optimal solution.

Above we showed that the search strategy, i.e. choosing the sub-problem to be solved, has an enormous effect on the efficiency of the B&B method. Unfortunately, no method guarantees that the 'golden' sub-problem is chosen. Many rules of thumb have been developed to speed up the searching process.

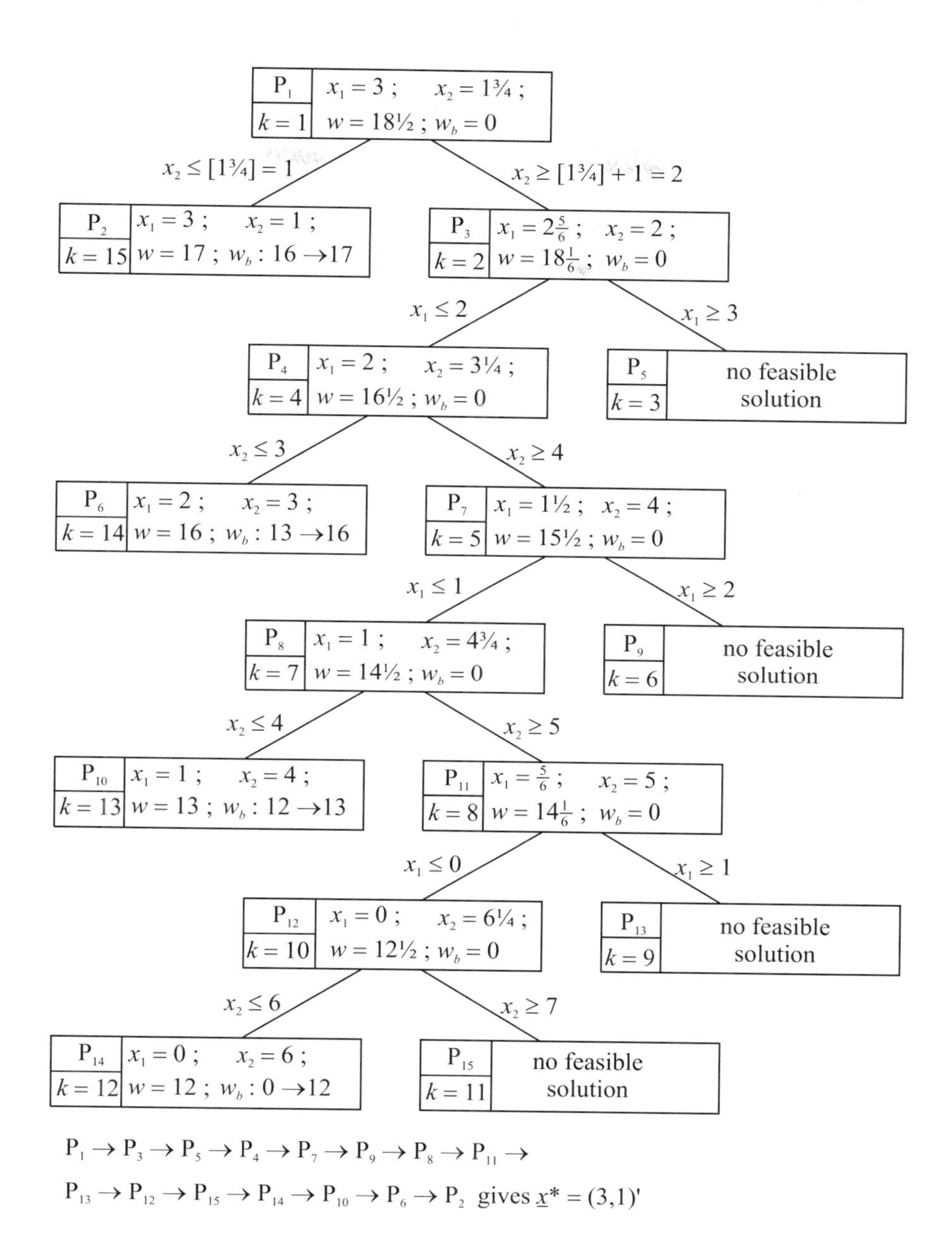

Figure 6.5. The B&B tree in case the 'right-hand side problems' ($x_j \geq$) are solved first.

Next we will present the B&B algorithm formally. Given the mixed integer LP problem (P):

$$\begin{aligned} &\max\{w = \underline{c}'\underline{x}\} \\ &\quad A\underline{x} \le \underline{b} \qquad\qquad\qquad (P) \\ &\quad \underline{x} \ge \underline{0} \\ &\quad x_j \text{ integer, } j \in I := \{1, 2, ..., p\},\ p \le n \end{aligned}$$

The order of the variables is such that the first p variables are integer. If all integer variables have an upper bound $0 \le x_j \le u_j$ ($j = 1, ..., p$), the optimal solution to (P) is found in a finite number of steps by the B&B method, provided that an optimal solution exists. If a feasible solution $\hat{\underline{x}}$ to (P) is known, then $w_b = \underline{c}'\hat{\underline{x}}$ is a lower bound for the optimal objective function value w^* of (P). If no feasible solution is known, initially $w_b = -\infty$ is taken.
The B&B-algorithm of Scheme 6.1 keeps a list L of continuous LP problems that still have to be solved.

Scheme 6.1. The Branch-and-Bound algorithm.

1. The set L initially consists of one continuous LP problem P_1 by relaxing (omitting) the integrality constraints of (P).
   Determine a lower bound w_b for the optimal value w^* of (P).
   Let $k = 0$.
2. STOP if $L = \varnothing$. $k := k + 1$.
   Choose one LP problem from L and call it P_k.
3. Solve problem P_k.
   If P_k has no feasible solution or the optimal value $w^{(k)}$ of P_k is lower than w_b ($w^{(k)} \le w_b$) then remove P_k from L and go back to 2. If $w^{(k)} > w_b$ then go to 4.
4. a. If the first p elements of optimal solution $\underline{x}^{(k)}$ of P_k are integer, let $\underline{x} := \underline{x}^{(k)}$ and $w_b := w^{(k)}$. Remove P_k and all problems P_i with $w^{(i)} \le w_b$ from L and go to 2.
   b. If at least one element $x_j^{(k)}$ ($j = 1, ..., p$) of the optimal solution $\underline{x}^{(k)}$ of P_k is not integer then go to 5.
5. Determine a $j \in \{1, 2, ..., p\}$ for which $x_j^{(k)} = \beta_j$ is not integer. Add two new LP problems to L :
   a. Problem P_k with extra restriction $x_j \le [\beta_j]$ (and upper bound w_b).
   b. Problem P_k with extra restriction $x_j \ge [\beta_j] + 1$ (and upper bound w_b).
   Go to 2.

Remarks

1. If we are interested in alternative optimal solutions, criterion $w^{(k)} \le w_b$ in step 3 can be replaced by $w^{(k)} < w_b$. We continue to step 4 if $w^{(k)} \ge w_b$. In step 4a criterion $w^{(i)} \le w_b$ should be replaced by $w^{(i)} < w_b$.
2. For the optimal objective function value w^* holds: $w_b \le w^* \le \max_i \{w^{(i)}\}$, in which $w^{(i)}$ is the upper bound of problem P_i in list L. The maximum of all $w^{(i)}$ values is an upper bound

for w^*. Usually the criterion $|\max_i \{w^{(i)}\} - w_b| < \varepsilon$ is applied as a stopping criterion for the B&B algorithm.

3. In step 5. a variable x_j with a non-integral value is selected as the branching variable. As can be seen in Example 6.2 the choice of the branching variable has a substantial effect on the efficiency of the B&B algorithm. In practice many concepts have been developed for this selection. However, a 'golden' rule does not exist.
4. Another selection mechanism in the algorithm is the choice of the LP problem from list L in step 2. Two aspects are relevant here:
   a. From a computational point of view it is attractive to elaborate the last solved problem P_k because the optimal basis is available. The optimal basis can be used when adding constraint $x_j \leq [\beta_j]$ or $x_j \geq [\beta_j] + 1$ by the so-called upper bound technique and the Dual Simplex Method.
   b. The problem with the highest upper bound $w^{(i)}$ seems most promising.

   Practical experience shows that it might be useful to proceed with branching first until an integer solution has been found. Next, the 'most promising' problem with the highest objective function value $w^{(i)}$ can be chosen.
5. Solving every continuous sub-problem completely requires an enormous computational effort. Therefore, finding efficient upper bound estimates for the objective function value of problem P_k is useful. If the estimate is already smaller than bound w_b, no further search in the corresponding branch is necessary.
6. Binary LP problems are integer LP problems in which variables are only allowed to take the values 0 or 1. An integer variable with an upper bound can be represented by a set of binary variables. For example, an integer variable x, $0 \leq x \leq 15$, can be represented by:

   $$x = 2^0 y_0 + 2^1 y_1 + 2^2 y_2 + 2^3 y_3, \text{ where } y_i \in \{0, 1\}, (i = 0, 1, 2, 3)$$

   The applicability of binary variables (see also Chapter 1) leads to an increasing interest in binary programming. Specific methods have been developed for this class of optimisation problems.
7. From a computational point of view (mixed) integer programming models are much harder to solve than similar sized (continuous) LP-models. In practice, integer LP problems are frequently solved by heuristic methods. Such methods don't guarantee to generate optimal solutions, but give reasonable results within an acceptable amount of time. We refer to Chapter 10 for an introduction to heuristic methods.
8. Usually the formulation of a mathematical model is not unique. In principle, the model builder should use as few integer variables as possible. For instance, variables with a considerable range (large values) can be approximated by continuous variables.
9. In practice it turns out that a wide class of problems can (or must) be modelled by using integer variables. It is beyond dispute that the flexibility in building (mixed) integer models is much greater than in building continuous linear programming models. In particular, binary variables, i.e. integer variables that are restricted to take only the values 0 or 1, are frequently used to model typical 'yes or no decisions'. As the flexibility in building (mixed) integer models is much greater than in building continuous LP models, the chance of formulating weak models increases too. We refer to Chapter 9 for a first introduction to modelling techniques in mathematical programming.

7. Networks

J.C. (Joke) van Lemmen-Gerdessen

In this chapter networks are introduced. Many problem situations can be modelled as a network. In the first paragraph notation and definitions from network theory are introduced that shall be used in subsequent paragraphs (Dolan and Aldous, 1993).

7.1. Notation and definitions

A *network* consists of a set of points connected by a set of lines. The points are called *nodes* (or vertices), and the lines are called *edges*. A synonym for network is *graph*.

When the nodes of a network are denoted $a, b, c, \dots$ then the edges can be written as $ab, ac, bc, \dots$ For a network the notation $G = (V, E)$ is often used, denoting that network G consists of the node (vertex) set V and the edge set E. The network of Figure 7.1 consists of the node set $V = \{a, b, c, d, e\}$ and the edge set $E = \{ab, ac, ad, bc, be, cd, ce, de\}$.
As a and b are the endpoints of edge $ab \in E$ we say that:

- edge ab is *incident* with nodes a and b,
- nodes a and b are incident with edge ab,
- a and b are *adjacent* nodes.

Networks can represent various situations.

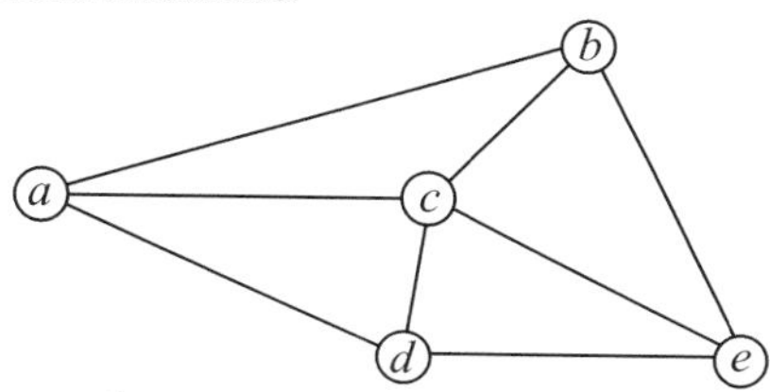

Figure 7.1. Example of a network.

Example 7.1.

Within a group of people at an office several people cooperate in several ways. We can summarize these relations in a table, such as given below. It is also possible to represent the cooperation with the network in Figure 7.2, where the node set consists of the different persons. Adjacent persons cooperate.

Person	Cooperates with
Hank	Ria, Liza, Peter
Ria	Hank, Peter, Liza
Liza	Hank, Peter, Ria, Mary
Mary	Liza, Peter
John	Peter
Peter	Ria, Hank, Liza, Mary, John

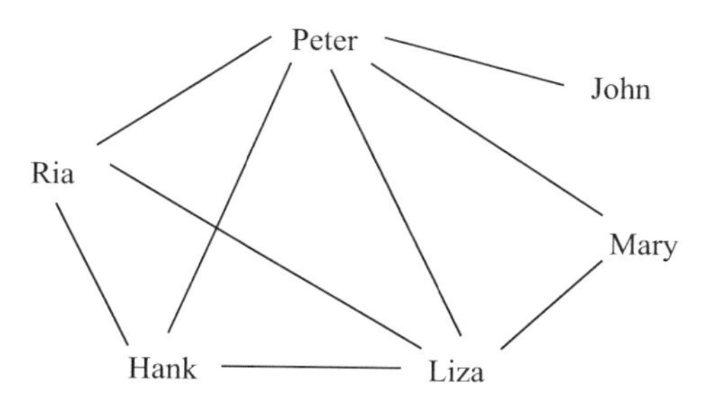

Figure 7.2. Two representations of the way in which people cooperate.

Example 7.2.

The left half of Figure 7.3 shows the floor plan of a house. For small-scale applications such a drawing can be useful in showing which rooms are connected. For large-scale applications a more schematic representation can be practical, such as the network in the right half of Figure 7.3. The node set consists of the rooms. The edge set represents the doors. Notice that the networks shows very clearly which rooms are connected (adjacent), but that other information, like the size and the shape of the rooms, is lost.

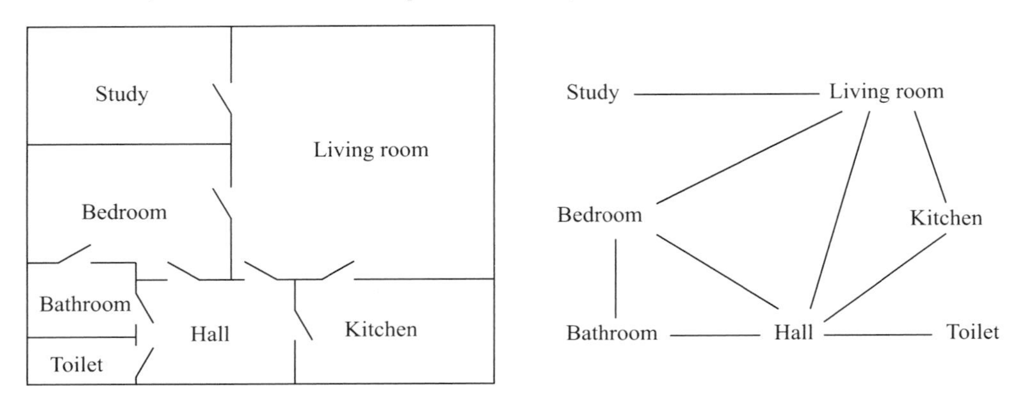

Figure 7.3. Two representations of the floor plan of a house.

The networks in Figures 7.1-7.3 are *undirected networks:* the edges of the networks do not represent a direction. If direction is important we can indicate this by adding a direction arrow to every edge. The resulting network is called a *directed network*, and its directed edges are called *arcs*. For a directed network the notation $G = (V, A)$ is used, with A being the set of arcs.

In an undirected network ab and ba represent the same edge. In a directed network ab and ba are arcs with an opposite direction. The arc ab is called an *outgoing arc* of a and an *incoming arc* of b.

The undirected network that arises when substituting the arcs of a directed network with edges is called the *underlying network* of that directed network. The undirected network of Figure 7.4 is the underlying network of the directed network in Figure 7.4.

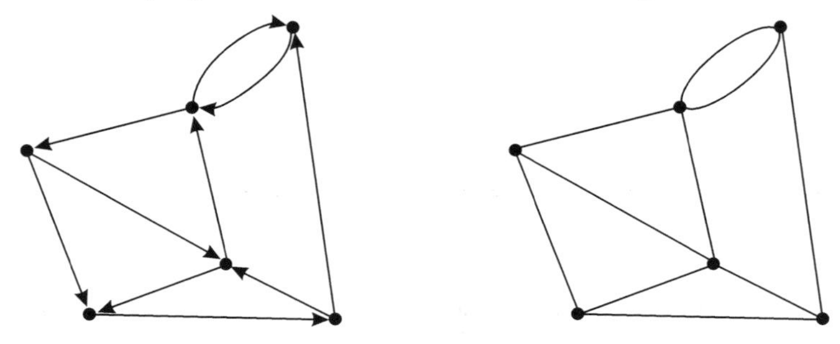

Figure 7.4. A directed network and its underlying undirected network.

Let G be a network with node set $V(G)$ and edge set $E(G)$. A *subnetwork* S of G is a network for which:

- $V(S) \subseteq V(G)$ and
- $E(S) \subseteq E(G)$ and
- if edge $ij \in E(S)$ then $i \in V(S)$ and $j \in V(S)$.

Example 7.3.

Every component of Figure 7.5b and Figure 7.5c is a subnetwork of the network in Figure 7.5a. Also $S = (\{a, c, d, e\}, \{ac, de\})$ and $S = (\{a, b, c\}, \varnothing)$ are subnetworks of G.

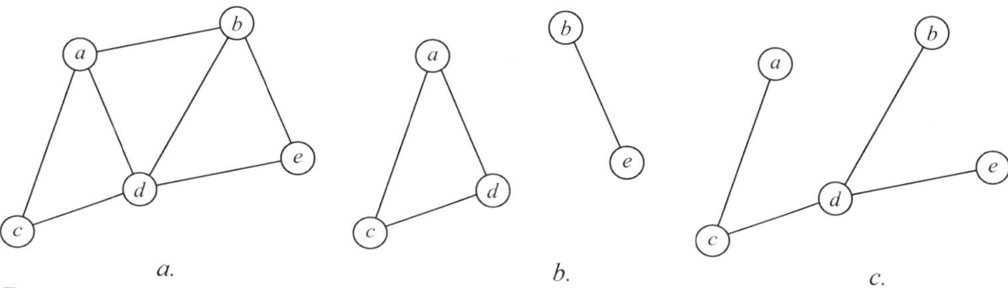

Figure 7.5. A network and two of its subnetworks.

The *degree* of a node represents the number of edges incident with that node.

In the network of Figure 7.3 (floor plan of a house) it can easily be seen how many doors each room has: the number of doors of a room is equal to the degree of the corresponding node. E.g. the number (= 5) of doors in the hall is equal to the number of edges that is incident with the node that represents the hall. The degrees of all the nodes add up to 18, which is twice the actual number of doors in the house, because we have counted every door twice. This is a general property of networks: *The sum of the degrees of all nodes in a network equals twice the number of edges in the network.*

This property has a number of consequences:

- In every network the sum of the degrees of the nodes is even.
- In every network the number of nodes with odd degree is even.

In the networks that were previously shown there was not always an edge between every pair of nodes.

A *complete network* is a network in which every pair of nodes is connected via an edge.

We denote the complete network of n nodes with K_n. In K_n every node has degree n-1. The total number of edges in K_n equals $n(n\text{-}1)/2$. In Figure 7.6 the complete networks K_1 to K_5 are given.

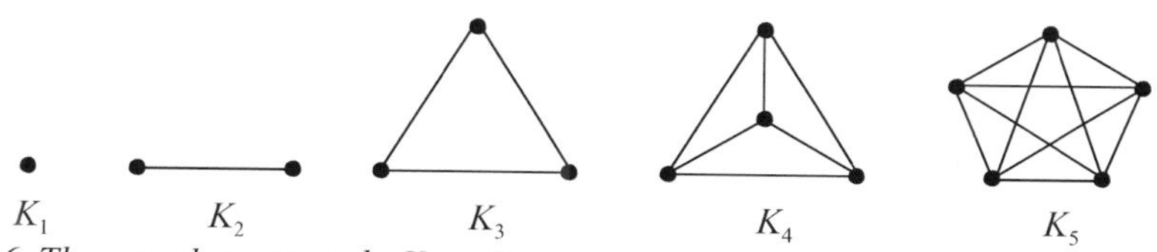

Figure 7.6. The complete networks K_1 to K_5.

We will now introduce the concepts of walk, path and cycle. These are applicable to directed networks as well as undirected networks.

A *walk* of length k is a succession of k edges in a network of the form *ab*, *bc*, *cd*, *de*, ... In a directed network a walk always follows the direction of the arcs. In a *closed walk* the first node is equal to the last node.

A *path* is a walk of which all nodes (and edges) are different. A *cycle* is a path of which the last node is equal to the first node.

A cycle that uses every node in a network is called a *Hamiltonian cycle*. A network that contains a Hamiltonian cycle is called a *Hamiltonian network*.

Example 7.4.

The map of a city can be represented by a directed network. In Figure 7.7 the node set consists of the points in the city where several streets come together. One-way streets are represented by one arc; two-way streets are represented by two arcs with opposite directions. For example, a motorcyclist starting in node *a* can drive the walk *acdebac*, or the path *acdbe*, or the (Hamiltonian) cycle *acdeba*.

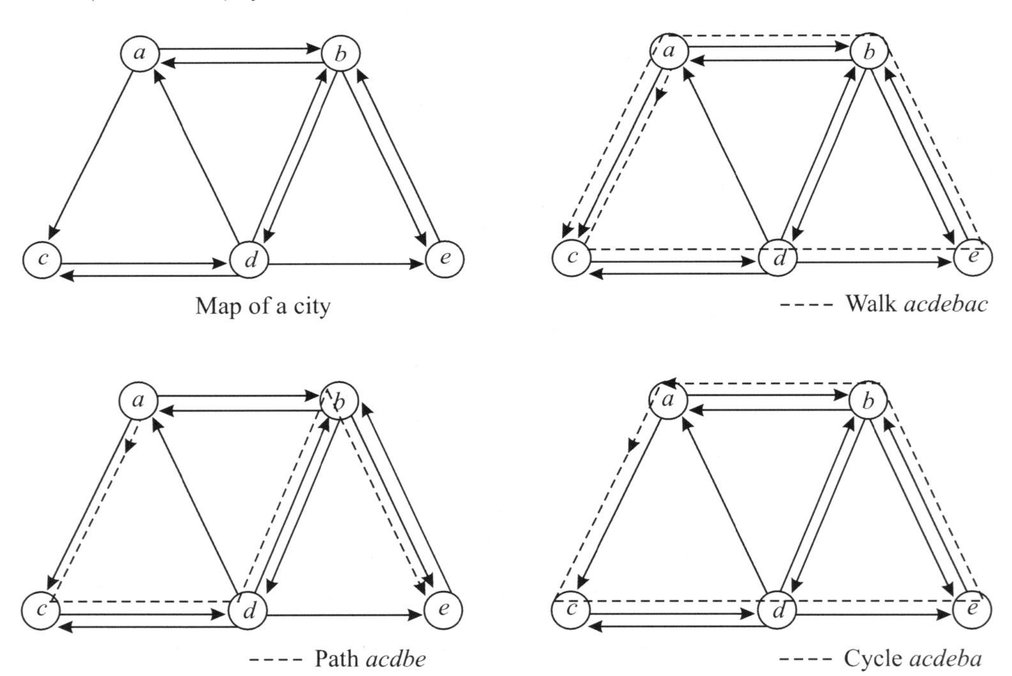

Figure 7.7. Map of a city with a walk, a path and a cycle.

Not every network contains a cycle, see e.g. Figure 7.8. A network that does not contain a cycle is called an *acyclic network*. Note that the underlying network of the middle network of Figure 7.8 contains a cycle.

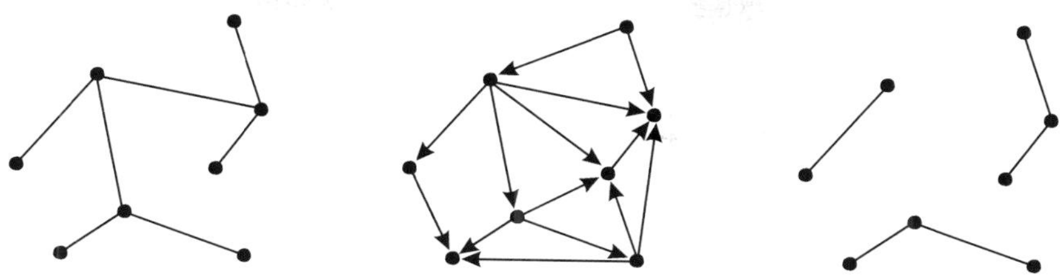

Figure 7.8. Acyclic networks.

Example 7.5.

A product developer is investigating which of five new fruit-flavoured desserts is liked most by the consumer. Members of a panel have to indicate which of two flavours they prefer.
The results of the investigation can be represented in a network. The node set of the network consists of the desserts. A directed arc *ab* is used to indicate that dessert *a* is preferred to dessert *b*. Two possible outcomes of the investigation are shown in Figure 7.9.

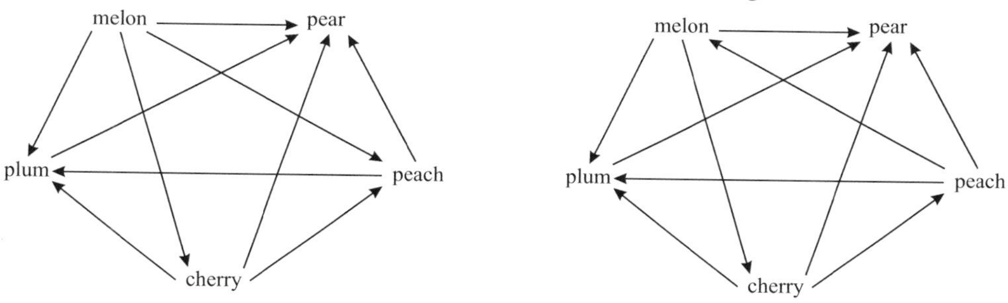

Figure 7.9. Preference networks.

The left-hand side network of Figure 7.9 is acyclic. In this network it can be seen which flavour is preferred by the members of the panel: *melon* has an outgoing arc to all other flavours. So *melon* is the preferred flavour.
The situation is more complicated in the right-hand side network of Figure 7.9: the network contains a cycle *melon, cherry, peach, melon*. No flavour has outgoing arcs to all other flavours. (Note that *pear* has incoming arcs from all other nodes. Apparently every flavour is preferred to *pear*.)
In this case the product developer will have to use a different way of investigating which flavour is liked most.

A network is *connected* if between any pair of nodes a path can be found. Otherwise the network is *disconnected*. The left-hand side network of Figure 7.10 is connected. The middle network of 7.10 is disconnected; it has *components acd* and *be*.
A *tree* is a connected undirected network that does not contain a cycle, or a connected directed network of which the underlying network does not contain a cycle.

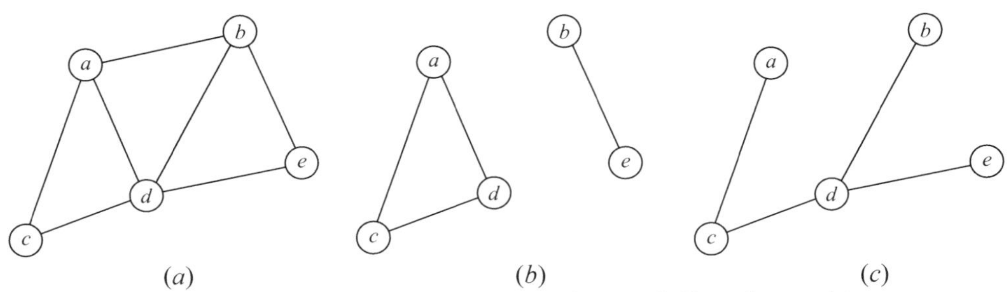

Figure 7.10. A connected network (a), a disconnected network (b) and a tree (c).

A *bipartite network* is a network of which the set of nodes can be divided into two subsets V and W such that every edge of the network has one node in V and one node in W.

In Figure 7.11 the nodes of V are white and the nodes of W are black. Another way to distinguish V and W is to draw the nodes of V on the left-hand side and those of W on the right-hand side, as in the third network of Figure 7.11.

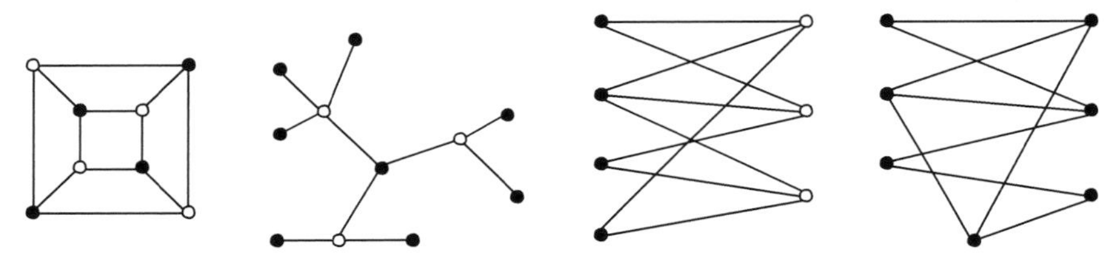

Figure 7.11. Three bipartite networks and a nonbipartite network.

Example 7.6.

A swimming team for a medley consists of four persons, one for each stroke. Every person has to swim exactly one stroke. The best result is obtained if every swimmer is assigned a stroke that he swims really fast.
This situation can be modelled with a bipartite network, see Figure 7.12. The nodes on the left-hand side denote the swimmers, the nodes on the right-hand side denote the strokes. A swimmer and a stroke are connected with an edge if the swimmer is really fast at that stroke.
The problem of finding a set of edges that represents a situation in which every swimmer is assigned to exactly one stroke is called a *matching problem*.

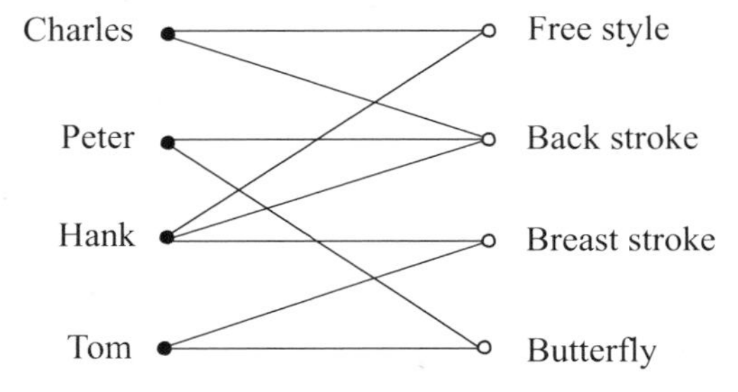

Figure 7.12. A bipartite network.

7.2. Shortest path, longest path

In this section we focus on topics like:
- What is the shortest route between two cities?
- What is the minimum duration of a large project?

The first question can be modelled as searching for the shortest path in a network. The second question involves searching for the longest path in a network.

Shortest path

Consider the network $G = (V, E)$. The length of edge ij is denoted d_{ij}. The length of a path in G is defined as the sum of the lengths of its edges. A path from s to t is called an . Now what is the shortest st-path? The length of the shortest st-path is called the *distance* from s to t.

The shortest st-path can be found with *Dijkstra's algorithm*. Dijkstra's algorithm assigns labels to the nodes of the network. A label is either temporary or permanent. A *temporary label* (L_i) of node i is an upper bound for the distance from s to i. A *permanent label* (L_i^*) of node i is the shortest distance from s to i.
At first only the starting node s gets a permanent label, all other nodes get a temporary label. Then in an iterative procedure all temporary labels are converted into permanent labels. The algorithm stops when the end node t gets a permanent label.

We present a stepwise description of Dijkstra's algorithm for finding the shortest st-path in an undirected network. This algorithm can also be used to find the shortest st-path in a directed network. The algorithm will be illustrated with an example.

Algorithm 7.1. Dijkstra's algorithm for finding the shortest st-path in a network $G = (V, E)$.

1. Assign a permanent label to the starting node s: $L_s^* = 0$. All other nodes are assigned a temporary label in the following way:
   - if $si \in E$, then $L_i := d_{si}$
   - if $si \notin E$, then $L_i := \infty$.

   Choose the smallest of the temporary labels, and make this label permanent.

2. Let node k be the node that was assigned a permanent label in the last iteration. Adjust the temporary labels of all nodes that are adjacent to k:
   $L_i := \min\{L_i, L_k + d_{ki}\}$
   Choose the smallest temporary label and make it permanent.
   If the node with the smallest temporary label is node t then the algorithm stops. Otherwise step 2 is executed again.

If in any iteration several nodes have the (same) smallest temporary label then any one of these can be chosen.

The shortest st-path is now given by the edges for which $d_{ij} = L_j^* - L_i^*$, with L_i^* and L_j^* permanent labels (starting at the endpoint t).

Example 7.7.

Dijkstra's algorithm is illustrated with the network of Figure 7.13. In this network the shortest st-path has to be found. The iterations of Dijkstra's algorithm are displayed in Table 7.1.

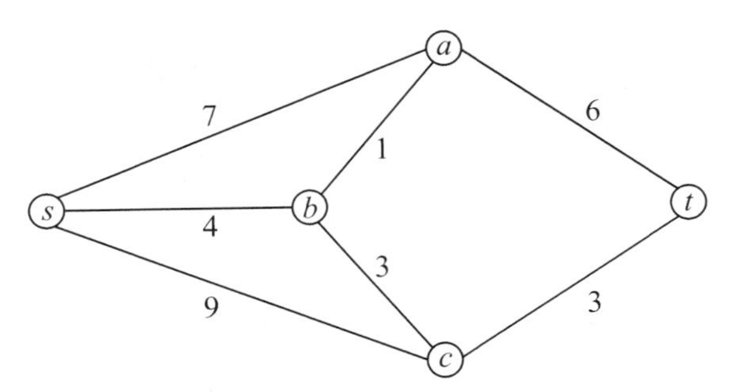

Figure 7.13.

Iteration 1

Node s is assigned a permanent label $L_s^* = 0$, because the shortest distance from s to s can never be less than 0. All other nodes get a temporary label that is an upper bound for the distance of s to these nodes. For the nodes that are adjacent to s this upper bound equals the length of the connecting edge. For nodes that are not adjacent to s the upper bound is ∞. So $L_s^* = 0$; $L_a = 7$; $L_b = 4$; $L_c = 9$; $L_t = \infty$.
The node with smallest temporary label is node $L_b = 4$, so this label is converted into a permanent label: $L_b^* = 4$.

Iteration 2

In iteration 0 node b was assigned a permanent label. For every node that is adjacent to b we check whether the path from s via b is shorter than the current temporary label. In this case the temporary label can be updated:

- $L_a := \min\{L_a ; L_b^* + d_{ba}\} = \min\{7; 4+1\} = 5$,
- $L_c := \min\{L_c ; L_b^* + d_{bc}\} = \min\{9; 4+3\} = 7$,
- $L_t = \infty$, no update because t has no incoming arc from b.

Node a has the smallest temporary label, so this label is converted into a permanent label: $L_a^* = 5$.

Iteration 3

In iteration 1 node a was assigned a permanent label. Only node t is adjacent to a, so only one temporary label has to be updated: $L_t := \min\{L_t ; L_a^* + d_{at}\} = \min\{\infty ; 5+6\} = 11$.
Node c has the smallest temporary label, so $L_c^* = 7$.

Iteration 4

Node t is adjacent to c, so the label of t is updated: $L_t := \min\{L_t; L_c + d_{ct}\} = \min\{11; 7+3\} = 10$.
Now there is only one node left with a temporary label, so this label can be made permanent: $L_t^* = 10$. So the shortest distance from s to t in this network is 10.

Table 7.1. Iterations of Dijkstra's algorithm.

	s	a	b	c	t
Iter. 1	$L_s^* = 0$	$L_a = d_{sa} = 7$	$L_b^* = d_{sb} = 4$	$L_c = d_{sc} = 9$	$L_t = \infty$
Iter. 2		$L_a^* := \min\{L_a; L_b^* + d_{ba}\}$ $= \min\{7; 4+1\} = 5$		$L_c := \min\{L_c; L_b^* + d_{bc}\}$ $= \min\{9; 4+3\} = 7$	$L_t = \infty$
Iter. 3				$L_c^* = 7$	$L_t := \min\{L_t; L_a^* + d_{at}\}$ $= \min\{\infty; 5+6\} = 11$
Iter. 4					$L_t^* := \min\{L_t; L_c^* + d_{ct}\}$ $= \min\{11; 7+3\} = 10$

Now that the length of the shortest st-path is known we can track down the edges of the shortest path. Starting in t we use the permanent labels to find out which edges are in the shortest path:

$$L_t^* - L_c^* = d_{ct}$$

so ct is in the shortest path. Also

$$L_c^* - L_b^* = d_{bc} \quad \text{and} \quad L_b^* - L_s^* = d_{sb}$$

so bc and sb are in the shortest st-path too.
The shortest path is $sbct$ with length 10.

Example 7.8.

Figure 7.14 shows a communication network. Through the edges of this network messages are transported from s to t. On the edges of the network failures can occur. If a failure occurs then the message is lost. Now the question is: which path has maximum reliability?
Figure 7.14 shows along every edge ij of the network the chance p_{ij} that no failure occurs at ij. The chance that a message that is sent via the path $sact$ actually arrives in t is equal to $p_{sa}p_{ac}p_{ct}$. The most reliable path is the path for which the arrival chance of the message is maximal.

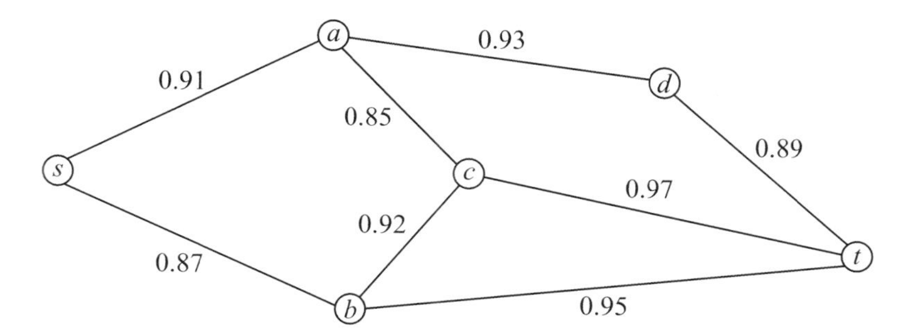

Figure 7.14. Finding the shortest route in a communication network.

At first sight it seems impossible to solve this problem with Dijkstra's algorithm, after all the reliability is calculated via a product, and not via a sum. But this objection can easily be overcome by using logarithms:

$$\log(p_{sa}p_{ac}p_{ct}) = \log(p_{sa}) + \log(p_{ac}) + \log(p_{ct})$$

All p_{ij} represent chances, so $0 < p_{ij} \leq 1$. So the logarithm of p_{ij} is negative. The path with maximum reliability has the maximum product $p_{si}p_{ij}p_{jt}$, so it has the least negative logarithm. So if we use c_{ij} = -log(p_{ij}) as the length of the edges we can find the most reliable path by searching for the shortest path with respect to measure c_{ij}.

Longest path

The longest path in a network is only defined for directed networks (V, A) that do not contain cycles of positive length.
The algorithm for finding a longest path closely resembles Dijkstra's algorithm for finding a shortest path. Only one type of label is used, representing the length of the longest path to that node. Iteratively all nodes are labelled.

Algorithm 7.2. Finding a longest path in a directed network G = (V, A) that does not contain a cycle of positive length.

1. Assign a label to the starting node s: $L_s := 0$.
2. Let P be the set of nodes with a label. Let U be the set of unlabelled nodes that can only be reached from nodes in P. Now for all nodes $u \in U$ the label L_u is calculated:

$$L_u := \max_{p \in P}\{L_p + d_{pu}\} \text{ if } \quad pu \in E$$

Repeat step 2 until t is labelled. The longest st-path is now given by the arcs for which $d_{ij} = L_j - L_i$.

In fact Algorithm 7.2 equals the Dynamic Programming algorithm of Chapter 8.

Example 7.9.
We search for the longest st-path in Figure 7.15.

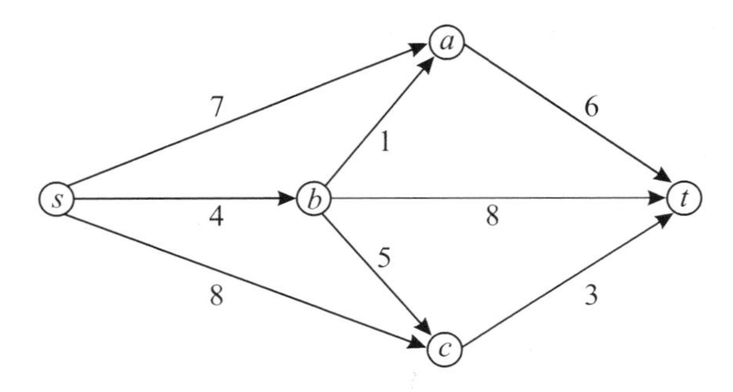

Figure 7.15.

Iteration 1

Node s is assigned a label $L_s = 0$.

Iteration 2

Only s is labelled so $P = \{s\}$. Node b is the only node that can only be reached from s, so $U = \{b\}$.
Now $L_b := L_s + d_{sb} = 4$.

Iteration 3

Now $P = \{s, b\}$, and $U = \{a, c\}$
$L_a := \max\{L_s + d_{sa}; L_b + d_{ba}\} = \max\{0+7; 4+1\} = 7$.
$L_c := \max\{L_s + d_{sc}; L_b + d_{bc}\} = \max\{0+8; 4+5\} = 9$.

Iteration 4

Now $P = \{s, b, a, c\}$, and $U = \{t\}$
$L_t := \max\{L_a + d_{at}; L_b + d_{bt}; L_c + d_{ct}\} = \max\{7+6; 4+8; 9+3\} = 13$.

Now we know that the longest st-path has length 13. To find out which arcs are in the longest path we check the labels: $L_t - L_a = d_{at}$ and $L_a - L_s = d_{sa}$, so sat is the longest path.

Project planning

An application of the longest path problem can be found in project planning. In a large project many activities have to be carried out. Some cannot start until others are finished. A project can be represented as a directed network. A longest path through this network represents the minimum duration of the project. We illustrate this with a small network. A more formal description can be found in (Hillier and Lieberman, 2005) and in (Winston, 2004).

Suppose a project has 5 activities: A, B, C, D, E. Activity i has duration t_i. Activities C and D can only be started after activity A is finished and activity E can only be started after both B and D are finished. These relations between the start and end times of activities are called *precedence constraints*.
Figure 7.16 shows a network model that can be constructed for this problem. The activities A to E are modelled by the arcs. The lengths of the arcs equal the durations of the activities. Nodes are inserted to represent the precedence constraints: the arcs representing activities C and D can only be entered after activity A is finished. A dummy activity F (finish) is added that can only be started after the whole project is finished.

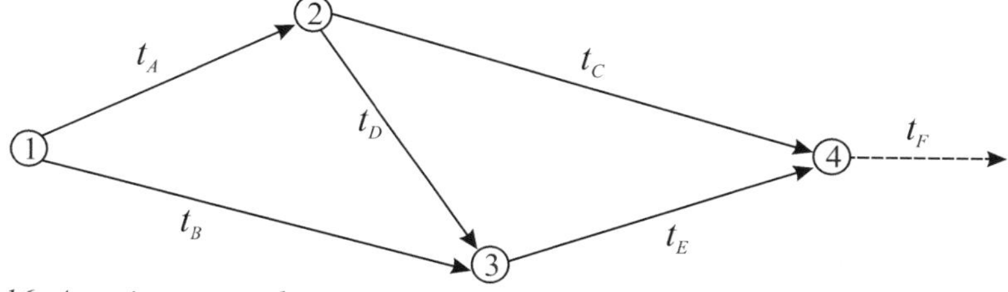

Figure 7.16. A project network.

We define y_i as the earliest starting time of activity i.
Suppose that A and B start a time $t = 0$, so $y_A = 0$ and $y_B = 0$. Activity C can start at the same time as activity D, so the variable $y_{CD} = y_C = y_D$ is defined as the earliest starting time of activity C and D. Variable y_F (the starting time of dummy activity F) denotes the time needed to complete the whole project (also called the duration of the project). The duration of the project can be modelled with the following LP model:

$$
\begin{aligned}
&\min \{ y_F \} \\
&\text{s.t.} \\
&\quad y_{CD} \geq t_A \\
&\quad y_E \geq t_B \\
&\quad y_F \geq t_C + y_{CD} \\
&\quad y_E \geq t_D + y_{CD} \\
&\quad y_F \geq t_E + y_E \\
&\quad y_{CD}, y_E, y_F \text{ free}
\end{aligned}
$$

The restrictions $y_E \geq t_B$ and $y_E \geq t_D + y_{CD}$ denote that activity E can only start after activities B and C have been finished. Activity B can finish at time t_B and activity D can finish at time $t_D + y_{CD}$.

This LP model can also be written as:

$$
\begin{aligned}
&\min \{ y_F \} \\
&\text{s.t.} \\
&\begin{array}{rrrl}
y_{CD} & & & \geq t_A \\
 & y_E & & \geq t_B \\
-y_{CD} & & + \; y_F & \geq t_C \\
-y_{CD} & + \; y_E & & \geq t_D \\
 & - \; y_E & + \; y_F & \geq t_E
\end{array} \\
&y_{CD}, y_E, y_F \text{ free}
\end{aligned}
$$

The dual problem of this model is

$$
\begin{aligned}
&\max \{ u_A t_A + u_B t_B + u_C t_C + u_D t_D + u_E t_E \} \\
&\text{s.t.} \\
&\begin{array}{rrrrrl}
u_A & & - u_C & - u_D & & = 0 \\
 & u_B & & + u_D & - u_E & = 0 \\
 & & u_C & & + u_E & = 1
\end{array} \\
&u_A, u_B, u_C, u_D, u_E \geq 0
\end{aligned}
$$

The latter model can be considered as maximising the length of a path through the network. It can also be considered as a model for sending a flow of value 1 through the network. This will be discussed in 7.3.

So the original LP model of minimising the duration y_F of this project is reformulated as the problem of determining the longest path in the network.

For controlling large projects the methods CPM (Critical Path Method) and PERT (Program Evaluation and Review Technique) are developed (Hillier and Lieberman, 2005). CPM is based on the model that was shown here. PERT includes the uncertainty about the duration of each activity, and calculates the probability that the entire project will be finished by a given target date.

Example 7.10.

Suppose the five activities of a project have durations and precedence constraints as shown in Figure 7.17.

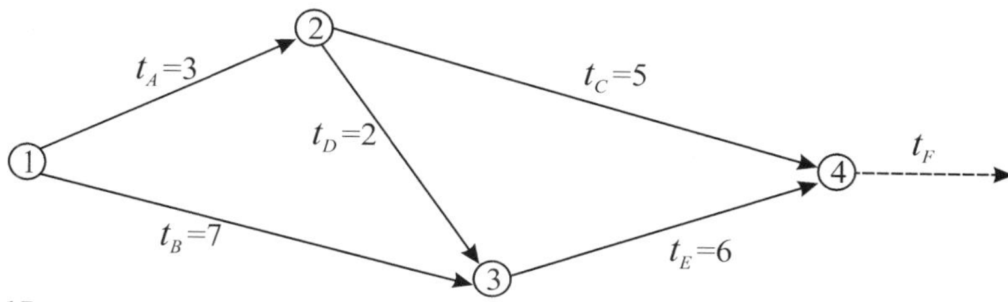

Figure 7.17.

The longest path in the project network is {1, 3, 4} with length 7+6=13. So the duration of this project is 13.
The earliest starting times of the activities are: $y_A = 0$, $y_B = 0$, $y_{CD} = y_A + t_A = 3+0 = 3$, $y_E = \max\{y_{CD} + t_D;\ y_B + t_B\} = \max\{3+2;\ 0+7\} = 7$. The duration of the entire project is $y_F = \max\{y_{CD} + t_C;\ y_E + t_E\} = \max\{3+5;\ 7+6\} = 13$.
Note that changes in the durations of B and E will influence the duration of the entire project: if a delay occurs in B or E then the duration of the entire project will increase. Therefore the longest path in a project network is called the *critical path*. If the duration of a project needs to be reduced then one should start to reduce the durations of the activities on the critical path.

7.3. Maximum flow problem, minimum cut problem

In this section we focus on topics like:

- How much oil can flow through a pipeline network in a day?
- How many vehicles can pass through a city in an hour?

These questions are examples of a class of network problems that are called *flow problems*.

The objective in a flow problem is to maximise the amount of flow from a source s to a sink t in a network without exceeding the capacity restrictions in the network.

In order to discuss flow problems the following topics will be described:

- LP formulation of the maximum flow problem,
- Solution method for the maximum flow problem,
- Constructing a residual network,
- Finding a flow-augmenting path,
- Minimum cut.

Example 7.11.

Consider an oilfield and a refinery. The oil that is extracted at the oil field is transported to the refinery via a pipeline network. The diameters of the pipelines restrict the flow through the network. How large is the maximum flow from the oilfield to the refinery? And how much oil flows through each of the pipes when the total flow is maximised?

This problem is represented in network G, see Figure 7.18. The oilfield s, the pumps $a...e$ and the refinery t are the nodes of the network. The arcs represent the pipelines. The numbers along the arcs denote the capacity of the pipelines (the flow is not restricted by the capacities of the pumps).

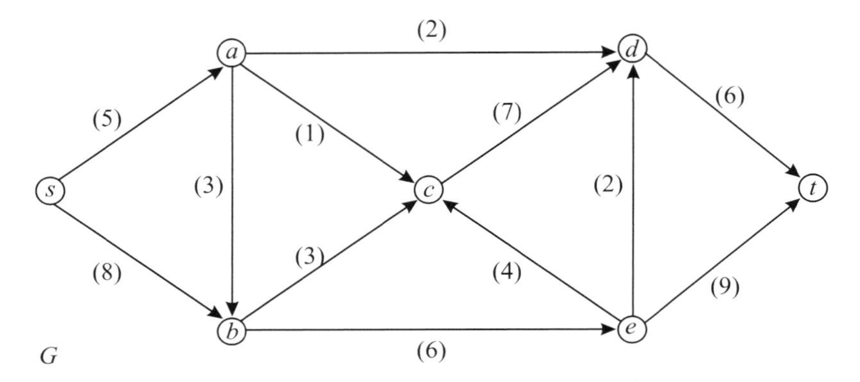

Figure 7.18.

The sum of the flows along the incoming arcs of a node is called the *incoming flow* of that node. The sum of the flows along the outgoing arcs of a node is called the *outgoing flow* of that node.

A feasible flow has the following characteristics:

- on every arc the flow does not exceed the capacity of the arc,
- on every arc the flow is non-negative,
- for all intermediate nodes (that is: for every node other than the source s and the sink t) the incoming flow equals the outgoing flow. We say there is *flow balance* in these nodes.

The size of a flow is defined as the total outgoing flow of the source s. Because no flow is created or lost at the intermediate nodes of the network the total flow is also equal to the total incoming flow of the sink t.

LP formulation of the maximum flow problem

Finding the maximum flow through a network $G = (V, A)$ can be formulated as an LP model. The total flow through the network is called v. The flow on arc ij is x_{ij}. The capacity of arc ij is c_{ij}. The LP model of the maximum flow problem can be formulated as:

$$\max \{v\}$$
s.t.

$$\sum_{\{i|si \in A\}} x_{si} = v$$

$$\sum_{\{i|ij \in A\}} x_{ij} = \sum_{\{i|ji \in A\}} x_{ji} \qquad \forall j \in V \setminus \{s,t\}$$

$$\sum_{\{i|it \in A\}} x_{it} = v$$

$$0 \le x_{ij} \le c_{ij} \qquad \forall\ ij \in A$$

The first and third restriction ensure that both the outgoing flow from s and the incoming flow in t equal the total flow through the network. The second restriction guarantees that in all intermediate nodes the incoming flow equals the outgoing flow (so-called balance constraints). So no flow is lost in the network. Therefore the first and third restriction are equivalent; one of them can be left out of the model. The fourth restriction ensures that the flow on an arc never exceeds the capacity.

Example 7.12.

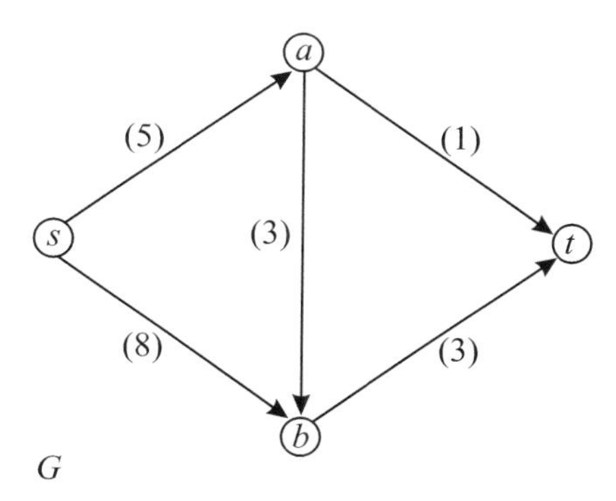

Figure 7.19.

For maximising the flow through the network of Figure 7.19 the following LP model can be formulated:

$$\max \{v\}$$
s.t.

$x_{sa} + x_{sb} = v$ (v equals the outgoing flow of s)

$x_{sa} = x_{ab} + x_{at}$ (in a the outgoing flow must equal the incoming flow)

$x_{sb} + x_{ab} = x_{bt}$ (in b the outgoing flow must equal the incoming flow)

$0 \le x_{sa} \le 5;\ 0 \le x_{sb} \le 8;\ 0 \le x_{ab} \le 3;\ 0 \le x_{at} \le 1;\ 0 \le x_{bt} \le 3$

It is possible to solve flow problems by solving the LP model. It can be proven that if all capacities are integer then the optimal solution will also be integer.

If a flow problem is very big, or if it has to be solved many times then it is better to use methods that are tailor-made for the flow problem: these are much more efficient for the flow problem than the simplex method (which was designed for general purposes).

The value of the LP formulation is in recognising classes of problems. In many cases network problems have the same structure as a flow problem. If the formulation of an arbitrary problem has the same structure as that of the flow problem then this arbitrary problem can be modelled and solved like a flow problem.

Solution method for the maximum flow problem

Now we return to network G of Figure 7.18 in order to illustrate a tailor-made method for solving the maximum flow problem.
In network G flow can be sent from source s to sink t along path *sadt*. The arcs of path *sadt* have capacities 5, 2 and 6. So the maximum amount of flow on this path is 2.
Flow can also be sent along path *sbet*. The minimum capacity of the arcs in this path is 6, so 6 units of flow can be sent along path *sbet*. The resulting solution is shown in Figure 7.20. The numbers without brackets along the arcs represent the current flow. The numbers between brackets represent the capacities of the arcs.

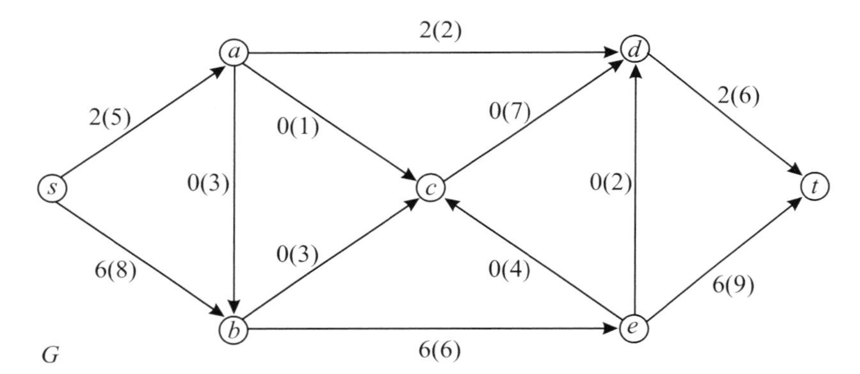

Figure 7.20.

In Figure 7.20 every arc of path *sabcdt* has an unused capacity of at least 3 units. Therefore the flow on this path is increased by 3 units. Now the total flow on the network is 2+6+3=11, see Figure 7.21. No more *st*-paths can be found that consist of only unsaturated arcs. At first sight there are no more possibilities to increase the flow in the network.

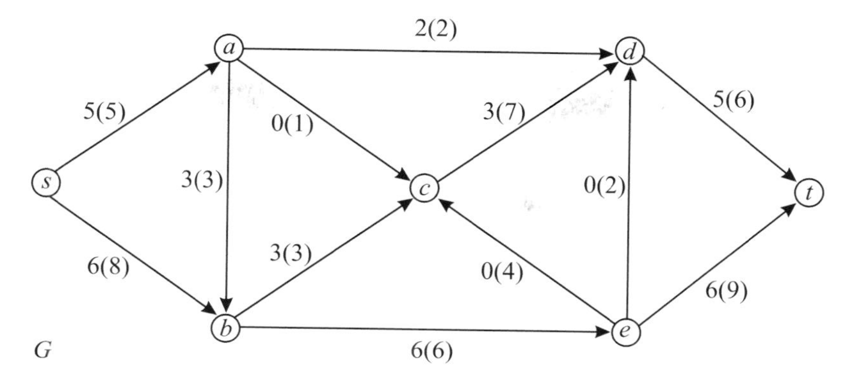

Figure 7.21.

The incoming flow in node a has size 5. Of this amount 3 units flow to b and 2 units flow to d. It would also be possible to guide a flow of 1 via ac. This would not lead to problems downstream, because cd and dt have enough slack capacity.
If we guide a flow of 1 via arc ac then the flow on ab can be reduced by 1 unit, resulting in an unbalanced situation in node b: the incoming flow is 6+2=8 and the outgoing flow is 6+3=9. The balance can be recovered by increasing the flow on sb with 1. The net result of these operations is that the total flow from s to t is increased by 1, so the total flow is 11+1=12, see Figure 7.22. The increase is realised by diverting part of the flow in order to create space for new flow.

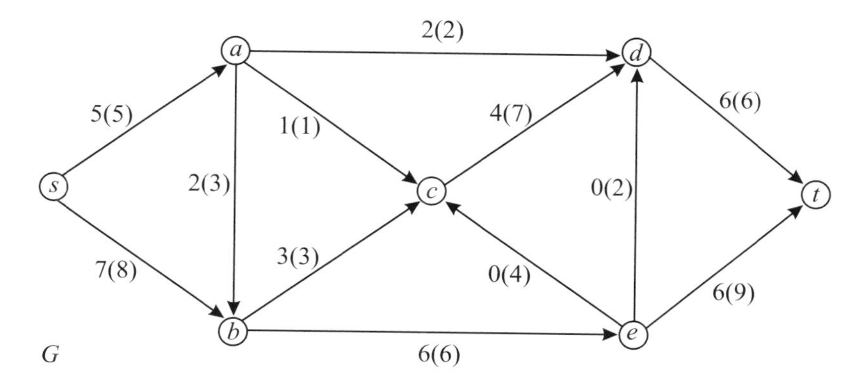

Figure 7.22.

The increased flow in Figure 7.22 was found by close inspection of Figure 7.22. It is also possible to use a more systematic approach (Ahuja, 1993) to find out if the flow through a network can be increased. We illustrate this for the network of Figure 7.21.

Constructing a residual network

For the network of Figure 7.21 we construct a so-called *residual network H*, see Figure 7.23. The nodes of H are equal to the nodes of G. The arcs of H depict in which way the flow in G can be changed:

- If $x_{ij} < c_{ij}$ then the flow on ij can be increased by an amount of $c_{ij} - x_{ij}$ or less. Therefore arc ij is added to H with capacity $c_{ij} - x_{ij}$.
- If $x_{ij} > 0$ then the flow on ij can be decreased to 0. In other words: the amount in the direction ij can be increased by $- x_{ij}$, or the flow in the direction ji can be increased with x_{ij}. Therefore arc ji is added to H with capacity x_{ij}.

So for arcs in G with $0 < x_{ij} < c_{ij}$ two arcs are added to H. E.g. for arc sb in G with flow $x_{sb} = 6$ and capacity $c_{sb} = 8$ two arcs are added in H: arc sb with capacity 2 and arc bs with capacity 6.

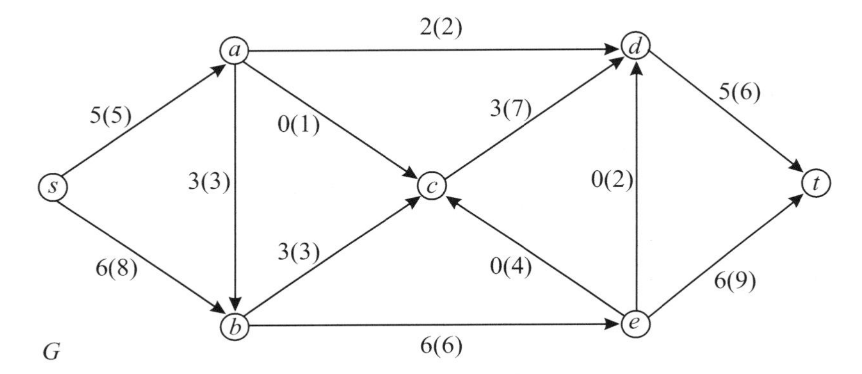

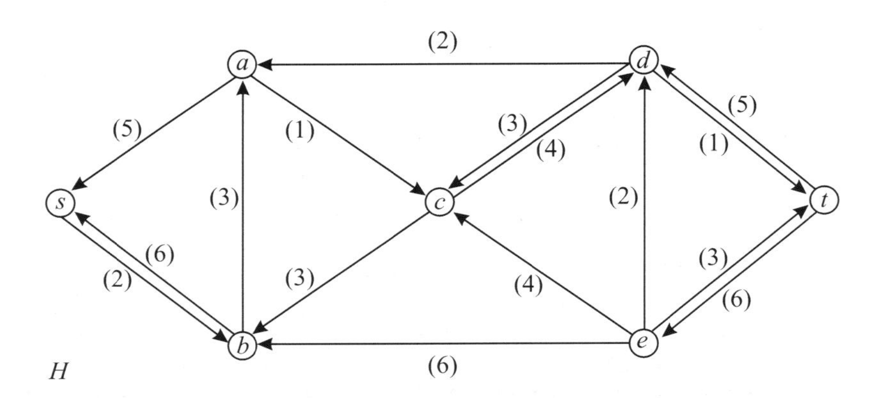

Figure 7.23. The network G (of Figure 7.21) and its residual network H.

Now consider the path *sbacdt* in H. The minimum capacity in H on path *sbacdt* is denoted with Δ. In this case $\Delta = c_{ac} = 1$. Suppose that on every arc of path *sbacdt* the flow in H is increased with Δ. Then the following happens in G:

- The flow on *sb* in G is increased by Δ.
- The flow on *ba* in G is increased by Δ. This means that the flow on *ab* is increased by $-\Delta$, resulting in a decrease of Δ.
- The flow on *ac*, *cd*, and *dt* in G is increased by Δ.

The resulting flow is still feasible, see Figure 7.24:

- The flow on *sb*, *ac*, *cd* and *dt* is increased by Δ. As Δ is equal to the minimum capacity along the path *sbacdt* in H none of these capacities shall be violated.
- The flow on *ab* is decreased by Δ. As Δ is equal to the minimum capacity along the path *sbacdt* in H the flow on *ab* shall not be lower than 0.
- In nodes c and d both the incoming and the outgoing flow have increased by Δ. So the incoming flow still equals the outgoing flow.
- In node b the incoming flow from s is increased by Δ, and the incoming flow from a is decreased by Δ, so the total incoming flow in b did not change.
- In node a the outgoing flow to b is decreased with Δ, and the outgoing flow to c is increased by Δ, so the total outgoing flow did not change.

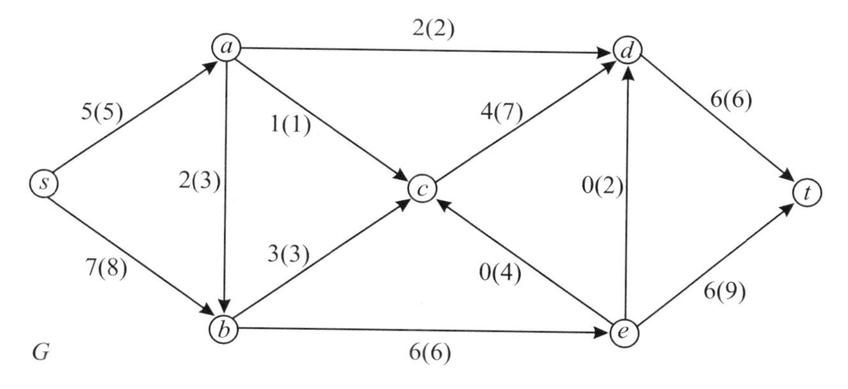

Figure 7.24.

The outgoing flow in s and the incoming flow in t are increased by Δ, so the flow through the network has been increased by Δ. Therefore the path *sbacdt* in H is called a *flow-augmenting* path.
Note that the flow along a flow-augmenting path cannot be changed by an amount larger than Δ, because then no feasible flow in G arises.

Now we can formulate an algorithm for finding a maximum flow in a network:

Algorithm 7.3. Finding a maximum flow in a network.

1. Find a feasible flow in G.
2. Draw the residual network H of this feasible flow in G.
3. If H contains an st-path then this is a flow-augmenting path. Find the minimum capacity Δ along this flow-augmenting path. Change the flow along the flow-augmenting path in G with Δ.
   If no st-path exists in H then the flow in G is maximal.

Steps 2 and 3 are repeated as long as the flow is not maximal.

Note that we did not prove that the absence of a flow-augmenting path in H implies that the flow in G is maximal. This will follow later.

Finding a flow-augmenting path

For convenience we assumed in the description of the maximum flow algorithm that it can easily be seen whether the residual network H contains an st-path. However, in large networks this is not always simple. We use the network of Figure 7.25 to show a systematic method of determining whether a network contains an st-path. Figure 7.25 equals Figure 7.23.

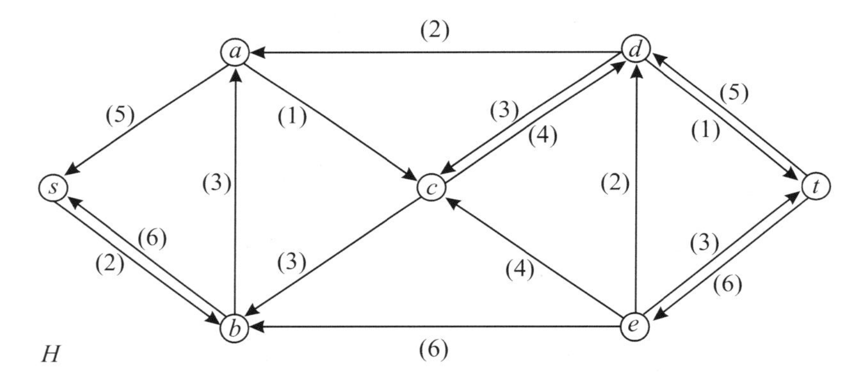

Figure 7.25.

In order to find out if H contains a flow-augmenting path we label all the nodes with an incoming arc from s. Next we label all the unlabelled nodes that have an incoming arc from a labelled node, etcetera. We continue this procedure until one of the following events takes place:

- t is labelled; in this case a flow-augmenting path does exist,
- no more new nodes can be labelled; in that case no flow-augmenting path exists.

Now we make an attempt to label the nodes in H (see Figure 7.25). Node b has an incoming arc from s, so b is labelled. From b we can reach a, so a is labelled. From a we can reach c so c is labelled. From c the nodes d and b can be labelled. Now we can label t from d. So H contains the flow-augmenting path *sbacdt*. The minimum capacity along this flow-augmenting path *sbacdt* in H is 1. So we can increase the flow along this path in H by 1. Then the total flow through G is 11+1=12 (see Figure 7.26).

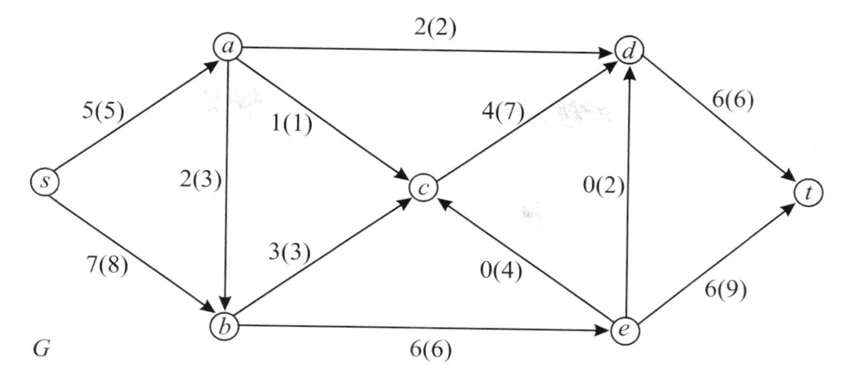

Figure 7.26.

The residual network that corresponds to the flow of Figure 7.26 is given in Figure 7.27. Again we label the nodes in H. Node b can be reached from s so b is labelled. Node a can be reached from b so a is labelled. No more nodes can be labelled from s, a and b. Apparently H does not contain a flow-augmenting path, so the flow 12 is maximal.

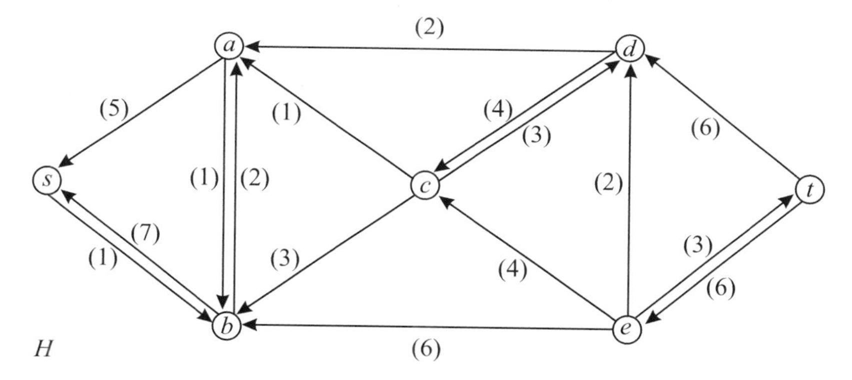

Figure 7.27.

Minimum cut

Closely related to the maximum flow problem is the *minimum cut problem*. Also this problem shall be introduced using the example of the pipeline network, see Example 7.11.

Suppose there is an emergency at the oil field, so that no oil flows from the oil field to the refinery. The cost of blocking a pipeline is supposed to be proportional to the capacity of the pipeline. Now what is the cheapest way to block the stream from the oil field to the refinery? In other words: which blockade comprises the minimum capacity? In the sequel we shall use the word 'cut' instead of 'blockade'. So a *cut* is a set of arcs whose removal from the network prevents any flow from s to t.

The pipeline network is given again in Figure 7.28. One of the ways to block the oil flow from s to t is to close arcs sa and sb. This cut has a capacity of 5+8=13. Another option is to close dt and et, with a capacity of 15. This is more expensive than blocking sa and sb. Another possible cut is ad, ac, bc, be with a capacity of 12. This cut is the cheapest one that we have found so far.

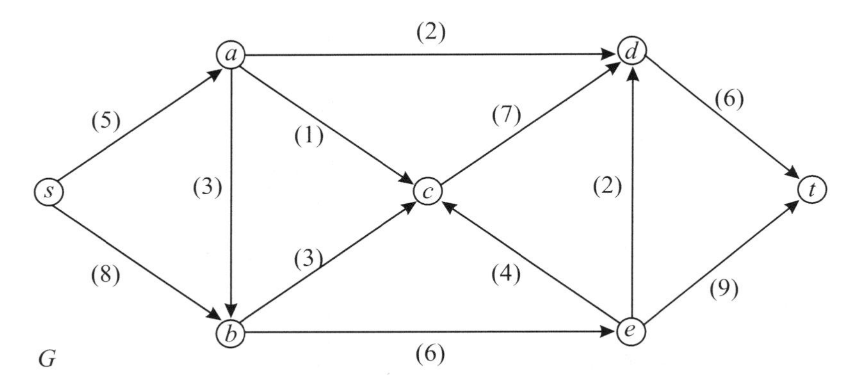

Figure 7.28.

Is it useful to continue the search? Does a cut with a capacity of less than 12 exist? The answer is: NO, such a cut does not exist. After all we have calculated that it is possible to send a flow with size 12 from s to t. All this flow has to pass the cut somewhere. So the capacity of every cut has to be at least 12. This is an important connection between flows and cuts: the capacity of the minimum cut cannot be smaller than the value of the maximum flow: capacity of minimum cut ≥ value of maximum flow. Later we shall prove that *the capacity of the minimum cut is equal to the value of the maximum flow.*

We will now introduce the concept of a cut in a more formal way. Let $S \subset V$ be a set of nodes such that $s \in S$ and $t \in \bar{S}$. ($\bar{S}$ is the complement of S so $S \cup \bar{S} = V$ and $S \cap \bar{S} = \varnothing$ hold.) The set of edges ij with $i \in S$ and $j \in \bar{S}$ is called a cut. (The word cut is used for the set of nodes S as well as for the set of arcs with one end in S and one end not in S. The context will clarify what is meant.) Elimination of the arcs of the cut blocks the flow from source s to sink t. The capacity $c(S)$ of cut S is defined as

$$c(S) = \sum_{i \in S, j \in \bar{S}} c_{ij}$$

The following result holds: the function $c(S)$ is minimal for the cut S^* that is defined by:

S^* = the set of nodes that were labelled in the last iteration of the maximum flow algorithm

Proof

Say that the maximum flow algorithm yields the maximum flow x^*. In the last iteration of the algorithm the labelling procedure was terminated because the set of labelled nodes in H did not contain an outgoing arc to the set of unlabelled nodes, so that the sink t could not be labelled.

The set of labelled nodes in H did not contain an outgoing arc because:

1. in G the flows on all arcs from the labelled nodes to the unlabelled nodes were equal to the capacities of these arcs (so $x_{ij}^* = c_{ij}$),
2. in G the flows on all arcs of the unlabelled nodes to the labelled nodes were 0 (so $x_{ji}^* = 0$).

We define the *value* $v(x)$ *of the flow* in the network as the size of the outgoing flow of s (which equals the incoming flow of t). For the value of an arbitrary flow x and an arbitrary cut S the following inequality has to hold:

$$v(x) \leq c(S)$$

In words: the value of an arbitrary flow never exceeds the value of an arbitrary cut (because all flow has to 'flow through the cut' somewhere).

Also for every flow x and cut S the following applies:

$$v(x) = \text{(total flow from nodes in } S \text{ to nodes in } \overline{S}) - \text{(total flow from nodes in } \overline{S} \text{ to nodes in } S).$$

For flow x^* and cut S^* we have $x_{ij}^* = c_{ij}$ for the arcs from S^* to $\overline{S^*}$ and $x_{ji}^* = 0$ for the arcs from $\overline{S^*}$ to S^*, so that

$$v(x^*) = c(S^*)$$

The inequality $v(x) \leq c(S)$ holds for every flow and every cut, so it also holds for the maximum flow and the minimum cut: $\max_x\{v(x)\} \leq \min_S\{c(S)\}$. So

$$v(x^*) \leq \max_x\{v(x)\} \leq \min_S\{c(S)\} \leq c(S^*)$$

As $v(x^*) = c(S^*)$ in any of these inequality signs the equality has to hold:

$$v(x^*) = \max_x\{v(x)\} = \min_S\{c(S)\} = c(S^*)$$

So we proved two things:

1. S^* is the minimum cut,
2. value of the maximum flow = capacity of the minimum cut.

Item 2 proves that the maximum flow algorithm found the maximum flow as soon as the residual network no longer contained a path from s to t. (We had not proved that previously.)

In the Figure 7.28 the minimum cut is $S^* = \{s, a, b\}$. The capacity of this cut is $c_{ad} + c_{ac} + c_{bc} + c_{be} = 2 + 1 + 3 + 6 = 12$, which indeed equals the value of the maximum flow.

7.4. Remarks

Everywhere we look in our daily lives, networks are apparent. Electrical power networks bring lighting and comfort into our homes. Telephone networks allow us to communicate with each another over large distances. Road networks, railway networks and airline networks provide us with the means to cross great distances. Manufacturing and distribution networks supply us with food and consumer products. And internet has a major impact on the way that we share information and conduct our business and personal life (Ahuja, 1985).
In each of these domains, and in many more, we wish to move some entity (electricity, a consumer product, a person, information) from one point to another in an underlying network, and to do so as efficiently as possible, both to provide good service to the users of the network and to use the underlying (typically expensive) facilities efficiently and effectively (Ahuja, 1985).
Network theory offers a vast range of problems. But in this chapter we confined ourselves to shortest path problems, longest path problems, and maximum flow-minimum cut problems. We did this because these problems arise frequently in practice, can be solved efficiently, and are frequently encountered as subproblems of more intricate networks problems.
Just like the combinatorial problems that will be discussed in Chapter 10 network problems might seem very easy to solve: with complete enumeration one can list all feasible solutions of the problem, evaluate their objective functions, and pick the best. However, in practice many problems of reasonable size have so many possible solutions that it will not be achievable to list them all in reasonable time, let alone to evaluate them. Fortunately very efficient algorithms have been developed in the past for problems that we introduced in this chapter.
Each of the problems that we introduced in this chapter could have been formulated and solved with either Dynamic Programming (Chapter 8) or Linear Programming (Chapters 2, 3, 4). Still we chose to present the network algorithms because they

1. provide a very efficient way of solving the problems by exploiting their characteristics,
2. give more insight into the structure of the problems,
3. can deepen the reader's understanding of LP and DP by seeing the problems in a different context.

8. Dynamic programming

Th.H.B. (Theo) Hendriks

8.1. Introduction

Dynamic Programming (DP) is a mathematical technique that can be used to solve many optimisation problems. It provides a systematic procedure for determining an optimal combination of linked decisions by breaking up the original problem into smaller, more tractable, sub-problems. In particular, problems related to a sequence of interrelated decisions, e.g. consecutive decisions in time, are often associated with DP. A situation where N decisions have to be made consecutively is also called an *N-step decision problem*. It should be mentioned that the presence of a time aspect in the problem formulation is not a condition for applying DP. Dynamic programming is due to the mathematician R. Bellman (1957). Bellman's contribution to the field is remembered in the name of the Bellman equation. This equation is the central result of DP, which reformulates the optimisation problem in recursive form.
In contrast to linear programming (Chapter 3) there is no standard mathematical formulation for DP. Dynamic programming is a general approach to problem solving and the necessary equations must be developed for each specific situation. This also explains why there is hardly any standard software for DP. A certain degree of creativity and insight into the general structure of DP is required to recognize when and how a problem can be solved effectively by dynamic programming.

Many practical decision problems can be solved by dynamic programming; some application areas are:

- production planning;
- inventory control;
- replacement decisions;
- personal management;
- investment decisions;
- allocation of scarce resources.

The most profound characteristics of DP are illustrated from Example 8.1. A more formal treatment of DP is dealt with in the other sections of this chapter.

Example 8.1.

A person living in city 1 intends to visit city 11 (see Figure 8.1). Three more or less parallel roads are available to travel from city 1 to 11; i.e. the north -, the middle - and the south route. Every route has three junctions or stages where the traveller can switch from one route to the other. As illustrated in Figure 8.1, the traveller may decide, for instance, to take the north route from the departure city 1 to the next crossing or decision moment in stage 1 (node 2). After he arrives in city 2 he may decide to change over to the middle route and travel to city 6 in stage 2, etc.
The essential idea of dynamic programming is that the possible routes can be depicted in a *network*. A network consists of *nodes* of which some are connected by *arcs* (arrows).

The nodes in the network, indicated by circled numbers in Figure 8.1, represent all decision situations divided over three decision moments or *stages*. The arcs indicate how one node can be reached from earlier nodes. The length of an arc is given in units that are relevant for the decision criterion. In this example the criterion is the travelling time in minutes or the distance in kilometres.
As in DP the decisions are made stepwise, the direction of travel is always fixed, e.g. from left to right. These types of networks are called *directed networks* (see also Chapter 7).

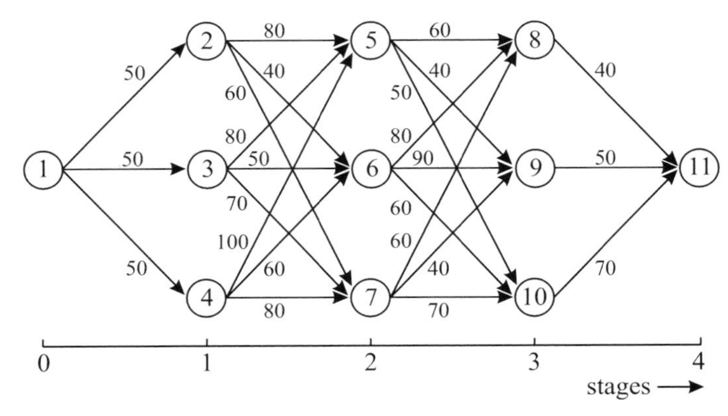

Figure 8.1. Determination of the shortest path between node 1 *and node* 11.

We refer to Section 8.2 for a more formal notation. On the horizontal axis in Figure 8.1 the stages are depicted. *Stage k* denotes the decision moment. In stage 0 ($k = 0$) the traveller is at the starting point (node 1) and he has to decide which direction to take (north, middle or south). In stage $k = 1$ the traveller is at the first crossing of either the north route (node 2), the middle route (node 3) or south route (node 4). In each node of stage $k = 1$ similar decisions have to be made. After the decisions in stage 1 are made the trip continues via the stages $k = 2$ and $k = 3$ until the final node (node 11) has been reached. The total 'costs' of the trip are given as distances in kilometres. The costs between subsequent nodes are given at the corresponding arrow in the network of Figure 8.1, e.g. the distance or costs between the nodes 2 and 6 are 40 kilometres.

The main question is: which road should the traveller take at the crossings he passes, such that the total travelling distance is at a minimum. What is generally noted as minimum costs, is given as a distance criterion in this example.
Starting in stage $k = 0$ (node 1), the traveller must choose between three directions for the first part of the trip. The corresponding costs (km) are given in Table 8.1:

Table 8.1.

from/to	2	3	4
1	50	50	50

All possible choices from node 1 span 50 km. From each of the nodes 2, 3 and 4 in stage $k = 1$ three possible directions are possible, leading to either node 5, 6 or 7 in stage $k = 2$. The costs in stage 1 are given in Table 8.2:

Table 8.2.

from/to	5	6	7
2	80	40	60
3	80	50	70
4	100	60	80

The nodes 5, 6 and 7 again give three possible choices for sub-trips towards the nodes 8, 9 and 10. The corresponding costs in stage $k = 2$ are given in Table 8.3:

Table 8.3.

from/to	8	9	10
5	60	40	50
6	80	90	60
7	60	40	70

Finally, from the nodes 8, 9 and 10 just one direction is possible to the final node 11. The corresponding costs in stage $k = 3$ are:

Table 8.4.

from/to	11
8	40
9	50
10	70

Verify that 27 possible routes exist for travelling from city 1 to city 11. Each route has it's own corresponding length. In this example, it is relatively simple to construct all routes and to calculate the length of each route. In practical problems, however, the number of possible routes in a network easily exceeds the number of one million. In our example this will happen when the number of stages increases from 4 to 15. Finding a solution by *complete enumeration* is in practice hardly realistic.
DP helps to do this work more efficiently; it guarantees that the stepwise choices are made in such a way that the overall trip is one of minimum cost. Next we will introduce the idea of working backwards and illustrate how Example 8.1 can be solved by DP. The notation makes use of the indices i and j to represent the start and end node of a sub trip (arrow), c_{ij} represents the corresponding costs (distance), e.g. $c_{3,7} = 70$ km and $c_{9,11} = 50$ km.

We introduce the function V_k:

$$V_k(i) = \text{costs of the shortest path from node } i \text{ in stage } k \text{ to the final node 11} \qquad (8.1)$$

So, determination of the shortest path from node 1 to node 11 corresponds to calculating the value function $V_0(1)$. The function $V_k(i)$ is called a *value function*. At the start only $V_4(11)$ is known: $V_4(11) = 0$. After all, the shortest path from node 11 to node 11 in stage 4 has a length of 0 km. The value functions are now calculated backwards (i.e. from right to left) in the network until $V_0(1)$ is known.

Remark:

The calculations start from the final stage of the decision problem. Usually the calculation procedure goes backwards. However, in this example it is also possible to determine the value functions from the start to the end. The interpretation of the value function in (8.1) is different in that case.

We started with $V_4(11) = 0$. Consider the value functions in stage 3 for the nodes 8, 9 and 10:

$$V_3(8) = 40$$
$$V_3(9) = 50$$
$$V_3(10) = 70$$

The values for V_3 can be derived directly from Table 8.4, as there is no choice in each of the nodes:

Table 8.5. Stage 3.

I	$V_3(i)$	j^*
8	40	11
9	50	11
10	70	11

In Table 8.5, column j^* indicates the optimal node j that leads to the shortest path from i to the final node ($j = 11$). In stage $k = 3$, j^* is always node 11, because the best path to reach node 11 does not depend on the point of departure in stage 3, i.e. node $i = 8$, $i = 9$, or $i = 10$.

Next the value functions $V_k(i)$ for the nodes $i = 5$, 6 and 7 in stage $k = 2$ are determined. We will demonstrate from an example, i.e. the value function $V_2(5)$, how these calculations are carried out. In node $i = 5$ of stage $k = 2$ there are three possible decisions:

- move to node 8 ('north')
- move to node 9 ('middle')
- move to node 10 ('south')

The costs of the arcs corresponding to all possible decisions are given in Table 8.3; $c_{5.8} = 60$, $c_{5.9} = 40$ and $c_{5.10} = 50$. We already know the minimal costs of travelling from node 8, 9 and 10 in stage 3 to the final node; $V_3(8)$, $V_3(9)$ and $V_3(10)$ respectively. The costs of the three options in node $i = 5$ of stage $k = 2$ are:

$$\begin{aligned} V_2(5) = c_{5.8} &+ V_3(8) = 60 + 40 = 100, \text{ or} \\ c_{5.9} &+ V_3(9) = 40 + 50 = 90, \text{ or} \\ c_{5.10} &+ V_3(10) = 50 + 70 = 120. \end{aligned}$$

The best option from node 5 to reach the final node 11 is to move towards node 9 (direction 'middle'). This decision leads to the shortest path of 90 km from node 5 to the final node 11. Generalising this way of reasoning gives:

$$V_2(5) = \underset{j=8,9,10}{\text{minimum}} \{c_{5j} + V_3(j)\}$$

such that:

$$V_2(5) = \min\{60 + V_3(8), 40 + V_3(9), 50 + V_3(10)\} = \min\{100, 90, 120\} = 90$$

In the same way $V_2(6)$ and $V_2(7)$ can be determined:

$$V_2(6) = \underset{j=8,9,10}{\text{minimum}} \{c_{6j} + V_3(j)\} = 120$$

and

$$V_2(7) = \underset{j=8,9,10}{\text{minimum}} \{c_{7j} + V_3(j)\} = 90$$

Summarising the results of stage 2:

Table 8.6. Stage 2.

I	$V_2(i)$	j^*	Decision
5	90	9	middle
6	120	8	north
7	90	9	middle

The interpretation of Table 8.6 is as follows. If the traveller arrives in node 5, the best direction to proceed is going 'middle' to node 9. The route from 5 via 9 to the final node 11 takes 90 km.

Next in stage $k = 1$, the value functions $V_1(i)$ are determined for the nodes $i = 2$, 3 and 4:

$$\begin{aligned} V_1(2) &= \underset{j=5,6,7}{\text{minimum}} \{c_{2j} + V_2(j)\} = \min\{c_{2,5} + V_2(5), c_{2,6} + V_2(6), c_{2,7} + V_2(7)\} \\ &= \text{minimum } \{80+90, 40+120, 60+90\} = 150 \quad \text{for } j^* = 7 \text{ ('south')} \end{aligned}$$

$$V_1(3) = \underset{j=5,6,7}{\text{minimum}} \{c_{3j} + V_2(j)\} = 160 \ \text{ for } j^* = 7 \text{ ('south')}$$

$$V_1(4) = \underset{j=5,6,7}{\text{minimum}} \{c_{4j} + V_2(j)\} = 170 \ \text{ for } j^* = 7 \text{ ('south')}$$

Summarising the results of stage 1:

Table 8.7. Value functions and optimal decisions in stage 1.

i	$V_1(i)$	j^*	Decision
2	150	7	south
3	160	7	south
4	170	7	south

Finally, the value function in stage $k = 0$ can be determined:

$$V_0(1) = \underset{j=2,3,4}{\text{minimum}} \{c_{1j} + V_1(j)\} = 200 \ \text{ for } j^* = 2 \text{ ('north')}$$

Summarising stage 0:

Table 8.8. Stage 0.

i	$V_0(i)$	j^*
1	200	2

So, the length of the shortest path is 200 km. This path follows the nodes 1, 2, 7, 9 and 11 (see Figure 8.2).

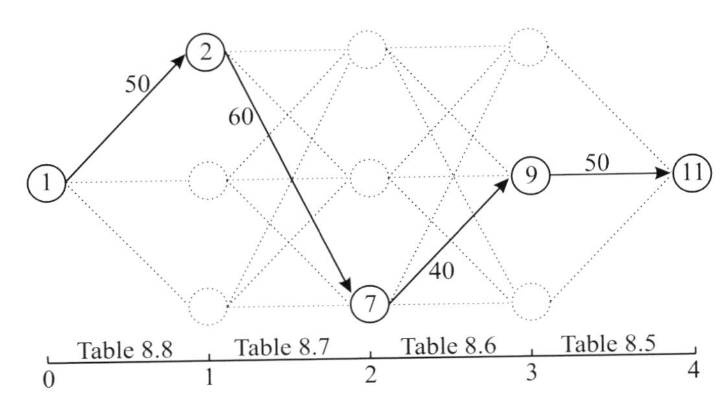

Figure 8.2. The shortest path.

The way the tables are constructed gives the opportunity to read the best directions in successive stages from Table 8.5 up to 8.8, (optimal next node j^*, in every node). This determines the optimal path from node 1 to final node 11 according to the so-called *forward procedure*.

The alternative method of complete enumeration implies generation and investigation of *27* routes. This requires $27 \times 3 = 81$ additions to be made and $27 - 1 = 26$ comparisons of two numbers. Solving the example with DP requires $8 \times 3 = 24$ additions and $7 \times 2 = 14$ comparisons of two numbers. The difference between enumeration and DP increases considerably when the number of possibilities and stages increases.

The relation between two consecutive value functions can be written as follows:

$$V_k(i) = \underset{j \in B_i}{\text{minimum}} \{c_{ij} + V_{k+1}(j)\} \qquad \text{for } k = 3, 2, 1, 0 \; ; \; i \in N_k \tag{8.2}$$

where B_i is the set of nodes reachable directly from node i and N_k is the set of nodes in stage k. In the example: $B_4 = \{5, 6, 7\}$, $B_9 = \{11\}$, $N_3 = \{8, 9, 10\}$.

Equation (8.2) is called the *recursive Bellman equation*. The minimisation in (8.2) determines the best next destination j^* in every node i of stage k and the corresponding length of the shortest path to the final node. The recursive Bellman calculation method for determining $V_k(i)$ for $k = 3, 2, 1, 0$ is called the *backward procedure*. After the starting node in stage $k = 0$ has been reached, the optimal path must be determined. This can be done by the so-called *forward procedure*. To implement the forward procedure, two methods exist:

1. Only the value functions $V_k(i)$ determined by the backward procedure are stored in a table. During the forward procedure the best next node is determined starting from the first node by taking $\min_{j \in B_1} \{c_{1j} + V_1(j)\}$. The next node is determined analogously using V_2.
2. The optimal j^* as well as the value functions $V_k(i)$ are stored for each state i in stage k during the backward procedure, as was done in the example. This requires more storage memory. On the other hand, it requires less calculation time.

In Section 8.2, the theory of DP is elaborated on further. In Section 8.3 we present a number of examples of deterministic DP. Probabilistic or Stochastic DP will not be discussed in this textbook.

8.2. Deterministic dynamic programming

In this section we introduce the general terminology of DP for solving N-step decision problems. The moments at which decisions are taken are called *stages*. In the network, stages are represented on the horizontal axis. In Example 8.1, the stages are the moments where the next direction of the path is chosen.

The nodes in the network represent the *states* in a certain stage. In Example 8.1 three possible states exist; at the first, second and third parallel route between the cities 1 and 11. The states are given on the vertical axis of the network (see Figure 8.3). The arrows in Figure 8.3 denote the decisions. Given a state i in any stage k, the outgoing decision clearly defines the state j in the next stage $k+1$.

A decision transforms the state i into a new state j in the next stage. This alternating sequence of state - decision - state - decision - ... continues up to a final state in the last stage. Almost all N-step decision problems can be transformed or reduced, into a problem of determining the shortest path in a network. A general N-step decision problem is illustrated in Figure 8.3.

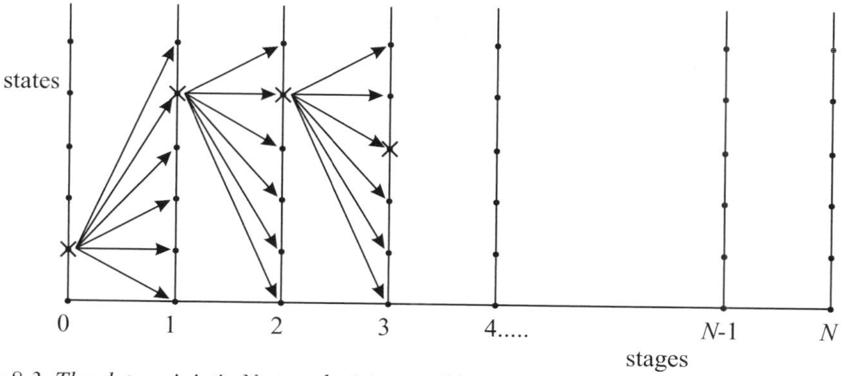

Figure 8.3. The deterministic N-step decision problem.

The dots in Figure 8.3 represent the feasible states in each stage 0, 1, ..., N. The arrows indicate the decisions. In each state of this example, six possible decisions are possible. The crosses represent the states when following a path of optimal decisions. The term *deterministic* N-step decision problem is used to express that the new state in the next stage is completely determined by the state and policy decision in the current stage; no probability or uncertainty is involved. When uncertainty is modelled via stochastic variables, the new states are reached with a certain probability and the term *stochastic N-step decision problem* is used. The stochastic case will not be discussed in this textbook. In Figure 8.3, the initial state corresponds to stage 0 and the final state to stage N. Next we discuss and formalize dynamic programming in more general terms.

State of a system: $\underline{s}_k$

The symbol $\underline{s}_k$ represents the state of a system in stage k. In Figure 8.1, $\underline{s}_0$ includes only one possibility, i.e. node 1. State $\underline{s}_1$ includes three possibilities: the nodes 2, 3 or 4. The symbol $\underline{s}_k$ is underlined to indicate that it is in general a vector in the n-dimensional space $\mathbb{R}^n$. So $\underline{s}_k \in \mathbb{R}^n$ where $\underline{s}_k = (s_{1k}, s_{2k}, ..., s_{nk})'$.
For example, in an inventory management problem for two products, it is usual to model the inventory of the two products at every decision moment k (= stage k) as the state variables. In that case, s_{1k} represents the inventory level of product one and s_{2k} represents the inventory level of product two at decision moment k. So, in that case $\underline{s}_k$ is a vector in the two dimensional space $\mathbb{R}^2$: $\underline{s}_k = (s_{1k}, s_{2k})'$. The symbol S_k represents the set of all feasible states $\underline{s}_k$ in stage k. In Figure 8.1, $S_0 = \{1\}$, $S_1 = \{2, 3, 4\}$ etc.

Decision variables: $\underline{x}_k$

Symbol $\underline{x}_k$ represents the vector of decision variables in stage k. In an inventory management problem, decision variables can be the quantities to be ordered of each product at decision moment k; x_{1k} denotes the order quantity of product 1 and x_{2k} the quantity for product 2. In case of two products, $\underline{x}_k$ is a vector in two-dimensional space $\mathbb{R}^2$; so $\underline{x}_k \in \mathbb{R}^2$; $\underline{x}_k = (x_{1k}, x_{2k})'$. In general, $\underline{x}_k \in \mathbb{R}^m$. Vector $\underline{x}_k$ gives the decisions a decision maker takes in stage $k-1$. Given a current state $\underline{s}_{k-1}$ in stage $k-1$, the decision $\underline{x}_k$ transforms the system into a new state $\underline{s}_k$ in stage k. The symbol X_k represents the set of all feasible decisions $\underline{x}_k$ in stage k.

In Example 8.1, X_1 contains three feasible decisions: travelling from node 1 to either node 2, 3 or node 4. In general X_k in stage k depends on state $\underline{s}_{k-1}$ of the system in stage $k-1$. This can be formalised by:

$$X_k = X_k(\underline{s}_{k-1})$$

Transformation function: T_k

Given a current state $\underline{s}_{k-1}$ and decision vector $\underline{x}_k$ the new state $\underline{s}_k$ is determined. In Example 8.1 for instance when the traveller is in node 6 and stage 2, the decision to go to node 8 determines the new state $\underline{s}_k$ = node 8. This is formalised in mathematical notation as follows. A *transformation function* T_k has as input the two vectors $\underline{s}_{k-1}$ and $\underline{x}_k$ and as output vector $\underline{s}_k$:

$$\underline{s}_k = T_k(\underline{s}_{k-1}, \underline{x}_k)$$

This equation is called the *balance equation* or *equation of motion*.
Suppose $\underline{s}_{k-1}$ represents the inventory level of a product at moment $k-1$. If $\underline{x}_k$ units are produced and the demand equals $\underline{d}_k$ units, then the following holds:

$$\underline{s}_k = \underline{s}_{k-1} + \underline{x}_k - \underline{d}_k$$

So, in this case the transformation function $T_k(\underline{s}_{k-1}, \underline{x}_k) = \underline{s}_{k-1} + \underline{x}_k - \underline{d}_k$

Costs: G_k

The transition from state $\underline{s}_{k-1}$ to $\underline{s}_k$ due to decision $\underline{x}_k$ involves costs. In Example 8.1 the costs are the distances between two nodes. Symbol G_k is used for the costs. The costs G_k depend on state $\underline{s}_{k-1}$ in stage $k-1$ and the decision $\underline{x}_k$ in stage k. So,

$$G_k = G_k(\underline{s}_{k-1}, \underline{x}_k)$$

In Example 8.1, in stage 2 at node 5 (= $\underline{s}_2$), the decision $\underline{x}_3$ to go to node 10 (direction 'south') implies $G_3 = 50$.
In the examples of this chapter, G_k represent costs, i.e. a minimisation problem. In other cases the optimisation criterion might, for example, be the profit. In such cases the value function (see below) must be maximised.

Value function: V_k

The value function V_k is defined for every state $\underline{s}_k$ in stage k ; $V_k = V_k(\underline{s}_k)$. In Example 8.1, $V_k(\underline{s}_k)$ represents the length of the shortest path from any node $\underline{s}_k$ in stage k to the final node $s_k = 11$ in stage $k = 4$. In general $V_k(\underline{s}_k)$ is defined as the minimal total cost from state $\underline{s}_k$ in stage k over all following stages.

Remark

In this section, X_k and S_k are assumed to be finite sets. The value function V_k for $k = 0, ..., N–1$ is defined as:

$$V_k(\underline{s}_k) = G_{k+1}(\underline{s}_k , \underline{x}_{k+1}) + G_{k+2}(\underline{s}_{k+1} , \underline{x}_{k+2}) + ... + G_N(\underline{s}_{N-1} , \underline{x}_N)$$

in which the decisions $\underline{x}_{k+1}$, $\underline{x}_{k+2}$, ..., $\underline{x}_N$ denote the optimal decisions or policy path from state $\underline{s}_k$ in stage k to the final stage. In this expression, $V_k(\underline{s}_k)$ only depends on $\underline{s}_k$ and the chosen decisions $\underline{x}_{k+1}$, $\underline{x}_{k+2}$, ..., $\underline{x}_N$. The intermediate states $\underline{s}_{k+1}$, $\underline{s}_{k+2}$, ..., $\underline{s}_N$ are not relevant in the expression. Due to the balance equation $\underline{s}_{i+1} = T_{i+1}(\underline{s}_i, \underline{x}_{i+1})$ for $i = k, k+1, ..., N–1$, every $\underline{s}_i$ for $i = k+1, ..., N$ can be derived from $\underline{s}_k$ and $\underline{x}_{k+1}, ..., \underline{x}_N$.

As in the final stage $k = N$ there are no additional costs, the value function $V_N(\underline{s}_N) = 0$. The value function $V_k(\underline{s}_k)$ can be written as:

$$V_k(\underline{s}_k) = \underset{\underline{x}_{k+1},\ldots,\underline{x}_N}{\text{minimum}} \left\{ \sum_{i=k+1}^{N} G_i(\underline{s}_{i-1}, \underline{x}_i) \right\} \tag{8.3}$$

$$= \underset{\underline{x}_{k+1},\ldots,\underline{x}_N}{\text{minimum}} \left\{ G_{k+1}(\underline{s}_k, \underline{x}_{k+1}) + \sum_{i=k+2}^{N} G_i(\underline{s}_{i-1}, \underline{x}_i) \right\} \quad \text{for } k = 0, 1, 2, ..., N-1$$

Expression (8.3) can be rewritten as:

$$V_k(\underline{s}_k) = \underset{\underline{x}_{k+1}}{\text{minimum}} \left\{ G_{k+1}(\underline{s}_k, \underline{x}_{k+1}) + \underset{\underline{x}_{k+2},\ldots,\underline{x}_N}{\text{minimum}} \left\{ \sum_{i=k+2}^{N} G_i(\underline{s}_{i-1}, \underline{x}_i) \right\} \right\} \quad \text{such that}$$

$$V_k(\underline{s}_k) = \underset{\underline{x}_{k+1}}{\text{minimum}} \left\{ G_{k+1}(\underline{s}_k, \underline{x}_{k+1}) + V_{k+1}(\underline{s}_{k+1}) \right\} \tag{8.4}$$

In expression (8.4) the balance equation or transformation function should be kept in mind:

$$\underline{s}_{k+1} = T_{k+1}(\underline{s}_k , \underline{x}_{k+1})$$

By this, the minimisation for fixed $\underline{s}_k$ in expression (8.4) only depends on $\underline{x}_{k+1}$. The relation between V_k and V_{k+1} in (8.4) is called the *Bellman equation*. A decision vector $\underline{x}_{k+1}$ which minimises (8.4) is noted by $\underline{x}^*_{k+1}$. So, $\underline{x}^*_{k+1}$ only depends on $\underline{s}_k$; $\underline{x}^*_{k+1} = \underline{x}^*_{k+1}(\underline{s}_k)$. The sequence

$x_1^*, x_2^*, ..., x_N^*$ is called an *optimal strategy* of an *N*-step decision problem. For the final decision $\underline{x}_N$, expression (8.4) can be rewritten into:

$$\begin{aligned} V_{N-1}(\underline{s}_{N-1}) &= \min_{\underline{x}_N}\{G_N(\underline{s}_{N-1}, \underline{x}_N) + V_N(\underline{s}_N)\} \\ &= \min_{\underline{x}_N}\{G_N(\underline{s}_{N-1}, \underline{x}_N)\} \end{aligned} \tag{8.5}$$

Note that we took $V_N(\underline{s}_N) = 0$ for all final states $\underline{s}_N \in S_N$ so far in this chapter. Expression (8.5) allows us to derive $V_{N-1}(\underline{s}_{N-1})$ for all $\underline{s}_{N-1} \in S_{N-1}$. From there, (8.4) shows the way to derive $V_{N-2}(\underline{s}_{N-2})$, ..., $V_0(\underline{s}_0)$, using the backward procedure. Finally, $V_0(\underline{s}_0)$ represents the costs to reach final state $\underline{s}_N$ starting in state $\underline{s}_0$ in stage 0 via *N* optimal decisions $\underline{x}_k^*$, $k = 1, 2, ..., N$.

Remark

Up to now we have taken $V_N(\underline{s}_N) = 0$ for all $\underline{s}_N \in S_N$. This is not a necessary condition, as we can model the preference of the final states via $V_N(\underline{s}_N)$. More specifically, if we prefer to end up in a subset $\hat{S}_N$ of S_N, so $\underline{s}_N \in \hat{S}_N$, then the value function $V_N(\underline{s}_N)$ in the final stage *N* can be defined as:

$$V_N(\underline{s}_N) = \begin{cases} 0 & \text{for } \underline{s}_N \in \hat{S}_N \\ \infty & \text{for } \underline{s}_N \notin \hat{S}_N \end{cases} \tag{8.6}$$

In this way, a penalty of infinite costs is incurred for ending up in any state $\underline{s}_N$ outside $\hat{S}_N$.

Using the balance equation $\underline{s}_{k+1} = T_{k+1}(\underline{s}_k, \underline{x}_{k+1})$, the formulas (8.4) and (8.5) can be rewritten as:

$$V_k(\underline{s}_k) = \underset{\underline{x}_{k+1}}{\text{minimum}}\{G_{k+1}(\underline{s}_k, \underline{x}_{k+1}) + V_{k+1}(T_{k+1}(\underline{s}_k, \underline{x}_{k+1}))\} \tag{8.7}$$

for each $\underline{s}_k \in S_k$ and $k = N-1, N-2, ..., 0$

Expression (8.7) implies the following:

If $\underline{x}_k^*$ for $k = 1, 2, ..., N$ is an optimal strategy for the original *N*-step decision problem, starting at initial state $\underline{s}_0$, then the decisions $\underline{x}_k^*$, ..., $\underline{x}_N^*$ are also optimal for the $(N-k+1)$-step decision problem starting at initial state $\underline{s}_{k-1}$. This phenomenon is called the *optimality principle of Bellman*. Verify what this means in Figure 8.4 taking for instance $k = 3$.

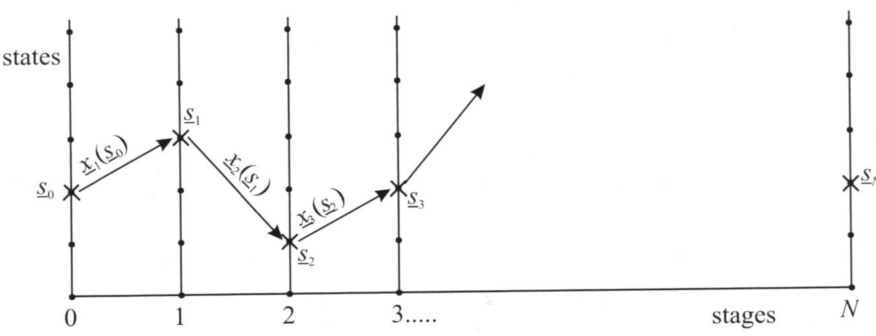

Figure 8.4. General deterministic N-step decision problem.

Many problems can be formulated as an N-step decision problem; usually planning problems. Solving such a problem using DP requires a formulation such that the network structure is visible as in Figure 8.4.

To recognise the network structure in a general problem formulation requires an unambiguous definition of the state vector $\underline{s}_k$, the decision vector $\underline{x}_k$ and the transformation function T_k. This may not be easy and calls in some way for a certain degree of ingenuity.

After the N-step decision problem has been translated into a network formulation, the Bellman equation (8.7) can be applied in order to find the optimal solution (shortest or longest path) of the problem.

In Section 8.3, examples of model formulation and model solution are discussed.

8.3. Applications of deterministic DP

This section further elaborates some typical problems that can be modelled and solved using DP. The focus will be on the network view of the problem. This makes it possible to handle the problems with the methodology that has been introduced in Section 8.2.

8.3.1. An allocation problem in workforce planning

Example 8.2.

A consultancy firm employs seven workers. Recently three projects, indexed by $k = 1, 2, 3$, have been acquired. Project 1 requires teams with a size of four persons. Project 2 requires teams of three persons. Any number of persons can be assigned to project 3.

The modelling proceeds as follows.
Variable x_1 represents the number of teams composed of four persons for project 1; x_2 the number of teams composed of three persons for project 2, and x_3 the number of workers on project 3. So the decision variables are x_1, x_2 and x_3.

Now the state variables are introduced: s_0, s_1, s_2 and s_3. State variable s_0 gives the total number of available workers that can be assigned to the projects 1, 2 and 3. So, $s_0 = 7$. State variable s_1 denotes the available number of workers for the projects 2 and 3 after x_1 has been fixed. State variables s_2 and s_3 are defined analogously. The profit to be maximised for project k is given by $G_k(s_{k-1}, x_k)$ and depends in this example only on x_k via:

$$\begin{aligned} G_1(s_0, x_1) &= 9x_1^3, \quad \text{where } s_0 = 7 \\ G_2(s_1, x_2) &= 2x_2^2, \quad \text{for all states } s_1 \\ G_3(s_2, x_3) &= x_3, \quad \text{for all states } s_2 \end{aligned} \tag{8.8}$$

As G_k is independent of the value of s_{k-1} with $k = 1, 2, 3$, notation $G_k(x_k)$ will be used here. The total profit G as a function of x_1, x_2 and x_3 is now:

$$G = 9x_1^3 + 2x_2^2 + x_3 \tag{8.9}$$

Seven workers are available that all have to be allocated, so:

$$4x_1 + 3x_2 + x_3 = 7 \tag{8.10}$$

The problem can be formulated as:

$$\begin{aligned} &\max\{G = 9x_1^3 + 2x_2^2 + x_3\} \\ &4x_1 + 3x_2 + x_3 = 7 \\ &x_1, x_2, x_3 \geq 0 \text{ and integer} \end{aligned} \tag{8.11}$$

A network formulation for problem (8.11) is given in Figure 8.5. In order to construct a network it is necessary to consider (8.11) as a 3-step decision problem. The state in each stage $k = 0, 1, 2$ and 3 denotes the number of workers that remain for the assignment to projects.

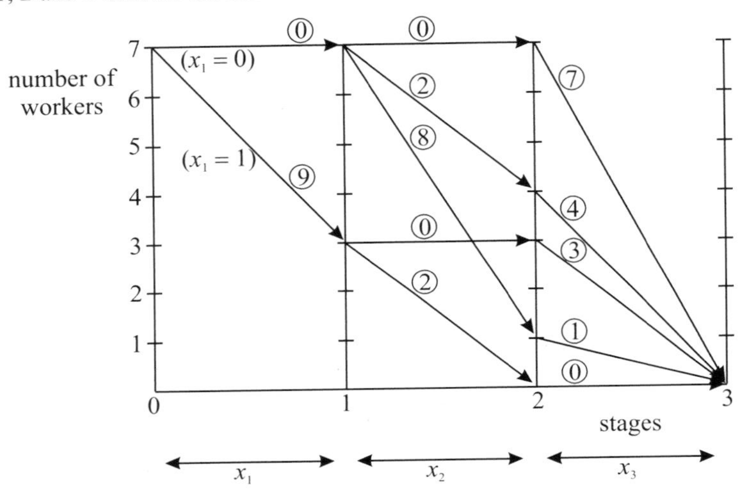

Figure 8.5. Workforce allocation.

In stage 0, seven workers ($s_0 = 7$) are available to be allocated to the projects and we decide on the number x_1 of teams (each team consists of four persons) for project 1. In stage 1, the number of teams x_2 (each team consists of three persons) are assigned to project 2 and finally in stage 2 the allocation of the remaining workers x_3 to project 3 is done. All possible allocations, i.e. decisions, are depicted in Figure 8.5 by arrows. Note that the order of the decisions influences the shape or density of the network. Starting with decision variable x_3 in stage 0, first allocating the number of workers to project 3, would imply more arrows in the network.

A profit indicator based on (8.9) is given to every arrow. In this way, the decision problem of allocating the workforce is translated into finding a path of maximum length in Figure 8.5. In the former examples as well as the theoretical elaboration of DP we were minimising. By replacing 'minimum'(min) by 'maximum' (max) and 'cost' by 'profit' in the formulas, the longest path can be found. Determination of value functions V_k for $k = 0, 1, 2, 3$ according to the *backward procedure*, goes as follows.

$k = 3$

Value function $V_3(s_3)$ is given as $-\infty$ for all states $s_3 \neq 0$. In this way no work force is left over. Furthermore, $V_3(0) = 0$

$k = 2$

As shown in Figure 8.5, s_2 can take the values 0, 1, 3, 4 and 7. The Bellman equation in stage $k = 2$ is:

$$V_2(s_2) = \underset{0 \leq x_3 \leq s_2}{\text{maximum}}\{G_3(x_3) + V_3(s_2 - x_3)\} \quad \text{for } s_2 = 0, 1, 3, 4, 7$$

The maximum is found for $x_3^* = s_2$; in this case no workers are left over. So,

$V_2(s_2) = G_3(s_2) = s_2$ for $s_2 = 0, 1, 3, 4, 7$

This means $V_2(7) = 7$, $V_2(4) = 4$, ..., $V_2(0) = 0$

$k = 1$

Now $V_1(s_1)$, can be determined for $s_1 = 7$ and $s_1 = 3$:

$$V_1(7) = \underset{x_2 = 0,1,2}{\text{maximum}}\{G_2(x_2) + V_2(7 - 3x_2)\}$$
$$= \max\{0 + V_2(7),\ 2 + V_2(4),\ 8 + V_2(1)\} = \max\{0+7, 2+4, 8+1\} = 9 \text{ for } x_2^* = 2$$

If in stage $k = 1$ seven workers are available ($s_1 = 7$), the best decision is to assign two teams of three persons to project 2. So, $x_2^* = 2$.

$$V_1(3) = \underset{x_2 = 0,1}{\text{maximum}}\{G_2(x_2) + V_2(3 - 3x_2)\}$$
$$= \max\{0 + V_2(3),\ 2 + V_2(0)\} = \max\{3,\ 2\} = 3.$$

If in stage $k = 1$ three workers remain for the assignment to the second project ($s_1 = 3$), the best decision is to allocate no workers to project 2. So, $x_2^* = 0$.

$k = 0$

Finally, the value function $V_0(7)$ is determined:

$$V_0(7) = \underset{x_1 = 0,1}{\text{maximum}}\{G_1(x_1) + V_1(7 - 4x_1)\}$$
$$= \max\{0 + V_1(7),\ 9 + V_1(3)\} = \max\{0+9, 9+3\} = 12 \text{ for } x_1^* = 2$$

Because all seven workers are available at stage 0, the best decision is to assign one team of four persons to project 1. So, $x_1^* = 1$

The calculations show that the longest path in the network has length 12. Next, the *forward procedure* must be applied:

$k = 0$, $s_0 = 7$

The best decision in stage 0 and state $s_0 = 7$ is $x_1^* = 1$. This implies that $s_k = 3$ for $k = 1$.

$k = 1$, $s_1 = 3$

The optimal decision in stage 1 and state $s_1 = 3$ is $x_2^* = 0$. This implies that $s_k = 3$ for k=2.

$k = 2$, $s_2 = 3$

The optimal decision in stage 2 and state $s_2 = 3$ is $x_3^* = 3$. This implies that $s_k = 0$ for k =3.

$k = 3$, $s_3 = 0$
The final node in stage $k = 3$ and state $s_3 = 0$ has been reached.

Problem (8.11) is an example of the so-called class of *knapsack problems* as introduced in Chapter 1. Typical is the structure of the objective function and the constraints. Both consist of a summation of functions in one variable. This property is called *separability*.

A function f of n variables $x_1, x_2, ..., x_n$ is called *separable* if it can be written as a summation of n functions g_i in one variable x_i, for $i = 1, 2, ..., n$ (see also Chapter 9).

$$f(x_1, x_2, ..., x_n) = g_1(x_1) + g_2(x_2) + ... + g_n(x_n)$$

The general (discrete) knapsack problem with one constraint can be formulated as:

$$\begin{aligned} &\max\left\{\sum_{j=1}^{n} G_j(x_j)\right\} \\ &\text{s.t.} \\ &\quad \sum_{j=1}^{n} g_j(x_j) \le B \\ &\quad x_j \ge 0 \ \text{ integer}, \quad \text{for } j = 1, 2, ..., n \end{aligned} \tag{8.12}$$

where G_j and g_j are functions of the integer variable x_j and B represents the available scarce resource, e.g. capacity. The *Bellman equation* for the *general knapsack problem* is:

$$V_{k-1}(s_{k-1}) = \max_{x_k}\{G_k(x_k) + V_k(s_{k-1} - g_k(x_k))\} \tag{8.13}$$

for $k = n, \ n-1, ..., 1$ and all $s_{k-1} \le B$
in which the maximisation is carried out for $g_k(x_k) \le s_{k-1}$ and $x_k \ge 0$.

8.3.2 Continuous knapsack problems

In this section, two examples are elaborated where the knapsack problem is extended to the case where the decision variables do not necessarily take an integer value.

Example 8.3.
Consider optimisation problem (8.14)

$$\begin{aligned} &\max\{3x_1 + x_2 - x_3\} \\ &\text{s.t.} \\ &\quad x_1^2 + x_2^2 + x_3 \le 16 \\ &\quad x_1, x_2, x_3 \ge 0 \end{aligned} \tag{8.14}$$

The objective function as well as the left-hand-side of the constraint are separable functions. In contrast to the former examples, the variables x_1, x_2 and x_3 are continuous in problem (8.14). This implies that the state variables s_k can also take infinitely many values. Although the complete network can not be drawn, we will show that the network formulation and dynamic programming can be applied in order to solve *continuous knapsack problems* by DP.

First we have to define the state variables. The initial state s_0 at the beginning of stage $k = 0$ is $s_0 = 16$ available units. In stage $k = 0$, decision variable x_1 is chosen. So, $s_1 = s_0 - (x_1)^2$. Consequently in $k = 1$ a value for x_2 is determined such that $s_2 = s_1 - (x_2)^2$. Finally, $s_3 = s_2 - x_3$.

In Figure 8.6 we can only give a schematic picture of the network corresponding to problem (8.14). The dotted arrows denote the infinite number of values for the decision variables from one of the possible states in each stage, pointing at the continuum of states for the infinite set of decisions in the next stage. In stage $k = 0$ there is only one state, i.e. $s_0 = 16$ as a starting point. In stage $k = 1$ and stage $k = 2$ the arrows can start in any state $0 \leq s_i \leq 16$, $i = 1, 2$. The arrows that have been drawn completely in Figure 8.6 represent the optimal strategy of the problem. Next, we follow the *backward procedure* of the Bellman equation.

$k = 3$

$V_3(s_3) = 0, \;\; 0 \leq s_3 \leq 16$

where s_3 can take any value between 0 and 16. In Example 8.2 only integer values were possible.

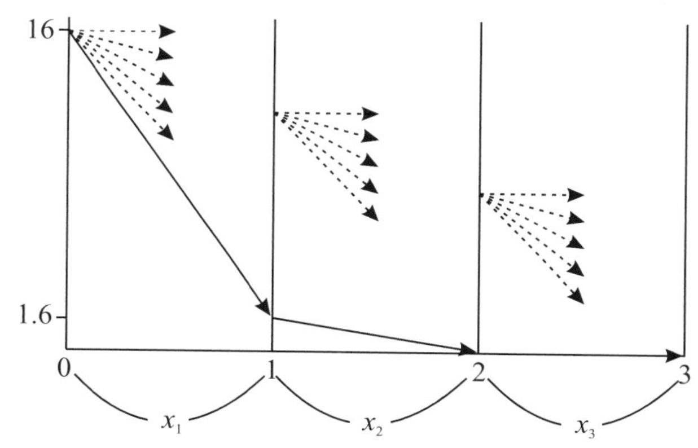

Figure 8.6. Continuous knapsack problem.

Following the backward procedure, we continue to determine the value functions for all states s_k in stage $k = 2$ (decision variable x_3):

$k = 2$

$$\begin{aligned} V_2(s_2) &= \underset{0 \leq x_3 \leq s_2}{\text{maximum}}\{-x_3 + V_3(s_2 - x_3)\} \\ &= \underset{0 \leq x_3 \leq s_2}{\text{maximum}}\{-x_3 + 0\} = 0, \quad 0 \leq s_2 \leq 16 \end{aligned} \tag{8.15}$$

As $V_3(s_3) = 0$ for all possible values of s_3, the maximum of (8.15) is reached for $x_3^* = 0$.

$k = 1$

$V_1(s_1)$ can be calculated according to:

$$V_1(s_1) = \underset{0 \le x_2^2 \le s_1}{\text{maximum}}\{x_2 + V_2(s_1 - x_2^2)\}$$
$$= \underset{0 \le x_2^2 \le s_1}{\text{maximum}}\{x_2 + 0\} = \sqrt{s_1}, \quad 0 \le s_1 \le 16 \qquad (8.16)$$

The best decision x_2^* in (8.16) is $x_2^* = \sqrt{s_1}$

$k = 0$

Finally, $V_0(s_0)$ is determined:

$$V_0(s_0) = \underset{0 \le x_1^2 \le s_0}{\text{maximum}}\{3x_1 + V_1(s_0 - x_1^2)\}$$
$$= \underset{0 \le x_1^2 \le s_0}{\text{maximum}}\{3x_1 + \sqrt{s_0 - x_1^2}\} \qquad (8.17)$$

The value function $V_0(s_0)$ in (8.17) should be determined only for $s_0 = 16$. Consider the expression between brackets in (8.17). This expression can be considered as a function f of variable x_1:

$$f(x_1) = 3x_1 + \sqrt{s_0 - x_1^2} = 3x_1 + \sqrt{16 - x_1^2}$$

Formula (8.17) requires finding the maximum of $f(x_1)$ on the interval $0 \le x_1 \le \sqrt{s_0} = \sqrt{16}$. Using the first order conditions of an interior maximum means that the derivative $f'(x_1)$ should be zero:

$$f'(x_1) = 3 - \frac{x_1}{\sqrt{16 - x_1^2}} = 0$$

Indeed $x_1^* = \sqrt{14.4}$ is an interior maximum point on the interval.

Now we can apply the forward procedure to determine the optimal decision policy.

$k = 0$

Knowing that $x_1^* = \sqrt{14.4}$ offers the possibility to determine x_2^* from $x_2^* = \sqrt{s_1}$ in stage $k = 1$.

$k = 1$

Given the transformation function $s_1 = s_0 - (x_1^*)^2$, we can see that $s_1 = 16 - 14.4 = 1.6$

Formula (8.16) gives $x_2^* = \sqrt{s_1} = \sqrt{1.6}$

$k = 2$

In stage $k = 2$, the new state $s_2 = s_1 - (x_1^*)^2 = 1.6 - (\sqrt{1.6})^2 = 0$. Finally (8.15) gives $x_3^* = 0$ and consequently in the next and final stage:

$k = 3$

The state $s_3 = s_2 - x_0^* = 0$

Example 8.3 showed that for applying the Bellman equation, the state space S_k does not necessarily have to be a finite set.
Of course, finding the solution for Example 8.3 does not require DP. Due to the coefficient -1 for x_3 in the objective function, the optimal solution for x_3 is obviously $x_3^* = 0$. In this way a function in two variables x_1 and x_2 remains that can be solved graphically. We refer to the Exercises chapter for an example of a (simple) two-dimensional knapsack problem for which it is less trivial to find the optimal solution.

The next example of a continuous knapsack problem illustrates a more general form of the Bellman equation.

Example 8.4.

Consider the following optimisation problem in three decision variables x_1, x_2 and x_3.

$$\begin{aligned} &\max\{x_1 x_2 x_3\} \\ &\text{s.t.} \\ &\quad x_1 + x_2 + x_3 \leq 10 \\ &\quad x_1,\ x_2,\ x_3 \geq 0 \end{aligned} \tag{8.18}$$

At first sight the objective function in (8.18) seems not to be separable, whereas this property is necessary to apply the Bellman equation. However, replacing '$\max\{x_1\ x_2\ x_3\}$' by '$\max\{\log(x_1\ x_2\ x_3)\}$' does not change the optimal strategy of (8.18), as 'log' is a strictly monotonous function on the interval $(0, \infty)$. Expression '$\max\{\log(x_1 x_2 x_3)\}$' can be written as:

'$\max\{\log x_1 + \log x_2 + \log x_3\}$'

such that between brackets a separable function of x_1, x_2 and x_3 appears.

Remark

In this way the concept of a separable function of $f(\underline{x})$ for $\underline{x}' = (x_1, \ldots, x_n) \in \mathbb{R}^n$ has been extended to a summation and as a product of functions g_k in one variable x_k :

$$f(x_1, x_2, \ldots, x_n) = g_1(x_1) \times g_2(x_2) \times \ldots \times g_n(x_n)$$

For problem (8.18) using a product instead of summation gives the following Bellman equation:

$$V_k(s_k) = \underset{0 \leq x_{k+1} \leq s_k}{\text{maximum}}\{x_{k+1} V_{k+1}(s_k - x_{k+1})\} \quad \text{for } k = 2, 1, 0 \text{ and } 0 \leq s_k \leq 10 \tag{8.19}$$

In the previous Example 8.3, the value function in the last stage $k = 3$ was set to zero. So, $V_3(s_3) = 0$, for $0 \leq s_3 \leq 16$. As we have a product of functions in Example 8.4, the value function for all states in the final stage becomes:

$k = 3$

$V_3(s_3) = 1, \quad 0 \leq s_3 \leq 10$

In other words: zero in an addition gives the same result as one in a multiplication!

The balance equation remains unaltered:

$$s_{k+1} = T_{k+1}(s_k, x_{k+1}) = s_k - x_{k+1}$$

Let the state s_k represent again the remaining units of the 'scarce resource' (e.g. the remaining space in the knapsack) in stage k for $x_{k+1}, \ldots, x_3$. Consequently, $s_0 = 10$. The *backward procedure* as formulated in (8.19), together with the value function $V_3(s_3) = 1$ implies:

$k = 2$

$$V_2(s_2) = \underset{0 \le x_3 \le s_2}{\text{maximum}}\{x_3 V_3(s_2 - x_3)\} = \underset{0 \le x_3 \le s_2}{\text{maximum}}\{x_3 \times 1\} = s_2$$

with corresponding optimal decision $x_3^* = s_2$. For the value functions in the next stage:

$k = 1$

$$V_1(s_1) = \underset{0 \le x_2 \le s_1}{\text{maximum}}\{x_2 V_2(s_1 - x_2)\} = \underset{0 \le x_2 \le s_1}{\text{maximum}}\{x_2(s_1 - x_2)\} \tag{8.20}$$

Maximisation in (8.20) implies that we need to find the maximum over a parabola; $x_2^* = \frac{1}{2}s_1$, which is in the interval $0 \le x_2 \le s_1$. In this way $V_1(s_1) = \frac{1}{2}s_1(s_1 - \frac{1}{2}s_1) = \frac{1}{4}s_1^2$.

Similarly,

$k = 0$

$$V_0(s_0) = \underset{0 \le x_1 \le s_0}{\text{maximum}}\{x_1 V_1(s_0 - x_1)\} = \underset{0 \le x_1 \le s_0}{\text{maximum}}\{x_1 \times \tfrac{1}{4}(s_0 - x_1)^2\} \tag{8.21}$$

Maximisation in expression (8.21) implies finding the maximum of a cubic function on the interval $0 \le x_1 \le s_0$. Verify that a maximum point is $x_1^* = \frac{1}{3}s_0$. Substitution in (8.21) gives $V_0(s_0) = \frac{1}{3}s_0 \times \frac{1}{4}(s_0 - \frac{1}{3}s_0)^2 = \frac{1}{27}s_0^3$

In order to find the optimal values x_1^*, x_2^* and x_3^*, we apply the *forward procedure*:

$k = 0$

In stage $k = 0$, $s_0 = 10$ such that $x_1^* = \frac{1}{3}s_0 = \frac{10}{3}$

$k = 1$

In stage $k = 1$, $s_1 = s_0 - x_1^* = 10 - \frac{10}{3} = \frac{20}{3}$ such that $x_2^* = \frac{1}{2}s_1 = \frac{10}{3}$

$k = 2$

Finally in stage $k = 2$, $s_2 = s_1 - x_2^* = \frac{20}{3} - \frac{10}{3} = \frac{10}{3}$ leading again to $x_3^* = s_2 = \frac{10}{3}$

8.3.3 Dynamic Programming with more than one state variable

In the examples presented so far only one state variable was sufficient to describe the behaviour of the system. Usually more than one state variable and decision variable are required. E.g.

A. In a production planning situation the inventory levels of several products can be the state vector. The decision vector is typically defined as the amounts to be produced of the products.

B. A farmer deciding on activities around the harvesting of grain, will also look at states like:
   - maturity of the grain;
   - area that still has to be harvested;
   - humidity of the grain;
   - availability of workforce and machines;
   - weather prediction.

   Furthermore, the decision vector will have more than one dimension too. Besides the area to be harvested, a decision might be related to the harvesting method, etc.

Solving a problem with a *more-dimensional state space* $\underline{s}_k$ requires additional storage capacity of the computer as well as more calculation time. Let $\#S_k$ be the number of elements in the set S_k. Applying the Bellman equation (8.7) requires $V_k(\underline{s}_k)$ to be determined for all $\#S_k$ vectors. With an n-dimensional state space, the set S_k is a finite subset of $\mathbb{R}^n$. With an increasing dimension n, the calculation time will increase considerably.
In a situation where every state variable in $\underline{s}_k$ can have 10 possible values, the number of values to be calculated and stored is $\#S_k = 10^n$. So when $n = 3$, the Bellman equation has to be determined 1000 times in each stage. Of course also $\#X_{k+1}(\underline{s}_k)$ influences the calculation time in (8.7). This explosion, due to the so-called *curse of dimensionality*, implies that N-step decision problems are in general hardly solvable for cases in which the dimension of the state space is more than four.
Next, we elaborate an example of a knapsack problem in which the state vector is two-dimensional and the decision vector is one-dimensional.

Example 8.5.

In a knapsack problem three products can be taken, each with a weight m_i, a volume l_i and importance b_i as given in Table 8.9 for $i = 1, 2, 3$.

Table 8.9. A two-dimensional knapsack problem.

product	weight m_i	volume l_i	importance b_i
1. tent	8	6	11
2. sleeping bag	4	8	10
3. cooker	1	2	3
at most	10	12	

At most one unit of every product can be taken along in the knapsack. The maximum allowed total weight is 10 kg and the maximum volume is 12 litres. The total importance should be maximised.

Mathematically, the knapsack problem can be formulated as:

$$\begin{aligned} &\max\{w = 11x_1 + 10x_2 + 3x_3\} \\ &\quad 8x_1 + 4x_2 + x_3 \le 10 \\ &\quad 6x_1 + 8x_2 + 2x_3 \le 12 \qquad \text{(P)} \\ &\quad x_1, x_2, x_3 \text{ binary} \end{aligned}$$

where $x_1 = 0$ represents that the tent is not included and $x_1 = 1$ means the tent is included in the knapsack. The problem can be represented in a network (see Figure 8.7).

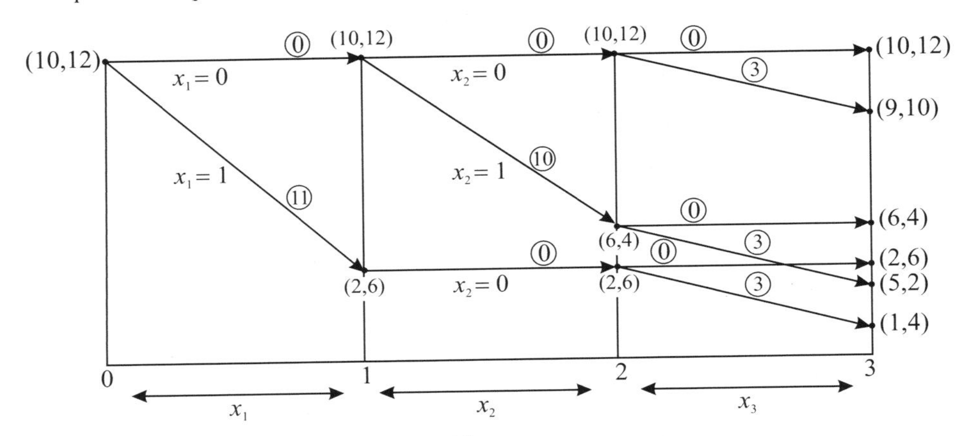

Figure 8.7. Two-dimensional knapsack problem.

State vector $\underline{s}_k = (M_k, L_k)$ is a two-dimensional vector, where M_k represents the available space (weight) in kg and L_k the available volume in litre in stage k. At stage $k = 0$, $\underline{s}_0 = (10, 12)$ such that 10 kg and 12 litre are available. The decision $x_1 = 0$ (take no tent) results into state $\underline{s}_1 = (10, 12)$ in stage 1. Taking the tent along implies $\underline{s}_1 = (2, 6)$ etc. The corresponding Bellman equation is:

$$V_0(10, 12) = \max_{x_1=0,1}\{11x_1 + V_1(10-8x_1, 12-6x_1)\}$$
$$= \max\{0 + V_1(10, 12), 11 + V_1(2, 6)\}$$

Now we elaborate the *backward procedure* for Figure 8.7:

$k = 3$

$V_3(\underline{s}_3) = 0$ for all states $\underline{s}_3$; $V_3(10, 12) = 0$, $V_3(9, 10) = 0$, ..., $V_3(1, 4) = 0$

$k = 2$

$$V_2(10, 12) = \max\{\underbrace{0 + V_3(10, 12)}_{x_3=0}, \underbrace{3 + V_3(9, 10)}_{x_3=1}\} = \max\{0, 3\} = 3 \text{ for } x_3^* = 1$$

Moreover, $V_2(6, 4) = V_2(2, 6) = 3$ for $x_3^* = 1$

$k = 1$

$$V_1(10, 12) = \max\{\underbrace{0 + V_2(10, 12)}_{x_2=0}, \underbrace{10 + V_2(6, 4)}_{x_2=1}\} = \max\{0+3, 10+3\} = 13 \text{ for } x_2^* = 1$$

$V_1(2, 6) = V_2(2, 6) = 3$ for $x_2^* = 0$

$k = 0$

$$V_0(10, 12) = \max\{\underbrace{0 + V_1(10, 12)}_{x_1=0}, \underbrace{11 + V_1(2, 6)}_{x_1=1}\} = \max\{0+13, 11+3\} = 14 \text{ for } x_1^* = 1$$

The *forward procedure* leads to the simple solution $x_1^* = 1$, (take the tent); $x_2^* = 0$, forget the sleeping bag and $x_3^* = 1$ (take the cooker). The optimal objective function value is $w^* = 14$.

A general knapsack problem with a two-dimensional state space can be formulated as:

$$\begin{array}{llllll} \max\{w = b_1x_1 + \ldots + b_{k+1}x_{k+1} + \ldots\ b_nx_n\} & & & & & \\ \quad m_1x_1 & + \ldots & + m_{k+1}x_{k+1} & + \ldots\ m_nx_n & \leq & M \\ \quad l_1x_1 & + \ldots & + l_{k+1}x_{k+1} & + \ldots\ l_nx_n & \leq & L \\ \quad x_j = 0 \text{ of } 1 \quad (j = 1, \ldots, n) & & & & & \end{array}$$

The general Bellman equation for this case is:

$$V_N(M_N, L_N) = 0$$

$$V_k(M_k, L_k) = \max_{x_{k+1}} \{b_{k+1}x_{k+1} + V_{k+1}(M_k - m_{k+1}x_{k+1},\ L_k - l_{k+1}x_{k+1})\} \qquad k = N-1, \ldots, 0$$

where the maximisation should fulfil the constraints for the binary decision variable x_{k+1}. So,

$$0 \leq x_{k+1} \leq \min\left\{1, \left[\frac{M_k}{m_{k+1}}\right], \left[\frac{L_k}{l_{k+1}}\right]\right\}$$

9. Modelling techniques for (non)linear and integer programming

G.D.H. (Frits) Claassen and Th.H.B. (Theo) Hendriks

9.1. Introduction

A major part of this book refers to an important class of optimisation models which are also known as *constrained optimisation models*. In these kinds of models the vector of decision variables: $\underline{x}' = (x_1, x_2, \dots, x_n)$ is restricted to taking only certain allowable values. In other words, the best possible result must be achieved while observing one or more constraints. Within this class of models we can distinguish three subclasses which are also known as linear, (mixed) integer and non-linear programming models. They all share the same general structure of optimising (maximising or minimising) an objective function subjected to at least one constraint.

The basic background relating to algorithms for solving models in each subclass has been covered in several chapters within this book, in particular linear programming as it is the simplest and in practice most commonly applied of the three. Since the fastest and most powerful solution methods are those for linear programming, it is advisable, where possible, to use this technique instead of solving big non-linear or integer programming models.

In Section 9.2 we offer some basic understanding on the question 'why are non-linear models usually far more difficult to solve than correspondingly sized linear models'. This insight inspired researchers to look for ways to transform models with special non-linear features into linear programming models. It turned out that these kinds of transformations are sometimes possible and efficient. In Section 9.3 we will focus on modelling techniques related to linear programming. Section 9.3 also includes some examples in which models with special non-linear features are transformed into (continuous) linear programming models. Quite often it is also possible to model and solve an approximation of the original non-linear model by an extension of linear programming called separable programming. The importance of separable programming in mathematical programming is that non-linear terms can be approximated by piecewise linear functions. This topic will be discussed in Section 9.3 too. However, it will become clear that the distinction between convex and non-convex programming (Section 9.2) is important for modelling and solving piecewise linear approximations of non-linear terms in mathematical programming.

As the emphasis in this chapter is on abstract modelling techniques in general, we will pay attention to (mixed) integer modelling techniques as well (see Section 9.4). In practice it turns out that a wide class of problems can (or has to) be modelled by using integer variables $y_j \in \{0, 1, 2 \dots J\}$ for all j. In cases where such a model consists solely of integer variables, the model is called an *integer programming* model (IP). More commonly there are both continuous variables $\underline{x}$ and integer variables $\underline{y}$. These models are called *mixed integer programming* models (MIP).

In Section 9.4 the focus is on modelling techniques that demonstrate the power of IP as a method of modelling. In particular, models in which the integer variables are restricted to take on binary values, so $y_j \in \{0, 1\}$ for all j, are frequently used to represent 'yes or no' decisions. Logical connections between such decisions can often be modelled by using linear constraints. Binary variables can also be used to model piecewise linear approximations of non-linear functions in non-convex separable programming. This and other (M)IP formulations will be explained in Section 9.4. An important consideration should be taken into account when building (M)IP models. From a computational point of view these kinds of models are much harder to solve than similar sized (continuous) linear programming models. As the flexibility in building (M)IP models is much greater than in building linear programming models, the chance of formulating bad and weak models increases too. However, it is frequently possible to reformulate a problem with another model which is much easier to solve.

9.2. Linear versus non-linear programming

The goal of this section is to give a basic understanding regarding the complexity of solving linear (Chapter 3) and non-linear programming models (Chapter 12). The main message will be that non-linear programming models are in general harder to solve than linear ones. The question is 'why?'. In order to answer this question it is convenient to classify models into *convex* and *non-convex programming* problems. Only convex programming problems have at most one optimal objective value, which makes them in general much easier to solve. It will be explained that linear programming is a special case of convex programming whereas non-linear programming is much wider. For non-linear programming it is not predetermined that all problems in this class are convex.

Definition convex set

An area of space or set $S \subset \mathbb{R}^n$ is called convex if the line segment joining any two points $\underline{p} \in S$ and $\underline{q} \in S$ is situated in the set S too. Mathematically: $\underline{p} + \lambda(\underline{q} - \underline{p}) \in S$; $0 \leq \lambda \leq 1$, or $\lambda\underline{q} + (1-\lambda)\underline{p} \in S$; $0 \leq \lambda \leq 1$.

For example, the interior of a circle is a *convex set S* as all points of a line segment between any two points of S will always belong to this set.

However, the set S' of the same circle without a part of its interior ($S' \subset S$) is an example of a non-convex set.

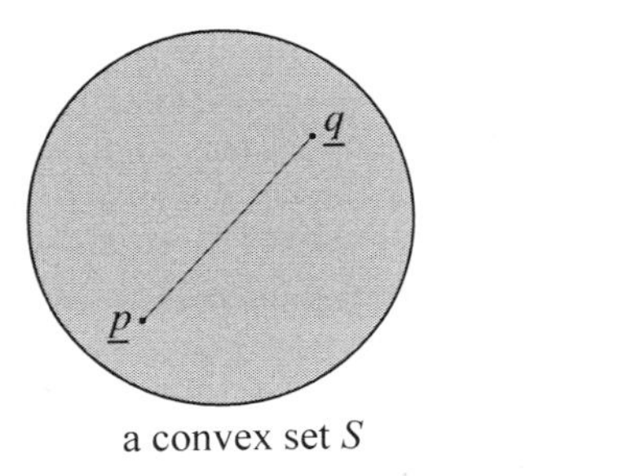

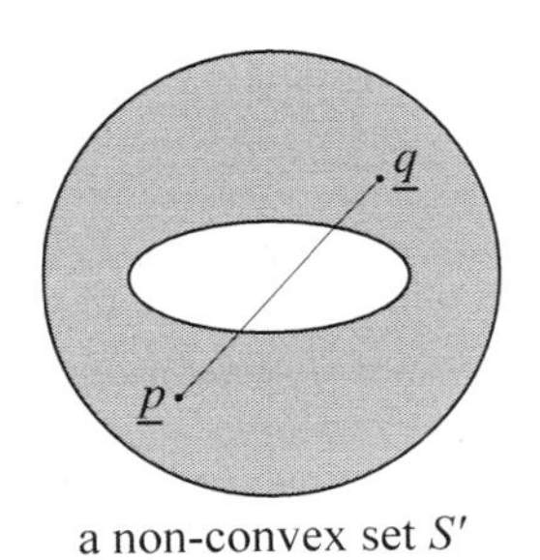

Figure 9.1. A convex set S and a non-convex set S'.

Definition convex and concave function

A function $f(x)$ is said to be *convex* if the line segment drawn between any two points on the graph of $f(x)$ never lies below the graph.
Mathematically: $f(\lambda \underline{a} + (1-\lambda)\underline{b}) \leq \lambda f(\underline{a}) + (1-\lambda) f(\underline{b})$; $0 \leq \lambda \leq 1$

A function $g(x)$ is said to be *concave* if the line segment drawn between any two points on the graph of $g(x)$ will never lie above the graph.
Mathematically: $g(\lambda \underline{a} + (1-\lambda)\underline{b}) \geq \lambda g(\underline{a}) + (1-\lambda) g(\underline{b})$; $0 \leq \lambda \leq 1$

So, for example, the function $f(x) = x^2$ in Figure 9.2 is convex and the function $g(x) = -x^2$ is concave.

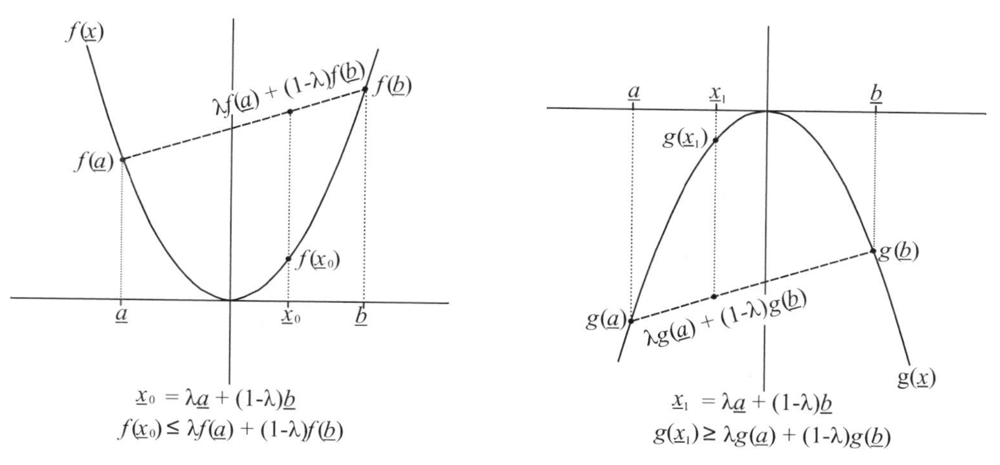

Figure 9.2. A convex function $f(x)$ and a concave function $g(x)$.

All other cases are called neither convex nor concave. Note that a linear function is both convex and concave. Now an important subset of mathematical programming models can be defined which are called *convex programming* problems.

Definition

A mathematical programming model is said to be *convex* if it involves either the minimisation of a convex function or the maximisation of a concave function over a convex set S (Bazaraa *et al.*, 1993).

Within this context it may be obvious that linear programming is a special case of convex programming. The linear objective function which should be either minimised or maximised is both convex and concave. Moreover, it can be shown that the feasible area of a linear programming problem is always convex (Bazaraa *et al.*, 1993). Figure 9.3 is a typical example of a convex set S in linear programming. We have already learned in Chapter 3 that if a linear programming model has an optimal solution, then there is always an optimal basic solution. A basic solution corresponds to a vertex.

If, for example, the convex feasible region is defined by the shaded area in Figure 9.3, then for any linear objective function the Simplex Algorithm determines the optimal vertex in an efficient way (see Chapter 3).

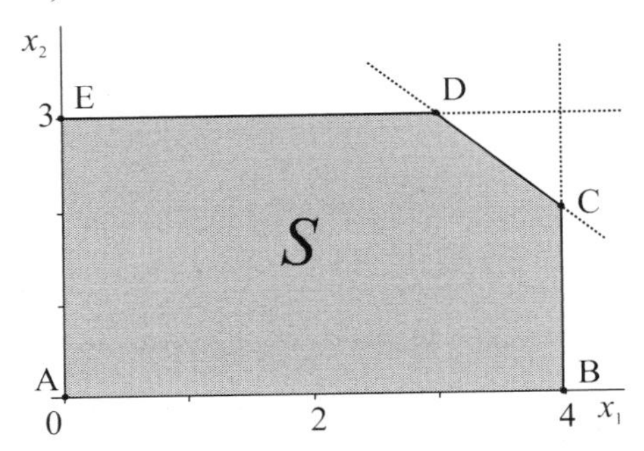

Figure 9.3. A convex feasible set S.

Now, let's consider an example in the class of convex, non-linear programming problems.

Example 9.1. A convex non-linear programming problem.

$$\max \{f(\underline{x}) = -(x_1 - 2)^2 - (x_2 - 2)^2 + 8\} \quad (9.1)$$

$$x_1 \leq 4 \quad (9.2)$$

$$x_2 \leq 3 \quad (9.3)$$

$$x_1 + x_2 \leq 6 \quad (9.4)$$

$$x_1, x_2 \geq 0 \quad (9.5)$$

The function $f(\underline{x}) = -(x_1 - 2)^2 - (x_2 - 2)^2 + 8$ is concave (see Figure 9.4). As the problem involves the maximisation of a concave function over convex set *S*, the problem is said to be convex.

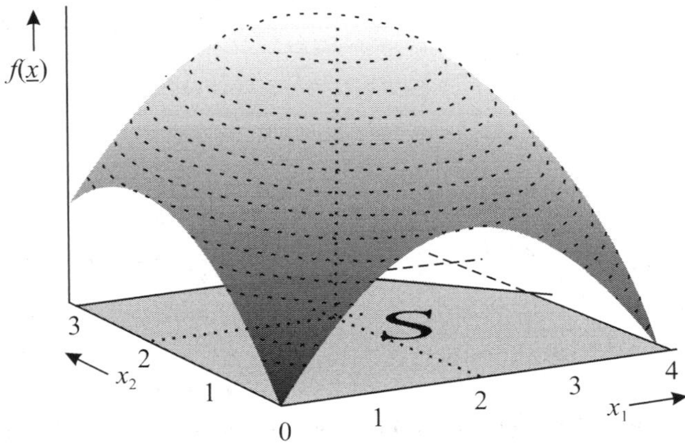

Figure 9.4. Maximisation of $f(\underline{x}) = -(x_1 - 2)^2 - (x_2 - 2)^2 + 8$ over the convex set S.

It is easy to see that the optimal solution for problem (9.1) to (9.5) is $(x_1^*, x_2^*) = (2,2)$. Note that, in contrast to LP, the optimal solution $(x_1^*, x_2^*) = (2,2)$ is not a vertex. In fact, the solution is not even on the boundary of the feasible region but an interior point (see Figure 9.4).

Apparently, if an optimal solution exists for a non-linear programming problem, it may lie 'everywhere' in the feasible area. In fact this observation has a substantial impact on solving non-convex programming problems in particular. If a solution is found, it may only be optimal with respect to the points in a small neighbourhood (local optimum), while the optimal solution (global optimum) is situated elsewhere. The terms *globally* and *locally optimal* are used to make the distinction. Non-linear programming problems appear in many different forms. As opposed to the simplex method for linear programming, no single algorithm exists for solving all these different types of problems. Although sophisticated algorithms have been developed for solving individual classes or special types of non-linear programming problems, the complexity of solving non-convex programming problems is incomparable with solving convex programming problems. The possibility of finding local optima in non-convex programming is what makes such models much more difficult to solve than convex programming models. An intuitive way of describing the situation is to consider the problem of a mountaineer on a range of mountains in a thick fog (Williams, 1990). It is easy for the mountaineer to determine when he is at a local optimum, i.e. a mountain peak. There will be a decrease in height no matter which direction he takes. This, however, does not guarantee that there is no other, higher mountain peak, hidden somewhere in the fog. It is beyond the scope of this chapter to discuss the class of non-linear programming problems extensively. For this we refer to Chapter 12 in this book or Bazaraa *et al.* (1993).

In the next Section 9.3 we focus on modelling techniques related to linear programming. This section also includes several examples of special non-linear models which can be transformed into (continuous) linear programming models. In Section 9.3 we also introduce an extension of linear programming called separable programming.
Separable programming can be used to approximate non-linear terms by piecewise linear functions. The modelling techniques in Section 9.3 are based on the assumption of convex programming. Some modelling techniques for non-convex programming are discussed in Section 9.4.

9.3. Special features for linear programming

In this section several modelling techniques will be discussed which make use of linear programming. Some of them can be helpful in transforming models with special (non-linear) structures into linear programming models. Each section will end with at least one example to which the discussed modelling technique can be applied.

9.3.1. Hard and soft constraints

The following constraint (9.6) is a typical example of a constraint in linear programming:

$$\sum_{i=1}^{I} a_i x_i \leq b \tag{9.6}$$

As a constraint like (9.6) excludes any solution in which the *left-hand-side* $\sum_i a_i x_i$ violates the *right-hand-side* value b, these kinds of constraints are sometimes called *hard constraints*. There are situations in which this is hardly realistic. For example, when it comes to the availability of a raw material in (9.6), it might be worthwhile violating the available inventory level b by purchasing additional raw material at a higher price. In such cases we may decide to replace (9.6) by constraint (9.7) in which a non-negative surplus variable s^+ is introduced.

$$\sum_{i=1}^{I} a_i x_i - s^+ \leq b \tag{9.7}$$

If we associate a positive cost coefficient c_1 with the surplus variable s^+ for a minimisation problem, the upper bound b can be expanded to $b + s^+$ at a cost of $c_1 s^+$ in the objective function. Note that the cost coefficient c_1 for the surplus variable s^+ should be negative in case of a maximisation problem.

In cases in which (9.6) is a greater than or equal constraint ($\geq$), a comparable approach should be followed by introducing a non-negative slack variable s^-

$$\sum_{i=1}^{I} a_i x_i + s^- \geq b \tag{9.8}$$

Once again we associate a positive cost coefficient c_2 with the slack variable s^- for a minimisation problem. Now the lower bound b can be decreased to $b - s^-$ at a cost of $c_2 s^-$ in the objective function. If it involves a maximisation problem the cost coefficient c_2 should be negative.

If (9.6) should be an equality constraint, the allowed deviation in both directions of the right-hand-side value b can be modelled by introducing both a non-negative surplus variable s^+ and a non-negative slack variable s^-.

$$\sum_{i=1}^{I} a_i x_i - s^+ + s^- = b \tag{9.9}$$

Both variables s^+ and s^- should be given appropriate coefficients in the objective function. Note that in the latter case either the surplus variable s^+ or the slack variable s^- or both variables are zero in the optimal solution. After all, the objective function value of any feasible solution in which s^+ and s^- are both positive can be improved by subtracting the smallest value of s^+ and s^- from the largest.

A well-known application of two-sided deviational variables like s^+ and s^- in (9.9) occurs, for example, in curve fitting. The objective function might be to minimise the sum of absolute 'errors' (see Section 9.3.2), but other objective functions are also possible (see Section 9.3.3).

Constraints like (9.7), (9.8) and (9.9) are also called *soft constraints* as the right-hand side value b may be violated at a certain cost.

An interesting application of soft constraints arises when problems with multiple, often conflicting, objectives are considered. For example, we might want to maximise the national income and minimise the emissions of greenhouse gasses simultaneously. One way of tackling these kinds of problems is to set target values for each of the objectives and replace the objectives by soft constraints. The overall objective may be to reach the target values as closely as possible. These types of models are known as *goal programming* models. We refer to Chapter 5 for a general introduction to multiple objective problems and goal programming in particular.

9.3.2. Absolute values

In addition to the preceding Section 9.3.1, it sometimes occurs that a variable x_j represents the deviations of the left-hand-side from the right-hand-side value b of a constraint. The deviations may be positive or negative. The objective might be to minimise the sum totals of the deviations in both directions simultaneously. In such cases *absolute values* can be used. A typical example can be found in *curve fitting* which will be illustrated at the end of Section 9.3.3. Only if absolute values appear as terms in the objective function, it is sometimes possible to reformulate the model as a linear program. When absolute values of variables appear in a general system of equations, it is not possible to reformulate the problem as a linear program (Dantzig, 1997). However, these situations can be modelled using integer programming.

Suppose we define the following problem in which the objective is to minimise the weighted sum of absolute values, subject to a set of linear constraints:

$$\min\left\{\sum_{j=1}^{J} c_j \left| x_j \right|\right\} \qquad (9.10)$$

s.t.

$$\sum_{j=1}^{J} a_{ij} x_j \le b_i \qquad \text{for all } i \qquad (9.11)$$

$$\text{in which } c_j \ge 0 \qquad \text{for all } j$$

Consider the term $w_j = c_j \left| x_j \right|$ of the objective function. With reference to the transformation rules in Section 3.3.3, the free variable x_j can be replaced by:

$$x_j = x_j^+ - x_j^-$$

with $x_j^+ = \max\{x_j, 0\}$ and $x_j^- = \max\{-x_j, 0\}$ in which $x_j^+ \ge 0$ and $x_j^- \ge 0$, implying that x_j^+ takes a value in case x_j is positive, while x_j^- takes a value when x_j is negative. By definition of the variables x_j^+ and x_j^- it follows that:

$$w_j = c_j \left| x_j \right| = c_j(x_j^+ + x_j^-)$$

As $c_j > 0$ it is not possible that the variables x_j^+ and x_j^- are both positive in the optimal solution:
For example, suppose $x_j^+ = 18$ and $x_j^- = 7$ So, $x_j = 18 - 7 = 11$ and $w_j = c_j(18+7) = 25\ c_j$. However, the solution $x_j^+ = 18 - 7 = 11$ and $x_j^- = 7 - 7 = 0$ also results in $x_j = 11$ but now the value of the objective function $w_j = c_j(11+0) = 11\ c_j \leq 25\ c_j$. So, any feasible solution in which x_j^+ and x_j^- are both positive can be improved by subtracting the smallest of x_j^+ and x_j^- from the largest and setting the smallest of x_j^+ and x_j^- to zero. Now, the equivalent linear programming formulation for (9.10) to (9.11) is:

$$\min \sum_{j=1}^{J} c_j (x_j^+ + x_j^-) \qquad (9.10b)$$

s.t.

$$\sum_{j=1}^{J} a_{ij} (x_j^+ - x_j^-) \leq b_i \qquad \text{for all } i \qquad (9.11b)$$

$$x_j^+, x_j^- \geq 0 \qquad \text{for all } j \qquad (9.11c)$$

$$\text{in which } c_j \geq 0 \qquad \text{for all } j$$

An application of absolute values arises, for example, in curve fitting which will be demonstrated in Example 9.2 at the end of Section 9.3.3.

Remarks

1. In case of $\max\{w_j = c_j |x_j|\}$ with $c_j > 0$ the solution $x_j^+ = (11+M)$ and $x_j^- = M$, shows that $w_j = c_j |x_j| = c_j(x_j^+ + x_j^-) = c_j(11+2M)$ is unbounded. So in case of $\max\left\{\sum_{j=1}^{J} c_j |x_j|\right\}$ the values c_j should be less then zero.
2. The transformation of absolute values into equivalent (continuous) linear programming models is restricted to cases in which these terms occur in the objective function (Dantzig, 1997). It should be pointed out that terms like $|x_j|$ in constraints, or more generally the transformation of absolute value functions of two or more variables (either in the objective function or in the constraints), cannot be dealt with by conventional linear programming. Generally speaking, modelling these terms require (mixed) integer linear programming.

9.3.3. Minimax or maximin problems

An objective which requires a maximum to be minimised is called a *minimax* objective. An application of a minimax objective arises, for example, in curve fitting which will be demonstrated at the end of Section 9.3.3 (see Example 9.2). A problem that requires a minimum to be maximised is called a *maximin* problem.

Consider the following minimax model:

$$\min\left\{\max_k \sum_{j=1}^{J} c_{kj}x_j\right\} \tag{9.12}$$

s.t.

$$\sum_{j=1}^{J} a_{ij}x_j = b_i \qquad \text{for all } i \tag{9.13}$$

$$x_j \geq 0 \qquad \text{for all } j \tag{9.14}$$

For example, if $k \in \{1,2\}$ and $j \in \{1,2,3\}$ then the objective function (9.12) is:

$$\min\{\max(c_{11}x_1 + c_{12}x_2 + c_{13}x_3\ ,\ c_{21}x_1 + c_{22}x_2 + c_{23}x_3)\}$$

Note that in Section 3.3.3 it has already been explained that any set of linear constraints can be transformed into the standard form, i.e. a constraint-set like (9.13) and (9.14).
The model (9.12) to (9.14) can be converted into a linear programming model by introducing a variable z which is defined by:

$$z = \max_k \sum_{j=1}^{J} c_{kj}x_j$$

In order to establish this relationship the following set of constraints should be added to the model:

$$\sum_{j=1}^{J} c_{kj}x_j \leq z \qquad \text{for all } k \tag{9.15}$$

The constraints in (9.15) guarantee that z will be greater than, or equal to $\sum_j c_{kj}x_j$ for all k.

By minimising z, the value for this variable will be forced down to the maximum value of all left-hand-sides in (9.15). So, the equivalent linear programming formulation for the model (9.12) to (9.14) is :

$$\min z \tag{9.12a}$$

s.t.

$$\sum_{j=1}^{J} a_{ij}x_j \leq b_i \qquad \text{for all } i \tag{9.13}$$

$$\sum_{j=1}^{J} c_{kj}x_j \leq z \qquad \text{for all } k \tag{9.15}$$

$$x_j \geq 0 \qquad \text{for all } j \tag{9.14}$$

The problem for a maximin objective can be transformed similarly. However, a maximax or minimin objective cannot be transformed by linear programming. This requires (mixed) integer programming.

An example of a minimax objective arises in curve fitting (see Example 9.2). These kinds of objective also occur in Game Theory problems. It should be pointed out that recognizing a minimax or maximin objective is not always trivial in practice.

Example 9.2. Curve fitting.

Linear regression is a well-known statistical method for fitting a straight line through observed data. Suppose we want to fit a straight line through a number of observed data expressed by the points (x_i, y_i) in which $i = 1 \ldots I$. (see Figure 9.5). It is obvious that the points in Figure 9.5 do not lie on a straight line. If a linear relationship is assumed, the question is which line best fits the observed data.

Suppose that an arbitrary line, given by the expression $y = ax + b$, is drawn through the data. The coefficient a is called the *slope* of the desired line and b is called the *intercept*. Now, the problem is to determine the coefficients a and b such that the line $y = ax + b$ fits the data best.

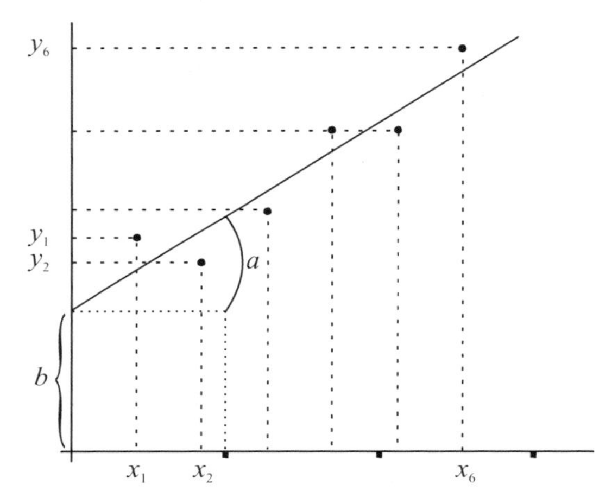

Figure 9.5. Linear regression.

The coefficients a and b can be determined by using the following general model:

$$\min\{f(\underline{z}) = f(z_1, z_2, \ldots, z_I)\} \tag{9.16}$$

s.t.

$$z_i = ax_i + b - y_i \qquad \text{for all } i$$

The variables z_i denote for each observed data point i the deviation between the value of $ax_i + b$ and the observed value y_i. In other words: z_i can be considered as the 'error' in the y-direction. Note that in the above model a and b are the decision variables, whereas x_i and y_i are given data.

Several options exist for the objective function $f(\underline{z})$ in (9.16).

One possible technique is to minimise the maximum 'error', which is also called '*least maximum deviation estimation*'. This in fact is an example of a minimax objective.

For a least maximum deviation estimation in linear regression, the objective function $f(\underline{z})$ in (9.16) should be replaced by the minimax objective (9.16a):

$$\min\left\{\max_i |z_i|\right\} \tag{9.16a}$$

An alternative for the objective function (9.16) is to minimise the sum of absolute 'errors'. This approach is called absolute deviation estimation. For the latter approach the objective function (9.16) takes the form (9.16b):

$$\min\left\{\sum_i |z_i|\right\} \tag{9.16b}$$

It should be pointed out that the fitted lines in both approaches (9.16a) or (9.16b), subject to the constraint-set $z_i = ax_i + b - y_i$ for all i, can be quite different. In case the data contain only a few extreme observations y_i, the objective (9.16b) is appropriate as it is less influenced by extreme points.
In practice, the most commonly used approach is called *least square estimation*. In this approach the objective function (9.16) is replaced by:

$$\min\left\{\sum_i z_i^2\right\} \tag{9.16c}$$

As the objective functions (9.16a) and (9.16b) consist of absolute values, it is not possible to apply linear programming directly. However, in Section 9.3.2 it was demonstrated that non-linear models with absolute value functions like in (9.16a) or (9.16b), can be easily transformed into an equivalent linear programming model. The objective function (9.16c) consists of a non-linear quadratic term. In case we want to solve these kinds of problems by using LP software, we introduce an approximation technique in Section 9.3.6 for solving non-linear terms by piecewise linear programming formulations.

9.3.4. Fractional objectives

Consider the following non-linear programming model (FP) in which the objective function consists of a ratio between two linear terms. These kind of objectives are also called *fractional* or *ratio objectives*. The solution space or set $S := \{\underline{x} \mid A\underline{x} \leq \underline{b}\,;\, \underline{x} \geq \underline{0}\}$ is assumed to be convex, not-empty and bounded, i.e. S is called a *compact set*.

$$\max_{\underline{x}}\left\{\frac{\underline{c}'\underline{x}+c_0}{\underline{d}'\underline{x}+d_0}\right\}$$

s.t. (FP)

$$\underline{x} \in S$$

Charnes and Cooper (1962) introduced an approach in which model (FP) can be transformed into an equivalent linear programming model provided that the denominator $\underline{d}'\underline{x} + d_0$ in the objective function is either positive or negative over the entire set S. So, $\underline{d}'\underline{x} + d_0 \neq 0$. Compared to model (FP) the equivalent linear programming model needs exactly one additional variable and only one additional constraint.

First we define a vector $\underline{w}$ and a scalar t of decision variables:

$$\underline{w} = \frac{\underline{x}}{\underline{d}'\underline{x} + d_0} \tag{9.17}$$

$$t = \frac{1}{\underline{d}'\underline{x} + d_0} \tag{9.18}$$

Note that we assume $\underline{d}'\underline{x} + d_0 \neq 0$. From (9.17) and (9.18) it follows that:

$$\underline{w} = t \cdot \underline{x}$$

and

$$\underline{x} = \frac{\underline{w}}{t}$$

Now, the basic idea is to reformulate model (FP) by means of (9.17) and (9.18) such that a linear programming model arises in terms of the variables $\underline{w}$ and t. If this new model can be solved for all relevant values of t, the solution of any fractional problem derived from (FP) can be found by $\underline{x} = \frac{\underline{w}}{t}$.

Next, the reformulation method will be illustrated for cases in which the denominator in the objective function of (FP) is positive over the entire set S. So, $\underline{d}'\underline{x} + d_0 > 0$ for all $\underline{x} \in S$.

The objective function of the model (FP) can be rewritten as follows:

$$\max_{\underline{x}} \left\{ \frac{\underline{c}'\underline{x} + c_0}{\underline{d}'\underline{x} + d_0} \right\} = \max_{\underline{x}} \left\{ \underline{c}' \cdot \frac{\underline{x}}{\underline{d}'\underline{x} + d_0} + c_0 \cdot \frac{1}{\underline{d}'\underline{x} + d_0} \right\}$$

Using definition (9.17) and (9.18) this can be reformulated as:

$$\max_{\underline{w},t} \left\{ \underline{c}'\underline{w} + c_0 t \right\} \tag{9.19}$$

Using $\underline{x} = \frac{\underline{w}}{t}$ the constraints $A\underline{x} \leq \underline{b}$ of the compact set S can be written as:

$A\underline{x} \leq \underline{b} \Rightarrow A \cdot \underline{w}/t \leq \underline{b} \Rightarrow A\underline{w} \leq \underline{b}\, t$. So,

$$A\underline{w} - \underline{b}t \leq \underline{0} \tag{9.20}$$

Definition (9.18) of the (new) variable t in (9.19) and (9.20) needs to be added:

$$t = \frac{1}{\underline{d}'\underline{x} + d_0} \Rightarrow t(\underline{d}'\underline{x} + d_0) = 1 \Rightarrow \underline{d}'t\underline{x} + t d_0 = 1. \text{ Using } \underline{w} = t \cdot \underline{x}:$$

$$\underline{d}'\underline{w} + t d_0 = 1 \tag{9.21}$$

Note that we assumed $\underline{d}'\underline{x} + d_0 > 0$. So, $t > 0$.

Reformulating the non-negative constraints $\underline{x} \geq \underline{0}$ of the set S in terms of the variables $\underline{w}$ and t using $\underline{w} = t \cdot \underline{x}$ and $t > 0$ gives:

$$\underline{x} \geq \underline{0} \Rightarrow \underline{w}/t \geq \underline{0}, \; t > 0. \quad \text{So,}$$

$$\underline{w} \geq \underline{0}, \; t > 0 \tag{9.22a}$$

In order to get a linear programming model we suppose $t \geq 0$. This is allowed because the set $S{:} = \{\underline{x} \mid A\,\underline{x} \leq \underline{b}\,;\, \underline{x} \geq \underline{0}\}$ is a compact set. Now, the following linear programming model is equivalent to the original fractional model (FP), provided that $\underline{d}'\underline{x} + d_0 > 0$.

$$\max_{\underline{w},t} \{\underline{c}'\underline{w} + c_0 t\} \tag{9.19}$$

s.t.

$$A\underline{w} - \underline{b}\,t \leq \underline{0} \tag{9.20}$$

$$\underline{d}'\underline{w} + t d_0 = 1 \tag{9.21}$$

$$\underline{w} \geq \underline{0}, \; t \geq 0 \tag{9.22b}$$

Note that $t \geq 0$ in (22.b) is just for the form's sake of linear programming. If $\underline{d}'\underline{x} + d_0 > 0$, $\underline{x} \in S$, S is compact and $t > 0$ (see 9.18), then the optimal value $t = t^* > 0$ for problem (9.19) - (9.21a) will also satisfy (9.22b).

Remark for $\underline{d}'\underline{x} + d_0 < 0$:

Until now we assumed the denominator $\underline{d}'\underline{x} + d_0$ in problem (FP) to be positive for all $\underline{x} \in S$. If this term is strictly negative ($\underline{d}'\underline{x} + d_0 < 0$ for all $\underline{x} \in S$), the transformation principle can be applied too. In that case, the vectors $(\underline{c}', c_0)$ and $(\underline{d}', d_0)$ of the fractional model (FP) can be replaced by $(-\underline{c}', -c_0)$ and $(-\underline{d}', -d_0)$. This in turn leaves the ratio $\frac{\underline{c}'\underline{x} + c_0}{\underline{d}'\underline{x} + d_0}$ unaltered for all $\underline{x} \in S$. Moreover, the replacement of $(\underline{d}', d_0)$ by its negative makes the denominator of the ratio positive.

In summary, any finite maximum of the non-linear fractional program (FP) can be found by solving one of the LP models in Table 9.1. For convenience we repeat the general model (FP):

$$\max_{\underline{x}} \left\{ \frac{\underline{c}'\underline{x} + c_0}{\underline{d}'\underline{x} + d_0} \right\}$$

s.t. (FP)

$$\underline{x} \in S$$

in which $S := \{\underline{x} \mid A\underline{x} \leq \underline{b}\ ;\ \underline{x} \geq \underline{0}\}$ is a compact set.
The sign of the denominator $\underline{d}'\underline{x} + d_0$ determines which of the LP models in Table 9.1 is equivalent to the model (FP).

Table 9.1. Transformation principle for fractional objectives.

Denominator $\underline{d}'\underline{x} + d_0 > 0$	Denominator $\underline{d}'\underline{x} + d_0 < 0$
$\max\limits_{\underline{w},t}\{\underline{c}'\underline{w} + c_0 t\}$	$\max\limits_{\underline{w},t}\{-\underline{c}'\underline{w} - c_0 t\}$
s.t.	s.t.
$A\underline{w} - \underline{b}t \leq \underline{0}$	$A\underline{w} - bt \leq \underline{0}$
$\underline{d}'\underline{w} + t d_0 = 1$	$-\underline{d}'\underline{w} - t d_0 = 1$
$\underline{w} \geq \underline{0}\ ,\ t \geq 0$	$\underline{w} \geq \underline{0}\ ,\ t \geq 0$

The term $\underline{d}'\underline{x} + d_0$ in problem (FP) should be either strictly positive or strictly negative for all possible values of $\underline{d}$ and d_0. If $(\underline{d}', d_0)$ is not either strictly positive or strictly negative, there exists a solution for the non-negative variables $\underline{x}$ for which the denominator $\underline{d}'\underline{x} + d_0 = 0$. In such cases the (transformed) problem is obviously not defined. A simple check can be introduced previous to any run of the (transformed) model in order to get an insight into the values of $\underline{d}'\underline{x} + d_0$.
It should be stated that the transformation principle can also be applied in case (FP) concerns a minimisation model.

Although fractional objectives can arise in several practical environments (for example, turnover ratios or productivity ratios) an interesting application of ratio objectives arises in *Data Envelopment Analysis* (*DEA*). DEA was developed by Charnes *et al.* (1978) and extended by Banker *et al.* (1984).

9.3.5. Range constraints

On a regular basis it is necessary, often even useful, to model both an upper bound U and a lower bound L on a constraint set $\underline{L} \leq A_2\underline{x} \leq \underline{U}$. The most obvious way to model this is to specify two separate subsets of constraints like (9.23) and (9.24) in the following general example (RP).

$$\max\{\underline{c}'\underline{x}\}$$

s.t. (RP)

$$A_1\underline{x} \leq \underline{b}$$

$$A_2\underline{x} \leq \underline{U} \qquad (9.23)$$

$$A_2\underline{x} \geq \underline{L} \qquad (9.24)$$

$$\underline{x} \geq \underline{0}$$

Together, the constraint-sets (9.23) and (9.24) are called *range constraints*. For reasons of efficiency and convenience we will discuss an alternative way to model range constraints.

A more compact and elegant formulation is to specify only the constraints in (9.23), together with new (slack) variables $\underline{y} \geq \underline{0}$:

$$A_2\underline{x} + \underline{y} = \underline{U} \tag{9.25}$$

Next the following bound is imposed upon $\underline{y}$:

$$\underline{0} \leq \underline{y} \leq \underline{U} - \underline{L} \tag{9.26}$$

It may be obvious that if, for some i, $y_i = 0$ in (9.26), the corresponding term $\underline{a}_i'\underline{x}$ in (9.25) is equal to the upper bound U_i. If $y_i = U_i - L_i$ in (9.26), the term $\underline{a}_i'\underline{x}$ in (9.25) is equal to the lower bound L_i. It should be pointed out that the (revised) simplex algorithm in most computer packages can cope with simple bounds on variables like (9.26) algorithmically.

So, an equivalent formulation for problem (RP) is the following model (RP_2):

$$\begin{aligned} &\max\ \{\underline{c}'\underline{x}\} \\ &\text{s.t.} \qquad\qquad\qquad\qquad (RP_2) \\ &\quad A_1\underline{x} \leq \underline{b} \\ &\quad A_2\underline{x} + \underline{y} = \underline{U} \\ &\quad \underline{y} \leq \underline{U} - \underline{L} \\ &\quad \underline{x},\ \underline{y} \geq \underline{0} \end{aligned}$$

Example 9.3.

Suppose $L_i \leq \underline{a}_i'\underline{x} \leq U_i$. takes the form: $2 \leq 2x_1 + 3x_2 \leq 7$. Instead of defining two separate constraints in problem (RP):

$$2x_1 + 3x_2 \leq 7$$
$$2x_1 + 3x_2 \geq 2$$

We can also specify just one constraint like (9.25) and set a simple bound on the slack variable y (see (9.26)).

$$2x_1 + 3x_2 + y = 7$$
$$0 \leq y \leq 7 - 2 = 5$$

Instead of expressing (9.26) as constraints in a model it is more efficient to regard these variables as having an upper bound. Other reasons in favour of model formulation (RP_2) are:

- The values of the slack variables $\underline{y}$ in (9.25) are limited to an upper bound of $\underline{U} - \underline{L}$. The slack variables in (9.23) of problem (RP), which will be automatically introduced after the problem is transformed into standard form, will have a (weaker) upper bound of $\underline{U}$.
- The problem size (i.e. the number of rows and columns) of problem (RP_2) in standard form is less than the problem size of problem (RP) in standard form.

- As the same type of constraints in (9.25) is specified just once, they must be modified only in one place of the model in case any change occurs.

9.3.6. Piecewise linear formulations in convex programming

The most essential feature of linear programming is that all relations between the variables are linear. In the preceding sections it was shown that some models with special non-linear features can be transformed into linear programming models. This Section 9.3.6 will focus on a surprisingly wide class of non-linear programming models which can be modelled by an extension of linear programming called *separable programming*. The main advantage of separable programming is that non-linear terms can be approximated by linear ones, resulting in either a linear programming model or a (mixed) integer linear programming model. It is the distinction between convex and non-convex programming (see Section 9.2) that determines which type of model formulation is applicable. The application of *piecewise linear model formulations* for non-linear terms in non-convex programming will be discussed in Section 9.4.6. In this Section 9.3.6 the focus is on piecewise linear formulations in convex programming. First we introduce *separable functions* as a prerequisite for separable programming. Next we introduce piecewise linear approximations for non-linear separable functions and the concept of separable programming. For modelling piecewise linear approximations there are two alternative formulation methods, i.e. the *λ-formulation)* and the *δ-formulation*. Both methods will be discussed.

Definition separable functions

A separable function is a function in which all terms consist of functions of single variables. So, $f(x_1, x_2, \ldots, x_n) = g_1(x_1) + g_2(x_2) + \ldots + g_n(x_n)$.

The function $f_1(\underline{x}) = \dfrac{1}{x_1} + \sqrt{x_2} + \ln x_3 - e^{x_3}$ is an example of a separable function since

$f_1(\underline{x}) = f_1(x_1, x_2, x_3) = g_1(x_1) + g_2(x_2) + g_3(x_3)$
for $g_1(x_1) = \dfrac{1}{x_1}$, $g_2(x_2) = \sqrt{x_2}$ and $g_3(x_3) = \ln x_3 - e^{x_3}$

However, the function $f_2(\underline{x}) = \dfrac{x_1}{1+x_2} + x_1x_3$ is not separable since

$h_1(x_1, x_2) = \dfrac{x_1}{1+x_2}$ and $h_2(x_1, x_3) = x_1 x_3$ are both functions of more than one variable.

The importance of separable functions in mathematical programming is that non-linear terms can be approximated by piecewise linear functions. Although the class of separable functions might seem to be restrictive, it should be pointed out that it is sometimes possible to transform non-separable terms into separable functions. A well-known example is the elimination of products of variables. For example: $f(x_1, x_2, x_3) = x_1x_2x_3^2$ can be rewritten into $h(x_1, x_2, x_3) = \log f = \log x_1 + \log x_2 + 2\log x_3$. An alternative way of dealing with product terms will be discussed in Section 9.6.

Piecewise linear approximations

Now we introduce the concept of separable programming. It will become apparent that only models within the class of convex programming problems can be solved satisfactorily by (continuous) piecewise linear approximations. Non-convex separable problems should be dealt with by (mixed) integer linear programming. These problems are (mostly) harder to solve. This will be described in Section 9.4.6.

Consider the convex programming problem (CP):

$$\max\{f(\underline{x}) = -x_1^2 + 4x_1 + 2x_2\}$$

$$\begin{aligned} x_1 \quad & \leq 4 \\ x_2 & \leq 3 \qquad\qquad \text{(CP)} \\ x_1 + x_2 & \leq 6 \\ x_1 \,,\, x_2 & \geq 0 \end{aligned}$$

In order to convert this non-linear problem into a suitable form for separable programming it is necessary to make a piecewise linear approximation to each non-linear (separable) function. Problem (CP) consists of only one non-linear function to be approximated: $f_1(x_1) = -x_1^2 + 4x_1$. Figure 9.6 illustrates a piecewise linear approximation of the function $f_1(x_1)$. The graph of the function $f_1(x_1)$ is approximated by the piecewise linear (concave) function $\varphi_1(x_1)$. For that purpose the curve $f_1(x_1)$ is divided into four 'pieces' by the points *O*, *A*, *B*, *C* and *D*. The approximation function $\varphi_1(x_1)$ consists of the line segments *OA*, *AB*, *BC* and *CD*.

The points *O*, *A*, *B*, *C* and *D* of the piecewise linear curve $\varphi_1(x_1)$ are called breakpoints. Note that the domain for the functions $f_1(x_1)$ and $\varphi_1(x_1)$ ends for $x_1 = 4$ according to the constraint $x_1 \leq 4$ in problem (CP). It may be obvious that the piecewise linear function $\varphi_1(x_1)$ introduces some inaccuracy. For example, if $x_1 = 2.5$ the function value $f_1(2.5) = 3.75$ while the transformed function has a value of $\varphi_1(2.5) = 3.5$. This inaccuracy can be reduced by introducing more breakpoints involving more (straight) line segments.

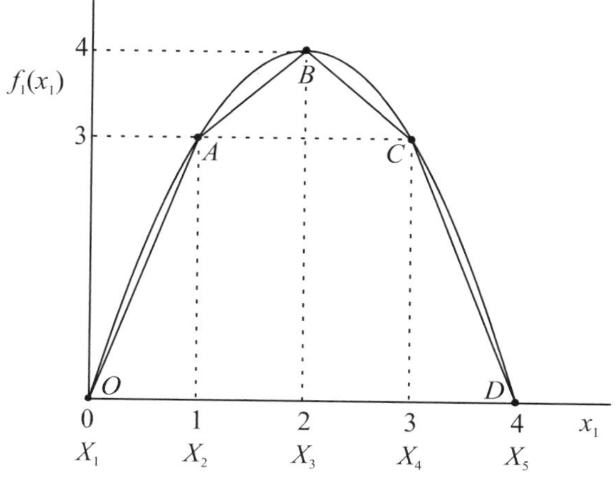

Figure 9.6. A piecewise linear approximation $\varphi_1(x_1)$ of the function $f_1(x_1) = -x_1^2 + 4x_1$.

λ-formulation

Now the model formulation (CP) has to be changed in order to express this approximation mathematically. A well-known method for doing so is called the '*λ-formulation')*. This method will be described below.

Let the capital letters X_1, X_2, X_3, X_4 and X_5 denote the x_1-coordinates of the five breakpoints and let $f_1(X_1)$, $f_1(X_2)$, $f_1(X_3)$, $f_1(X_4)$ and $f_1(X_5)$ denote the corresponding function values $f_1(X_i)$ for all i. So, $f_1(X_1) = \varphi_1(0) = 0$, $f_1(X_2) = \varphi_1(1) = 3$, $f_1(X_3) = \varphi_1(2) = 4$, $f_1(X_4) = \varphi_1(3) = 3$ and $f_1(X_5) = \varphi_1(4) = 0$. Now, any point x_1 and the corresponding function value $\varphi_1(x_1)$ can be represented as a weighted sum of two adjacent points X_i and $\varphi_1(X_i)$ respectively. For instance $x_1 = 0.6$ between $X_1 = 0$ and $X_2 = 1$ can be written as $x_1 = 0.4X_1 + 0.6X_2$. The corresponding function value $\varphi_1(x_1) = \varphi_1(0.6) = 0.4\varphi_1(0) + 0.6\varphi_1(1) = 1.8$.

Now, define λ_1, λ_2, λ_3, λ_4 and λ_5 as the (nonnegative) weighting or interpolation variables such that their sum is equal to 1. Then the piecewise linear approximation $\varphi_1(x_1)$ of the function $f_1(x_1)$ in Figure 9.6 can be formulated by (9.27) to (9.30):

$$x_1 = \lambda_1 X_1 + \lambda_2 X_2 + \lambda_3 X_3 + \lambda_4 X_4 + \lambda_5 X_5 \quad (9.27)$$

$$\varphi_1(x_1) = \lambda_1 f_1(X_1) + \lambda_2 f_1(X_2) + \lambda_3 f_1(X_3) + \lambda_4 f_1(X_4) + \lambda_5 f_1(X_5) \quad (9.28)$$

$$\lambda_1 + \lambda_2 + \lambda_3 + \lambda_4 + \lambda_5 = 1, \text{ moreover} \quad (9.29)$$

it is necessary to add another constraint related to the new variables λ_i :

at most two adjacent λ_i may become non-zero (9.30)

Restriction (9.30) guarantees that an approximation with the coordinates $(x_1, \varphi_1(x_1))$ will always lie on one of the line segments *OA, AB, BC* or *CD* and not on a line segment between any other pair of breakpoints. The necessity of (9.30) for $x_1 = 1.5$ is illustrated in Figure 9.7.

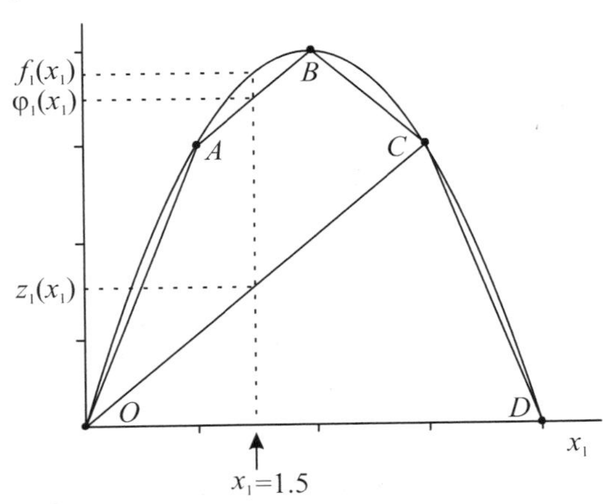

Figure 9.7. Piecewise linear approximation of $f_1(x_1)$ by $\varphi_1(x_1)$ and $z_1(x_1)$.

It is obvious that the approximation for $f_1(x_1)$ should be $\varphi_1(x_1)$ on the line segment AB ($\lambda_2 = \frac{1}{2}$; $\lambda_3 = \frac{1}{2}$; adjacent λ_i) and $f_1(x_1)$ should not be approximated by, for example, $z_1(x_1)$ on the line segment OC for $\lambda_1 = \frac{1}{2}$ and $\lambda_4 = \frac{1}{2}$ (not adjacent λ_i).

Unfortunately restriction (9.30) cannot be modelled using linear programming. It takes the introduction of binary variables to rewrite this verbal restriction mathematically. This will be described in Section 9.4.6.
However, there are situations in which constraint (9.30) may be omitted because the properties of a specific class of problems are such that the optimal solution will automatically fulfil (9.30). To be more precise, if the separable model formulation belongs to the class of convex programming problems, i.e.:

- the minimisation of a convex function; or
- the maximisation of a concave function.

over a convex region of space, condition (9.30) need not be imposed explicitly.

The property of convexity holds for our problem (CP). In this problem the concave function $f_1(x_1) = -x_1^2 + 4x_1$ needs to be maximised. Figure 9.7 illustrates that for any value of x_1 the optimisation automatically tends to choose for one of the line segments OA, AB, BC or CD of $\varphi_1(x_1)$ instead of a line segment between any other pair of breakpoints.
To summarise: the piecewise linear model formulation for the convex programming problem (CP) is:

$$\max \{\varphi_1(x_1) + 2x_2\}$$

s.t.

$$\begin{aligned}
x_1 \quad\quad & \leq 4 \\
x_2 & \leq 3 \\
x_1 + x_2 & \leq 6 \\
x_1 &= \sum_{i=1}^{5} \lambda_i X_i \\
\varphi_1(x_1) &= \sum_{i=1}^{5} \lambda_i f_1(X_i) \\
\sum_{i=1}^{5} \lambda_i &= 1 \\
x_1, x_2, \lambda_1, \lambda_2, \ldots, \lambda_5 &\geq 0
\end{aligned}$$

In case of convex programming it does not matter whether the non-linearities appear in the objective function, in the constraints or both. If the original non-linear problem is convex and the non-linear functions are separable functions, a piecewise linear approximation can be modelled by linear programming. However, it might not be easy to decide whether a problem is convex or not.

In case the problem turns out to be non-convex, a piecewise linear approximation of the original problem cannot be modelled by linear programming but (mixed) integer linear programming should be used (see Section 9.4.6).

δ-formulation

It should be mentioned that an approximation of a separable function like $f_1(x_1)$ by a piecewise linear curve $\varphi_1(x_1)$ can be modelled in several ways. From a computational point of view the so-called δ-*formulation* is usually considered to be more efficient (Williams, 1990). This explains why computer packages with a module for separable programming are mostly designed to work with the δ-formulation.

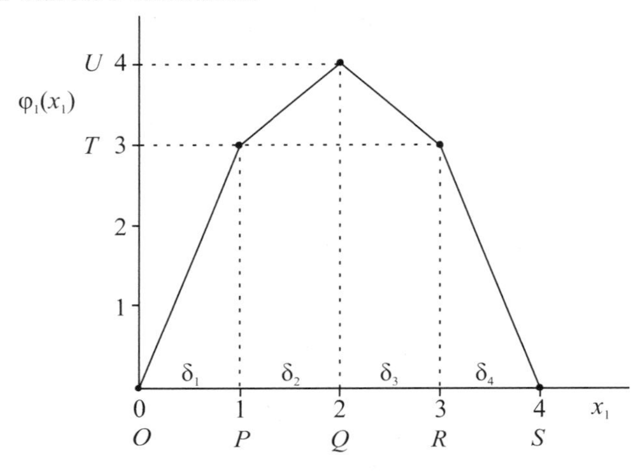

Figure 9.8. Piecewise linear function $\varphi_1(x_1)$.

The δ-formulation will be demonstrated for the original, non-linear problem (CP). The piecewise linear approximation $\varphi_1(x_1)$ of the function $f_1(x_1) = x_1^2 + 4x_1$ is redrawn in Figure 9.8.

Now the variables δ_1, δ_2, δ_3 and δ_4 are defined to represent proportions of the intervals *OP, PQ, QR* and *RS* . The length of each interval in the x_1 direction (in this example all 1) can be used to construct any value for x_1. So,

$$x_1 = 1\delta_1 + 1\delta_2 + 1\delta_3 + 1\delta_4 \qquad (9.31)$$

in which

$$0 \leq \delta_1, \delta_2, \delta_3, \delta_4 \leq 1 \qquad (9.32)$$

Note that in this example the line segments *OP, PQ, QR* and *RS* are all of length 1 which implies that the coefficients in (9.31) for each of the variables δ_i for $i = 1... 4$ are 1. The values of these coefficients may of course be different in other cases.

Similarly, the slope of $\varphi_1(x_1)$ on these intervals are 3, 1, −1 and −3 respectively (see Figure 9.8). Looking in the $\varphi_1(x_1)$ direction, the slope of $OT = 3$ and $TU = 1$. Once we reach the maximum value $\varphi_1(x_1) = 4$, the value of $\varphi_1(x_1)$ will decrease in the interval $UT = -1$ and $TO = -3$. So, any value of $\varphi_1(x_1)$ can be found by:

$$\varphi_1(x_1) = 3\delta_1 + 1\delta_2 - 1\delta_3 - 3\delta_4 \qquad (9.33)$$

Moreover, we have to ensure that every possible approximation for $f_1(x_1)$ lies on the piecewise linear function $\varphi_1(x_1)$ (see Figure 9.8). This can be done by adding the following condition (correspondingly condition (9.30) in the λ-formulation):

if any $0 < \delta_i < 1$ is non-zero all preceding δ_i must take the value of 1
and all succeeding δ_i must take a value of 0 (9.34)

Example 9.4.

Suppose $x_1 = 2.5$ then according to (9.31), (9.32) and (9.34):

$$\delta_1 = \delta_2 = 1,\quad \delta_3 = \tfrac{1}{2} \text{ and } \delta_4 = 0$$

According to (9.33):

$$\varphi_1(2.5) = (3 \cdot 1) + (1 \cdot 1) - (1 \cdot ½) - (3 \cdot 0) = 3\tfrac{1}{2}$$

Note that all variables δ_i in (9.32) have an upper bound of 1. Instead of expressing (9.32) as constraints in a model, most computer packages can cope with these kinds of (simple) bounds on variables, algorithmically.

The purpose and necessity of condition (9.34) in the δ-formulation corresponds to condition (9.30) in the λ-formulation. This also implies that (9.34) may be omitted for convex programming problems.

If the problem turns out to be non-convex, we refer to Section 9.4.6.

9.4. Special features for (mixed) integer programming

It turns out that a wide class of problems can be modelled by using integer variables $y_j \in \{0, 1, 2, \ldots, n\}$ for all j. Such problems are called integer (linear) programming problems. They can be divided into the class of:

1. pure integer programming models (*IP*), in which all the decision variables are restricted to integer values, and
2. the class of mixed integer programming models (*MIP*) consisting of both continuous and integer decision variables.

It is beyond dispute that the flexibility in building (*M*)*IP* models is much greater than in building linear programming models. The versatility of (*M*)*IP* in applications originates from

the fact that in many practical problems activities and resources like transportation vehicles or production facilities are indivisible. Moreover, many problems require typical 'yes or no' decisions. These kinds of decisions can be modelled by binary variables, i.e. integer variables that are restricted to take only 0 – 1 values.
In the first part of Section 9.4 (i.e. 9.4.1 to 9.4.5) the emphasis is on modelling techniques that demonstrate the power of using integer variables as a method of modelling.
The last part of Section 9.4.6 discusses modelling techniques which can be helpful to transform some models with special (non-linear) structures into (mixed) integer linear programming models.

9.4.1. Semi-continuous variables

Consider a linear programming model with the additional condition that the value of a decision variable x should be either zero or if x is positive, the value should be greater than or equal to some specified lower bound L. The variable x is called a *semi-continuous variable*. So,

$$x = 0 \quad \text{or} \quad x \geq L \tag{9.35}$$

Condition (9.35) can be interpreted as two constraints which cannot hold simultaneously.

Modelling condition (9.35) requires the introduction of a binary variable $y \in \{0, 1\}$. Sometimes these kinds of variables y are also called *indicator variables*. This appellation appears from the next definition:

$$y = \begin{cases} 0 & \text{if} \quad x = 0 \\ 1 & \text{if} \quad x \geq L \end{cases}$$

After defining a sufficiently large upper bound M for x, the binary variable y can be linked to x by the following two constraints (9.36) and (9.37):

$$x \leq M \cdot y \tag{9.36}$$

and

$$x \geq L \cdot y \tag{9.37}$$

$$y \in \{0, 1\}$$

Constraint (9.36) and (9.37) implies that if $y = 0$, the variable $x = 0$. From both constraints (9.36) and (9.37) we can see that if $y = 1$, it implies that $L \leq x \leq M$.

Note that the value of M should be sufficiently large. However, for numerical reasons this value should not be 'too' large. From a computational point of view it is wise to set the value of M as small as possible.

It should be mentioned that in (most) computer packages condition (9.35) can also be dealt with algorithmically through the method of integer programming. The two possibilities of condition (9.35) correspond to a branch in a solution tree as demonstrated in Figure 9.9 in which P_k is defined as sub-problem P in node k of the search-tree:

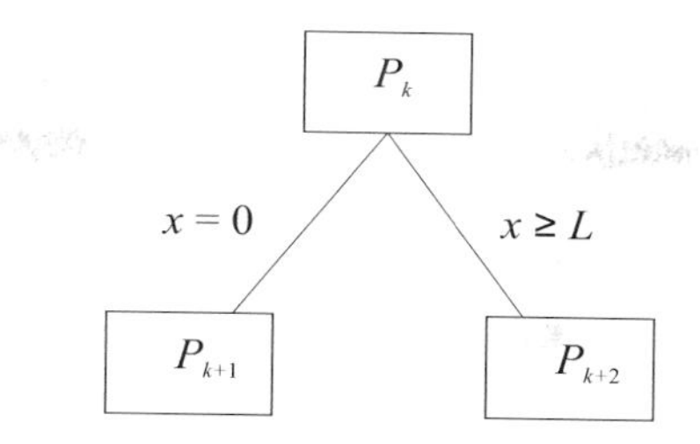

Figure 9.9. Branching scheme for a semi-continuous variable.

The use of semi-continuous variables may arise, for example, in batch-size problems. In these kinds of problems the model includes variables that represent the amounts of used raw material(s). The transportation of raw materials to a factory may require that whenever a raw material is ordered, a minimum batch size is delivered to a specified lower bound. Another example occurs when high set-up costs are associated with manufacturing products. A production batch can be produced but only at amounts greater than or equal to the specified lower bound L.

9.4.2. Discrete valued variables

In addition to the preceding Section 9.4.1, it sometimes occurs that variables are restricted to take values in a specified discrete set $\{\beta_1, \ldots, \beta_k)$ in which $\beta_j \in \mathbb{N}^+$. These kinds of variables are called *discrete valued variables*. Modelling the condition that a variable $x \in \{\beta_1, \ldots, \beta_k\}$ needs the introduction of binary variables $y_j \in \{0, 1\}$ for all j:

$$y_j = \begin{cases} 1 & \text{if } \; x = \beta_j \in \{\beta_1, \ldots, \beta_k\} \\ 0 & \textit{otherwise} \end{cases}$$

The binary variables y_j can be linked to the variable x as follows:

$$x - (\beta_1 y_1 + \ldots + \beta_k y_k) = 0 \tag{9.38}$$

$$y_1 + \ldots + y_k = 1 \tag{9.39a}$$

$$y_j \in \{0, 1\} \qquad \text{for all } j \tag{9.39b}$$

Constraint (9.39a) together with the conditions (9.39b) ensure that only one variable y_j of the set $\{y_1, \ldots, y_k\}$ will get a value of 1. Constraint (9.38) implies that the value for the variable x takes one value within the set $\{\beta_1, \ldots, \beta_k\}$.

The use of discrete valued variables may arise, for example, in designing pipeline distribution networks. One (or more) of the variables denotes the diameter of the pipes used. Pipes are usually only available in a restricted number of diameters, for example, the set (6 mm, 8 mm, 12 mm, 24 mm}. Another example might be in manufacturing car engines. The cubic capacity of engines is often restricted to a predefined set, for example, the set {1300 cc, 1600 cc, 1800 cc, 2000 cc}.

9.4.3. Either-or constraints

In the preceding Section 9.4.2 we discussed semi-continuous variables in which $x_j = 0$ or $x_j \geq L$ has to hold. In fact these kinds of conditions can be considered as the simplest example of a much wider class of conditions which are also known as '*either-or constraints*'. In this section we will generalize the concept of either-or constraints to subsets of constraints.

Consider the following model:

$$\max\left\{\sum_{j=1}^{J} c_j x_j\right\}$$

s.t.

$$\sum_{j=1}^{J} a_{1j} x_j \leq b_1 \qquad (1)$$

$$\vdots \qquad \vdots \qquad \vdots \qquad\qquad \vdots$$

$$\sum_{j=1}^{J} a_{rj} x_j \leq b_r \qquad (r)$$

at least k of the constraints (1) to (r) must hold ($k < r$) $\qquad (r+1)$

$x_j \geq 0$ for all j

Constraint (r + 1) implies that not all constraints of the model have to hold. It is crucial to realise that in a general linear programming model all constraints must hold. As a result we need integer programming to model constraint (r + 1) mathematically.

Again, we introduce the binary (indicator) variables $y_i \in \{0, 1\}$ for all i and sufficiently large upper bounds $M_1, \ldots, M_r$ for the left-hand sides of each of the constraints (1) to (r). In other words: the left-hand side ($\sum_j a_{ij} x_j$) of a constraint i is always smaller than $b_i + M_i$.

Now the requirement in which at least k of the constraints (1) to (r) must hold is equivalent to:

$$\sum_{j=1}^{J} a_{1j} x_j \leq b_1 + M_1\, y_1$$

$$\vdots \qquad \vdots \qquad \vdots$$

$$\sum_{j=1}^{J} a_{rj} x_j \leq b_r + M_r\, y_r$$

$$\sum_{i=1}^{I} y_i = r - k$$

$$x_j \geq 0; \quad y_i \in \{0, 1\} \qquad \text{for all } i, j$$

An example of either-or constraints might occur in situations where a product has to be produced on at least k different places. Another example might be manufacturing processes where two (or more) ways of operation are available.

9.4.4. If – Then constraints

In many applications, situations occur in which we want to ensure that if a constraint $f(\underline{x}) > 0$ is satisfied, then the constraint $g(\underline{x}) \geq 0$ must be satisfied too. On the other hand, if $f(\underline{x}) > 0$ is not satisfied, then constraint $g(\underline{x}) \geq 0$ may or may not be satisfied. These types of requirements can be modelled using integer programming and are called *If – Then constraints* (Winston, 2004).

Modelling these conditions requires the introduction of a binary (indicator) variable $y \in \{0, 1\}$ and a sufficiently large upper bound M. The following constraints must be added to the model formulation:

$$f(\underline{x}) \leq M(1-y) \qquad (9.40a)$$

$$-g(\underline{x}) \leq My \qquad (9.40b)$$

$$y \in \{0, 1\}$$

The value for M must be chosen such that the left-hand-sides $f(\underline{x})$ and $-g(\underline{x})$ of the constraints (9.40a) and (9.40b) are always smaller than M for all possible values of $\underline{x}$ that satisfy the other constraints in the problem.
If $f(\underline{x}) > 0$ is satisfied, then according to (9.40a) the binary variable $y = 0$. Consequently, if $y = 0$ constraint (9.40b) implies that $-g(\underline{x}) \leq 0$ or $g(\underline{x}) \geq 0$ which is in fact the desired result. If $f(\underline{x}) > 0$ is not satisfied (i.e. $f(\underline{x}) \leq 0$), then constraint (9.40a) allows $y = 1$ or $y = 0$. These two possible outcomes for the binary variable y also imply that $g(\underline{x}) < 0$ and $g(\underline{x}) \geq 0$ are both possible (see (9.40b)).

9.4.5. Selection problems

Another broad class of frequently occurring problems are related to *selection problems*. The diversity of the *plant location problem* (also called the *facility location* or *warehouse location problem*) is a well-known example of this problem class.

Multiple choice problems

Consider a problem in which various activities j are considered. Suppose x_j denotes the level at which activity j is performed. The variables x_j have to be non-negative and satisfy some linear constraints. Suppose, from the set of activities $j = 1 \ldots J$ at most k activities may be performed.

Define the variables

$$y_j = \begin{cases} 1 & \text{if} \quad x_j > 0 \qquad \text{(i.e. activity } j \text{ is performed)} \\ 0 & \text{otherwise} \end{cases}$$

Moreover, let M be a large positive number that represents an upper bound on the level of any activity. Now, the constraints that at most k of the $j = 1 \ldots J$ activities may be performed can be modelled as follows:

$$x_j \leq M\,y_j \qquad \text{for all } j$$

$$\sum_{j=1}^{J} y_j \leq k$$

$$x_j \geq 0,\; y_j \in \{0,1\} \qquad \text{for all } j$$

Fixed costs problems

Another very common application and extension of multiple choice problems are *fixed costs or fixed charge problems*.

The decision variables in linear programming models usually represent the level at which an activity j is performed ($x_j \geq 0$ for all j). It is assumed that the objective function, which will be either minimised or maximised, is a linear function $f(x_j) = c_j\, x_j$ of the continuous variables x_j. However, in many practical problems the costs of performing an activity involves a fixed cost F_j too. The threshold costs $F_j > 0$ are also called set-up or fixed charge costs. Now, the total costs $Q(x_j)$ of performing activity j are defined as:

$$Q(x_j) = \begin{cases} 0 & \text{if } \; x_j = 0 \\ F_j + c_j x_j & \text{if } \; x_j > 0 \end{cases} \qquad \text{for all } j$$

The discontinuous function $Q(x_j)$ is illustrated in Figure 9.10 and can be regarded as non-linear. Hence, it is not possible to model this situation using linear programming.

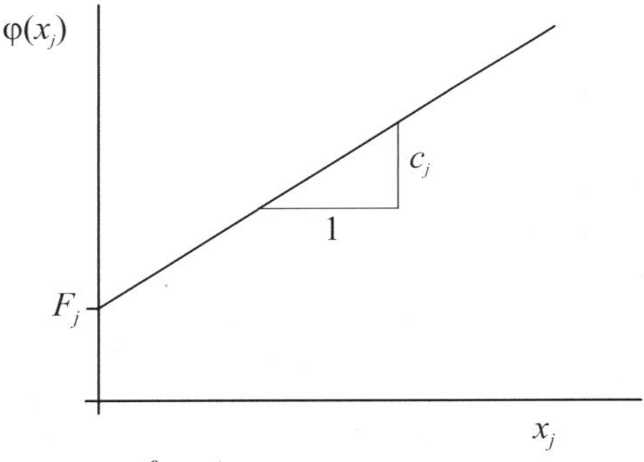

Figure 9.10. A discontinuous cost function.

However, the fixed cost problem can be modelled quite easily using integer programming. Just as in the previous section we define binary indicator variables:

$$y_j = \begin{cases} 1 & \text{if } \; x_j > 0 \\ 0 & \text{if } \; x_j = 0 \end{cases}$$

and a large positive number M that represents an upper bound on the level of any activity j. Now, the cost function can be specified by:

$$Q(x_j, y_j) = c_j x_j + F_j y_j \tag{9.41}$$

Once again the continuous variables x_j can be linked to the binary variables y_j by the constraints:

$$x_j \leq M y_j \quad \text{for all } j \tag{9.42}$$
$$x_j \geq 0\,,\ y_j \in \{0,1\} \quad \text{for all } j \tag{9.43}$$

From (9.42) follows that when x_j is positive, the binary variable y_j will be forced to 1. So, $x_j > 0$ implies $y_j = 1$.
Alternatively, if $x_j = 0$ the corresponding binary variable may be $y_j = 0$ or $y_j = 1$. Although a solution $x_j = 0$, $y_j = 1$ does not violate (9.42), it should be mentioned that this is clearly not an optimal solution. As $F_j > 0$, the cost function (9.41) will have a lower value for the combination $x_j = 0$, $y_j = 0$.

In general the fixed cost problem can be formulated by the following mixed integer linear programming model (FP):

$$\min\left\{\sum_{j=1}^{J} c_j x_j + \sum_{j=1}^{J} F_j y_j\right\}$$
$$\text{s.t.} \qquad \text{(FP)}$$
$$A\underline{x} \leq \underline{b}$$
$$x_j \leq M y_j \quad \text{for all } j$$
$$x_j \geq 0,\ y_j \in \{0,1\} \quad \text{for all } j$$

Examples of fixed cost problems occur, for instance, when set-up costs are charged for starting a new production run on a machine or fixed costs for building new plants at one or more sites (i.e. plant location, facility location or warehouse location problems).

Plant location problems are an important class of practical problems. Problem (FP) is one of the simplest examples, i.e. the *uncapacitated plant location problem*. Note that the large positive number M in (9.42) might equally be replaced by the (production) capacity at site j; in such cases the problem is called the *capacitated plant location problem*.

In most practical cases the problem is not only to determine a subset of sites for locating the plants based on the cost function $\varphi(x_j, y_j)$ at each site. Usually the problem is to determine an optimal subset of sites for locating the plants and a shipping schedule such as to minimise the total costs of building the plants, setting up the routes and transporting the product(s) to all customers. Usually this problem is decomposed into easier-to-solve sub-problems. The first problem is to solve a plant location problem in which not only the location costs $\varphi(x_j, y_j)$ are taken into consideration but also the costs for transporting the products from location j to distribution centre (DC) k. At this stage it is important to find a balance between (the fixed) location costs on the one hand and the (variable) transportation costs between the selected locations and the DC's on the other hand. Opening less locations will reduce the total fixed costs. However, this also implies that the average distance (i.e. the total transportation costs) between the selected locations and the DC's will increase. The second sub-problem, the transportation of products from a DC to several customers, is a vehicle routing problem.

It should be mentioned that the assumed linearity of the variable costs $c_j x_j$ in the function $\varphi(x_j, y_j)$ might have a non-linear structure in practice. The unit cost contribution c_j regularly depends on the quantity x_j. For example in public transport the transportation costs per unit usually decrease with increasing loading capacities of (different) trucks, transport by railway or even (inland) vessels. The total costs $\varphi(x_j, y_j)$ might be described by a curve as illustrated in Figure 9.11. The principle in which the cost per unit decrease as the output x_i increases, is also called *Economies of scale*.

As the minimisation of a concave function $\varphi(x_j, y_j)$ is an example of non-convex programming (see Section 9.2), any piecewise linear approximation of this function should be modelled by using integer programming. This will be demonstrated in the next section.

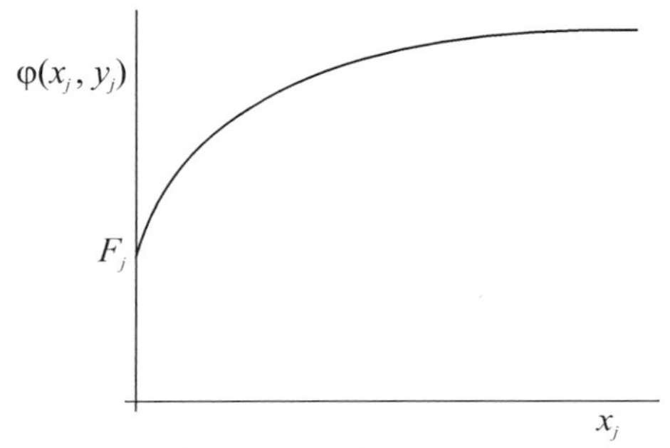

Figure 9.11. A non-linear cost function $\varphi(x_j, y_j)$.

9.4.6. Piecewise linear formulations in non-convex programming

As stated in Section 9.3.6 piecewise linear approximations for non-convex, separable, problems need to be modelled by using integer programming. Consider the non-convex programming problem (NC):

$$\begin{aligned} &\min\{f(\underline{x}) = -x_1^2 + 4x_1 + 2x_2\} \\ &\text{s.t.} \qquad\qquad\qquad\qquad\qquad\qquad \text{(NC)} \\ &\quad A\underline{x} \le b \\ &\quad \underline{x} \ge \underline{0} \end{aligned}$$

The objective function $f(\underline{x}) = f_1(x_1) + f_2(x_2)$ of problem (NC) consists of the non-linear function $f_1(x_1) = -x_1^2 + 4x_1$ to be approximated.. In Section 9.3.6 we presented two piecewise linear model formulations for the approximation of non-linear functions, namely the λ-formulation and the δ-formulation. We will use the λ-formulation for demonstrating piecewise linear approximations in non-convex programming. The λ-formulation for problem (NC) is (see Section 9.3.6):

$$\min_{\underline{x}}\{\varphi_1(x_1)+2x_2\} \tag{9.44}$$

s.t.

$$A\underline{x} \leq \underline{b} \tag{9.45}$$

$$\underline{x} \geq \underline{0} \tag{9.46}$$

$$x_1 = \sum_{i=1}^{I}\lambda_i X_i \tag{9.47}$$

$$\varphi_1(x_1) = \sum_{i=1}^{I}\lambda_i f_1(X_i) \tag{9.48}$$

$$\sum_{i=1}^{I}\lambda_i = 1 \tag{9.49}$$

$$\lambda_i \geq 0 \quad \text{for all } i \tag{9.50}$$

at most two adjacent λ_i may become non-zero (9.51)

Note that the capital letter X_i in (9.47) denotes the coordinates of the chosen breakpoints (O, A, B and C) along the x_1-axis (see Figure 9.12). Any point on the piecewise linear function $\varphi_1(x_1)$ can be found as a convex combination of the breakpoints (X_i, $f_1(X_i)$).
An example of a piecewise linear approximation for the function $f_1(x_1)$ is illustrated in Figure 9.12. For convenience we assume that the constraints in (9.45) and (9.46) restrict the domain of x_1 to $0 \leq x_1 \leq 3$. Restriction (9.51) guarantees that any approximation for the function $f_1(x_1)$ should lie on one of the line segments between two consecutive breakpoints O, A, B and C. For example, if $x_1 = 1.5$ then the approximation of $f_1(x_1) = 3.75$ should be $\varphi_1(x_1) = 3.5$ instead of $z_1(x_1) = 1.5$ (see Figure 9.12).

Note that in contrast to the model formulation in Section 9.3.6 the concave objective function (9.44) should now be minimised, i.e. problem (NC) is a non-convex programming problem (see Section 9.2). So, without condition (9.51) the approximation $z_1(x_1)$ will be selected instead of $\varphi_1(x_1)$. This observation implies that the solution of the piecewise linear approximation (9.44) to (9.50) for problem (NC) does not automatically fulfil constraint (9.51) (see Figure 9.12). Integer programming is needed to guarantee that (9.51) is satisfied too.

We demonstrate the integer programming approach as described by Dantzig (1960).

A constraint like (9.51) can be modelled by introducing additional binary variables δ_j. To be more precise, for each line segment j of the piecewise linear function $\varphi_1(x_1)$ a binary variable δ_j is defined. So, in our case: $\delta_j \in \{0,1\}$ for j =1, 2, 3. We associate the binary variable δ_1 with line segment OA; line segment AB will be associated with the variable δ_2 and the binary variable δ_3 is associated with line segment BC (see Figure 9.12).

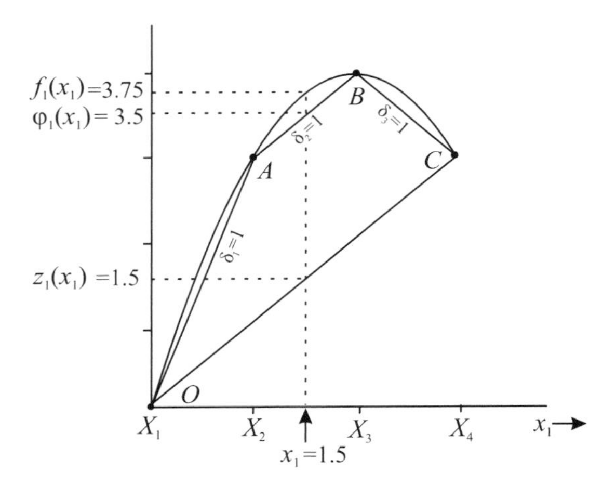

Figure 9.12. Piecewise linear approximation of $f_1(x_1)$ *by* $\varphi_1(x_1)$.

In general the binary variables δ_j are defined as:

$$\delta_j = \begin{cases} 1 & \text{a point with the coordinates } (x_i, \varphi_1(x_i)) \text{ lies on line segment } j \\ 0 & \text{otherwise} \end{cases}$$

Now, any point $(x_i, \varphi_1(x_i))$ should lie either on line segment OA $(\delta_1 = 1)$, on AB $(\delta_2 = 1)$ or on the line segment BC $(\delta_3 = 1)$. Mathematically:

$$\delta_1 + \delta_2 + \delta_3 = 1 \qquad (9.52)$$

$$\delta_j \in \{0,1\} \qquad \text{for all } j \qquad (9.53)$$

In the λ-formulation of Section 9.3.6 the interpolation variables $\lambda_i \geq 0$ are defined as weights to be attached to the breakpoints *O, A, B* and *C*. As the breakpoints *O, A, B* and *C* are part of the line segments, the related weighting or interpolation variables $\lambda_i (i = 1,2,3,4)$ should also be associated with the line segments represented by $\delta_j (j = 1,2,3)$.

For any value x_i in the domain $(X_1, X_4]$ of our example we can distinguish three possible cases:

- $X_1 \leq x_i \leq X_2$
  This implies that $(x_i, \varphi_1(x_i))$ should be a point on line segment OA $(\delta_1 = 1)$. If $\delta_1 = 1$ then at least one of the weights related to the breakpoints *O* and *A* is positive ($\lambda_1, \lambda_2 \geq 0$). The weights related to the breakpoints *B* and *C* are both equal to zero ($\lambda_3 = \lambda_4 = 0$).
- $X_2 \leq x_i \leq X_3$
  The point $(x_i, \varphi_1(x_i))$ should lie on line segment AB $(\delta_2 = 1)$. If $\delta_2 = 1$ then $\lambda_2, \lambda_3 \geq 0$ (weights related to *A* and *B*) and $\lambda_1 = \lambda_4 = 0$ (weights related to *O* and *C*).

- $X_3 \leq x_i \leq X_4$
  The point $(x_i, \varphi_1(x_i))$ should lie on line segment BC $(\delta_3 = 1)$. If $\delta_3 = 1$ then $\lambda_3, \lambda_4 \geq 0$ (weights related to B and C) and $\lambda_1 = \lambda_2 = 0$ (weights related to O and A).

The above-mentioned relations between the continuous variables λ_i $(i = 1,2,3,4)$ and the binary variables $\delta_j (j = 1,2,3)$ can be modelled using linear constraints. For example, the interpolation variable λ_1 is associated with breakpoint O and belongs to the line segment $OA(\delta_1 = 1)$. In other words:

$$\lambda_1 - \delta_1 \leq 0 \tag{9.51a}$$

The interpolation variable λ_2 is associated with breakpoint A and belongs both to the line segments OA $(\delta_1 = 1)$ and AB $(\delta_2 = 1)$. So,

$$\lambda_2 - \delta_1 - \delta_2 \leq 0 \tag{9.51b}$$

The interpolation variable λ_3 is associated with breakpoint B and belongs both to the line segments $AB(\delta_2 = 1)$ and $BC(\delta_3 = 1)$:

$$\lambda_3 - \delta_2 - \delta_3 \leq 0 \tag{9.51c}$$

Finally, the interpolation variable λ_4 is associated with breakpoint C and belongs to the line segment $BC(\delta_3 = 1)$:

$$\lambda_4 - \delta_3 \leq 0 \tag{9.51d}$$

Note that $\delta_1 + \delta_2 + \delta_3 = 1$ and $\delta_j \in \{0,1\}$ for all j. Now, the complete mixed integer linear programming model for our example is:

$$\min_{\underline{x}}\{\varphi_1(x_1) + 2x_2\} = \min_{\underline{x}}\left\{\sum_{i=1}^{4}\lambda_i f_1(X_i) + 2x_2\right\} \tag{9.44}$$

s.t.

$$A\underline{x} \leq \underline{b} \tag{9.45}$$

$$\underline{x} \geq \underline{0} \tag{9.46}$$

$$x_1 = \sum_{i=1}^{4}\lambda_i X_i \tag{9.47}$$

$$\varphi_1(x_1) = \sum_{i=1}^{4}\lambda_i f_1(X_i) \tag{9.48}$$

$$\sum_{i=1}^{4}\lambda_i = 1 \tag{9.49}$$

$$\lambda_i \geq 0 \qquad \text{for all } i \tag{9.50}$$

$$\lambda_1 - \delta_1 \leq 0 \tag{9.51a}$$
$$\lambda_2 - \delta_1 - \delta_2 \leq 0 \tag{9.51b}$$
$$\lambda_3 \quad - \delta_2 - \delta_3 \leq 0 \tag{9.51c}$$
$$\lambda_4 \quad\quad - \delta_3 \leq 0 \tag{9.51d}$$
$$\delta_1 + \delta_2 + \delta_3 = 1 \tag{9.52}$$
$$\delta_j \in \{0,1\} \quad \text{for all } j \tag{9.53}$$

In general we can state that a piecewise linear approximation based on n different breakpoints for a non-linear, separable function in non-convex programming requires $n + 1$ additional constraints and $n - 1$ additional binary variables.

In the next Section 9.5.2 we will discuss an alternative approach called 'Special Ordered Sets of type 2' for modelling and solving problem (NC) i.e. (9.44) to (9.51).

9.5. Special ordered sets of variables

Some particular types of restrictions are quite common in mathematical programming. For two of them Beale and Tomlin (1969) introduced a concept which is referred to as *Special Ordered Sets type 1* (SOS_1) and *Special Ordered Sets type 2* (SOS_2). Nowadays most commercial computer packages offer the facility to apply the concept of SOS_1 and SOS_2 in mathematical programming. Next we will demonstrate the use and benefits of both concepts.

9.5.1. Special ordered sets type 1 (SOS_1)

The concept of SOS_1 is related to very common restrictions in which from a set of variables, continuous or integer, at most one variable may be non-zero.
Assignment problems or (un)capacitated facility location problems (see Section 9.4.5) are typical examples in which SOS_1 sets can be applied. But also restriction (9.52) together with (9.53) in the preceding section can be regarded as an example of SOS_1 sets. Suppose we define the following uncapacitated facility location problem:

$$\min\left\{w = \sum_{i=1}^{I}\sum_{j=1}^{J} c_{ij}x_{ij} + \sum_{i=1}^{I} F_i\, y_i\right\} \tag{9.54}$$

s.t.

$$x_{ij} \leq y_i \quad \text{for all } i,j \tag{9.55}$$
$$\sum_{i=1}^{I} x_{ij} = 1 \quad \text{for all } j \tag{9.56}$$
$$0 \leq x_{ij} \leq 1 \quad \text{for all } i,j \tag{9.57a}$$
$$y_i \in \{0,1\} \quad \text{for all } i \tag{9.58}$$

The variables y_i indicate whether at location i a facility will be constructed ($y_i = 1$) or not ($y_i = 0$). The variables x_{ij} represent the fraction of the total demand of customer j supplied from facility i (9.57a).

From (9.55) it follows that when any x_{ij} is positive, the corresponding binary variable y_i will be forced to one. So, $x_{ij} > 0$ for any j implies $y_i = 1$. The constraints in (9.56) imply that the total demand for every client j must be fulfilled. Although (9.54) to (9.58) is a very common formulation for the uncapacitated facility location problem, we initially assume for convenience that every customer j must be supplied by a single facility i. In this case condition (9.57a) is replaced by (9.57b):

$$x_{ij} \in \{0,1\} \qquad \text{for all } i,j \qquad (9.57b)$$

Suppose there exist I different locations ($i = 1 \ldots I$) and J different clients ($j = 1 \ldots J$). Instead of defining $I \times J$ different binary variables in (9.57b) and subsequently branch on individual variables in a branch-and-bound (B&B) tree, the integrality constraints (9.57b) can be relaxed and it is possible to apply the concept of special ordered sets type 1.

Definition special ordered set type 1

A special ordered set of type 1 (SOS_1) is a set of variables within which at most one variable may be non-zero.

In problem (9.54) to (9.58) we may define for each customer j an SOS_1 set of (continuous) variables:

$SOS_{1j} := \{ x_{1j}, x_{2j}, \ldots, x_{(I-1)j}, x_{Ij} \}$, moreover:
at most one of the variables within SOS_{1j} can be non-zero $\qquad$ for all j $\qquad$ (9.59)

According to the definition the sets {0, 0, 0, 0, 0, 0}, {0, 0, 0, 0, 0, 0.8} or {0, 1, 0, 0, 0, 0, 0} are all examples of *feasible* SOS_1 *sets*.
On the other hand, SOS_1-sets like {0, 0, 0, 0.4, 0, 0.4} or {0.5, 0.3, 0, 0, 0, 0.2} are called *infeasible* SOS_1 *sets*.

Note that it is not necessary to treat the variables x_{ij} in (9.57) as binary variables since the SOS_{1j} conditions in (9.59), together with the constraints in (9.56) ensure that exactly one (continuous) variable in each SOS_{1j} -set will get a final value of one.
As an alternative to define the variables x_{ij} as binary integers for all i, j in (9.57b), it is convenient to consider each SOS_{1j} -set as a generalisation of a 0–1 variable. The additional conditions (9.59) can be dealt with algorithmically which will be shown next.
Treating each set as an entity instead of a collection of individual variables makes it possible to branch in a branch-and-bound algorithm on entities rather than on individual (integer) variables. After all, a non-zero variable in a feasible SOS_{1j} -set of (9.59) will lie either to the left, or to the right, of any marker m between two consecutive variables within a set:

$$\{x_{1j}, \ldots, x_{mj}, x_{(m+1)j}, \ldots, x_{Ij}\}$$
$$\uparrow$$
$$m$$

So,

either $\{x_{1j}, x_{2j}, \ldots, x_{mj}\}$ are all zero
or $\{x_{(m+1)j}, x_{(m+2)j}, \ldots, x_{Ij}\}$ are all zero

These two possibilities correspond to a branch in a search tree as demonstrated in Figure 9.13 in which P_{kj} is defined as a sub-problem P for an SOS_{1j} -set in node k of the search tree (Williams, 1993).

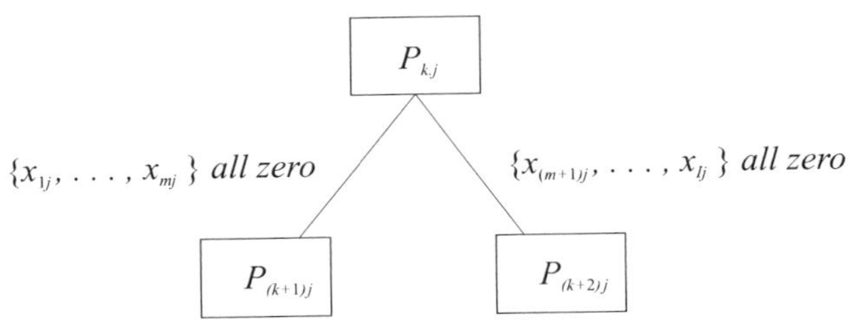

Figure 9.13. The branching procedure in an SOS_1 search tree.

For any node in the search tree, for example, problem $P_{(k+1)j}$, one of the following situations holds:

- problem $P_{(k+1)j}$ is infeasible which implies that the search-tree stops below node $P_{(k+1)j}$.
- problem $P_{(k+1)j}$ is feasible. Now, two possibilities are left:
  - the subset $\{x_{(m+1)j}, \ldots, x_{lj}\}$ is feasible i.e. at most one of the variables in the set is non-zero. If the value of the objective function w for problem $P_{(k+1)j}$ is better than the best bound w_b found so far, the value for w_b is updated ($w_b := w$). The search tree stops below node $P_{(k+1)j}$.
  - the subset $\{x_{(m+1)j}, \ldots, x_{lj}\}$ is infeasible i.e. at least two variables in the subset are non-zero. If the value of the objective function w in problem $P_{(k+1)j}$ is worse or equal to the bound w_b found so far, the search tree below node $P_{(k+1)j}$ stops. If the value of the objective function w for problem $P_{(k+1)j}$ is better than the best bound w_b, the branching procedure is to be continued on the subset $\{x_{(m+1)j}, \ldots, x_{lj}\}$. Note that in any node below problem $P_{(k+1)j}$, at least the variables $\{x_{1j}, \ldots, x_{mj}\}$ are all zero.

Now the question remains: where to put the marker m and 'break' an infeasible SOS_1-(sub)set into two subsets as demonstrated in Figure 9.13. A simple and intuitive rule might be to split an infeasible SOS_1 (sub)set at node P_k into two subsets of comparable sizes. Such a strategy is hardly efficient because the solution in one of the sub problems P_{k+1} or P_{k+2} might be the same as the one already found in node P_{kj} (see Example 9.5).

Example 9.5.

Suppose there are six potential locations ($I = 6$) for the facilities in problem (9.54) to (9.58) and for some client c in node P_{kc} of the SOS_1 search tree we found the solution:
$SOS_{1c} := \{x_{1c}, x_{2c}, x_{3c}, x_{4c}, x_{5c}, x_{6c}\} = \{0.6, 0.4, 0, 0, 0, 0\}$. The set SOS_{1c} in node P_{kc} is called infeasible. Suppose we branch the infeasible set into two subsets of comparable sizes. This might give a solution in the nodes $P_{(k+1)c}$ and $P_{(k+2)c}$ as demonstrated in Figure 9.14.

Obviously, the solution in node $P_{(k+2)c}$ is equal to the one found in node P_{kc}.

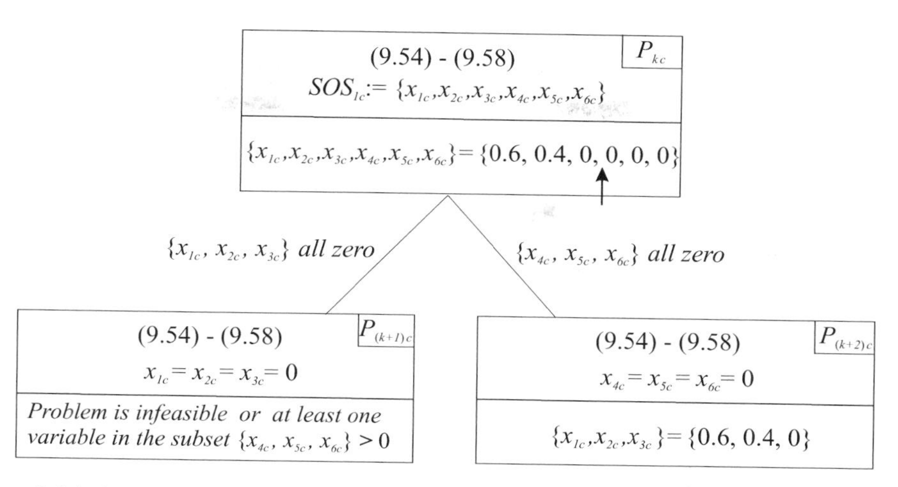

Figure 9.14. An example of inefficient branching in an SOS_1 search tree.

So the question is:' Where to put the branching marker?' A more sensible way to split an SOS_1 is to break the set in node P_{kj} (see Figure 9.13) in such a way that each branching subset, i.e. $\{x_{1j}, x_{2j}, \ldots, x_{mj}\}$ and $\{x_{(m+1)j}, x_{(m+2)j}, \ldots, x_{Ij}\}$ contains at least one non-zero variable of node P_{kj}. Condition (9.59) implies that if more than one variable of an SOS_1-(sub)set takes a non-zero value, the (sub)set is infeasible. In order to measure this infeasibility the variables in an SOS_1 can be associated with some monotonic, increasing or decreasing, set of numbers (a_1, a_2, …, a_I), known as the reference row. For example, the index numbers of the variables can be used as a reference row in order to associate each variable in the set with its place in the ordering, so $a_1 = 1$, $a_2 = 2$, …, $a_I = I$. So, in Example 9.5 we should define in node P_{kc}:

$$\begin{aligned} SOS_{1c} \quad &:= \{ x_{1c}, x_{2c}, x_{3c}, x_{4c}, x_{5c}, x_{6c} \} \\ \textit{Ref. row}_c &:= \{ 1, 2, 3, 4, 5, 6 \} \end{aligned}$$

Now, the position of the branching marker m of an infeasible SOS_1-set can be calculated by:

$$\sum_{i=1}^{I} a_i x_{ij} \Big/ \sum_{i=1}^{I} x_{ij} \qquad \text{for all } j \qquad (9.60)$$

in which x_{ij} are the solution values of the variables in the relevant node of the branch-and-bound tree. As the numbers a_i are monotonic, there will be some a_m such that

$$a_m \leq \left(\sum_{i=1}^{I} a_i x_{ij} \Big/ \sum_{i=1}^{I} x_{ij} \right) < a_{m+1} \qquad \text{for all } j \qquad (9.61)$$

indicating that the 'centre of gravity' of a set has come out between the index numbers m and $m+1$ (Williams, 1993). So, if the set is infeasible the branching marker will be placed between the variables $x_{m,j}$ and $x_{(m+1)j}$.

Example 9.6.

Suppose $I = 6$ and for some client c in node P_{kc} of the SOS_1 search tree we found the solution: $SOS_{1c} := \{x_{1c}, x_{2c}, x_{3c}, x_{4c}, x_{5c}, x_{6c}\}=\{0.6, 0.4, 0, 0, 0, 0\}$. The index numbers $i = 1...6$ of the variables x_{ic} are used as a reference row. Using (9.60), the breakpoint of the infeasible set is: $\sum_i a_i x_{ic} = (1\times 0.6) + (2\times 0.4) + (3\times 0) + (4\times 0) + (5\times 0) + (6\times 0) = 1.4$ Note that the constraints (9.56) of problem (9.54) to (9.58) imply that the denominator in (9.60) can be omitted. According to (9.61) the value $m = 1.4$ is located between $a_1 = 1$ and $a_2 = 2$ (see Figure 9.15).

Computational advantage of the *SOS* formulation

It can be proved by complete induction that the upper bound for the number of branches in an *SOS*-based search tree is equal to the potential number of branches in a 'conventional' branch-and-bound tree (Claassen, 2007), i.e. branching on individual binary variables x_{ij} for problem (9.54) to (9.58). As a consequence, any (computational) advantage of the *SOS*-formulation must be based on finding strong and/or earlier bounds in the *SOS*-based search tree.

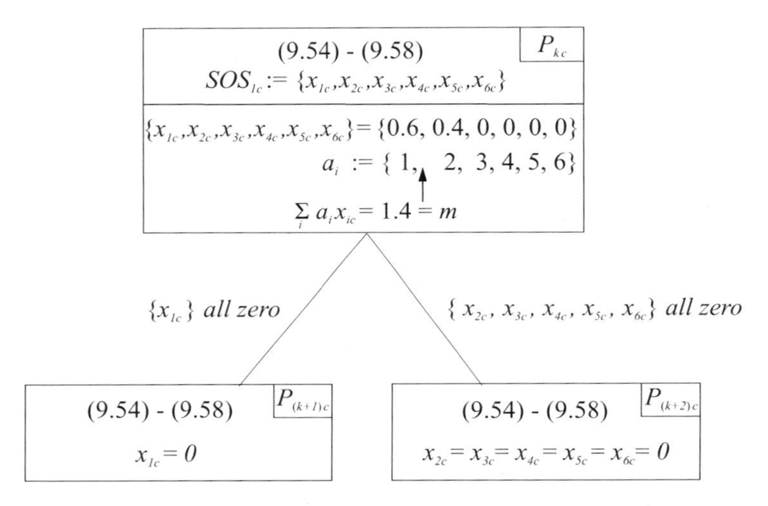

Figure 9.15. An example of efficient branching in an SOS_1 search tree.

It is obvious that finding strong bounds in an early stage of the branch-and-bound (B&B) search tree will have a significant effect on the efficiency of a B&B algorithm. Although a general and robust strategy for strong bounds in any B&B procedure may be hard to find, the SOS_1 branching principle offers an interesting opening to that end. If the actual position of the branching marker m is such that the subsets corresponding to each of the branches in Figure 9.13 are unequal in size, the potential depth of the branch related to the largest subset will be less than the depth of the opposite branch. So, it is likely that the chances of finding an early solution (i.e. bound) will be larger in a node beneath the branch on the largest subset. As the actual ordering of the variables within any SOS_1 (sub)set is free (in our examples the index numbers i of the variables x_{ij} are only used as an example in order to associate each variable in the set with its place in the ordering), any computational advantage of the SOS_1 formulation is based on the ordering of the variables within the SOS_1 (sub)sets. For that purpose Williams

(1990) mentioned that there is a great computational advantage to be gained in the *SOS* formulation only if the variables have a natural ordering within the sets. Next we will give an example of a natural ordering of variables in an SOS_1.

The ordering of variables within an SOS_1

In the formulation of some applications the set of monotonic increasing or decreasing numbers (a_1 , a_2 , ..., a_l), i.e. the reference row, arises from a constraint. For example, the choice between different (increasing) production capacities (e.g. the levels l =1...L) could be modelled easily by adding the restrictions (9.62) and (9.63) to the uncapacitated location problem (9.54) to (9.58) at the start of Section 9.5.1. After some minor adjustments in (9.54) to (9.58), the *capacitated facility location problem* can be formulated as:

$$\min\{\sum_{i=1}^{I}\sum_{j=1}^{J} c_{ij}\, x_{ij} + \sum_{i=1}^{I}\sum_{l=1}^{L} F_{il}\, y_{il}\} \qquad (9.54b)$$

s.t.

$$x_{ij} \leq \sum_{l=1}^{L} y_{il} \qquad \text{for all } i,j \qquad (9.55b)$$

$$\sum_{i=1}^{I} x_{ij} = 1 \qquad \text{for all j} \qquad (9.56)$$

$$0 \leq x_{ij} \leq 1 \qquad \text{for all } i,j \qquad (9.57)$$

$$y_{il} \in \{0,1\} \qquad \text{for all } i,l \qquad (9.58b)$$

$$\sum_{j=1}^{J} D_j\, x_{ij} \leq \sum_{l=1}^{L} Cap_l\, y_{il} \qquad \text{for all } i \qquad (9.62)$$

$$\sum_{l=1}^{L} y_{il} \leq 1 \qquad \text{for all } i \qquad (9.63)$$

The parameters D_j and Cap_l in (9.62) denote the quantity of the demand of client j and the capacity level l (l = 1 .. L) at location i, respectively. The binary variables y_{il} indicate whether at location i a facility with production level l wil be constructed (y_{il} = 1) or not (y_{il} = 0). The restrictions in (9.63) imply that at every location i at most one production level l is allowed. We assume $Cap_l > 0$ for all l.

Note that instead of the 'coupling constraints' in (9.55b) the capacity constraints in (9.62) may be used. If any x_{ij} in (9.62) is positive, one of the binary variables y_{i1}, y_{i2} ..., y_{iL} in the right-hand-side of (9.62) will be forced to one. Although the constraints (9.55b) are redundant and can be omitted, these constraints are a simple example of so-called variable upper bound (VUB) constraints or valid inequalities. Valid inequalities can have a great, positive impact on the efficiency in a B&B search tree. They may yield a much tighter LP-relaxation than a formulation without them. We refer to Schrage (1975) for additional information on VUB constraints and Wolsey (1998) or Pochet and Wolsey (2006) for more information on valid inequalities in (mixed) integer programming.

It may be obvious that the sets:

$$SOS_{1i} := \{ y_{i.1}, y_{i.2} \ldots, y_{i.L} \} \qquad \text{for all } i \tag{9.64}$$

can be regarded as an SOS_1 too. In contrast with the SOS_{1j} as defined in (9.59) the reference row for the SOS_{1i} in (9.64) arise naturally from the constraints in (9.62). Now, the set of (monotonic increasing) values Cap_{i1}, Cap_{i2}, …, Cap_{iL} in (9.62) can be used to associate each variable in (9.64) with its place in the ordering.

There is a great computational advantage to be gained from the *SOS* formulation (Williams, 1990) in particular for cases in which the variables in the SOS_1 can be ordered in a natural way (e.g. the increasing capacity levels in 9.62).

9.5.2. Special ordered sets type 2 (SOS_2)

In Section 9.3.6 we introduced separable programming for the transformation of a certain class of non-linear programming models. It was shown that non-linear terms can be approximated by linear ones, resulting in either:

- a linear programming model for convex separable problems (Section 9.3.6)
- or a (mixed) integer linear programming model for non-convex separable problems Section 9.4.6).

The application of *special ordered sets type 2* (SOS_2) is most commonly related to the latter, i.e. the approximation of non-convex functions by piecewise linear functions.

Definition Special ordered set type 2

An SOS_2 is defined as a set of non-negative variables in which at most two variables may be non-zero. In addition, the non-zero variables must be adjacent to each other in the ordering of the set members.

Considering the definition of an SOS_2, their application in the approximation of non-convex functions by piecewise linear functions may be obvious (see constraint (9.51) in Section 9.4.6). The use of an SOS_2 will be demonstrated with an example. Suppose we define the non-convex problem (NC):

$$\begin{aligned} &\min \{ f(\underline{x}) = -x_1^2 + 4x_1 + 2x_2 \} \\ &\text{s.t.} \qquad\qquad\qquad\qquad\qquad\qquad \text{(NC)} \\ &\quad A\underline{x} \leq \underline{b} \\ &\quad \underline{x} \geq \underline{0} \end{aligned}$$

In Section 9.4.6, the λ-formulation or piecewise linear approximation for problem (NC) is given:

$$\min_{\underline{x}} \{ \varphi_1(x_1) + 2x_2 \} \tag{9.44}$$

s.t.

$$A\underline{x} \leq \underline{b} \tag{9.45}$$

$$\underline{x} \geq \underline{0} \tag{9.46}$$

$$x_1 = \sum_{i=1}^{I} \lambda_i X_i \tag{9.47}$$

$$\varphi_1(x_1) = \sum_{i=1}^{I} \lambda_i f_1(X_i) \tag{9.48}$$

$$\sum_{i=1}^{I} \lambda_i = 1 \tag{9.49}$$

$$\lambda_i \geq 0 \qquad \text{for all } i \tag{9.50}$$

$$\text{at most two adjacent } \lambda_i \text{ may become non-zero} \tag{9.51}$$

Instead of introducing additional binary variables δ_j and the constraints (9.51a) to (9.52) in Section 9.4.6 to formulate (9.51) in mathematical terms we may apply SOS_2 and treat (9.51) algorithmically. The SOS_2 consist of the complete set of interpolation variables λ_i:

$$SOS_2 := \{\lambda_1, \lambda_2, \ldots, \lambda_{I-1}, \lambda_I\}$$

The branching scheme for an SOS_2 is quite similar to the SOS_1 branching principle. By analogy the branching marker which splits an infeasible SOS_2 into two subsets, can be determined by:

$$\sum_{i=1}^{I} r_i \lambda_i$$

in which r_i denotes the reference number and λ_i the solutions of the weighing variables in a node of the search tree.
The reference numbers r_i are used to associate each variable λ_i in the set with its place in the ordering (compare the reference values a_i in (9.60) for an SOS_1). As for an SOS_1, their values should be monotonic too (increasing or decreasing). In contrast with an SOS_1, the ordering of the set members λ_i within an SOS_2 is fixed. The interpolation variables λ_i are associated with predefined breakpoints X_i , which implies that their sequence within the SOS_2 is fixed (see Figure 9.17). Obviously, with SOS_2 applied to non-linear functions, there is a natural ordering of the set members. The set of monotonic (increasing) reference values $X_1, X_2, \ldots, X_{I-1}, X_I$ in (9.47) reflects a natural ordering of the variables λ_i and can be used as reference numbers (see Figure 9.17).

So, $r_i = X_i$ for all i. As in an SOS_1, the reference values r_i of an SOS_2 are monotonic which implies that there will be some value r_n such that:

$$r_n \leq \sum_{i=1}^{I} r_i \lambda_i < r_{n+1}$$

If the calculated value for $\sum_{i=1}^{I} r_i \lambda_i$ is between r_n and r_{n+1} a rule must be chosen in order to decide on the exact branching point in the set (either λ_n or λ_{n+1}). In a feasible SOS_2 at most two adjacent set members can be non-zero. Suppose, in some node P_k of the search-tree the branching point for an infeasible SOS_2 is found at the set-member λ_n. Now, the problem in node P_k can be split into two sub problems:

$$P_{k+1} : \text{either } \{\lambda_1, \lambda_2, ..., \lambda_{n-1}\} \quad \text{are all zero}$$
$$P_{k+2} : \text{or } \{\lambda_{n+1}, \lambda_{n+2}, ..., \lambda_l\} \quad \text{are all zero}$$

Note that the interpolation variable λ_n is not part of the subsets in the nodes P_{k+1} and P_{k+2}. These two possibilities correspond to a branch in the solution tree as demonstrated in Figure 9.16.

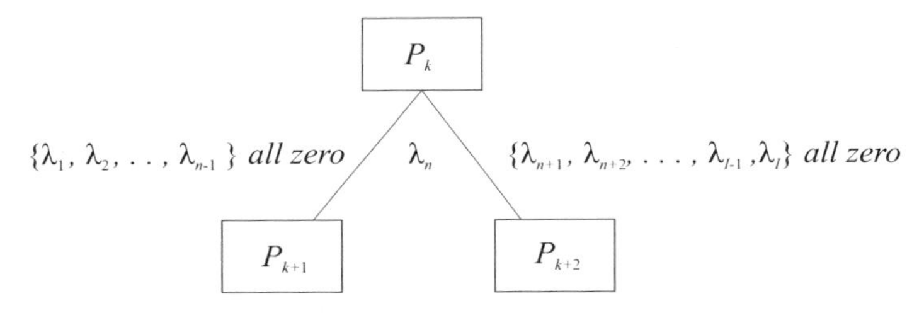

Figure 9.16. The SOS_2 branching scheme (Williams, 1993).

Next we will explain why the interpolation variable λ_n is left out of the branching subsets.

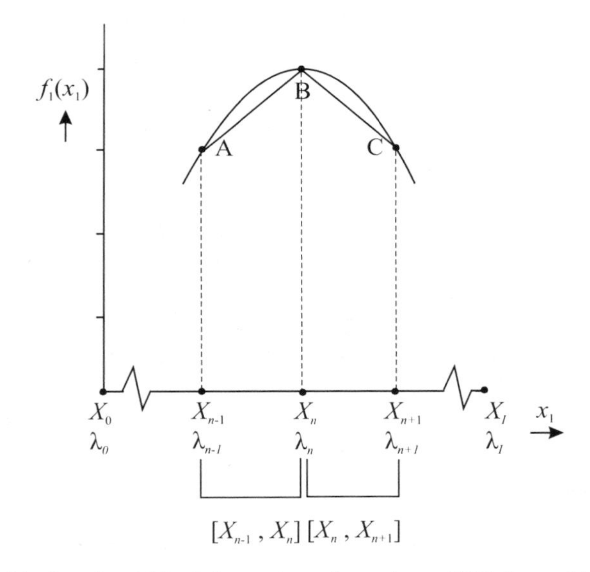

Figure 9.17. Variable λ_n should be left out the subsets in an SOS_2 branching scheme.

From the definition of an SOS_2 we know that any feasible SOS_2 consists of at most two adjacent, non-zero variables λ_i. If the interpolation variable λ_n is a member of either the branching subset $\{\lambda_1, \lambda_2, ..., \lambda_{n-1}\}$ or $\{\lambda_{n+1}, \lambda_{n+2}, ..., \lambda_I\}$ in Figure 9.16, it implies that solutions for any value of x_1 in the domain $\langle X_n, X_{n+1}\rangle$, i.e. on line segment *BC*, and any value for x_1 in the domain $\langle X_{n-1}, X_n\rangle$, i.e. on line segment *AB*, are excluded (see Figure 9.17). Only if the variable λ_n is excluded from both subsets, is any solution in the domain $\langle X_{n-1}, X_{n+1}\rangle$ possible ($\lambda_n > 0$ and either the adjacent variable λ_{n-1} or λ_{n+1} is non-zero).

9.6. Products of variables

There are numerous applications that give rise to non-linear constraints and objectives. In this Section 9.6 we focus on the linearization of mathematical programming models in which *product terms of variables* are incorporated. We distinguish three different cases related to product terms of two variables. In all cases the linearization is based on the following concept:

1. define one or more new variables z_i and replace the product term by a new function $g(z)$
2. add a number of additional constraints.

9.6.1. Product terms of two continuous variables

In Section 9.3.6 separable programming was introduced as a method for the approximation of a specific class of non-linear programming models by piecewise linear functions. Obviously, a function of two continuous variables like $f(x_1, x_2) = x_1 \cdot x_2$ for $x_1, x_2 \geq 0$, is a non-separable function. Next we will show an alternative way of converting a product term of two continuous variables into a separable form.

1. Define two new (continuous) variables $z_1 \geq 0$ and $z_2 \geq 0$ and replace the non-separable function $f(x_1, x_2) = x_1 \cdot x_2$ by the separable function $g(z_1, z_2) = z_1^2 - z_2^2$.
2. Add the following equalities or definitions of z_1 and z_2 to the problem:

$$z_1 = \tfrac{1}{2}(x_1 + x_2)$$
$$z_2 = \tfrac{1}{2}(x_1 - x_2)$$

If the definitions of $z_1 = \frac{1}{2}(x_1 + x_2)$ and $z_2 = \frac{1}{2}(x_1 - x_2)$ are substituted in $g(z_1, z_2)$, it becomes clear that $g(z_1, z_2) = \frac{1}{4}(x_1 + x_2)^2 - \frac{1}{4}(x_1 - x_2)^2 = x_1 \cdot x_2 = f(x_1, x_2)$.

Now the resulting non-linear term $z_1^2 - z_2^2$ is a separable function that can be approximated by a piecewise linear function, either in convex programming (Section 9.3.6) or in non-convex programming (Sections 9.4.6 and 9.5.2).

9.6.2. Product terms of one continuous and one binary variable

If a product term such as $x \cdot \delta$ for $x \geq 0$ and $\delta \in \{0, 1\}$ appears anywhere in a model, the term can be linearised by the following steps:

1. define a continuous variable $z = x \cdot \delta$ and replace the product term by z.

2. add the following constraints in order to force z to take the possible values of $x \cdot \delta$. So, if $\delta = 0$ then $z = 0$, if $\delta = 1$ then $z = x$:

$$z \leq M\delta \quad (9.65)$$
$$z \leq x \quad (9.66)$$
$$z \geq x + M(\delta - 1) \quad (9.67)$$
$$z \geq 0 \quad (9.68)$$

in which M is an upper bound for the continuous variable x (consequently for z too).

The validity of the constraints (9.65) to (9.68) is demonstrated in Table 9.2 in which all possible situations are listed. Note that $\tilde{x}$ denotes any value for $0 < x \leq M$.

Table 9.2. Validity of the constraints (9.65) to (9.68).

possible values		(9.65) $z \leq M \cdot \delta$	(9.66) $z \leq x$	(9.67) $z \geq x + M(\delta - 1)$	(9.68) $z \geq 0$	outcome z
$\delta = 0$	$x = 0$	$z \leq 0$	$z \leq 0$	$z \geq -M$	$z \geq 0$	$z = 0$
	$x = \tilde{x}$	$z \leq 0$	$z \leq \tilde{x}$	$z \geq \tilde{x} - M$	$z \geq 0$	$z = 0$
$\delta = 1$	$x = 0$	$z \leq M$	$z \leq 0$	$z \geq 0$	$z \geq 0$	$z = 0$
	$x = \tilde{x}$	$z \leq M$	$z \leq \tilde{x}$	$z \geq \tilde{x}$	$z \geq 0$	$z = \tilde{x}$

9.6.3. Product terms of two binary variables

If a product term such as $\delta_1 \cdot \delta_2$ for $\delta_1, \delta_2 \in \{0, 1\}$ appears anywhere in a model, the term can be linearized by the following steps:

1. define an additional binary variable z and replace $\delta_1 \cdot \delta_2$ by z in which $z \in \{0, 1\}$
2. in order to force z to take the possible values of $\delta_1 \cdot \delta_2$, the following constraints should be added:

$$z \leq \delta_1 \quad (9.69)$$
$$z \leq \delta_2 \quad (9.70)$$
$$z \geq \delta_1 + \delta_2 - 1 \quad (9.71)$$
$$z \in \{0, 1\}$$

The validity of the constaints (9.69) to (9.71) is demonstrated in Table 9.3.

Table 9.3. Validity of the constraints (9.69) to (9.71).

Possible values		(9.69) $z \leq \delta_1$	(9.70) $z \leq \delta_2$	(9.71) $z \geq \delta_1 + \delta_2 - 1$	Outcome z
$\delta_1 = 0$	$\delta_2 = 0$	$z \leq 0$	$z \leq 0$	$z \geq -1$	$z = 0$
	$\delta_2 = 1$	$z \leq 0$	$z \leq 1$	$z \geq 0$	$z = 0$
$\delta_1 = 1$	$\delta_2 = 0$	$z \leq 1$	$z \leq 0$	$z \geq 0$	$z = 0$
	$\delta_2 = 1$	$z \leq 1$	$z \leq 1$	$z \geq 1$	$z = 1$

10. Heuristics

J.C. (Joke) van Lemmen-Gerdessen

10.1. Introduction

A major part of this book is dedicated to problems that can be modelled with linear and/or integer programming techniques. These methods place restrictions on the form of the objective function and the constraints: they have to be linear.
In many cases problems are too big or too difficult to find an optimal solution in a reasonable amount of time. Often it's not even possible to find a feasible solution in reasonable time. In those situations we can try to find a reasonably good feasible solution in a reasonable amount of time by applying so-called *heuristics*. Heuristics are methods that try to find a reasonably good solution in a reasonable amount of time. They cannot give us the guarantee that the optimal solution will be found, but we can hope that they provide us with a satisfying outcome.
Heuristics can be classified in several ways, e.g. by distinguishing common sense heuristics, metaheuristics, and heuristics based on mathematical programming. In this book common sense heuristics and metaheuristics are described. For heuristics based on mathematical programming we refer to (Wolsey, 1998).
This chapter starts with a short introduction to combinatorial optimisation (a problem field where heuristics are often applied). Then examples of some classical common sense heuristics are provided. We have chosen to illustrate these common sense heuristics mainly with relatively easy problems. In the last paragraphs three metaheuristics are introduced: simulated annealing, tabu search, and genetic algorithm.

10.2. Combinatorial optimisation

Many optimisation problems can be structured as a function of decision variables $\underline{x}$, in the presence of some constraints. Such problems can be formulated generally:

$$\begin{aligned} &\max \{c(\underline{x})\} \\ &\text{s.t.} \\ &\quad a_i(\underline{x}) \leq b_i \quad \text{for all } i \\ &\quad \underline{x} \in \mathbb{R}^n \end{aligned}$$

Here, $\underline{x}$ is an n-dimensional vector of decision variables, and $c(\underline{x})$ and $a_i(\underline{x})$ are functions $\mathbb{R}^n \rightarrow \mathbb{R}$. This formulation has assumed that the problem is one of maximisation. It can be modified to represent a minimisation problem.
There are many specific classes of such problems, obtained by placing restrictions on the type of functions under consideration, and on the values that the decision variables can take. Perhaps the most well known of these classes is that obtained by restricting $c(\underline{x})$ and $a_i(\underline{x})$ to linear functions of continuous decision variables, which leads to problems of linear programming.

The term *combinatorial* is usually reserved for problems in which the decision variables are discrete - i.e. where the solution is a set, or a sequence, of integers or other discrete objects. The problem of finding optimal solutions to such problems is known as *combinatorial optimisation*. We present four examples of combinatorial problems. The first three are relatively simple problems, the fourth is notoriously difficult. In Section 10.4 we will present a common sense heuristic for each example.

Example 10.1. Knapsack problem (see also Chapter 8. Dynamic programming).

Given I items, each having a volume v_i and a value w_i ($i = 1 \ldots I$), and a knapsack with volume V. The issue in the *knapsack problem* is to determine which items should be packed in the knapsack in order to maximise the total value $w = \sum w_i$ of the packed items, subject to the volume restriction of the knapsack.

This problem can be formulated as an integer-programming problem (IP problem) by introducing binary variables:

$$x_i = \begin{cases} 1 & \text{if item } i \text{ is packed,} \\ 0 & \text{otherwise.} \end{cases} \qquad i = 1, \ldots, I$$

With these variables the knapsack problem can be formulated as

$$\max \left\{ \sum_{i=1}^{I} w_i x_i \right\} \tag{10.1}$$

s.t.

$$\sum_{i=1}^{I} v_i x_i \leq V \tag{10.2}$$

In the objective function (10.1) the total value of the packed items is maximised. Restriction (10.2) guarantees that the volume of the packed items does not exceed the available capacity V of the knapsack.

Example 10.2. Bin packing problem.

Given J items, each having a volume v_j ($j = 1, \ldots, J$), and I identical bins each of a given volume V, the *bin packing problem* calls for packing all the items into the minimum number of bins, subject to the volume capacity constraints.

This problem can be formulated as an IP problem by introducing binary variables:

$$y_i = \begin{cases} 1 & \text{if bin } i \text{ is in use (i.e. bin } i \text{ is not empty)} \\ 0 & \text{if bin } i \text{ is empty} \end{cases}$$

$$x_{ij} = \begin{cases} 1 & \text{if item } j \text{ is packed in bin } i \\ 0 & \text{if item } j \text{ is not packed in bin } i \end{cases}$$

With these variables the bin-packing problem can be formulated as

$$\min\left\{\sum_{i=1}^{I} y_i\right\} \quad (10.3)$$

s.t.

$$\sum_{j=1}^{J} v_j x_{ij} \leq V \qquad \text{for all } i \quad (10.4)$$

$$\sum_{i=1}^{I} x_{ij} = 1 \qquad \text{for all } j \quad (10.5)$$

$$x_{ij} \leq y_i \qquad \text{for all } i, j \quad (10.6)$$

$$x_{ij}, y_i \in \{0, 1\}$$

In the objective function (10.3) the number of used bins is minimised. Restriction (10.4) guarantees that the volume of the packed items in bin i does not exceed the volume V of bin i. Restriction (10.5) implies that every item is assigned to a bin, and restriction (10.6) prevents items being assigned to a bin that is not in use.

Example 10.3. Uncapacitated facility location problem (UFLP).

Given J potential depots and I clients, suppose there is a fixed cost f_j associated with the use of depot j, and a transportation cost c_{ij} if all of client i's order is delivered from depot j. The problem is to decide which depots to open, and which depot serves which client so as to minimise the sum of the fixed cost and the transportation cost.
We introduce depot opening variable $y_j = 1$ if depot j is used, and $y_j = 0$ otherwise. The binary variable $x_{ij} = 1$ if the total demand of client i is satisfied from depot j, and $x_{ij} = 0$ otherwise. Now the *uncapacitated facility location problem* can be stated as:

$$\min\left\{\sum_{i=1}^{I}\sum_{j=1}^{J} c_{ij} x_{ij} + \sum_{j=1}^{J} f_j y_j\right\} \quad (10.7)$$

s.t.

$$\sum_{j=1}^{J} x_{ij} = 1 \qquad \text{for all } i \quad (10.8)$$

$$x_{ij} \leq y_j \qquad \text{for all } i, j \quad (10.9)$$

$$x_{ij}, y_j \in \{0, 1\} \qquad \text{for all } i, j$$

The objective function (10.7) minimises the sum of the transportation cost and the fixed cost of the open depots. Constraint (10.8) guarantees that each client's demand is satisfied. Constraint (10.9) ensures that clients are only served from depots that are opened.

Example 10.4. The Travelling Salesman Problem (TSP).

The *Travelling Salesman Problem* is a notorious problem in Operations Research because it is easy to explain, and tempting to try and solve. A salesman must visit each of J cities exactly once and then return to his starting point. The distance from city i to city j is c_{ij} $(i,j = 1 \ldots J)$. Find the shortest route for the travelling salesman.
This problem arises in a multitude of forms: a truck driver has a list of clients he must visit on a given day, or a machine must place modules on printed circuit boards, or a stacker crane

must pick up and depose containers. In the objective function, for example, distance, time, and money can be minimised.

Now we formulate the TSP as a binary programming problem with the help of $x_{ij} = 1$ if the salesman goes directly from city i to city j, and $x_{ij} = 0$ otherwise:

$$\min\left\{\sum_{i=1}^{J}\sum_{j=1}^{J} c_{ij}x_{ij}\right\} \qquad (10.10)$$

s.t.

$$\sum_{i=1}^{J} x_{ij} = 1 \qquad \text{for all } j \qquad (10.11)$$

$$\sum_{j=1}^{J} x_{ij} = 1 \qquad \text{for all } i \qquad (10.12)$$

'subtour breaking constraint' (10.13)

$$x_{ij} \in \{0, 1\} \qquad \text{for all } i, j$$

In the objective function (10.10) the total distance is minimised. Constraints (10.11) and (10.12) ensure that the salesman enters and leaves every city exactly once. The optimal solution of (10.10) to (10.12) might be of the form as shown in Figure 10.1b (i.e. a set of disconnected subtours). To eliminate these infeasible solutions, we need a so-called '*subtour breaking constraint*' (10.13). Examples of subtour breaking constraints can be found in (Lawler *et al.*, 1985).

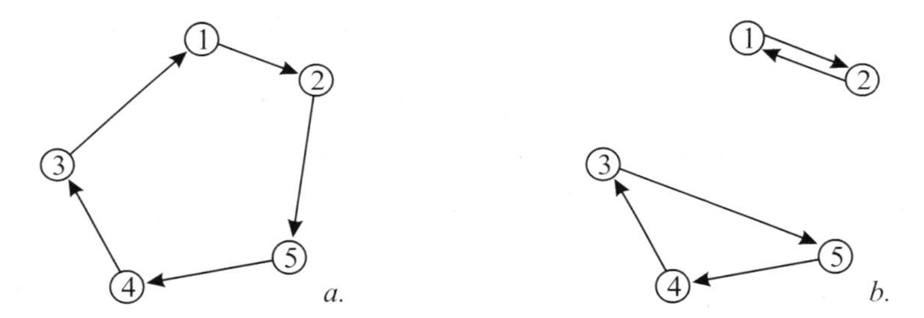

Figure 10.1. A feasible solution (a) and an infeasible solution (b, two disconnected subtours).

10.3. Why do we need heuristics?

It may sound relatively easy to solve an instance of a combinatorial optimisation problem: with *complete enumeration* one can list all feasible solutions to the problem, evaluate their objective functions, and pick the best. However, in practice many problems of reasonable size have so many possible solutions that it will not be possible to list them all in reasonable time, let alone evaluate them.

We illustrate this with the Travelling Salesman Problem (TSP) of Example 10.4. For a TSP along J cities a number of $(J - 1)!$ different solutions exist. If the distance between every pair of cities is the same regardless of the direction of travel (symmetry) we have to evaluate $\frac{1}{2}(J-1)!$ solutions in order to find the best one.

Suppose a computer is able to enumerate 1000 solutions per second. (Note that we ignore the fact that solutions to a bigger problem require (much) more time to be enumerated than solutions to smaller problems!) With that computer the $9!/2 = 1.8 \cdot 10^5$ solutions of a 10-city problem can be enumerated in 3 minutes. The $2 \cdot 10^7$ different solutions to a 12-city problem take 5.5 hours. It takes 331 years to enumerate the 10^{13} solutions to the 17-city problem. Buying a computer that is 1000 times faster (10^6 solutions per second) does not really help: it would still take 4 months to enumerate the 17-city problem, and almost twenty centuries to enumerate the 20-city problem. Because the computing time grows exponentially with the size of the problem, complete enumeration is clearly not a promising way to solve the TSP. The same goes for many other combinatorial problems.

For many problems no good (i.e. efficient) algorithm has yet been found, and nowadays it is strongly suspected that in many cases no such algorithm exists. However, this has not been proved yet! In this book we will not discuss the theory behind these 'difficult' problems. More on this topic can be found in (Papadimitriou and Steiglitz, 1982).

Because of the fact that it is often not possible to find an optimal solution for an optimisation problem OR practitioners have taken refuge in heuristics to find reasonably good solutions in reasonable time, instead of optimal solutions in 'infinite' time. Heuristics however can not guarantee that an optimal (or even a feasible) solution will be found. It may not even be possible to state how close to optimality a particular heuristic solution is.

This may sound as if there are serious problems inherent in the use of heuristics. Nevertheless, it should be emphasised that many modern heuristic techniques do give high-quality solutions in practice.

Furthermore, we do not always need the real optimum. In many situations we might be very happy with a solution that is guaranteed to be within 5% of the optimum, or with a method that generates several reasonably good solutions in a relatively short time. Heuristics might also be very valuable in these cases.

10.4. Common sense heuristics

Many common sense heuristics have been developed. They can be categorised in various ways. In this chapter we distinguish between *construction heuristics* and *improvement heuristics*. In 10.4.1 and 10.4.2 we shall use the Travelling Salesman Problem to illustrate these concepts. In 10.4.3 construction and improvement heuristics shall be given for the knapsack problem, the bin-packing problem, and the location problem.

10.4.1. Construction heuristics

Construction heuristics try to find a feasible solution by using a relatively simple method (rules-of-thumb). Some methods are *deterministic*: they always produce the same solution. Others are *stochastic*: the values of stochastic variables determine which solution is found.

Example 10.5. A deterministic construction heuristic for the TSP.

A travelling salesman has to visit 5 cities. His trip starts and ends in city A. The salesman has to find a trip that minimises the total distance travelled. His heuristic is to continue his trip by visiting the nearest non-visited city (this heuristic is called '*nearest neighbour heuristic*'). So the salesman starts in A, then visits closest city B, then E, D, C and returns to A. Figure 10.2a shows the resulting trip. Note that it is not optimal: it has crossings.
The optimal solution is shown in Figure 10.2b.

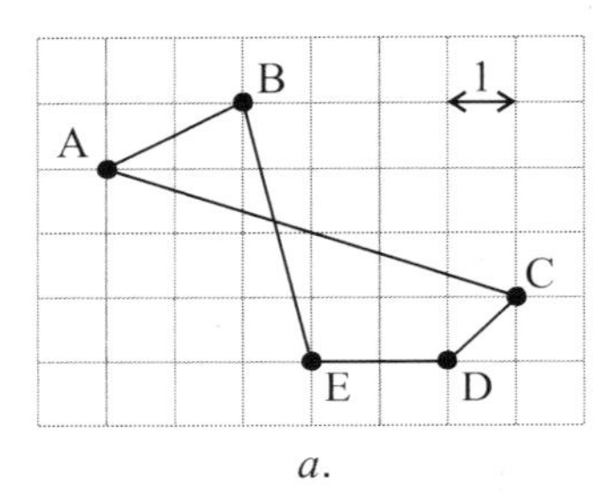

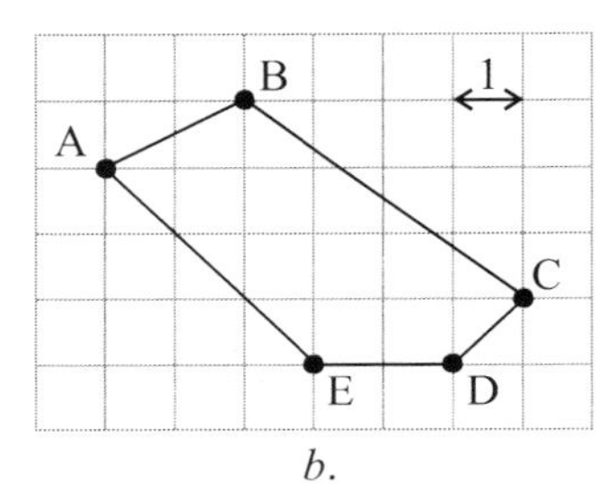

Figure 10.2.a. The result of the nearest neighbour heuristic, length = 16.1.
Figure 10.2.b. The optimal route for this TSP, length = 14.9.

The nearest neighbour heuristic is an example of a *greedy heuristic*, i.e. at any moment in time the decision that minimises the cost of the current step is taken, without taking into consideration the following steps.

For the TSP much better heuristics than nearest neighbour exist; see e.g. (Lawler, 1985) and (Laporte, 1992).

Example 10.6. A stochastic construction heuristic for the TSP.

The travelling salesman decides to apply a stochastic version of his not-very-successful nearest neighbour heuristic. Now in each step he will randomly choose from the three nearest cities. Figure 10.3a shows the resulting trip.
The salesman is not very content. He applies his stochastic heuristic for a second time, and finds a better route, see Figure 10.3b.
Also for this heuristic the chance of finding a very good solution is rather small.

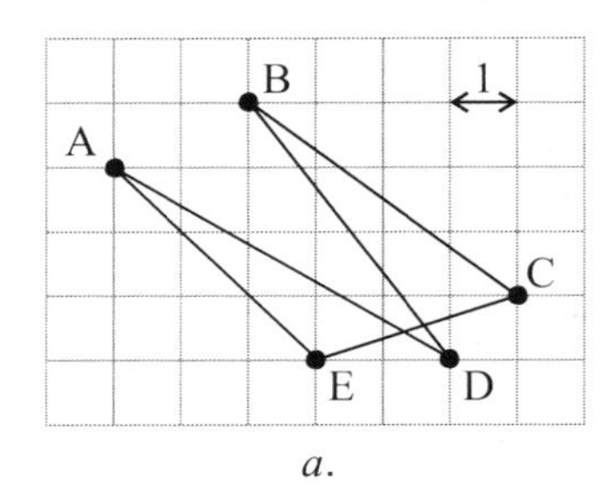

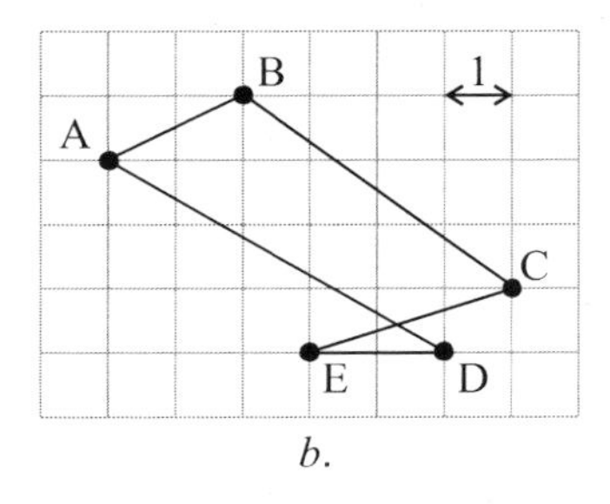

Figure 10.3.a. First result of the stochastic nearest neighbour heuristic, length = 23.2.
Figure 10.3.b. Second result of the stochastic nearest neighbour heuristic, length = 18.2.

10.4.2. Improvement heuristics, local search

Improvement heuristics try to improve an existing solution, usually in an iterative way. Another name for improvement heuristics is *local search* (also called *neighbourhood search*). Once an initial feasible solution has been found it is natural to try to improve this solution. The best feasible solution that has been found so far is called the *incumbent solution*. The idea of a local search heuristic is to define a neighbourhood of solutions close to the incumbent. Then the best solution in the neighbourhood is tracked down. If it is better than the incumbent, it replaces it, and the procedure is repeated. If no better solution can be found in the neighbourhood then the incumbent solution is called '*locally optimal*' with respect to the neighbourhood, and the heuristic terminates.

Example 10.7. Improvement heuristic for the TSP.

The salesman can try to improve his route by defining the following neighbourhood: the neighbourhood of a solution S contains all solutions that have n–2 of the n connections of S, and 2 'new' connections. Therefore this heuristic is called the *2-opt heuristic*.
In Figure 10.4 it is shown how a route in the neighbourhood of the route of Figure 10.2a can be found. The dotted lines in Figure 10.4b represent the 'new' connections. Careful inspection of the route shows that AC and BE have to be deleted. We say that the heuristics *moves* from solution ABEDCA to its neighbour ABCDEA. The result is shown in Figure 10.4c. In this example the optimal route was found.

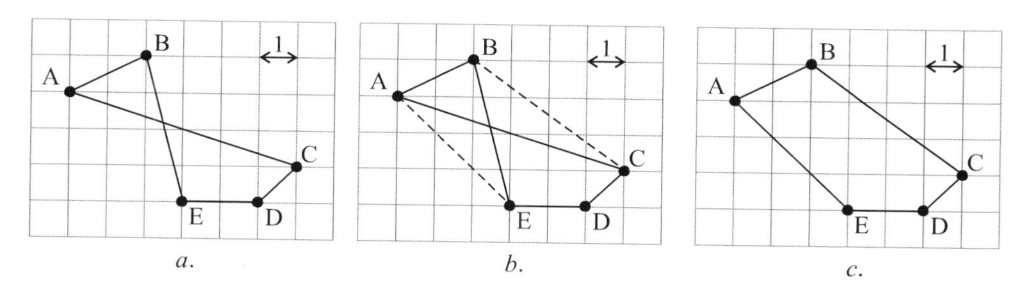

Figure 10.4.a. The route of Figure 10.2a.
Figure 10.4.b. Construction of a solution in the neighbourhood of Figure 10.2a.
Figure 10.4.c.The optimal solution.

In order to give a more formal description of local search we shall assume that a solution to our optimisation problem can be represented as a vector $\underline{x}$, that is a point in the feasible area X of our optimisation problem, so $\underline{x} \in X$. We assume that we are minimising an objective function $c(\underline{x})$. The neighbourhood $N(\underline{x})$ of $\underline{x}$ is made up of all solutions (neighbours) that can be reached from $\underline{x}$ in one *move*. Algorithm 10.1 describes a very general version of local search (Rayward-Smith *et al.*, 1996):

Algorithm 10.1. Local Search.

1. Initialisation
   1.1. Set $k := 0$. Select a starting solution $\underline{x}_0 \in X$.
   1.2 Record the incumbent solution by setting $\underline{x}_{\text{best}} := \underline{x}_k$ and define *bestcost* $:= c(\underline{x}_{\text{best}})$.
2. Choice and termination.
   2.1. Choose a solution $\underline{x}_{k+1} \in N(\underline{x}_k)$. If no valid solution can be found in $N(\underline{x}_k)$ (hence no solution qualifies to be $\underline{x}_{k+1}$), or if termination criteria apply (such as a limit on the total number of iterations), then the method stops.
3. Update.
   3.1. Reset $\underline{x}_k := \underline{x}_{k+1}$; and if $c(\underline{x}_k) <$ *bestcost*, perform Step 1.2.
   Set $k := k+1$. Then return to Step 2.

In Algorithm 10.1 moves are allowed both to solutions with lower cost and solutions with higher cost. A solution that is better (i.e. has lower cost) than all previously found solutions is stored in $\underline{x}_{\text{best}}$ (the incumbent solution), until the next improvement is found.
The procedures for step 2 and step 3 can be specified in many ways, thus influencing the results of the heuristic. The art of designing a local search heuristic is in tuning the procedures in such a way that solutions of a satisfying quality are found in an acceptable amount of time.

A well-known version of local search is the *descent method* in which only moves to better solutions are allowed. A descent method stops in case no improved solution can be found. In the descent method the step 2 and step 3 are integrated, see Algorithm 10.2 (Rayward-Smith *et al*., 1996):

Algorithm 10.2. Descent method.

1. Initialisation
   1.1. Set $k := 0$. Select a starting solution $\underline{x}_0 \in X$.
   1.2. Record the incumbent solution by setting $\underline{x}_{\text{best}} := \underline{x}_k$ and define bestcost $:= c(\underline{x}_{\text{best}})$.
2. Choice and termination
   2.1. Choose $\underline{x}_{k+1} \in N(\underline{x}_k)$ such that $c(\underline{x}_{k+1}) < c(\underline{x}_k)$. Terminate if no such $\underline{x}_{k+1}$ can be found.

The failing of the descent method is easy to see: *when the descent method comes into a local optimum it is not able to get out of this local optimum* (and find the global optimum), see Figure 10.5. In Figure 10.5 the search method has found the solution that is marked with the black dot. It is a local minimum. In the neighbourhood of this local minimum the local search method cannot find a better solution. The neighbourhood is not big enough to 'climb' over the 'hill' (to the right), and see the global minimum.

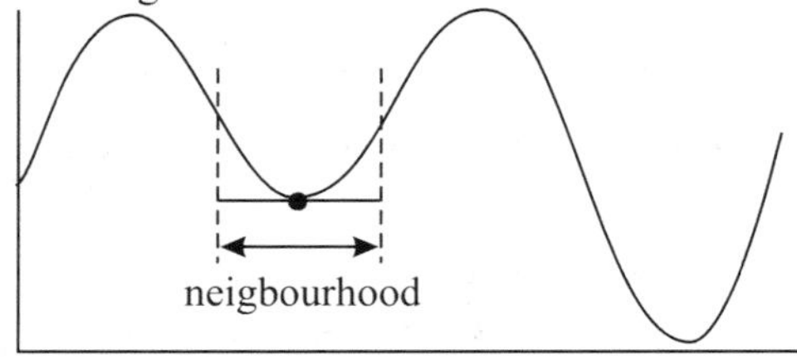

Figure 10.5. A local search method can get stuck in a local minimum.

The problem of getting stuck in a local optimum can for example be tackled by specifying step 2 in a randomised way, see Algorithm 10.3 (based on Rayward-Smith *et al.*, 1996):

Algorithm 10.3. Randomisation method.

1. Initialisation
   1.1. Set $k := 0$. Select a starting solution $\underline{x}_0 \in X$.
   1.2. Record the incumbent solution by setting $\underline{x}_{\text{best}} := \underline{x}_k$ and define bestcost := $c(\underline{x}_{\text{best}})$.
2. Choice
   2.1. Randomly select $\underline{x}_{k+1}$ from $N(\underline{x}_k)$.
   2.2. Calculate $\Delta c = c(\underline{x}_{k+1}) - c(\underline{x}_k)$.
   2.3. If $\Delta c < 0$ then accept $\underline{x}_{k+1}$.
   If $\Delta c > 0$ then accept $\underline{x}_{k+1}$ with a probability that decreases with increasing Δc.
3. Update and termination
   3.1. If in 2.3 $\underline{x}_{k+1}$ is accepted then reset $\underline{x}_k := \underline{x}_{k+1}$; and if $c(\underline{x}_k) < bestcost$, then reset $\underline{x}_{\text{best}} := \underline{x}_k$ and $bestcost := c(\underline{x}_{\text{best}})$.
   3.2. Terminate by a chosen termination rule.
   3.3. Set $k := k+1$. Then return to Step 2.

In Algorithm 10.3 the termination rule is still to be specified by the user. For example, it can be formulated as 'stop after 1000 iterations', or 'stop if in the n previous iterations no improvement was observed'.
An example of a randomised procedure is *simulated annealing*, see Section 10.6.
Other heuristics that are designed to avoid local minima are *tabu search* and *genetic algorithm*, see Sections 10.7 and 10.8. But unfortunately also these heuristics cannot guarantee that they will find the global optimum.

10.4.3. Heuristics for specific problems

Many heuristics are problem-specific (*tailor-made heuristics*), i.e. they are designed to find a solution for a certain type of problem. An advantage of well-designed tailor-made heuristics is that they often find better solutions in a shorter time than general heuristics (like simulated annealing, tabu search, genetic algorithm), because tailor-made heuristics are able to exploit the characteristics of a problem. The reverse side of the picture is that a small change in the characteristics of a problem might require major changes in a tailor-made heuristic, or even make it useless.
In this section we present common sense heuristics for the knapsack problem, the bin-packing problem and the uncapacitated facility location problem.

Example 10.8. A heuristic for the knapsack problem.

A very straightforward construction heuristic for the knapsack problem uses the observation that items with low volume v_i and high value w_i are preferred to items with high volume and low value. Therefore the heuristic proceeds as follows: the items are sorted by w_i/v_i, in descending order. Items are packed into the knapsack, starting with the first item on the sorted list, and proceeding with the next item that fits in, until none of the remaining items fits in.

An attempt can be made to improve the resulting solution. Therefore, remove the last one or two or three items that were packed into the knapsack. Proceed with the heuristic in the usual way.

Table 10.1 shows an example of the knapsack problem. The knapsack has a volume of 100. This knapsack problem can be written as an integer-programming problem:

$$\max \{ 6x_1 + 4x_2 + \ldots + x_{25} \}$$
$$x_1 + x_2 + \ldots + 3x_{25} \leq 100$$
$$x_i \in \{0, 1\}, i \in \{1, 2, \ldots, 25\}$$

where $x_i = 1$ if item i is packed, and $x_i = 0$ otherwise.

The items have been numbered beforehand in such a way that $w_i/v_i \geq w_{i+1}/v_{i+1}$, see Table 10.1. So in the left-hand section of Table 10.1 the items are already sorted by w_i/v_i, in descending order. The column 'first solution' shows which items are packed (they have value 1) and which are not packed (they have value 0). The items 1-12 fit in the knapsack. There is not enough space to pack item 13. The following items that fit in are 18 and 19. The total value of the packed items is 223.

In the right-hand section of Table 10.1 an attempt is made to improve the solution by leaving out items 19, 18 and 12 (their values are fixed to 0). The heuristic is applied in the usual way, without the possibility to pack items 19, 18 and 12. A better solution is found, with value 226.

Table 10.1. Example of the knapsack heuristic.

Data				First solution			Better solution		
item i	value w_i	volume v_i	w_i/v_i	x_i	cumulative volume	cumulative value	x_i	cumulative volume	cumulative value
1	6	1	6.00	1	1	6	1	1	6
2	4	1	4.00	1	2	10	1	2	10
3	17	5	3.40	1	7	27	1	7	27
4	18	6	3.00	1	13	45	1	13	45
5	8	3	2.67	1	16	53	1	16	53
6	27	11	2.45	1	27	80	1	27	80
7	16	7	2.29	1	34	96	1	34	96
8	24	11	2.18	1	45	120	1	45	120
9	27	13	2.08	1	58	147	1	58	147
10	18	9	2.00	1	67	165	1	67	165
11	28	15	1.87	1	82	193	1	82	193
12	24	13	1.85	1	95	217	0	82	193
13	22	12	1.83	0	95	217	1	94	215
14	11	6	1.83	0	95	217	1	100	226
15	18	10	1.80	0	95	217	0	100	226
16	14	9	1.56	0	95	217	0	100	226
17	14	9	1.56	0	95	217	0	100	226
18	3	2	1.50	1	97	220	0	100	226
19	3	2	1.50	1	99	223	0	100	226
20	20	14	1.43	0	99	223	0	100	226
21	12	9	1.33	0	99	223	0	100	226
22	12	9	1.33	0	99	223	0	100	226
23	6	5	1.20	0	99	223	0	100	226
24	4	5	0.80	0	99	223	0	100	226
25	1	3	0.33	0	99	223	0	100	226

This very straightforward heuristic is often very successful. But, as with all heuristics, optimality cannot be guaranteed.

Example 10.9. A heuristic for the bin-packing problem.

The construction heuristic that we present here for the bin-packing problem uses the common knowledge that it is harder to pack a big item than a small item. Therefore, the items are sorted by volume v_i, in descending order. The first (biggest) item on the list is packed into the first bin. Every following item is packed into the first bin that still has enough space for it.

In Table 10.2 the bin-packing heuristic is demonstrated for a problem with bins of volume 15 and items with volumes as shown in the table. The items have been numbered in such a way that $v_i \geq v_{i+1}$. The entries in row i in the column bin1, …, bin8 denote the free volume in bin1, …, bin8 after item i has been packed. The first item is assigned to bin 1. After that the free volume in bin 1 is 3 (in row '1', under 'bin1'). As items 1 to 5 all have a volume of more than half a bin it is obvious that each of these items will be assigned to a new bin.
Then item 6 has to be assigned. Bin 1 and bin 2 do not have enough space, so item 6 is assigned to bin 3 (which has no space left after that). In this way every item is assigned to the first bin that has enough space.

Table 10.2. Example of the bin-packing heuristic.

			free volume in							
item	volume	assigned to	bin1	bin2	bin3	bin4	bin5	bin6	bin7	bin8
1	12	1	3	15	15	15	15	15	15	15
2	10	2	3	5	15	15	15	15	15	15
3	9	3	3	5	6	15	15	15	15	15
4	8	4	3	5	6	7	15	15	15	15
5	8	5	3	5	6	7	7	15	15	15
6	6	3	3	5	0	7	7	15	15	15
7	6	4	3	5	0	1	7	15	15	15
8	5	2	3	0	0	1	7	15	15	15
9	5	5	3	0	0	1	2	15	15	15
10	4	6	3	0	0	1	2	11	15	15
11	4	6	3	0	0	1	2	7	15	15
12	4	6	3	0	0	1	2	3	15	15
13	3	1	0	0	0	1	2	3	15	15
14	3	6	0	0	0	1	2	0	15	15
15	1	4	0	0	0	0	2	0	15	15
	88									

In Table 10.2 the bin-packing heuristic finds a solution with 6 bins. In these six bins only two volume units are not used (in bin 5). This is less than the volume of one bin, so the heuristic has found an optimal solution.

In general this simple bin-packing heuristic is very successful. But there is no guarantee that it will find the optimal solution, as is shown in Table 10.3. In Table 10.3 the heuristic finds a solution that uses 7 bins. The optimal solution (found by another method) comprises only six bins, containing items 1 + 11; 2 + 12 + 15; 3 + 9; 4 + 10; 5 + 7 + 13; 6 + 8 + 14.

Table 10.3. Situation where the bin-packing heuristic does not find the optimal solution.

item	volume	placed in	free volume in bin1	bin2	bin3	bin4	bin5	bin6	bin7	bin8
1	12	1	3	15	15	15	15	15	15	15
2	11	2	3	4	15	15	15	15	15	15
3	10	3	3	4	5	15	15	15	15	15
4	10	4	3	4	5	5	15	15	15	15
5	7	5	3	4	5	5	8	15	15	15
6	7	5	3	4	5	5	1	15	15	15
7	6	6	3	4	5	5	1	9	15	15
8	6	6	3	4	5	5	1	3	15	15
9	5	3	3	4	0	5	1	3	15	15
10	5	4	3	4	0	0	1	3	15	15
11	3	1	0	4	0	0	1	3	15	15
12	3	2	0	1	0	0	1	3	15	15
13	2	6	0	1	0	0	1	1	15	15
14	2	7	0	1	0	0	1	1	13	15
15	1	2	0	0	0	0	1	1	13	15
	90									

Example 10.10. The add/drop heuristic for the Uncapacitated Facility Location Problem.

Two very simple heuristics for the uncapacitated facility location problem (UFLP, see Example 10.3) are the *add heuristic* and the *drop heuristic*. These heuristics try to find a solution with lower costs by adding/dropping facilities.
Consider an instance of the UFLP with 5 depots (A to E) and 8 clients, and costs as shown in Table 10.4.

Table 10.4. Data for an UFLP with 5 depots and 8 clients.

transportation costs	depot A	B	C	D	E
client 1	5	38	22	43	49
2	6	8	5	6	18
3	18	3	13	23	10
4	5	6	1	7	7
5	29	14	25	21	29
6	19	22	18	2	29
7	13	15	8	6	19
8	24	12	18	36	3
fixed costs	30	25	22	20	24

Example 10.10.a. The add heuristic.

The *add heuristic* starts with a solution of one open facility. To determine which facility is open in the starting solution the costs of every possible solution with one depot are calculated. E.g. supplying all clients from depot C (and keeping the other depots closed) will cost: transportation costs + fixed costs = (22 + 5 + … + 18) + 22 = 132.

The costs of all possible solutions with one open depot are shown in Table 10.5a. The solution with only depot C has the lowest costs, so in the starting solution only depot C is opened. Then heuristic tries to find a cheaper solution by opening an extra depot, keeping C open. All combinations of depot C and an extra depot are tried, see Table 10.5b. Opening both C and B is cheaper than just opening C.

Table 10.5.a.

open depot	costs of solution
A	149
B	143
C	132
D	165
E	188

Table 10.5.b.

open depot	costs of solution
C + A	145
C + B	130
C + D	131
C + E	138

Table 10.5.c.

open depot	costs of solution
CB + A	143
CB + D	132
CB + E	145

With B and C opened it's cheaper to service the clients 3,5,8 from B. Clients 1,2,4,6,7 are still serviced from C. The total costs can be calculated as:

transportation costs from B + transportation costs from C + fixed costs =
(3+14+12) + (22+5+1+18+8) + 25 + 22 = 130

Now an attempt is made to see if the costs can be further reduced by opening A or D or E, keeping C and B open, see Table 10.5c. The best solution found in this iteration is CBD with costs 132, which is more than the costs of the previous iteration. Therefore the heuristic stops. The final solution is to open depots B and C, with total costs of 130.

Example 10.10.b. The drop heuristic.

The *drop heuristic* constructs an initial solution by opening all depots. If all depots are opened every client can be served from its closest depot: $x_{1A} = x_{2C} = x_{3B} = x_{4C} = x_{5B} = x_{6D} = x_{7D} = x_{8E} = 1$. The total costs are calculated as:

transportation costs + fixed costs =
(5 + 5 + 3 + 1 + 14 + 2 + 6 + 3) + (30 + 25 + 22 + 20 + 24) = 160

Then an attempt is made to reduce the costs by dropping (i.e. closing) one depot. The best solution is obtained by dropping depot C. If depot C is closed then clients 2 and 4 can no longer be served from C. Client 2 is now served from depot A or depot D and client 4 is served from depot A. The total costs are reduced to 143, see Table 10.6a.

Table 10.6.a.

open depot	costs of solution
BCDE	147
ACDE	150
ABDE	143
ABCE	158
ABCD	145

Table 10.6.b.

open depot	costs of solution
BDE	147
ADE	133
ABE	147
ABD	128

Table 10.6.c.

open depot	costs of solution
BD	132
AD	138
AB	132

Then an attempt is made to reduce the costs by dropping a second depot. The solution with depots A, B and D has the lowest costs, see Table 10.6b.

The solution with depots A, B and D has lower costs than the previous one, therefore an attempt is made to drop one more depot, see Table 10.6c. All of the resulting solutions are more expensive than the previous one, so the drop heuristic stops here.

Note that the add heuristic finds a different solution from the drop heuristic. In this case the drop heuristic (accidentally) found the optimal solution.

10.5. Common sense heuristics versus metaheuristics

In 10.4 several common sense heuristics were presented. Often common sense heuristics are very problem specific: they are tailor-made for one specific problem. The art in designing a common sense heuristic is in finding a way (an algorithm) to produce high-quality solutions in an efficient way. If one succeeds, then this is a very good solution to precisely this problem.
The disadvantage of developing a common sense heuristic is that a small change in the problem characteristics can make the heuristic useless; a new tailor-made heuristic has to be designed for the new problem.

Other types of heuristics have been developed called metaheuristics. Metaheuristics share the characteristic that they are applicable on a relatively wide range of problems. The art in applying metaheuristics is to formulate the optimization problem in such a way that it can be tackled by the metaheuristic, and to choose the right parameters for the metaheuristic.
Many metaheuristics were inspired by natural processes. Of those we present Simulated Annealing (10.6) and Genetic Algorithm (10.8). Another metaheuristic that we describe is Tabu Search (10.7).
One advantage of using metaheuristics is that they are relatively robust in the face of changes to the problem characteristics. The drawback is that a metaheuristic is often less effective and efficient than a tailor-made heuristic.

10.6. Simulated annealing

Simulated annealing (SA) is a heuristic that tries to avoid getting stuck in local optima by introducing the probability to accept a move to a worse solution (see also Table 10.3). We start with a very short description of the heuristic. Then we explain the idea of SA, and provide a more detailed description.

SA starts from a feasible solution x_k. Then randomly a solution x_{k+1} is picked from the neighbourhood of x_k. If x_{k+1} is better than x_k the move from x_k to x_{k+1} is always accepted. If x_{k+1} is worse than x_k the move from x_k to x_{k+1} has a probability of being accepted. The size of this acceptance probability depends on

- the size of the deterioration Δ. If Δ is small then the acceptance probability is big, and vice versa,
- the effort that has been invested to obtain x_k. If you have already been searching for a long time then the acceptance probability tends to be small, and vice versa.

When the algorithm has just started moves to inferior solutions have a large probability of being accepted. As the algorithm progresses the probability of accepting a move to an inferior solution is gradually decreased until finally even small deteriorations have hardly any probability of being accepted.

Why would this heuristic be effective? To answer this question we describe the physical annealing process after which SA is modelled (Eglese, 1990) and its analogy with finding the optimal solution of a combinatorial optimisation problem.

Physical annealing refers to the process of finding low energy states of a solid by initially melting the substance, and then lowering the temperature slowly, spending a long time at temperatures close to the melting point. An example would be producing a crystal from the molten substance. In a liquid, the particles are arranged randomly. But the ground state of the solid, which corresponds to the minimum energy configuration, will have a particular structure, such as seen in a crystal. If the cooling is not done slowly, the resulting solid will not attain the ground state, but will be frozen into a metastable, locally optimal structure, such as a crystal with several defects in the structure.

In the analogy, the different states of the substance correspond to the different feasible solutions to the combinatorial optimisation problem, and the energy of the system corresponds to the function to be minimised. Lowering the 'temperature' of the combinatorial optimisation problem corresponds to reducing the probability of accepting inferior solutions. If this acceptance probability is reduced slowly enough then we might find a reasonably good (i.e. low-cost) solution to our problem, which corresponds to an (almost) perfect crystal. Lowering the temperature too fast (like in the descent method, where only moves to better neighbours are accepted) will probably yield an inferior local optimum (a crystal with defects in it).

Assume that we are minimising $c(x)$, and that the move from the current solution x_k to the neighbouring solution x_{k+1} results in a deterioration $\Delta = c(x_{k+1}) - c(x_k)$. In SA the acceptance probability of this move is calculated as (Eglese, 1990)

$$P_{\text{acc}} = e^{-\frac{\Delta}{T}}$$

where T is a control parameter that has the role of the temperature in the physical annealing process. When the heuristic starts the parameter T is high (so that the acceptance probability tends to be high), and during the execution of the heuristic T is lowered (which results in a decreasing acceptance probability).

For every temperature T a small deterioration Δ is more easily accepted than a large Δ. In other words, a small deterioration is always more acceptable than a large deterioration. Better solutions are always accepted. Figure 10.6 and Table 10.7 illustrate the way in which P_{acc} depends on T and on Δ.

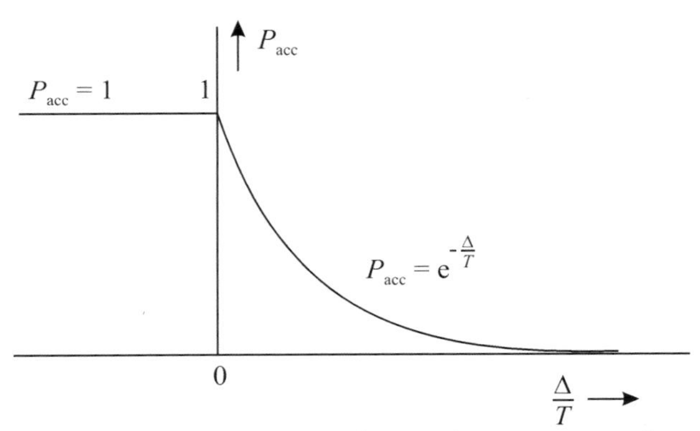

Figure 10.6. The probability of accepting an inferior solution, expressed as a function of Δ / T.

In order to determine whether a move with an acceptance probability of $P_{acc} = e^{-\Delta/T}$ will be accepted we compare $P_{acc} = e^{-\Delta/T}$ with a uniform random number $R \sim U(0,1)$. If $e^{-\Delta/T} > R$ then the move is accepted, otherwise it is rejected.
Table 10.7 summarises the effect of T and Δ on the acceptance rate of the algorithm.

Table 10.7. Overview of the effect of T and Δ on the acceptance rate of simulated annealing.

T	Δ	Δ/T	$e^{-\Delta/T}$	How frequently is $e^{-\Delta/T} > R$?	Acceptance rate
high	small	Close to 0	close to 1	very often	nearly all moves are accepted
⋮	⋮	⋮	⋮	⋮	⋮
low	large	high	close to 0	hardly ever	hardly any moves are accepted

Simulated annealing can be described as a special case of the randomisation method of Algorithm 10.3, see Algorithm 10.4.

Algorithm 10.4. Simulated annealing.

1. Initialisation
   1.1. Set $k := 0$. Select a starting solution $\underline{x}_0 \in X$.
   Select a starting temperature T_0 and a rule for calculating T_{k+1} from T_k.
   1.2. Record the incumbent solution by setting $\underline{x}_{best} := \underline{x}_k$ and define bestcost $:= c(\underline{x}_{best})$.
2. Choice
   2.1. Randomly select $\underline{x}_{k+1}$ from $N(\underline{x}_k)$.
   2.2. Calculate $\Delta = c(\underline{x}_{k+1}) - c(\underline{x}_k)$.
   2.3. If $\Delta < 0$ then accept $\underline{x}_{k+1}$.
   If $\Delta > 0$ then accept $\underline{x}_{k+1}$ with probability $P_{acc} = e^{-\Delta/T}$.
3. Update and termination
   3.1. If in 2.3 $\underline{x}_{k+1}$ is accepted then reset $\underline{x}_k := \underline{x}_{k+1}$; and if $c(\underline{x}_k) < bestcost$, then reset $\underline{x}_{best} := \underline{x}_k$ and $bestcost := c(\underline{x}_{best})$.
   3.2. Calculate T_{k+1} from T_k. Update $T_k := T_{k+1}$.
   3.2. Terminate by a chosen termination rule.
   3.3. Set $k := k+1$. Then return to Step 2.

For applying Simulated Annealing we need:

- a suitable representation of the solutions (often a binary string),
- a cost function,
- a neighbourhood structure,
- an initial value of T, and a procedure for decreasing T (a so-called *cooling schedule*).

These have to be chosen with care, as they may affect both the effectiveness and the efficiency of the algorithm. In SA the solutions are often represented with binary strings because binary strings can easily be manipulated.

Example 10.11.

We consider the problem:

$$\min\{c(x) = \tfrac{1}{2}x^2 + \cos(x) + \sin(6x-3)\} \text{ with } x \in \{0, 0.01, \ldots, 2.55\}$$

On $X = \{0, 0.01, \ldots, 2.55\}$ function c has a global minimum 0.000196 at $x = 0.24$, see Figure 10.7.

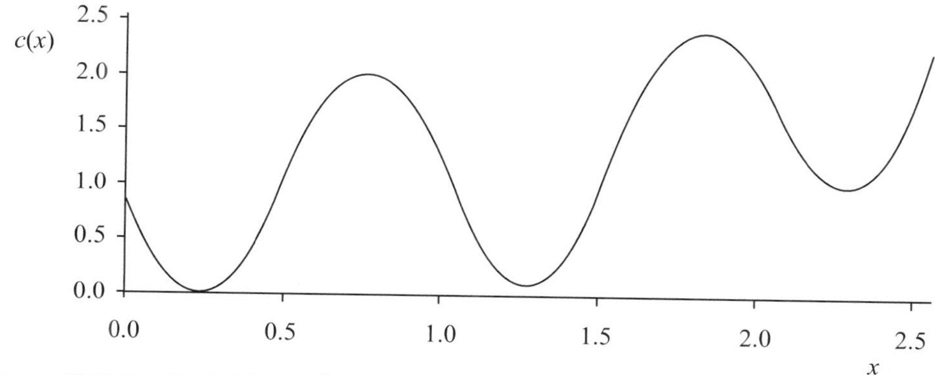

Figure 10.7. Graph of $c(x) = \tfrac{1}{2}x^2 + \cos(x) + \sin(6x-3)$ *with* $x \in \{0, 0.01, \ldots, 2.55\}$.

To illustrate SA every element of $\{0, 0.01, \ldots, 2.55\}$ is coded as an 8-bit binary string $\underline{b}$, e.g.

$$\underline{b}' = (0\ 0\ 1\ 0\ 1\ 1\ 0\ 0) \text{ corresponds to}$$

$$x = \tfrac{1}{100}\sum_{i=1}^{8} b_i \cdot 2^{8-i} =$$

$$\tfrac{1}{100}(0\cdot 2^7 + 0\cdot 2^6 + 1\cdot 2^5 + 0\cdot 2^4 + 1\cdot 2^3 + 1\cdot 2^2 + 0\cdot 2^1 + 0\cdot 2^0) =$$

$$\tfrac{1}{100}(0\cdot 128 + 0\cdot 64 + 1\cdot 32 + 0\cdot 16 + 1\cdot 8 + 1\cdot 4 + 0\cdot 2 + 0\cdot 1) = 0.44$$

We define the neighbourhood of a solution as the set of all solutions that differ in one bit from the current solution, so the 8 neighbours of $\underline{b}' = (0\ 0\ 1\ 0\ 1\ 1\ 0\ 0)$, $x = 0.44$ are

neighbour 1:	$\underline{b}'$ = (**1** 0 1 0 1 1 0 0),	$x = 1.72$
neighbour 2:	$\underline{b}'$ = (0 **1** 1 0 1 1 0 0),	$x = 1.08$
neighbour 3:	$\underline{b}'$ = (0 0 **0** 0 1 1 0 0),	$x = 0.12$
neighbour 4:	$\underline{b}'$ = (0 0 1 **1** 1 1 0 0),	$x = 1.28$
neighbour 5:	$\underline{b}'$ = (0 0 1 0 **0** 1 0 0),	$x = 0.36$
neighbour 6:	$\underline{b}'$ = (0 0 1 0 1 **0** 0 0),	$x = 0.40$
neighbour 7:	$\underline{b}'$ = (0 0 1 0 1 1 **1** 0),	$x = 0.46$
neighbour 8:	$\underline{b}'$ = (0 0 1 0 1 1 0 **1**),	$x = 0.45$

Note that for every neighbour the shaded element shows which bit differs from $\underline{b}'$ = (0 0 1 0 1 1 0 0). In this neighbourhood structure the neighbouring values of x are not all in a small interval around x.

The progress of SA is summarised in Table 10.8. We arbitrarily start from the string $\underline{b}_0$ = (0 0 1 0 1 1 0 0)', binary code for $x = 0.44$, with $c(0.44) = 0.649$. The neighbourhood of a solution is defined as the set of all solutions that differ in one bit from the current solution. We choose neighbours (i.e. the bits to change) randomly. The objective is to minimise $c(x)$. If $c(x_{k+1}) < c(x_k)$ a better solution is found, which is always accepted. If $c(x_{k+1}) > c(x_k)$ we have found a worse solution, so the deterioration $\Delta = \| c(x_{k+1}) - c(x_k) \|$ is calculated, and the corresponding acceptance probability $P_{acc} = e^{-\Delta/T}$. As a starting temperature $T = 1000$ is chosen. The temperature is lowered in every iteration according to the following cooling schedule: $T_{k+1} = 0.75T_k$.

Table 10.8. First attempt, starting $T_0 = 1000$, fast cooling: $T_{k+1} = 0.75T_k$.

k	T_k	$\underline{b}'_k$	x_k	$c(x_k)$	bit	$\underline{b}'_{k+1}$	x_{k+1}	$c(x_{k+1})$	Δ	$e^{-\Delta/T}$	R	move?
0	1000	0 0 1 0 1 1 0 0	0.44	0.649	5	0 0 1 0 0 1 0 0	0.36	0.256	better			Y
1	750.0	0 0 1 0 0 1 0 0	0.36	0.256	1	1 0 1 0 0 1 0 0	1.64	1.804	1.548	0.9979	0.1401	Y
2	562.5	1 0 1 0 0 1 0 0	1.64	1.804	3	1 0 0 0 0 1 0 0	1.32	0.141	better			Y
3	421.9	1 0 0 0 0 1 0 0	1.32	0.141	2	1 1 0 0 0 1 0 0	1.96	2.158	2.017	0.9952	0.1871	Y
4	316.4	1 1 0 0 0 1 0 0	1.96	2.158	7	1 1 0 0 0 1 1 0	1.98	2.081	better			Y
5	237.3	1 1 0 0 0 1 1 0	1.98	2.081	7	1 1 0 0 0 1 0 0	1.96	2.158	0.078	0.9997	0.9256	Y
6	178.0	1 1 0 0 0 1 0 0	1.96	2.158	4	1 1 0 1 0 1 0 0	2.12	1.434	better			Y
7	133.5	1 1 0 1 0 1 0 0	2.12	1.434	1	0 1 0 1 0 1 0 0	0.84	1.912	0.478	0.9964	0.6619	Y
8	100.1	0 1 0 1 0 1 0 0	0.84	1.912	1	1 1 0 1 0 1 0 0	2.12	1.434	better			Y
9	75.08	1 1 0 1 0 1 0 0	2.12	1.434	8	1 1 0 1 0 1 0 1	2.13	1.390	better			Y
10	56.31	1 1 0 1 0 1 0 1	2.13	1.390	1	0 1 0 1 0 1 0 1	0.85	1.884	0.494	0.9913	0.5414	Y
11	42.24	0 1 0 1 0 1 0 1	0.85	1.884	1	1 1 0 1 0 1 0 1	2.13	1.390	better			Y
12	31.68	1 1 0 1 0 1 0 1	2.13	1.390	6	1 1 0 1 0 0 0 1	2.09	1.573	0.183	0.9942	0.9972	N
13	23.76	1 1 0 1 0 1 0 1	2.13	1.390	2	1 0 0 1 0 1 0 1	1.49	0.854	better			Y
14	17.82	1 0 0 1 0 1 0 1	1.49	0.854	5	1 0 0 1 1 1 0 1	1.57	1.370	0.515	0.9715	0.1779	Y
15	13.36	1 0 0 1 1 1 0 1	1.57	1.370	5	1 0 0 1 0 1 0 1	1.49	0.854	better			Y
16	10.02	1 0 0 1 0 1 0 1	1.49	0.854	7	1 0 0 1 0 1 1 1	1.51	0.979	0.125	0.9876	0.1551	Y
17	7.517	1 0 0 1 0 1 1 1	1.51	0.979	3	1 0 1 1 0 1 1 1	1.83	2.410	1.431	0.8267	0.5580	Y
18	5.638	1 0 1 1 0 1 1 1	1.83	2.410	1	0 0 1 1 0 1 1 1	0.55	1.299	better			Y
19	4.228	0 0 1 1 0 1 1 1	0.55	1.299	1	1 0 1 1 0 1 1 1	1.83	2.410	1.111	0.7689	0.8003	N
20	3.171	0 0 1 1 0 1 1 1	0.55	1.299	2	0 1 1 1 0 1 1 1	1.19	0.239	better			Y

k	T_k	$\underline{b}'_k$	x_k	$c(x_k)$	bit	$\underline{b}'_{k+1}$	x_{k+1}	$c(x_{k+1})$	Δ	$e^{-\Delta/T}$	R	move?
21	2.378	0 1 1 1 0 1 1 1	1.19	0.239	5	0 1 1 1 1 1 1 1	1.27	0.107	better			Y
22	1.784	0 1 1 1 1 1 1 1	1.27	0.107	1	1 1 1 1 1 1 1 1	2.55	2.158	2.051	0.3167	0.4400	N
23	1.338	0 1 1 1 1 1 1 1	1.27	0.107	2	0 0 1 1 1 1 1 1	0.63	1.710	1.603	0.3018	0.7665	N
24	1.003	0 1 1 1 1 1 1 1	1.27	0.107	3	0 1 0 1 1 1 1 1	0.95	1.460	1.353	0.2596	0.6581	N
25	0.753	0 1 1 1 1 1 1 1	1.27	0.107	1	1 1 1 1 1 1 1 1	2.55	2.158	2.051	0.0655	0.1123	N
26	0.564	0 1 1 1 1 1 1 1	1.27	0.107	1	1 1 1 1 1 1 1 1	2.55	2.158	2.051	0.0264	0.9773	N
27	0.423	0 1 1 1 1 1 1 1	1.27	0.107	8	0 1 1 1 1 1 1 0	1.26	0.111	0.004	0.9901	0.6203	Y
28	0.317	0 1 1 1 1 1 1 0	1.26	0.111	7	0 1 1 1 1 1 0 0	1.24	0.130	0.019	0.9411	0.9624	N
29	0.238	0 1 1 1 1 1 1 0	1.26	0.111	8	0 1 1 1 1 1 1 1	1.27	0.107	better			Y
30	0.179	0 1 1 1 1 1 1 1	1.27	0.107	2	0 0 1 1 1 1 1 1	0.63	1.710	1.603	0.0001	0.3947	N
31	0.134	0 1 1 1 1 1 1 1	1.27	0.107	5	0 1 1 1 0 1 1 1	1.19	0.239	0.132	0.3731	0.5070	N
32	0.100	0 1 1 1 1 1 1 1	1.27	0.107	3	0 1 0 1 1 1 1 1	0.95	1.460	1.353	0.0000	0.5888	N
33	0.075	0 1 1 1 1 1 1 1	1.27	0.107	4	0 1 1 0 1 1 1 1	1.11	0.565	0.458	0.0022	0.0418	N
34	0.057	0 1 1 1 1 1 1 1	1.27	0.107	3	0 1 0 1 1 1 1 1	0.95	0.146	1.353	0.0000	0.7834	N
35	0.042	0 1 1 1 1 1 1 1	1.27	0.107	4	0 1 1 0 1 1 1 1	1.11	0.565	0.458	0.0000	0.6821	N
36	0.032	0 1 1 1 1 1 1 1	1.27	0.107	5	0 1 1 1 0 1 1 1	1.19	0.239	0.132	0.0157	0.5534	N
37	0.024	0 1 1 1 1 1 1 1	1.27	0.107	1	1 1 1 1 1 1 1 1	2.55	2.158	2.051	0.0000	0.8095	N
38	0.018	0 1 1 1 1 1 1 1	1.27	0.107	7	0 1 1 1 1 1 0 1	1.25	0.119	0.012	0.5098	0.5470	N
39	0.013	0 1 1 1 1 1 1 1	1.27	0.107	3	0 1 0 1 1 1 1 1	0.95	1.460	1.353	0.0000	0.1544	N
40	0.010	0 1 1 1 1 1 1 1	1.27	0.107	4	0 1 1 0 1 1 1 1	1.11	0.565	0.458	0.0000	0.0103	N

The first iteration starts with string $\underline{b}_0 = (0\ 0\ 1\ 0\ 1\ 1\ 0\ 0)'$, so $x_0 = 0.44$, $c(0.44) = 0.649$. In the column 'bit' it is randomly chosen which bit is changed in order to pick a neighbour from the neighbourhood of $\underline{b}_0$. In this case bit 5 is randomly chosen, so solution $\underline{b}_1 = (0\ 0\ 1\ 0\ 0\ 1\ 0\ 0)'$, with $x_1 = 0.36$, $c(0.36) = 0.256$ is picked from the neighbourhood. Obviously $c(0.36) < c(0.44)$, so the move from $x_0 = 0.44$ to $x_1 = 0.36$ is accepted.

In the second iteration $\underline{b}_1 = (0\ 0\ 1\ 0\ 0\ 1\ 0\ 0)'$, with $x_1 = 0.36$. Now bit 1 is changed, so $\underline{b}_2 = (1\ 0\ 1\ 0\ 0\ 1\ 0\ 0)'$, with $x_2 = 1.64$ and $c(1.64) = 1.804$. So we have a deterioration of $\Delta = 1.804 - 0.256 = 1.548$. The probability of accepting this deterioration at the current temperature $T = 750$ is calculated as exp(-1.548/750) = 0.9979. This probability exceeds the random number R = 0.1401, so the move to the inferior solution $\underline{b}_2 = (1\ 0\ 1\ 0\ 0\ 1\ 0\ 0)'$ is accepted.

In iteration 22 the current solution is $\underline{b}_{22} = (0\ 1\ 1\ 1\ 1\ 1\ 1\ 1)'$, so $x_{22} = 1.27$, $c(1.27) = 0.107$. Bit 1 is changed so $\underline{b}_{23} = (1\ 1\ 1\ 1\ 1\ 1\ 1\ 1)'$, with $x_{23} = 2.55$, $c(2.55) = 2.158$ is picked from the neighbourhood. This results in a deterioration $\Delta = 2.158 - 0.107 = 2.051$. At temperature $T = 1.784$ this deterioration has a probability of being accepted of exp(-2.051/1.784) = 0.3167. This acceptance probability is lower than the random number R = 0.4400 so the move from $x_{22} = 1.27$ to $x_{23} = 2.55$ is rejected, and $\underline{b}_{23} = \underline{b}_{22}$.
After iteration 22 the algorithm is not able to get out of the local optimum $x_k = 1.27$.

In the preceding example the temperature was lowered so quickly that there was not enough time to get out of the local minimum $x = 1.27$. Table 10.9 shows a summary of a second attempt, with slower cooling ($T_{k+1} = 0.90T_k$) and more iterations. This time the optimum solution $x = 0.24$ is found. Note that many iterations were left out, for sake of brevity.

Table 10.9. Second attempt: starting T_0 = 1000, slower cooling: $T_{k+1} = 0.90T_k$.

k	T_k	$\underline{b}'_k$	x_k	$c(x_k)$	bit	$\underline{b}'_{k+1}$	x_{k+1}	$c(x_{k+1})$	Δ	$e^{-\Delta/T}$	R	move?
0	1000	0 0 1 0 0 0 1 1	0.35	0.217	5	0 0 1 0 1 0 1 1	0.43	0.594	0.376	0.9996	0.2912	Y
1	900	0 0 1 0 1 0 1 1	0.43	0.594	1	1 0 1 0 1 0 1 1	1.71	2.152	1.558	0.9983	0.0836	Y
2	810	1 0 1 0 1 0 1 1	1.71	2.152	5	1 0 1 0 0 0 1 1	1.63	1.746	better			Y
3	729	1 0 1 0 0 0 1 1	1.63	1.746	5	1 0 1 0 1 0 1 1	1.71	2.152	0.406	0.9994	0.7703	Y
4	656.1	1 0 1 0 1 0 1 1	1.71	2.152	8	1 0 1 0 1 0 1 0	1.70	2.110	better			Y
5	590.5	1 0 1 0 1 0 1 0	1.70	2.110	5	1 0 1 0 0 0 1 0	1.62	1.686	better			Y
:	:	:	:	:	:	:	:	:	:			:
41	13.30	0 0 0 0 0 1 0 1	0.05	0.573	5	0 0 0 0 1 1 0 1	0.13	0.203	better			Y
42	11.97	0 0 0 0 1 1 0 1	0.13	0.203	8	0 0 0 0 1 1 0 0	0.12	0.241	0.038	0.9969	0.4006	Y
43	10.78	0 0 0 0 1 1 0 0	0.12	0.241	7	0 0 0 0 1 1 1 0	0.14	0.169	better			Y
44	9.698	0 0 0 0 1 1 1 0	0.14	0.169	2	0 1 0 0 1 1 1 0	0.78	2.009	1.841	0.8271	0.8592	N
45	8.728	0 0 0 0 1 1 1 0	0.14	0.169	8	0 0 0 0 1 1 1 1	0.15	0.137	better			Y
:	:	:	:	:	:	:	:	:	:			:
81	0.197	0 0 0 1 0 1 0 0	0.20	0.026	3	0 0 1 1 0 1 0 0	0.52	1.123	1.097	0.0038	0.1230	N
82	0.177	0 0 0 1 0 1 0 0	0.20	0.026	5	0 0 0 1 1 1 0 0	0.28	0.032	0.005	0.9704	0.8779	Y
83	0.159	0 0 0 1 1 1 0 0	0.28	0.032	7	0 0 0 1 1 1 1 0	0.30	0.068	0.037	0.7939	0.8843	N
84	0.143	0 0 0 1 1 1 0 0	0.28	0.032	3	0 0 1 1 1 1 0 0	0.60	1.570	1.538	0.0000	0.5192	N
85	0.129	0 0 0 1 1 1 0 0	0.28	0.032	2	0 1 0 1 1 1 0 0	0.92	1.611	1.580	0.0000	0.8507	N
:	:	:	:	:	:	:	:	:	:			:
121	0.003	0 0 0 1 1 0 0 0	0.24	0.000	2	0 1 0 1 1 0 0 0	0.88	1.783	1.783	0.0000	0.7848	N
122	0.003	0 0 0 1 1 0 0 0	0.24	0.000	8	0 0 0 1 1 0 0 1	0.25	0.003	0.002	0.3888	0.7810	N
123	0.002	0 0 0 1 1 0 0 0	0.24	0.000	5	0 0 0 1 0 0 0 0	0.16	0.108	0.108	0.0000	0.3459	N
124	0.002	0 0 0 1 1 0 0 0	0.24	0.000	5	0 0 0 1 0 0 0 0	0.16	0.108	0.108	0.0000	0.6174	N
125	0.002	0 0 0 1 1 0 0 0	0.24	0.000	1	1 0 0 1 1 0 0 0	1.52	1.044	1.043	0.0000	0.8575	N

As mentioned before the following is needed for applying SA:

- coding: a suitable representation of the solutions,
- a cost function,
- a neighbourhood structure,
- an initial temperature, a cooling schedule, and a final temperature (stopping criterion).

Coding

The real art of implementing SA is in designing an effective and efficient encoding for the solutions to the problem. In Example 10.11 the coding was very straightforward: numerical (real) values x were encoded as binary strings b. In many situations the encoding will be more difficult, e.g. for the Travelling Salesman Problem or the Capacitated Facility Location Problem. Literature study, research, and experience are needed to find a suitable encoding for a problem.

Neighbourhood structure

Closely related to the coding of the solution is the definition of the neighbourhood structure of a problem. Among others the following aspects have to be taken into consideration:

- *reachability,* i.e. every feasible solution can be reached from every other feasible solution via a finite series of neighbourhood moves. Reachability is needed in order to ensure that the optimal solution can be reached from the starting solution.

- *feasibility.* In Example 10.11 any neighbour of any solution is a feasible solution itself. In many problems this will not be the case. Two ways of coping with infeasible solutions are:
  - prohibit moves to infeasible solutions. In this way the algorithm cannot end up with an infeasible solution. But at the same time we are cutting off a possibly very short route from the current solution, via an infeasible solution, to the optimal solution.
  - incur penalties for infeasible solutions. In this way the procedure is discouraged from moving to infeasible solutions, especially at lower temperatures. But it is still possible that the algorithm ends up with an infeasible solution.
- *economic calculation.* It can be an advantage to formulate the neighbourhood in such a way that the costs of moving from x_k to x_{k+1} can be calculated efficiently. E.g. in the 2-opt heuristic (Example 10.7) the costs of going to the next solution can be calculated as the difference between the length of the old connections and the length of the new connections. It is not necessary to calculate the total lengths of the routes.

Initial temperature

In the initial phase of SA the heuristic has to get the opportunity to explore all the regions of the feasible space. The initial temperature has to be so high that most moves are accepted. What value is sufficiently high depends on the magnitude of the objective function: a problem with an objective function in the range [1000; 2000] will need a much higher initial temperature than a problem with the objective function in [0.100; 0.200]. It might take some experimentation to find a suitable initial temperature.

Cooling schedule

Basically two types of cooling schedule can be distinguished:

1. The temperature is reduced by a small amount after every move (as in Example 10.11). A very straightforward schedule is $T_{k+1} = \alpha T_k$, with α a constant close to 1. Often α is chosen in the range [0.8; 0.99].
2. At a fixed temperature many iterations are carried out. After equilibrium is reached the temperature is lowered considerably, and the procedure is repeated. This schedule is more complicated than the previous one. The problem is in deciding whether or not equilibrium has been reached.

Final temperature, stopping criterion

When the temperature is very low even small deteriorations will no longer be accepted, and the heuristic will move to a local optimum (which is hopefully the global optimum). Then there is no use proceeding. In practice such a stopping criterion can be formulated as 'if in n consecutive iterations no move has been made to a better solution then the heuristic stops'. Other types of stopping criteria can also be used, e.g. stop after 1000 iterations, or stop after 5 minutes.

The choice of a coding, a neighbourhood structure, a cooling schedule and a stopping criterion are very problem-specific, and they have a major impact on the efficiency and effectiveness of the heuristic. Therefore it's worthwhile paying considerable attention to tuning the heuristic.

10.7. Tabu search

When one wants to escape from a local minimum x_1 it might be a good idea to move to the best solution x_2 in the neighbourhood, even though its value is worse. Ideally the best neighbour of x_2 is not x_1, but another solution. In that case we have escaped from local minimum x_1 and found a better solution.
However, in many cases the best neighbour of x_2 is x_1, which causes the algorithm to alternate between two solutions: $x_1 \rightarrow x_2 \rightarrow x_1 \rightarrow x_2 \rightarrow x_1$. This alternating can be avoided by declaring x_1 *tabu* for some iterations, i.e. one is not allowed to visit x_1 in those iterations. In this way the algorithm is forced to move away from x_1 and explore other regions of the feasible area. This search principle is applied in *Tabu search*.

Tabu search (TS) tries to explore the feasible area by declaring certain moves or solutions forbidden or *tabu*. These moves or solutions are stored in a *tabu list*. The tabu list contains the moves that are tabu for a certain number of iterations or the solutions that are tabu for a certain number of iterations.

Using the terminology of the local search method of Table 10.1 we describe tabu search (TS) in terms of the way it modifies the neighbourhood. The basic local search method is thus modified as in Algorithm 10.5.

Algorithm 10.5. Tabu search.

1. Initialisation
   - 1.1. Set $k := 0$. Select a starting solution $x_0 \in X$.
   - 1.2. Record the current best-known solution by setting $x_{\text{best}} := x_k$ and define bestcost $:= c(x_{\text{best}})$.
   - 1.3. Initialise an empty tabu list.
2. Choice and termination
   - 2.1. Determine the best neighbour of x_k that is not forbidden by the tabu list. That neighbour is chosen to be x_{k+1}.
3. Update
   - 3.1. Reset $x_k := x_{k+1}$.
   - 3.2. If $c(x_k) < bestcost$, then $x_{\text{best}} := x_k$ and $bestcost := c(x_{\text{best}})$.
   - 3.3. If termination criteria apply, then the method stops.
   - 3.4. Update the tabu-list. Set $k := k + 1$. Then return to step 2.

The objective in TS is to encourage exploration of parts of the solution space that have not been visited before. As stated before, this can be aimed for in two ways:
1. by prohibiting the reversal of previous moves,
2. by prohibiting the revisiting of previously visited solutions.

So either moves or solutions become *tabu*. The *tabu tenure* specifies during how many iterations a move cannot be reversed (or a solution cannot be revisited). In the following examples both ways of applying TS are demonstrated (prohibiting moves and prohibiting solutions).

Example 10.12.a. Tabu search; reversal of the most recent moves is tabu.

Minimise $c(x) = \frac{1}{2}x^2 + \cos(x) + \sin(6x - 3)$ with $x \in \{0, 0.01, \ldots, 2.55\}$

For the application of TS every element of $\{0, 0.01, \ldots, 2.55\}$ is coded as an 8-bit binary string $\underline{b}$, as in Example 10.11. We define the neighbourhood of a solution as the set of all solutions that differ in one bit from the current solution (see Example 10.11). In this example the reversal of the most recent moves is tabu and the tabu tenure is 3 iterations: if in some iteration bit i is changed then during the next 3 iterations bit i cannot be changed again. Six iterations are shown in Table 10.10.

The starting string is arbitrarily chosen to be $\underline{b}_0 = (1\ 0\ 1\ 0\ 0\ 0\ 0\ 0)'$, with value $x_0 = 1.60$ and $c(x_0) = 1.562$. Initially $x_{\text{best}} = 1.60$ and $c(x_{\text{best}}) = 1.562$.

Table 10.10. Minimising c(x), the reversal of the most recent moves is prohibited.

Iteration		Bit	Neighbours $\underline{b}'$	x	$c(x)$	Tabu bits	Best
0	$\underline{b}_0 = (1\ 0\ 1\ 0\ 0\ 0\ 0\ 0)'$	1	(0 0 1 0 0 0 0 0)	0.32	0.118	-	$x_{\text{best}} = 1.28$
	$x_0 = 1.60$	2	(1 1 1 0 0 0 0 0)	2.24	1.039		$c(x_{\text{best}}) = 0.106$
	$c(x_0) = 1.562$	3	***(1 0 0 0 0 0 0 0)***	***1.28***	***0.106***		
		4	(1 0 1 1 0 0 0 0)	1.76	2.318		
		5	(1 0 1 0 1 0 0 0)	1.68	2.017		
		6	(1 0 1 0 0 1 0 0)	1.64	1.804		
		7	(1 0 1 0 0 0 1 0)	1.62	1.686		
		8	(1 0 1 0 0 0 0 1)	1.60	1.562		
1	$\underline{b}_1 = (1\ 0\ 0\ 0\ 0\ 0\ 0\ 0)'$	1	(0 0 0 0 0 0 0 0)	0.00	0.859	3	$x_{\text{best}} = 1.28$
	$x_1 = 1.28$	2	(1 1 0 0 0 0 0 0)	1.92	2.287		$c(x_{\text{best}}) = 0.106$
	$c(x_1) = 0.106$	3	tabu	-	-		
		4	(1 0 0 1 0 0 0 0)	1.44	0.567		
		5	(1 0 0 0 1 0 0 0)	1.36	0.233		
		6	(1 0 0 0 0 1 0 0)	1.32	0.141		
		7	(1 0 0 0 0 0 1 0)	1.30	0.116		
		8	***(1 0 0 0 0 0 0 1)***	***1.29***	***0.110***		
2	$\underline{b}_2 = (1\ 0\ 0\ 0\ 0\ 0\ 0\ 1)'$	1	(0 0 0 0 0 0 0 1)	0.01	0.800	8, 3	$x_{\text{best}} = 1.28$
	$x_2 = 1.29$	2	(1 1 0 0 0 0 0 1)	1.93	2.259		$c(x_{\text{best}}) = 0.106$
	$c(x_2) = 0.110$	3	tabu	-	-		
		4	(1 0 0 1 0 0 0 1)	1.45	0.621		
		5	(1 0 0 0 1 0 0 1)	1.37	0.264		
		6	(1 0 0 0 0 1 0 1)	1.33	0.159		
		7	***(1 0 0 0 0 0 1 1)***	***1.31***	***0.127***		
		8	tabu	-	-		
3	$\underline{b}_3 = (1\ 0\ 0\ 0\ 0\ 0\ 1\ 1)'$	1	(0 0 0 0 0 0 1 1)	0.03	0.684	7, 8, 3	$x_{\text{best}} = 1.28$
	$x_3 = 1.31$	2	(1 1 0 0 0 0 1 1)	1.95	2.194		$c(x_{\text{best}}) = 0.106$
	$c(x_3) = 0.127$	3	tabu	-	-		
		4	(1 0 0 1 0 0 1 1)	1.47	0.734		
		5	(1 0 0 0 1 0 1 1)	1.39	0.336		
		6	***(1 0 0 0 0 1 1 1)***	***1.35***	***0.204***		
		7	tabu	-	-		
		8	tabu	-	-		

Iteration		Bit	Neighbours $\underline{b}'$	x	$c(x)$	Tabu bits	Best
4	$\underline{b}_4$ = (1 0 0 0 0 1 1 1)'	1	***(0 0 0 0 0 1 1 1)***	***0.07***	***0.467***	6, 7, 8	x_{best} = 1.28
	x_4 = 1.35	2	(1 1 0 0 0 1 1 1)	1.99	2.039		$c(x_{best})$ = 0.106
	$c(x_4)$ = 0.204	3	(1 0 1 0 0 1 1 1)	1.67	1.967		
		4	(1 0 0 1 0 1 1 1)	1.51	0.979		
		5	(1 0 0 0 1 1 1 1)	1.43	0.516		
		6	tabu	-	-		
		7	tabu	-	-		
		8	tabu	-	-		
5	$\underline{b}_5$ = (0 0 0 0 0 1 1 1)'	1	tabu	-	-	1, 6, 7	x_{best} = 0.23
	x_5 = 0.07	2	(0 1 0 0 0 1 1 1)	0.71	1.963		$c(x_{best})$ = 0.001
	$c(x_5)$ = 0.467	3	(0 0 1 0 0 1 1 1)	0.39	0.388		
		4	***(0 0 0 1 0 1 1 1)***	***0.23***	***0.001***		
		5	(0 0 0 0 1 1 1 1)	0.15	0.137		
		6	tabu	-	-		
		7	tabu	-	-		
		8	(0 0 0 0 0 1 1 0)	0.06	0.519		

In iteration 0 the algorithm moves from x_0 = 1.60 to x_1 = 1.28, thus improving the solution. In iteration 1 it turns out that all neighbours of x_1 = 1.28 have a higher function value. In other words: x_1 = 1.28 is a local optimum. The algorithm chooses the neighbour with lowest $c(x)$, which is x_2 = 1.29.
Also, in iterations 2, 3, and 4 a move to a worse solution has to be made.
In iteration 5 a better solution is found: x_5 = 0.23. This solution is the best solution found so far. So by (temporarily) prohibiting previous moves the algorithm has managed to escape from the local optimum x_1 = 1.28.

Example 10.12.b. Tabu search; revisiting previously visited solutions is tabu.

Minimise $c(x) = \frac{1}{2}x^2 + \cos(x) + \sin(6x - 3)$ with $x \in \{0, 0.01, \ldots, 2.55\}$

As in the previous example the neighbourhood of a solution is defined as the set of all solutions that differ in one bit from the current solution. In this example revisiting the most recent solutions is tabu and the tabu tenure is 3 iterations: if in some iteration a solution x_i is visited then in the next 3 iterations this solution x_i cannot be visited again.
Again starting string is $\underline{b}_0$ = (1 0 1 0 0 0 0 0)', with value x_0 = 1.60 and $c(x_0)$ = 1.562. Six iterations are shown in Table 10.11.

Table 10.11. Minimising c(x), revisiting the most recent solutions is prohibited.

Iteration		Bit	Neighbours $\underline{b}'$	x	$c(x)$	Tabu values	Best
0	$\underline{b}_0$ = (1 0 1 0 0 0 0 0)'	1	(0 0 1 0 0 0 0 0)	0.32	0.118	-	x_{best} = 1.28
	x_0 = 1.60	2	(1 1 1 0 0 0 0 0)	2.24	1.039		$c(x_{best})$ = 0.106
	$c(x_0)$ = 1.562	3	***(1 0 0 0 0 0 0 0)***	***1.28***	***0.106***		
		4	(1 0 1 1 0 0 0 0)	1.76	2.318		
		5	(1 0 1 0 1 0 0 0)	1.68	2.017		
		6	(1 0 1 0 0 1 0 0)	1.64	1.804		
		7	(1 0 1 0 0 0 1 0)	1.62	1.686		
		8	(1 0 1 0 0 0 0 1)	1.61	1.562		

Iteration	Bit	Neighbours $\underline{b}'$	x	$c(x)$	Tabu values	Best
1 $\underline{b}_1 = (1\,0\,0\,0\,0\,0\,0\,0)'$	1	(0 0 0 0 0 0 0 0)	0.00	0.859	1.60	$x_{best} = 1.28$
$x_1 = 1.28$	2	(1 1 0 0 0 0 0 0)	1.92	2.287		$c(x_{best}) = 0.106$
$c(x_1) = 0.106$	3	(1 0 1 0 0 0 0 0)	1.60	tabu		
	4	(1 0 0 1 0 0 0 0)	1.44	0.567		
	5	(1 0 0 0 1 0 0 0)	1.36	0.233		
	6	(1 0 0 0 0 1 0 0)	1.32	0.141		
	7	(1 0 0 0 0 0 1 0)	1.30	0.116		
	8	***(1 0 0 0 0 0 0 1)***	***1.29***	***0.110***		
2 $\underline{b}_2 = (1\,0\,0\,0\,0\,0\,0\,1)'$	1	(0 0 0 0 0 0 0 1)	0.01	0.800	1.60	$x_{best} = 1.28$
$x_2 = 1.29$	2	(1 1 0 0 0 0 0 1)	1.93	2.259	1.28	$c(x_{best}) = 0.106$
$c(x_2) = 0.110$	3	(1 0 1 0 0 0 0 1)	1.61	1.625		
	4	(1 0 0 1 0 0 0 1)	1.45	0.621		
	5	(1 0 0 0 1 0 0 1)	1.37	0.264		
	6	(1 0 0 0 0 1 0 1)	1.33	0.159		
	7	***(1 0 0 0 0 0 1 1)***	***1.31***	***0.127***		
	8	(1 0 0 0 0 0 0 0)	1.28	tabu		
3 $\underline{b}_3 = (1\,0\,0\,0\,0\,0\,1\,1)'$	1	(0 0 0 0 0 0 1 1)	0.03	0.684	1.60	$x_{best} = 1.28$
$x_3 = 1.31$	2	(1 1 0 0 0 0 1 1)	1.95	2.194	1.28	$c(x_{best}) = 0.106$
$c(x_3) = 0.127$	3	(1 0 1 0 0 0 1 1)	1.63	1.746	1.29	
	4	(1 0 0 1 0 0 1 1)	1.47	0.734		
	5	(1 0 0 0 1 0 1 1)	1.39	0.336		
	6	(1 0 0 0 0 1 1 1)	1.35	0.204		
	7	(1 0 0 0 0 0 0 1)	1.29	tabu		
	8	***(1 0 0 0 0 0 1 0)***	***1.30***	***0.116***		
4 $\underline{b}_4 = (1\,0\,0\,0\,0\,0\,1\,0)'$	1	(0 0 0 0 0 0 1 0)	0.02	0.741	1.28	$x_{best} = 1.28$
$x_4 = 1.30$	2	(1 1 0 0 0 0 1 0)	1.94	2.228	1.29	$c(x_{best}) = 0.106$
$c(x_4) = 0.116$	3	(1 0 1 0 0 0 1 0)	1.62	1.686	1.31	
	4	(1 0 0 1 0 0 1 0)	1.46	0.677		
	5	(1 0 0 0 1 0 1 0)	1.38	0.299		
	6	***(1 0 0 0 0 1 1 0)***	***1.34***	***0.180***		
	7	(1 0 0 0 0 0 0 0)	1.28	tabu		
	8	(1 0 0 0 0 0 1 1)	1.31	tabu		
5 $\underline{b}_5 = (1\,0\,0\,0\,0\,1\,1\,0)'$	1	(0 0 0 0 0 1 1 0)	0.06	0.519	1.29	$x_{best} = 1.28$
$x_5 = 1.34$	2	(1 1 0 0 0 1 1 0)	1.98	2.081	1.31	$c(x_{best}) = 0.106$
$c(x_5) = 0.180$	3	(1 0 1 0 0 1 1 0)	1.66	1.915	1.30	
	4	(1 0 0 1 0 1 1 0)	1.50	0.916		
	5	(1 0 0 0 1 1 1 0)	1.42	0.467		
	6	(1 0 0 0 0 0 1 0)	1.30	tabu		
	7	***(1 0 0 0 0 1 0 0)***	***1.32***	***0.141***		
	8	(1 0 0 0 0 1 1 1)	1.35	0.204		

In this example TS did not get out of the local minimum at $x = 1.28$. Probably a larger tabu tenure is needed to force the algorithm out of the local minimum.

In Example 10.12a TS came very close to optimum solution $x = 0.24$ in the sixth iteration. However, as in the case of SA, the heuristic does not recognise that a near optimal solution has been found and carries on exploring (this is not shown in the table).

In these examples prohibiting the most recent moves worked out better than prohibiting the most recent solutions. This is not always the case. In applying TS the user should experiment with various neighbourhood structures and tabu tenures in order to find a good solution.

As mentioned before the following is needed for applying TS:
- coding: a suitable representation of the solutions;
- a cost function;
- a neighbourhood structure;
- a tabu tenure;
- a stopping criterion.

For coding, cost function, neighbourhood structure and stopping criterion we refer to Section 10.7.

Tabu tenure

A suitable tabu tenure usually has to be found by experimenting. If the tabu tenure is relatively small, the heuristic might not escape from local minima. If the tabu tenure is relatively large the heuristic might be forced out of an auspicious area of the search space.

10.8. Genetic algorithm

Until now algorithms have been presented that try to improve one solution at a time. In contrast to this the *Genetic Algorithm* (GA) seeks to improve a whole group of solutions.
GA was inspired by population genetics. In a population the fittest individuals have the biggest probability of surviving and passing their chromosomes to the next generation. In this way the average fitness of the next generation is maximised. This principle is simulated in GA to improve a population of solutions.

The application of a GA requires that a solution is written as a string of variables, e.g. (a_1, a_2, a_3, a_4, a_5). In the genetic analogy this string is called a *chromosome*, and the variables are called *genes*. In many GA applications a chromosome is simply a string of 0's and 1's.
For every chromosome the *fitness* can be calculated, which is a measure of the quality of the solution (e.g. in maximisation problems the objective value can be used as a fitness function).
The GA starts with a group of chromosomes, which is called a *population*. The chromosomes in the population are potential *parents*.
The process of generating new solutions (called *offspring* or *children*) is called *reproduction*. In a reproduction step two parents are chosen. The probability of a parent being selected is determined by his fitness: fitter parents have a bigger probability of being selected than weaker parents. The selected parents generate offspring, thus creating individuals of a new *generation*. The aim of GA is to maximise the fitness of the offspring by giving high-quality chromosomes a relatively large probability of being selected as a parent.

Two often-used methods for generating offspring from two selected parents are *crossover* and *mutation*.

For crossover we need 2 parent chromosomes, e.g. $\underline{a}$ and $\underline{b}$. Suppose that each parent's chromosome consists of 5 variables or genes: $\underline{a}' = (a_1\ a_2\ a_3\ a_4\ a_5)$ and $\underline{b}' = (b_1\ b_2\ b_3\ b_4\ b_5)$. A *crossover point* is chosen at random from the numbers 1, ..., 4, and offspring is produced by combining the pieces of the parents. For instance, if the crossover point is 2, then crossover leads to the following offspring

parents	crossover	offspring
$(a_1\ a_2\ a_3\ a_4\ a_5)$	$a_1\ a_2 \mid a_3\ a_4\ a_5$	$(a_1\ a_2\ b_3\ b_4\ b_5)$
$(b_1\ b_2\ b_3\ b_4\ b_5)$	$b_1\ b_2 \mid b_3\ b_4\ b_5$	$(b_1\ b_2\ a_3\ a_4\ a_5)$

If parents have the same value for a variable (i.e. $a_i = b_i$) then all their offspring will also have this value for variable i. This limits the search space. *Mutation* can be used to diversify the search.

Mutation provides the opportunity to produce offspring that cannot be produced by crossover alone. For mutation only one chromosome is needed. Each gene of a chromosome is examined in turn, and with a small probability its current value is changed. For example, a chromosome

$$(0\ 1\ \underline{1}\ 0\ 1) \text{ could become } (0\ 1\ \underline{0}\ 0\ 1)$$

if its 3rd gene is mutated.
By using mutation parts of the search space can be reached that might not be reached by crossover alone.

Crossover and mutation are the basic tools for generating a new generation out of the individuals of the old generation. In term of solutions: crossover and mutation can be used as tools to generate new solutions from old ones. If applied in a useful way crossover and mutation can help to improve the quality of a population of solutions. Their effect on the population is influenced by the way in which parents are chosen.

The GA procedure is summarised in Algorithm 10.6.

Algorithm 10.6. Genetic algorithm.

1. Obtain a population of parents. This is the first generation.
2. Produce a new generation
   - 2.1. *Evaluation*: evaluate the fitness of the individuals.
   - 2.2. *Parent selection*: select pairs of parents. The probability of a chromosome being selected as a parent depends on its fitness.
   - 2.3. *Crossover*: produce one or two new individuals from each pair of parents.
   - 2.4. *Mutation*: some genes of the offspring are modified randomly.
   - 2.5. *Population selection*: a new generation is selected replacing some or all of the original population by an identical number of offspring.

Repeat step 2 until some termination criteria apply.

When designing a GA a lot of choices have to be made, for example:

- How will the initial population be created?
- How many individuals does a generation contain?
- How will it be accomplished that fitter chromosomes have a bigger probability of being selected as a parent?
- Will the crossover be carried out at 1 crossover point, or at 2 or even more crossover points?
- Will 1 or 2 'children' of a crossover be used in the next generation?
- In what way will one generation be replaced by the next: gradually (each time adding 1 chromosome and removing 1 chromosome) or completely (the whole generation is replaced at once, as in Example 10.13)?
- How many generations will be calculated?
- What stopping rule will be used?

As in SA and TS the real 'art' of implementing GA is in designing an effective and efficient encoding for the solutions to the problem, i.e. in defining the chromosomes. In the example the values of x where encoded as binary strings. This approach is not always possible or effective. Much literature study and research might be needed to find a good encoding for a problem.

Example 10.13.

To make GA more transparent, we again use the problem of minimising the function
$c(x) = \frac{1}{2}x^2 + \cos(x) + \sin(6x - 3)$ with $x \in \{0, 0.01, \ldots, 2.55\}$

Table 10.12 shows the first 3 generations of a GA for this problem. In implementing this GA several choices are made:

- Every generation has 10 chromosomes.
- The 10 chromosomes of the first generation are generated randomly.
- The parents selection procedure is designed in such a way that parents with low $c(x)$ have a larger probability of being chosen than parents with high $c(x)$. The selection procedure is not shown in the table. Columns p_1 and p_2 indicate which parents are selected to produce new solutions.
- The crossover point is chosen randomly. Every potential crossover point has an equal chance of being selected. In the table the crossover point is written as '|'.
- Of the crossover result of two parents only the chromosome that consists of the first genes of p_1 and the last bits of p_2 is used. The chromosome that consists of the first genes of p_2 and the last bits of p_1 is not used.
- The mutation rate is 0.03: every gene has a 3% probability of being mutated. All mutated genes are underlined. The 10 chromosomes in column 'mutation' are the offspring of generation 1.
- After generating 10 offspring chromosomes all parents of the current generation are deleted. The 10 offspring chromosomes are used as the parents of the next generation.

For example, the first row of 'Producing offspring of generation 1' shows that chromosomes 6 and 1 are selected to be a parent ($p_1 = 6$ and $p_2 = 1$). Crossover takes place at 6: the first six genes of chromosome 6 are combined with the last two genes of chromosome 1, thus creating (0 0 1 0 1 1 | 1 0). Of this chromosome one gene (the 2nd) is mutated. The resulting chromosome (0 1 1 0 1 1 1 0) is used as the first parent of the second generation.

Table 10.12. Example genetic algorithm.

Generation 1			
i	Chromosomes b_i	x	$c(x)$
1	0 0 1 0 0 1 1 0	0.38	0.34
2	1 1 0 0 1 0 1 1	2.03	1.86
3	0 1 0 0 1 0 0 0	0.72	1.98
4	1 1 0 1 0 0 0 1	2.09	1.57
5	0 1 0 1 1 1 1 1	0.95	1.46
6	0 0 1 0 1 1 0 0	0.44	0.65
7	0 0 1 1 1 1 0 0	0.60	1.57
8	0 1 0 0 0 0 0 1	0.65	1.79
9	1 1 0 1 0 1 1 1	2.15	1.31
10	0 0 1 1 1 0 1 1	0.59	1.52

Producing offspring of generation 1						
Parents		Crossover result		Mutation result	x	$c(x)$
p_6	p_1	0 0 1 0 1 1	1 0	0 1 1 0 1 1 1 0	1.10	0.62
p_{10}	p_4	0 0 1 1 1 0 1	1	0 0 1 1 1 0 1 1	0.59	1.52
p_9	p_1	1 1 0 1 0	1 1 0	1 1 0 1 0 1 1 0	2.14	1.35
p_4	p_5	1 1	0 1 1 1 1 1	1 1 0 1 1 1 1 0	2.22	1.08
p_9	p_8	1 1 0 1 0 1	0 1	1 0 0 1 0 1 0 1	1.49	0.85
p_4	p_1	1 1	1 0 0 1 1 0	1 1 1 0 0 1 1 0	2.30	1.00
p_1	p_7	0 0	1 1 1 1 0 0	0 0 1 1 1 1 0 0	0.60	1.57
p_6	p_9	0 0 1 0 1 1 0	1	0 0 1 0 1 1 0 1	0.45	0.71
p_5	p_9	0 1	0 1 0 1 1 1	0 1 0 1 0 0 1 1	0.83	1.94
p_6	p_5	0 0 1	1 1 1 1 1	0 0 1 1 1 1 1 1	0.63	1.71

Generation 2			
i	Chromosomes b_i	x	$c(x)$
1	0 1 1 0 1 1 1 0	1.10	0.62
2	0 0 1 1 1 0 1 1	0.59	1.52
3	1 1 0 1 0 1 1 0	2.14	1.35
4	1 1 0 1 1 1 1 0	2.22	1.08
5	1 0 0 1 0 1 0 1	1.49	0.85
6	1 1 1 0 0 1 1 0	2.30	1.00
7	0 0 1 1 1 1 0 0	0.60	1.57
8	0 0 1 0 1 1 0 1	0.45	0.71
9	0 1 0 1 0 0 1 1	0.83	1.94
10	0 0 1 1 1 1 1 1	0.63	1.71

Producing offspring of generation 2						
Parents		Crossover result		Mutation result	x	$c(x)$
p_8	p_6	0 0 1 0 1	1 1 0	0 0 1 0 1 1 1 0	0.46	0.76
p_4	p_8	1 1 0	0 1 1 0 1	1 1 0 0 1 0 0 1	2.01	1.95
p_6	p_4	1 1 1 0 0	1 1 0	1 1 1 0 0 1 1 0	2.30	1.00
p_4	p_1	1 1 0 1 1 1 1	0	1 0 0 1 1 1 1 0	1.58	1.43
p_6	p_6	1 1 1	0 0 1 1 0	1 1 1 0 0 1 1 0	2.30	1.00
p_2	p_3	0 0	0 1 0 1 1 0	0 0 0 1 0 1 1 0	0.22	0.01
p_6	p_4	1 1 1	1 1 1 1 0	1 1 1 1 1 1 1 0	2.54	2.08
p_8	p_3	0 0	0 1 0 1 1 0	0 0 0 1 0 1 1 0	0.22	0.01
p_8	p_8	0 0 1 0 1 1	0 1	0 0 1 0 1 1 0 0	0.44	0.65
p_5	p_8	1 0 0 1 0 1 0	1	1 1 0 1 0 1 0 1	2.13	1.39

Generation 3			
i	Chromosomes b_i	x	$c(x)$
1	0 0 1 0 1 1 1 0	0.46	0.76
2	1 1 0 0 1 0 0 1	2.01	1.95
3	1 1 1 0 0 1 1 0	2.30	1.00
4	1 0 0 1 1 1 1 0	1.58	1.43
5	1 1 1 0 0 1 1 0	2.30	1.00
6	0 0 0 1 0 1 1 0	0.22	0.01
7	1 1 1 1 1 1 1 0	2.54	2.08
8	0 0 0 1 0 1 1 0	0.22	0.01
9	0 0 1 0 1 1 0 0	0.44	0.65
10	1 1 0 1 0 1 0 1	2.13	1.39

Producing offspring of generation 3						
Parents		Crossover result		Mutation result	x	$c(x)$
p_6	p_6	0 0 0	1 0 1 1 0	0 0 0 1 0 1 1 0	0.22	0.01
p_2	p_8	1 1 0 0 1 0 0	0	1 1 0 0 1 0 0 0	2.00	2.00
p_1	p_6	0 0	0 1 0 1 1 0	0 0 0 1 0 1 1 0	0.22	0.01
p_6	p_4	0 0 0 1	1 1 1 0	0 0 0 1 1 1 1 0	0.30	0.07
p_1	p_5	0 0 1	0 0 1 1 0	0 0 1 0 0 1 1 0	0.38	0.34
p_4	p_5	1 0 0 1 1 1	1 0	1 0 0 1 1 1 1 0	1.58	1.43
p_5	p_5	1 1 1 0 0 1	1 0	1 1 1 0 0 1 1 0	2.30	1.00
p_5	p_8	1 1	0 1 0 1 1 0	1 0 0 1 0 1 1 0	1.50	0.92
p_3	p_9	1 1 1 0 0	1 0 0	1 1 1 0 0 1 0 0	2.28	1.00
p_6	p_5	0 0 0 1 0 1	1 0	0 0 0 1 0 1 1 0	0.22	0.01

In this example 3 generations were shown. In practice many more generations are used.

11. Inventory management

K.G.J. (Karin) Pauls-Worm

11.1. Introduction

Every person or organisation has products in stock waiting to be used. People may have pencils or napkins on the shelf, and most probably have food products in the refrigerator for meals to come. Organisations like retailers have complete outlets filled with stored products waiting to be sold to customers and industrial producers have raw materials in stock waiting to be processed. Storing products costs money and has impact on performance criteria such as price, product quality, delivery reliability, and lead time. Inventory costs are an important part of the logistics costs in a production or business organisation. Therefore, optimising the inventory levels can save a lot of money and improve the service level of an organisation.

This chapter discusses inventory management with an emphasis on inventory management models. To discuss inventory management first a few questions have to be answered:

- What is inventory?
- Why inventory?
- What is inventory management?

11.1.1. What is inventory?

Inventory is goods and materials held available in stock. In the manufacturing industry inventories are called *stockkeeping units (SKU)*, which are held at a storage point. An SKU is a unique combination of all the components that are assembled into the purchasable item. Therefore any change in the packaging or product results in a new SKU.

Within a *production chain* starting from raw materials to final products, different production stages, separated by different kinds of inventories, can be distinguished. See Figure 11.1. A production chain can relate to successive production stages within one company or can relate to production stages that are conducted by a number of companies. When more than one company is involved in a production chain, it is called a *supply chain*. In a supply chain, the end product of a certain company, is an input to the next company.

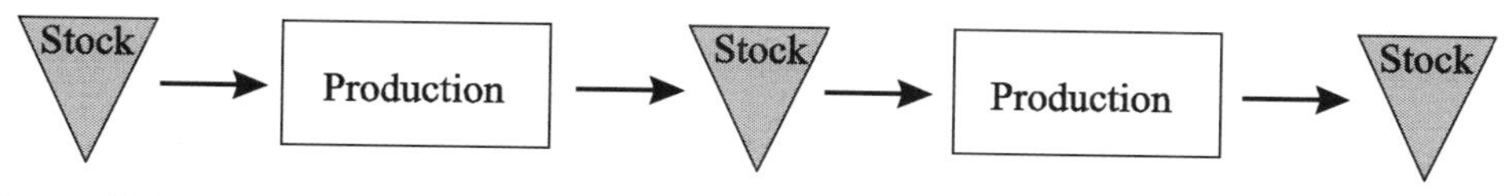

Figure 11.1. Production chain with items in stock.

In the beginning of the chain under consideration there are raw materials in stock that are needed for production. For example, milk collected at the farms or pigs delivered to the slaughterhouse. Between two production stages there can be an inventory of half-(semi-finished) products, for example, a tank with sterilised milk or a batch of slaughtered animals. Depending on the processing steps that take place successively more inventories of half-products can be identified. For example, yoghurt, cheese, or processed meat. The chain ends with the final (packaged) product ready for the customer. In conclusion, there are three main types of stockkeeping units: raw materials, half-products (often called work-in-process), and finished products.

11.1.2. Why inventory?

The most fundamental reason for having inventory is that in most cases it is not possible to have every item arrive exactly when and where it is needed. This is often physically impossible or economically infeasible. Some other reasons for having inventory are (Slack *et al*., 2006):

- *To deal with uncertainties in supply and demand*
  There may be unexpected delays in the delivery of supplies, then a buffer or safety stock is needed to be able to continue with production or fulfil demand. In practice demand is never completely certain. When there is a *safety stock*, one can deal with fluctuations in demand.
- *For decoupling purposes*
  Inventory can decouple operations when there is an inventory of half-products between two production stages. The purpose of this inventory is to reduce dependency between production stages. It prevents production being stopped due to a lack of half-products caused by machine breakdowns, material shortages or other production fluctuations. Furthermore, it allows production units to schedule their own operations. This type of inventory is called decoupling inventory.
- *To overcome seasonalities in supply and demand (production smoothing)*
  Sometimes raw materials are only available during a short period of the year, like fresh soft fruits. Then they can be processed into half-products to be able to produce the end-products throughout the year. In this case the inventory is held to overcome seasonalities in production. Finished products (or end products) can also be kept in stock to overcome seasonalities in demand, like ice-skates. This type of inventory is called anticipatory inventory or *seasonal inventory*.
- *To reduce costs or compensate for process inflexibility*
  Large order quantities can reduce costs in two ways. Firstly, the number of orders placed is less, so the fixed cost per order of administration and materials handling has to be paid less often. Secondly, sometimes a quantity discount can be obtained when ordering a large amount. These savings may be greater than the extra cost of holding inventory. The same holds for production. When a company manufactures a wide range of items, then the same equipment will be used to produce several items. Producing larger batches will improve productivity and reduce costs, although it results in higher inventories. This type of inventory is called *cycle inventory* indicating that the number of units purchased or produced is greater than the firm's immediate needs.
- *To fill the distribution pipeline*
  When goods are delivered to a supermarket, this is mostly a full truckload of many different items. Those items are brought together in a warehouse where many suppliers deliver their products. Products in transport and in warehouses are not available for customers, so it is in-transit or *pipeline inventory*.

Though there are many good reasons to have inventory, there are also many good reasons not to have inventory. Some of them are (Slack *et al.*, 2006):
- Inventory ties up money or working capital that cannot be spent otherwise to make a profit;
- Inventory takes up storage space that can be more valuable than the stored items (packaging materials);
- Storage of items can be hazardous (chemicals);
- High levels of inventory between production stages hides problems because it decouples activities of adjacent production stages;
- Inventory can become obsolete (when changes in production are made, or in the case of fashion items);
- Inventory can be damaged or can deteriorate (fresh food).

In inventory management it is important to find a balance between allowing for inventory to build up, in order to maintain the continuation of production and to meet a certain service level for customers, and reducing the inventory to save costs.

11.1.3. What is inventory management?

Inventory management is the activity of planning and controlling resources (inventories) as they move through supply chains, operations and processes (Slack *et al.*, 2006). In inventory management the basic questions are:
- When should an order be released?
- How large should each order be?
- How to control the inventory?

These questions should be answered on the conditions that costs will be minimised and a certain service level will be met, supported by inventory models. Section 11.2 will elaborate on the main cost factors that need to be taken into account in inventory models. We will classify the inventory models according to the demand patterns of the items under consideration.

Section 11.3 presents a classification of demand patterns. In the Sections 11.4, 11.5 and 11.6 different inventory models will be discussed according to the classification of demand patterns. The chapter is concluded with some final remarks.

11.2. Inventory costs

In inventory management there are four cost components to consider: (1) the purchasing or production cost of the items, (2) the ordering cost in case of buying items, or the set-up cost in case of producing items, (3) the inventory holding costs, and (4) the cost of having a shortage of items (so-called stock outs).

The *purchasing cost* is the unit value of an item, at the moment the item enters the company. When the supplier calculates the factory gate price, the value of the item for the receiving company is the factory gate price increased by the transportation cost. In case of producing items, the unit value is the *production cost* per item.

The *ordering cost* is the fixed cost of placing an order. That can be a cost that is accounted for by the supplier of the items, as a payment for administration, handling or transportation. It can also be the fixed cost of the company's own purchasing department, or any combination of the two. In case of the production of an item, there is a fixed *set-up cost* when the production of a batch is started, for instance, because of the adjustment of the equipment to produce a different item.

The *inventory holding costs* consist of the cost of the warehouse and the cost of the capital invested in inventory. The cost of the warehouse is the cost of the space needed, but also the handling in the warehouse. This cost also depends, among other things, on the volume of the product, but it is usually calculated as a percentage of the value of the items. The capital tied up in inventory cannot be used in an alternative way, for instance to put it on a bank account to gain interest. Therefore, an interest rate is used to calculate part of the inventory holding costs. In practice, the inventory holding costs are usually difficult to calculate, and a kind of estimate for an inventory carrying charge is used, partly based on the annual banking interest rate, increased by a raise to include the warehouse cost. The *carrying cost* is expressed in euros per Euro value of the item. Therefore, the carrying cost multiplied by the unit value of an item gives the annual holding costs per item.

Shortage cost can occur when there are not enough items in stock to fulfil current demand. Then two scenarios can occur. Firstly there is *lost sales*. The customer does not come back for the item, so a certain profit is lost, for the moment, but maybe also for the future when the customer finds an alternative. Secondly there is *backlogging*. The customer still gets the item, but at a later moment in time. A cost is incurred for delivering later.

11.3. Classification of demand patterns

In order to be able to manage the inventory it is important to know the demand pattern for the items under consideration. Demand patterns can be classified in three ways:
Firstly, demand is either deterministic or stochastic. When demand, prices and lead times of products are known with (almost) certainty, demand is *deterministic*. In a *stochastic model* some of the data are uncertain. Secondly, demand is constant in time or varying in time (fluctuating). Thirdly, demand is *continuous* (nondenumerable) or *discrete* (denumerable).

These three directions can be combined into eight different demand patterns. A continuous demand is, for instance, the demand for electricity. The demand for electricity goes on 24 hours a day, 7 days a week. For some applications the electricity use is constant and deterministic because certain equipment works day and night at a constant known level (for example, a radio station), for other equipment the demand for electricity varies with time, and is uncertain (for example, air conditioning in a building varies with temperature and time of day). A discrete demand is, for instance, the demand for bread. There is a daily demand for bread, during the opening hours of the bakery. The demand is expressed as a demand per discrete time unit, in this case the demand for bread per day. The demand for regular bread is more or less constant and certain, the demand for special breads varies and is usually uncertain (stochastic).

In Figure 11.2 the demand patterns are represented for the combined directions deterministic – stochastic and constant – variable. Variable deterministic demand is reflected as a discrete demand, the other demand patterns are reflected as a continuous demand. Often when demand is discrete in reality, it is approached as continuous in an inventory model.

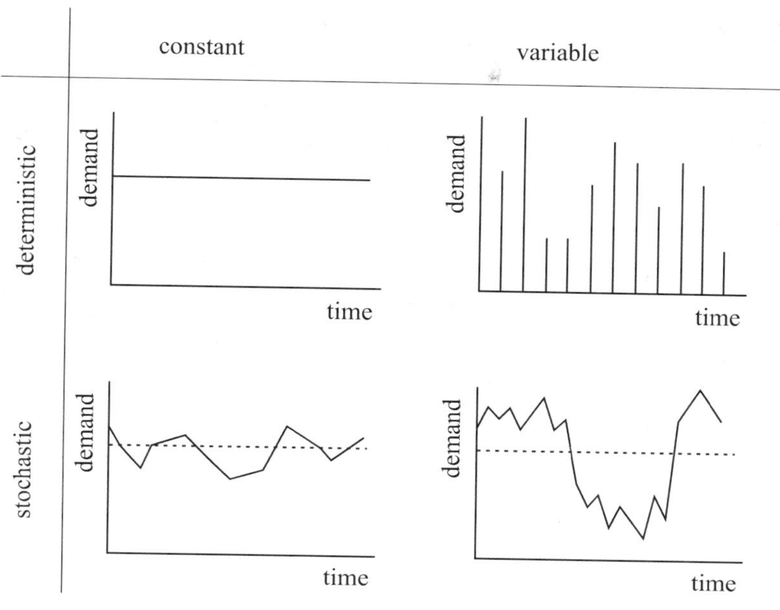

Figure 11.2. Demand patterns.

When demand is constant and deterministic and some other assumptions hold, the classical *Economic Order Quantity* can be derived. This will be discussed in Section 11.4, along with some extensions of the Economic Order Quantity. In case demand is variable and deterministic a replenishment strategy is provided in Section 11.5. When demand is stochastic and the mean can be estimated and the standard deviation is small, demand can be considered as constant and stochastic, with a known probability distribution function of demand. In Section 11.6 this situation is discussed as a one-period problem known as the Newsboy Problem, for the discrete case and the continuous case. Also, some multi-period order policies that are commonly used by practitioners are discussed in Section 11.6.
Variable, stochastic demand has a great standard deviation and often an unknown probability distribution function of demand. It is hard to use analytical models in this case. Simulation tools are often used to determine an order policy when demand is variable and stochastic with an unknown probability distribution function of demand. Simulation is discussed in Chapter 13.

The symbols that are used in this chapter are listed in the List of notations, elsewhere in this book, along with some basic relations between them. Note that variables to be calculated by the model are in capitals, coefficients are in lower case except for the order point s, which is a variable too. In this chapter the unit of time is a year, if not denoted otherwise. Other time units are also possible, for instance a month or a week.

11.4. Constant deterministic demand

11.4.1. Economic Order Quantity (EOQ)

The classical economic lot size model, also called the *Economic Order Quantity* (*EOQ*), was first derived in 1913 by F.W. Harris. It is also well known as Wilson's lot size formula or Camp's formula. It is a simple but nevertheless very useful approach for calculating the order quantity. In the case of the classical Economic Order Quantity a number of assumptions are made:

- only one product at a time is considered;
- demand d is deterministic and occurs at a constant rate (units per time);
- demand d is independent of the demand for other products;
- the order quantity Q is free, no restrictions;
- unit variable cost c is independent of the order size;
- a fixed reorder cost k is incurred every time an order is placed;
- lead time is zero;
- instantaneous re-supply: the entire quantity is delivered at the same time;
- no shortage allowed;
- all parameters remain constant for a long time.

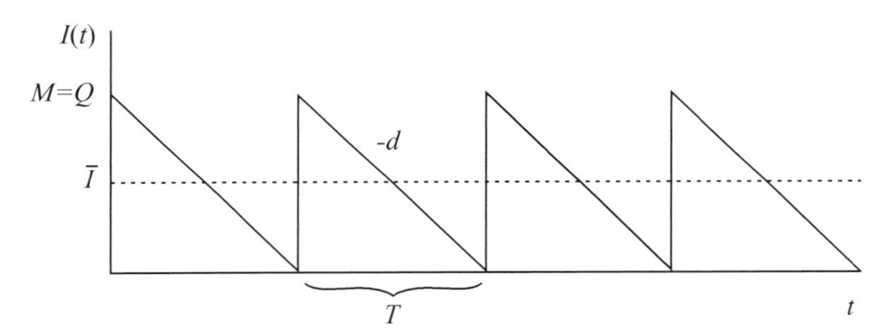

Figure 11.3. The classical economic lot size model.

Now the basics of the EOQ model will be explained. In the EOQ model the involved cost components, explained in Section 11.2, are minimised to derive the optimal Economic Order Quantity.
The inventory level as a function of time $I(t)$ has a characteristic pattern looking like saw teeth, see Figure 11.3. T is the time lapse between two consecutive orders. The moment an order with quantity Q is placed and replenished, the inventory level is Q. With a demand rate of d, the inventory level reaches zero after a time lapse of Q/d, so

$T = \frac{Q}{d}$ and the number of times an order should be placed per year is $\frac{d}{Q} = \frac{1}{T}$

The total costs per year $TC(Q)$ consist of the ordering cost, the inventory holding cost and the purchasing cost. The ordering cost k occurs every time an order is placed. This means, k has to be paid d/Q times a year.

So the ordering cost is $k \cdot \frac{d}{Q}$

The inventory holding cost h is paid over the average inventory level. The maximum inventory level is M, which equals Q. The average inventory level $\bar{I} = \frac{1}{2}M = \frac{1}{2}Q$.

So the inventory holding cost is $\frac{1}{2}Q \cdot h$.

The yearly purchasing cost is the demand d per year multiplied by the unit cost c per item.

So the purchasing cost is $c \cdot d$.

This gives the following relationship:

Total costs per year = Ordering cost + Inventory holding cost + Purchasing cost

$$TC(Q) = k \cdot \frac{d}{Q} + \frac{1}{2}Q \cdot h + c \cdot d \qquad (11.1)$$

The optimal value of Q can be derived by first order analysis of $TC(Q)$, such that $\frac{dTC}{dQ} = 0$.

$$\frac{dTC}{dQ} = -1 \cdot kdQ^{-2} + \tfrac{1}{2}h = 0 \quad \Leftrightarrow \quad \frac{kd}{Q^2} = \frac{h}{2} \quad \Leftrightarrow \quad Q^2 h = 2kd$$

$$\Rightarrow \quad Q^* = \sqrt{\frac{2kd}{h}} \quad \text{since } Q \geq 0 \qquad (11.2)$$

It can be shown that the second derivative $\frac{d^2TC}{dQ^2} > 0$, so Q^* is a minimum (see Section 12.3).

$Q^* = \sqrt{\frac{2kd}{h}}$ is called the *Economic Order Quantity*

From the optimal order quantity Q^*, the optimal time lapse T^* between two consecutive orders can be derived:

$$T^* = \frac{Q^*}{d} = \frac{1}{d}\sqrt{\frac{2kd}{h}} = \sqrt{\frac{2k}{hd}} \qquad (11.3)$$

Substituting Q^* in Equation (11.1) gives, after some rewriting, the total costs per time unit at the Economic Order Quantity:

$$TC(Q^*) = \sqrt{2kdh} + cd \qquad (11.4)$$

Note that the purchasing cost $c \cdot d$ does not depend on Q, so as long as the unit value of an item c is constant, the purchasing cost is not relevant for the determination of the optimal order quantity. Therefore, the Total Relevant Costs per year $TRC(Q)$ are often used in calculations.

$$TRC(Q) = k \cdot \frac{d}{Q} + \tfrac{1}{2}Q \cdot h \quad \text{or} \quad TRC(Q^*) = \sqrt{2kdh} \qquad (11.5)$$

Example 11.1. Economic order quantity.

A small liquor store sells wine from a special vineyard. The store has a group of enthusiastic customers who have a constant demand throughout the year for their special wine. Together they have a yearly demand of 1000 bottles of wine. The unit cost of one bottle of wine is € 12 and to keep one bottle in stock for a year costs € 0.96. Every time an order is placed, there is an ordering cost of € 20. How large should each order of the liquor store be, and how often must an order be placed?

Using Equation (11.2) gives $Q^* = \sqrt{\dfrac{2kd}{h}} = \sqrt{\dfrac{2 \cdot 20 \cdot 1000}{0.96}} = 204.12$

The number of orders to be placed is $\dfrac{d}{Q} = \dfrac{1000}{204.12} = 4.90$ orders a year.

Naturally, one can not order 0.12 bottles of wine, so Q^* is rounded off to 204.
Equation (11.1) or (11.4) gives the total costs $TC(Q^*) =$ € 12195.96
Equation (11.5) gives the total relevant costs $TRC(Q^*) =$ € 195.96

In Figure 11.4 the graph of the cost functions of Example 11.1 is drawn. The ordering cost, the inventory holding cost and the total relevant costs $TRC(Q)$ are depicted. Figure 11.4 shows that the value of $TRC(Q)$ increases only slightly on both sides around the optimal value of Q^*, which means that the value of Q^* is rather robust. Ordering a quantity close to the EOQ still leads to almost the same total (relevant) costs. When for efficiency reasons the order quantity has to be adapted to a quantity different from the optimal value, but still close, the costs increase less than the change in the order quantity.

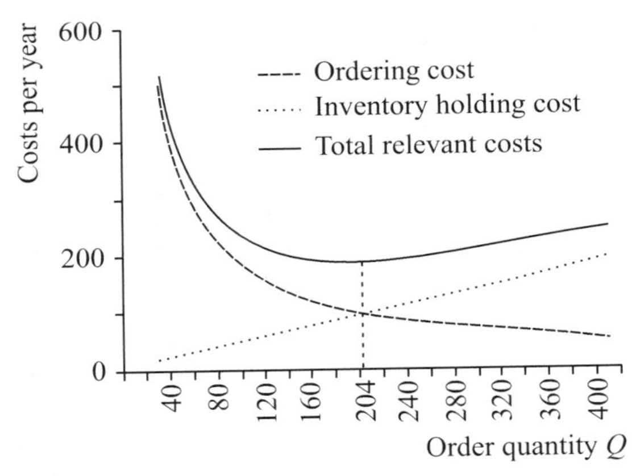

Figure 11.4. The cost functions of the classical economic lot size model.

Example 11.2.

In Example 11.1, the Economic Order Quantity $Q^* = 204$. The total relevant costs at Q^* are $TRC(204) =$ € 195.96. When the order quantity Q is diminished by about 10% to 180, then $TRC(180) =$ € 197.51, which is an increase of 0.8% with respect to the minimal costs. Increasing the order quantity Q by about 10% to 220, gives $TRC(220) =$ € 196.51, which is an increase of 0.3% with respect to the minimal costs.

The robustness of the Economic Order Quantity shown in Figure 11.4 and in Example 11.2 is the reason that the EOQ formula is used so often. Gathering the data needed to calculate it can be difficult, because the data are not always available within companies. Often the data have to be estimated. Even when the estimates are not completely accurate, the EOQ is a good approach for deciding on the order quantity.

11.4.2. Economic Production Quantity (EPQ).

In the case of the Economic Order Quantity an instantaneous re-supply is assumed. When determining the *Economic Production Quantity* that assumption is relaxed. The replenishment arrives gradually when production proceeds. There is a finite production rate p and the time to produce a replenishment of size Q is T_p. It holds that $p > d$, otherwise production can not keep up with demand. Instead of an ordering cost, in this case there is a set-up cost k to start the production of the item. Q is the amount of items to be produced in every production cycle. The inventory level as a function of time $I(t)$ in this case is represented in Figure 11.5.

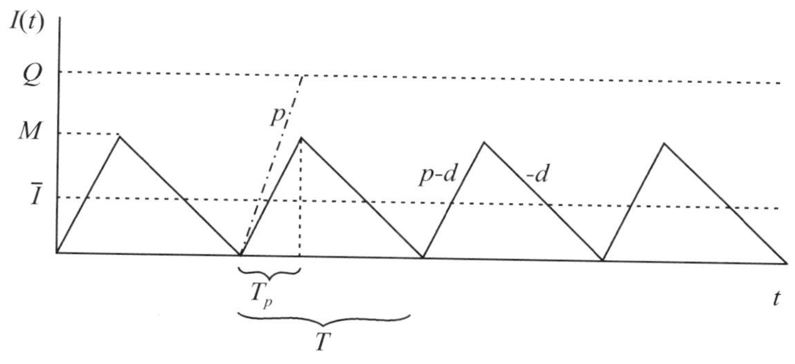

Figure 11.5. The inventory level as a function of time when the production rate is finite.

In order to formulate the total costs per year, the formulation of the average inventory holding cost per year is needed. In Figure 11.5 it is clear that the maximum inventory level is M. During the production time T_p there is a gross increase in the inventory level with production rate p, and also a decrease in the inventory level with demand rate d. So during T_p there is a net increase in the inventory level with rate $(p-d)$. The maximum inventory level M is the net increase in inventory $(p-d)$ multiplied by the time T_p needed to produce Q units at rate p. So,

$$M = T_p \cdot (p-d).$$

Production time $T_p = \frac{Q}{p}$, so M can be expressed as

$$M = \frac{Q}{p} \cdot (p-d)$$

The average inventory level is

$$\bar{I} = \tfrac{1}{2}M = \tfrac{1}{2}\cdot\frac{Q}{p}(p-d) = \frac{Q(1-\frac{d}{p})}{2}$$

T still is the time lapse between two consecutive orders (the cycle time), $T = \dfrac{Q}{d}$

The number of times production should be started per time unit is $\dfrac{1}{T} = \dfrac{d}{Q}$

Total costs per time unit = Set-up cost + Inventory holding cost + Production cost

$$TC(Q) = k\cdot\frac{d}{Q} + \frac{Q(1-\frac{d}{p})h}{2} + c\cdot d \quad (11.6)$$

The optimal value of $TC(Q)$ is obtained when $\dfrac{dTC}{dQ} = 0$. So

$$\frac{dTC}{dQ} = -\frac{kd}{Q^2} + \frac{(1-\frac{d}{p})h}{2} = 0 \quad \Rightarrow \quad Q^* = \sqrt{\frac{2kd}{(1-\frac{d}{p})h}} \quad \text{since } Q \geq 0$$

(11.7)

It can be shown that the second derivative $\dfrac{d^2TC}{dQ^2} > 0$, so Q^* is a minimum (see Section 12.3).

$Q^* = \sqrt{\dfrac{2kd}{(1-\frac{d}{p})h}}$ is called the *Economic Production Quantity*

Example 11.3. Economic production quantity.

A toy factory produces a special kind of radio-control toy car. There is a constant demand throughout the year. The yearly demand is 10,000 toy cars. The unit cost of one car is € 12 and to keep one radio-control toy car in stock for a year costs € 0.96. Every time a production cycle is started, there is a set-up cost of € 20. The production rate is 50,000 cars a year. How large should each production lot of the toy factory be for this car and how often should the production start for this car?

Equation (11.7) shows $Q^* = \sqrt{\dfrac{2kd}{(1-\frac{d}{p})h}} = \sqrt{\dfrac{2\cdot 20\cdot 10000}{(1-\frac{10000}{50000})\cdot 0.96}} = 721.69$

The number of times to start production is $\dfrac{d}{Q} = \dfrac{1000}{721.69} = 13.85$ times a year

Equation (11.6) results in total costs of € 120554.26

11.4.3. Shortages

When considering the assumptions made for the classical economic lot size model, there is another assumption that can be released. It is possible to allow for *shortages* in inventory, so that demand can not be fulfilled directly from stock. Then two possibilities occur: either there are *lost sales*, or demand can be *backlogged*. Both cases can be included in the model with instantaneous re-supply (purchasing) and in the model with non-instantaneous re-supply (production). The latter, with a production rate p, is the most general lot-sizing model with a constant, deterministic demand rate. All other models are special cases of this general model. The interested reader is referred to (Ghiani *et al.*, 2004).

11.4.4. Quantity discounts

In practice it is quite common for suppliers to offer *quantity discounts*, or for *economies of scale* to occur in a production environment. In a production environment, the unit value c of an item may vary with the production lot size Q. In a purchasing environment suppliers can offer discounts in two ways: quantity-discounts-on-all-units and incremental discounts. In the first case when the size of Q exceeds a certain value, i.e. the discount break, every item gets a discount. In the second case, the case of incremental discounts, when the size of Q exceeds the discount break, then the items below the discount break are bought at the standard price and the items above the discount break get the discount price. Quantity-discounts-on-all-units is the most common discount structure in practice, so this one will be discussed here. As will be shown, the EOQ model can be expanded with quantity discounts, only relaxing the assumption that the unit variable cost c is independent of the order size (Hax and Candea, 1984); (Ghiani *et al.*, 2004).

Quantity-discounts-on-all-units

The function $f(Q)$ gives the purchasing cost of Q items, it is assumed to be piecewise linear as shown in Figure 11.6.

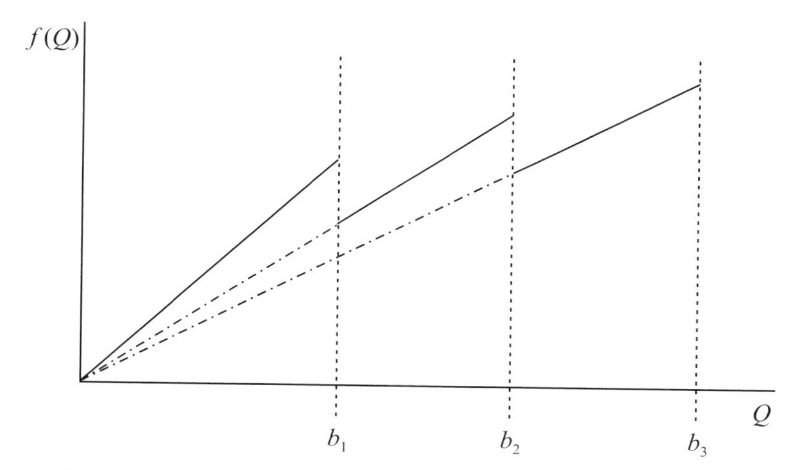

Figure 11.6. The purchasing cost function f(Q).

The purchasing cost function can be denoted as $f(Q) = c_i Q \quad b_{i-1} \leq Q < b_i \quad i = 1, 2, ...$ Then $b_0 = 0$ and b_1, b_2, etc. are the discount breaks set by the supplier of the product and $c_i > c_{i+1}$ for i = 1, 2, …. If the order size Q is between $b_{i-1} \leq Q < b_i$, then every item has the same unit value c_i.

The total costs function $TC(Q)$ can be written as

$$TC(Q) = TC_i(Q) \qquad b_{i-1} \le Q < b_i \qquad i = 1, 2, ... \quad \text{with } h_i = rc_i \qquad i = 1, 2, ...$$

$$TC_i(Q) = k \cdot \frac{d}{Q} + \tfrac{1}{2} Q \cdot h_i + c_i \cdot d \qquad i = 1, 2, ...$$

Compared to Equation (11.1), the total costs function appears to be the same as in the classical EOQ model, except now c_i and h_i depend on the discount level, so the total costs also depend on the discount level. Because $h_i = rc_i$, the value of h_i changes when c_i changes. Therefore, the following formulation of the total costs function is used:

$$TC_i(Q) = k \cdot \frac{d}{Q} + \tfrac{1}{2} Q \cdot rc_i + c_i \cdot d \qquad i = 1, 2, ... \qquad (11.8)$$

The optimal order size Q^* can be obtained in three steps using first order analysis of $TC_i(Q)$.

Algorithm 11.1. Order quantity with discounts-on-all-units.

Step 1.

Determine the economic order quantity for every $TC_i(Q)$ and call it Q_i:

calculate $Q_i = \sqrt{\dfrac{2kd}{rc_i}}$ for every value of c_i.

Step 2.

Determine for each discount interval i, $[b_{i-1}, b_i]$ the value of Q_i^* which minimises $TC_i(Q)$ in that interval.

So $Q_i^* = \begin{cases} b_{i-1} & \text{if } Q_i \le b_{i-1} \\ Q_i & \text{if } b_{i-1} \le Q_i \le b_i \qquad i = 1,2,\ldots \\ b_i & \text{if } Q_i > b_i \end{cases}$

Step 3.

Calculate $TC_i(Q_i^*) = k \cdot \dfrac{d}{Q_i^*} + \tfrac{1}{2} Q_i^* \cdot rc_i + c_i \cdot d$ for every Q_i^* $\quad i = 1, 2, ...$

Take the minimum of $TC_i(Q_i^*)$ over i. This gives the optimal Q^* and corresponding optimal discount level.

Example 11.4. Quantity discounts-on-all-units.

A small liquor store sells wine from a special vineyard. The store has a group of enthusiastic customers who have a constant demand throughout the year for their special wine. Together they have a yearly demand of 1000 bottles of wine. The unit cost of one bottle of wine is € 12 and to keep one bottle in stock for a year costs $h = rc = 0.08 \cdot 12 = € \ 0.96$. Every time an order is placed, there is an ordering cost of € 20. Suppose that the store can obtain the wine at a discount price of € 11 a bottle when an order of more than 200 bottles is placed. When more than 300 bottles are ordered a price of € 10 is in force. How large should each order of the liquor store be, at what price?

Step 1.

$c_1 = 12$ $\quad Q_1 = \sqrt{\frac{2kd}{rc_1}} = \sqrt{\frac{2 \cdot 20 \cdot 1000}{0.08 \cdot 12}} = 204.12$

$c_2 = 11$ $\quad Q_2 = 213.20$

$c_3 = 10$ $\quad Q_3 = 223.61$

Step 2.

$b_0 = 0 \quad b_1 = 200 \quad b_2 = 300 \quad b_3 = \infty$

$$Q_1 = 204.12 > b_1 \Rightarrow Q_1^* = 200$$
$$Q_2 = 213.20 \Rightarrow b_1 < Q_2 < b_2 \Rightarrow Q_2^* = 213.20$$
$$Q_3 = 223.61 < b_2 \Rightarrow Q_3^* = 300$$

Step 3.

$$TC_1(Q_1^*) = TC_1(200) = 20 \cdot \frac{1000}{200} + \tfrac{1}{2} \cdot 200 \cdot 0.96 + 12 \cdot 1000 = €12196$$
$$TC_2(Q_2^*) = TC_2(213.20) = €11187.62$$
$$TC_3(Q_3^*) = TC_3(300) = €10186.67$$

The optimal discount level is 3, with $Q^* = 300$ and a total cost of € 10186.67

The steps are graphically represented in Figure 11.7.

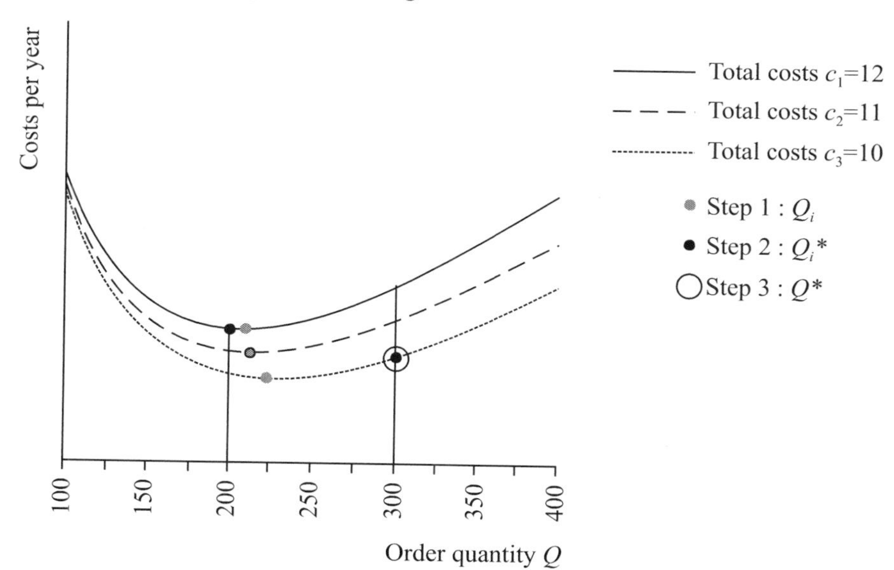

Figure 11.7. Quantity discounts-on-all-units.

11.5. Time-varying deterministic demand

The classical economic lot-size model was developed for the situation with a constant, deterministic demand. The demand was expressed continuously. There are also cases where demand is deterministic but time-varying. Here the demand is expressed in discrete time units, for instance a demand per week, or per month. Examples are:

- Items with a seasonal demand pattern, sometimes due to promotional activities;
- Replacement parts for preventive maintenance;
- Production to contract;
- Items for multi-echelon assembly operations, with a dependent demand. There is a customer demand for assembled end products. To produce them there is a dependent demand for components which have to be assembled into this end product. This kind of demand is often calculated with the help of a *Materials Requirement Planning* (MRP) system.

When the average deterministic demand rate varies with time, it is no longer optimal to order always the same replenishment quantity, with the same period between two consecutive orders. In practice one can just order the demand for every period (lot-for-lot) but it is also possible to calculate an order strategy that is more cost effective. In the latter case, there are essentially three approaches for calculating an order strategy (Silver *et al.*, 1998):

1. *Ignore the variability of demand and use the basic EOQ*. Based on an average demand the EOQ is calculated and used as a fixed order quantity. This approach can only be used when the variability of the demand is low.
2. *Calculate the optimal solution*. Under a specific set of assumptions the Wagner – Whitin algorithm can be used. This algorithm is an application of Dynamic Programming, which is discussed in Chapter 8. The Wagner – Whitin algorithm gives the optimal solution under restrictive assumptions. The algorithm has a complex structure and is therefore difficult to understand for practitioners. The interested reader can study the algorithm in (Silver *et al.*, 1998).
3. *Use a heuristic approach*. A heuristic approach does not guarantee finding the optimal solution, but aims to find a reasonable solution, see also Chapter 10. Many heuristic approaches have been developed to determine an order strategy for the case of variable deterministic demand. Most heuristics outperform the use of the basic EOQ with fixed order quantities. Baker (1989) found that the Silver – Meal heuristic in test problems only incurred an average cost penalty of less than 1% compared to the optimal solution obtained by the Wagner – Whitin algorithm (Silver *et al.*, 1998). As the Silver – Meal heuristic performs better than other tested heuristics it is discussed in this chapter.

Silver – Meal heuristic

For the Silver – Meal heuristic the following assumptions are made:

- demand d_t in period t is deterministic, but time-varying $t = 1, .., T$;
- there is a rolling and discrete time horizon $t = 1, .., T$ (a rolling-time horizon means that during calculations, periods can be added at the end of the current time horizon);
- the entire requirement for each period must be available at the beginning of that period;
- there is a fixed reorder cost k;
- the unit variable cost c does not depend on the replenishment quantity;
- the costs remain constant within the time horizon;

- the item is treated independently of other items;
- lead time is known with certainty;
- no backlogging allowed;
- instantaneous resupply;
- holding cost h only to inventory that is carried from one period to the next, not for inventory that is used during the running period.

From the assumptions above it follows that there is only replenishment when the inventory level is zero, and the replenishment quantity is for an integer amount of periods. The Silver – Meal heuristic is also called the Least Period Cost heuristic because the average cost per period is minimised. The concept is the following. If a replenishment arrives at the beginning of period $t = i$ and it covers supply to the end of $t = i + N - 1$, where N is the number of periods covering the replenishment, then

$$TRCP(N) = \frac{k + \text{total holding cost to the end of } i + N - 1}{N} = \frac{k + h\sum_{j=0}^{N-1} jd_{t+j}}{N}$$

So,

$$TRCP(1) = \frac{k}{1}$$

$$TRCP(2) = \frac{k + h \cdot d_{t+1}}{2}$$

$$TRCP(3) = \frac{k + h \cdot d_{t+1} + 2 \cdot h \cdot d_{t+2}}{3}, \text{ etc.}$$

The replenishment quantity is $Q_i = \sum_{t=i}^{i+N-1} d_t$

The criterion function is:
Minimise the Total Relevant Costs per Period (TRCP) for the duration of the replenishment quantity.

Algorithm 11.2. Silver – Meal heuristic.

Given demand data $d_1, d_2, .., d_T$ it determines values for $Q_1, Q_2, .., Q_T$ that are initially set to 0

Start with $i = 1$

While $i \leq T$ do

Starting with $N = 1$, evaluate $TRCP(N) = \frac{k + h\sum_{j=0}^{N-1} jd_{i+j}}{N}$ for increasing values of N until, for the first time $TRCP(N+1) > TRCP(N)$

N is the number of periods the replenishment should cover, so $Q_i = \sum_{t=i}^{i+N-1} d_t$

$i := i + N$

Example 11.5. Silver – Meal heuristic.

A toy shop Puzzle House faces a time-varying demand for one of their mind-challenging puzzles, according to the following:

Period t	1	2	3	4	5	6	7	8
Demand d_t	10	14	12	18	80	50	12	120

Ordering cost k is € 20 and holding cost h is € 1 per item, per period.

When should Puzzle House order, and what quantity, according to the Silver – Meal heuristic?

$i = 1$: $t = 1$:

$N = 1$: $TRCP(1) = \frac{k}{1} = \frac{20}{1} = 20$

$N = 2$: $TRCP(2) = \frac{k + h \cdot d_2}{2} = \frac{20 + 1 \cdot 14}{2} = 17$

$N = 3$: $TRCP(3) = \frac{k + h \cdot d_2 + 2 \cdot h \cdot d_3}{3} = \frac{20 + 1 \cdot 14 + 2 \cdot 12}{3} = 19.3 > TRCP(2)$

In period 1, order $Q_1 = 10 + 14 = 24$. In period 2, order $Q_2 = 0$
$i = t + N = 1 + 2 = 3$

$i = 3$: $t = 3$:

$N = 1$: $TRCP(1) = \frac{k}{1} = \frac{20}{1} = 20$

$N = 2$: $TRCP(2) = \frac{k + h \cdot d_4}{2} = \frac{20 + 1 \cdot 18}{2} = 19$

$N = 3$: $TRCP(3) = \frac{k + h \cdot d_4 + 2 \cdot h \cdot d_5}{3} = \frac{20 + 1 \cdot 18 + 2 \cdot 80}{3} = 66 > TRCP(2)$

In period 3, order $Q_3 = 12 + 18 = 30$. In period 4, order $Q_4 = 0$
$i = t + N = 3 + 2 = 5$

The reader can verify $Q_5 = 80$, $Q_6 = 50 + 12 = 62$, $Q_7 = 0$ and $Q_8 = 120$.

11.6. Constant stochastic demand

In the case of constant stochastic demand the average demand is approximately constant, but uncertain. Two cases are discussed with constant stochastic demand: the *single-period inventory problem* and the *inventory problem with infinite time horizon*. The following assumptions hold, for both cases:

- demand $\tilde{d}$ is stochastic and occurs approximately at a constant rate (units per year);
- demand $\tilde{d}$ is independent of the demand for other products;
- the order quantity is free, no restrictions;
- unit variable cost c is independent of the order size;

- only one item at a time is considered;
- a fixed reorder cost k is incurred every time an order is placed;
- instantaneous resupply;
- all parameters remain constant for a long time.

The order decision has to be made before the demand is known with certainty.

11.6.1. One-period stochastic demand: the Newsboy Problem.

The *Newsboy Problemn* is the popular name for the single-period inventory problem with stochastic demand. In this case there is only one opportunity to order a certain amount of product that will be sold in the upcoming period. The period in which the product is sold is relatively short: varying from one day to three or four months at the most. Examples for this problem are: newspapers, Christmas trees, seasonal pot plants, fashion items. After the selling period, the items that are not sold have a salvage value. The salvage value can be zero, or in the case of newspapers the value of wastepaper and in the case of fashion items the price the outlet store is willing to pay. In case the demand exceeds the amount ordered, there occurs a shortage, the sale is lost, backlogging is not possible. In determining the order quantity for the Newsboy Problem the purchasing cost, the selling price and the salvage value are cost factors to consider.

Newsboy Problem: Discrete probability distribution function

The Newsboy Problem is first discussed on the basis of an example with a discrete probability distribution function.

Example 11.6. Newsboy Problem: Discrete probability distribution function.

A newsboy has to decide how many newspapers to order at the beginning of the day. He faces the following costs:

c	purchasing cost	€ 1
re	selling price	€ 2
u	salvage value	€ 0.25

where $re > c > u$.

The demand is a discrete random variable $\tilde{d}$ with probability function $f_{\tilde{d}}(d) = \text{P}(\tilde{d} = d)$ and discrete cumulative distribution function $F_{\tilde{d}}(d)$, as shown in Table 11.1.

Table 11.1. Discrete demand distribution for the Newsboy Problem.

d	50	70	90	110	130	150
$f_{\tilde{d}}(d)$	$\frac{1}{6}$	$\frac{1}{6}$	$\frac{1}{6}$	$\frac{1}{6}$	$\frac{1}{6}$	$\frac{1}{6}$
$F_{\tilde{d}}(d)$	$\frac{1}{6}$	$\frac{1}{3}$	$\frac{1}{2}$	$\frac{2}{3}$	$\frac{5}{6}$	1

The expected demand $E(\tilde{d}) = \frac{1}{6}(50+70+90+110+130+150) = 100$ newspapers.
How many papers should the newsboy order to maximise his profit?

When he orders $Q = 90$, the expected profit $\rho(Q)$ will be:
$\rho(Q) = \rho(90) = re \cdot \frac{1}{6}(50+70+90) + re \cdot \frac{1}{6}(90+90+90) + u \cdot \frac{1}{6}(40+20) - c \cdot 90 = 72.5$

In words: the expected profit is the sum of four terms. The first two terms are the expected sales. The first term counts for the values of demand lower than the order quantity $Q = 90$: the amount sold (the demand) times the selling price multiplied by the probability of that demand. The second term counts when demand is greater than the order quantity. One cannot sell more than $Q = 90$. The third term applies when demand is smaller than Q. For the amount of Q not sold, the salvage value is obtained. The last term is the cost to purchase Q items which has to be subtracted.

For $Q = 110$:
$\rho(110) = re \cdot \frac{1}{6}(50+70+90+110) + re \cdot \frac{1}{6}(110+110) + u \cdot \frac{1}{6}(60+40+20) - c \cdot 110 = 75.01$
Taking $Q = 110$, the newsboy increases his expected profit compared to $Q = 90$. Can he further increase his expected profit by ordering $Q = 130$?

For $Q = 130$:
$\rho(130) = re \cdot \frac{1}{6}(50+70+90+110+130) + re \cdot \frac{1}{6} \cdot 130 + u \cdot \frac{1}{6}(80+60+40+20) - c \cdot 130 = 71.6$
The expected profit has decreased compared to the previous calculation. An order quantity of $Q = 110$ seems to be the best of these three options for the newsboy.

In the above example the expected profit is calculated for a few of the possible order quantities to obtain the optimal one. There should be a way to obtain the optimal order quantity without calculating the whole range.

The general function of the expected profit for the Newsboy Problem can be written as:

$$\rho(Q) = re \cdot \sum_{d=0}^{Q} d \cdot f_{\tilde{d}}(d) + re \cdot \sum_{d=Q+1}^{\infty} Q \cdot f_{\tilde{d}}(d) + u \cdot \sum_{d=0}^{Q} (Q-d) \cdot f_{\tilde{d}}(d) - c \cdot Q \tag{11.9}$$

Adding and subtracting $re \cdot \sum_{d=Q+1}^{\infty} d \cdot f_{\tilde{d}}(d)$:

$$\rho(Q) = re \cdot \sum_{d=0}^{Q} d \cdot f_{\tilde{d}}(d) + re \cdot \sum_{d=Q+1}^{\infty} d \cdot f_{\tilde{d}}(d) - re \cdot \sum_{d=Q+1}^{\infty} d \cdot f_{\tilde{d}}(d) + re \cdot \sum_{d=Q+1}^{\infty} Q \cdot f_{\tilde{d}}(d) + u \cdot \sum_{d=0}^{Q} (Q-d) \cdot f_{\tilde{d}}(d) - c \cdot Q$$

The first two terms sum to $re \cdot E(\tilde{d})$, so after rewriting:

$$\rho(Q) = re \cdot E(\tilde{d}) + re \cdot \sum_{d=Q+1}^{\infty} (Q-d) \cdot f_{\tilde{d}}(d) + u \cdot \sum_{d=0}^{Q} (Q-d) \cdot f_{\tilde{d}}(d) - c \cdot Q \tag{11.10}$$

To find the order quantity Q for which the expected profit $\rho(Q)$ has the maximum value, consider the approximate derivative (Section 12.3) of $\rho(Q)$: $\Delta\rho(Q) = \rho(Q+1) - \rho(Q)$

$$\rho(Q+1) = re \cdot E(\tilde{d}) + re \cdot \sum_{d=Q+2}^{\infty} ((Q+1)-d) \cdot f_{\tilde{d}}(d) + u \cdot \sum_{d=0}^{Q+1} ((Q+1)-d) \cdot f_{\tilde{d}}(d) - c \cdot (Q+1)$$

$$\rho(Q) = re \cdot E(\tilde{d}) + re \cdot \sum_{d=Q+1}^{\infty} (Q-d) \cdot f_{\tilde{d}}(d) + u \cdot \sum_{d=0}^{Q} (Q-d) \cdot f_{\tilde{d}}(d) - c \cdot Q$$

$$\Delta\rho(Q) = re \cdot \sum_{d=Q+1}^{\infty} f_{\tilde{d}}(d) + u \cdot \sum_{d=0}^{Q} f_{\tilde{d}}(d) - c$$

Because $\sum_{d=Q+1}^{\infty} f_{\tilde{d}}(d) = 1 - \sum_{d=0}^{Q} f_{\tilde{d}}(d)$,

$$\Delta\rho(Q) = re - re \cdot \sum_{d=0}^{Q} f_{\tilde{d}}(d) + u \cdot \sum_{d=0}^{Q} f_{\tilde{d}}(d) - c \text{, such that}$$

$$\Delta\rho(Q) = (re-c) - (re-u) \cdot \sum_{d=0}^{Q} f_{\tilde{d}}(d)$$

Because $re > c > u$, $(re - u) > (re - c)$. Note that, $\sum_{d=0}^{Q} f_{\tilde{d}}(d) \leq 1$. This means that when Q is small and therefore $\sum_{d=0}^{Q} f_{\tilde{d}}(d)$ is close to zero, the approximate derivative $\Delta\rho(Q)$ first has a positive value and later, when Q gets greater and $\sum_{d=0}^{Q} f_{\tilde{d}}(d)$ closer to one, $\Delta\rho(Q)$ gets a negative value. In other words, $\rho(Q)$ is first increasing in Q and later decreasing, so $\rho(Q)$ has a maximum. The optimal order quantity for the newsboy is at the maximum value for the expected profit.

Take as Q^* the smallest value of Q for which $\Delta\rho(Q)$ is negative, then $\rho(Q)$ has its maximum at Q^*.

$$\Delta\rho(Q) \leq 0 \quad \Leftrightarrow \quad (re-c) - (re-u) \cdot \sum_{d=0}^{Q} f_{\tilde{d}}(d) \leq 0$$

$$\Leftrightarrow \quad \sum_{d=0}^{Q} f_{\tilde{d}}(d) \geq \frac{re-c}{re-u} \quad \text{or} \quad \mathrm{P}(\tilde{d} \leq Q^*) \geq \frac{re-c}{re-u} \qquad (11.11)$$

Example 11.6 continued: Newsboy Problem: Discrete probability distribution function

Returning to the example and using $\sum_{d=0}^{Q} f_{\tilde{d}}(d) \geq \frac{re-c}{re-u}$ gives $\frac{re-c}{re-u} = \frac{2-1}{2-0.25} = 0.57$

The first value of Q for which $\sum_{d=0}^{Q} f_{\tilde{d}}(d) \geq 0.57$ is $Q^* = 110$. This is also the value that was found by trying the three possibilities in the example.

Newsboy Problem: Continuous probability distribution function

In the continuous case, the derivation resembles the line followed in the discrete case. Now $f_{\tilde{d}}(d)$ is a density function.

$$\rho(Q) \;= re \cdot \int_0^Q d \cdot f_{\tilde{d}}(d)\,\mathrm{d}d \;+\; re \cdot \int_Q^\infty Q \cdot f_{\tilde{d}}(d)\,\mathrm{d}d + \; u \cdot \int_0^Q (Q-d) \cdot f_{\tilde{d}}(d)\,\mathrm{d}d \;-\; c \cdot Q \quad (11.12)$$

In words: the expected profit is the sum of four terms. The first two terms are the expected sales. The first term holds for the values of demand lower than the order quantity Q: the amount sold (the demand) times the selling price multiplied by the probability of that demand. The second term holds when demand is greater than the order quantity. One cannot sell more than the amount Q ordered. The third term applies when demand is smaller than Q. For the amount not sold, the salvage value is obtained. The last term is the cost to purchase Q items.

After adding and subtracting $re \cdot \int_Q^\infty d \cdot f_{\tilde{d}}(d)\mathrm{d}d$ this expression can be rewritten as:

$$\rho(Q) \;= re \cdot E(\tilde{d}) \;+\; re \cdot \int_Q^\infty (Q-d) \cdot f_{\tilde{d}}(d)\mathrm{d}d + \; u \cdot \int_0^Q (Q-d) \cdot f_{\tilde{d}}(d)\,\mathrm{d}d \;-\; c \cdot Q \qquad (11.13)$$

It can be shown that $\rho(Q)$ is concave for $Q \geq 0$ and $\rho(Q) \to -\infty$ for $Q \to \infty$. Therefore the maximum expected profit is achieved when the first derivative of $\rho(Q)$ with respect to Q is zero. It can be shown that for the optimal order quantity Q^* the following equation holds:

$$P(\tilde{d} \leq Q^*) = \frac{re - c}{re - u} \qquad (11.14)$$

This result resembles the result obtained in the discrete case in Equation (11.11).

Important is the case where the demand $\tilde{d}$ has a Normal distribution with expected value μ and standard deviation σ or $\tilde{d} \sim N(\mu, \sigma^2)$. See also Appendix B. For calculation purposes it is common to transform a variable $\tilde{d} \sim N(\mu, \sigma^2)$ to a Standard Normal variable $\tilde{x} \sim N(0,1)$, with $\tilde{x} = \dfrac{\tilde{d} - \mu}{\sigma}$. Then

$$P(\tilde{d} \leq Q^*) = P\left(\frac{\tilde{d} - \mu}{\sigma} \leq \frac{Q^* - \mu}{\sigma}\right) = \Phi\left(\frac{Q^* - \mu}{\sigma}\right) = \Phi(x) \qquad (11.15)$$

where $\Phi(x)$ is the cumulative distribution function of $\tilde{x}$. Equation (11.15) is substituted in Equation (11.14) to determine the optimal order quantity. The value of x can be found in special tables for the Standard Normal distribution function or calculated in Excel (see Appendix B). Consequently $Q^* = \mu + x\sigma$ can be derived.

Example 11.7. The Newsboy problem with a Normal distribution.

Pete Peters is a famous designer of lady's shoes. Once a year he picks one of his unique pairs of boots from his collection of designer boots to take into production. There is a group of female customers who buy his special boots almost every year. He sells the boots at a price $re = €\ 500$ per pair. The unit value c of one pair of boots is € 100. Pete wants his boots to be exclusive, so boots not sold will be kept in his atelier, therefore the salvage value $u = €\ 0$. Pete estimates that the demand $\tilde{d}$ for his boots has a Normal distribution with expected demand $\mu = 200$ and a standard deviation $\sigma = 25$ or $\tilde{d} \sim N(200, 625)$. How many pairs of boots should he bring into production to maximise his profit?

To calculate the optimal order quantity Q^* Equation (11.14) holds:

$$P(\tilde{d} \leq Q^*) = \frac{500-100}{500-0} = 0.8$$

Combining Equation (11.14) and (11.15): $P(\tilde{d} \leq Q^*) = \Phi(x) = 0.8$. In the table of the Standard Normal distribution the value of x can be found: $\Phi^{-1}(0.8) = 0.85$.

So, $Q^* = \mu + x\sigma \Rightarrow Q^* = 200 + 0.85 \cdot 25 = 221.25$

Pete should take 221 pairs of boots into production.

11.6.2. Multiple period stochastic demand: Control systems

In the Newsboy problem there is only one period, so only one opportunity to order. The most common inventory problems with stochastic demand have an infinite planning horizon and therefore multiple opportunities to reorder. The assumption is made that the demand is fairly constant over time, and the other parameters remain unchanged during the time horizon. Then, just as in the sections before, the questions when to reorder (the reorder point) and how much to reorder (the order quantity) have to be answered.

There are a number of possible *control systems* to deal with uncertain demand. The most common ones will be discussed in this chapter. In the discussion the *inventory position* is considered. The inventory position consists of the *on-hand-inventory* and the *on-order-inventory*. The on-order-inventory is already ordered, but has not yet arrived. The on-hand-inventory can be used to fulfil demand immediately. When an item is out-of-stock, there can be lost sales, or backlogging. In the following discussion *backlogging* is assumed. This means that the demand still has to be fulfilled, as soon as a replenishment arrives.

There are two ways of reviewing the inventory level: continuous review and periodic review.

- In the *continuous review* the inventory level is monitored continuously, and as soon as the inventory level decreases to or below a certain level, the reorder point s, a new order is placed. See Figure 11.8. The reorder point s is expressed in units of the product.
- In the *periodic review* the inventory level is monitored every R units of time ($R > 0$). At time R an order is placed or a decision has to be made whether or not to place an order. The quantity to order can be a fixed quantity Q: every time an order is placed, the order quantity is Q. The order quantity can also vary, in that case an order-up-to-level S is placed.

The above-mentioned points lead to the following *order policies*, which will be discussed in the remainder of this section:

(s, Q) order point, order quantity system;
(s, S) order point, order-up-to-level system;
(R, S) periodic review, order-up-to-level system;
(R, s, S) periodic review, order point, order-up-to-level system.

(s, Q) order point, order quantity system

In the (s, Q) order point, order quantity system, there is a continuous review of the inventory level ($R = 0$). A fixed order quantity Q is ordered whenever the inventory level drops to the reorder point s or lower. The order quantity is delivered after the lead time t_l, so during t_l demand has to be fulfilled from the inventory level still in stock which has level s or lower. This is a so called *two-bin system*. When the first bin is empty, the second bin is opened, and an order is placed. During lead time t_l the second (smaller) bin of size s is used. When the order arrives, first the second bin is refilled, then the rest of Q goes to the first bin.

The (s, Q) system is a simple system to operate, but there is a problem when individual transactions are large. When the last transaction before reordering is large, Q may not be enough to pass the reorder point s.

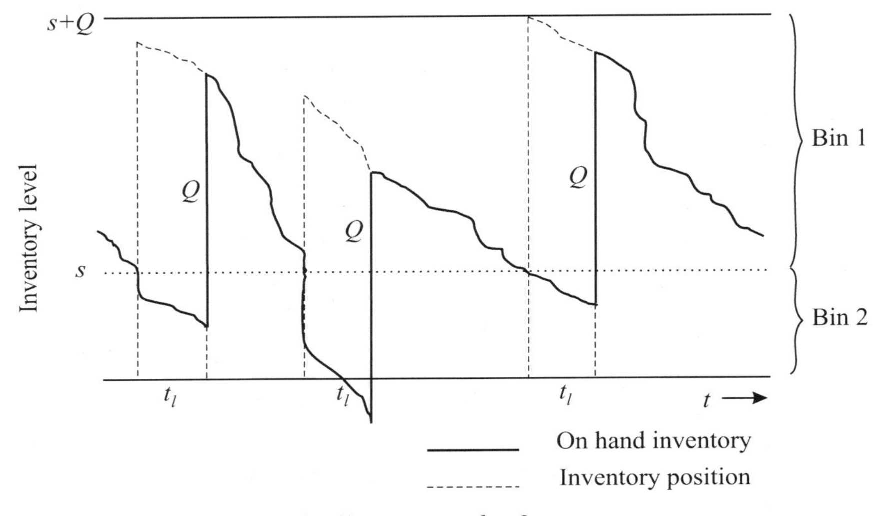

Figure 11.8. Continuous Review: (s, Q) system: order Q.

So how to calculate the optimal order quantity Q^* and the reorder point s?
Suppose the following assumptions hold:

- the demand rate $\tilde{d}$ has a Normal distribution with as expected value the average demand rate $\bar{d}$ and the standard deviation σ_d (the average demand rate $\bar{d}$ and σ_d can be estimated using techniques from forecasting theory (Chopra and Meindl, 2007). The demand during the lead time $\tilde{d}t_l$ has the following distribution: $\tilde{d}t_l \sim N(\bar{d}t_l,\ \sigma_d^2 t_l)$.
- $\bar{d}$ and σ_d are constant in time;
- the lead time t_l is deterministic and constant;
- the demand rate and the lead time are statistically independent;
- when a shortage occurs, demand will be backlogged.

In this case the order quantity Q^* and the reorder point s can be approached with the formula for the Economic Order Quantity that was derived for the deterministic case with constant demand according to Equation (11. 2).

First calculate $Q^* = \sqrt{\dfrac{2k\bar{d}}{h}}$

To calculate the reorder point s a service level α has to be set, e.g. a service level of $\alpha = 99\%$, which means that 99% of the demand should be fulfilled directly from stock. To meet the preset service level the inventory level has to be nonnegative during lead time with a probability of α. In other words: the probability that the demand $\tilde{d}$ during lead time t_l is smaller than or equal to the reorder point s has to be greater than or equal to the service level α. In formula: $\mathrm{P}(\tilde{d}t_l \leq s) \geq \alpha$.

Because $\tilde{d}$ has a Normal distribution (see Appendix B) this can be rewritten as $P\left(\dfrac{\tilde{d}t_l - \bar{d}t_l}{\sigma_d\sqrt{t_l}} \leq \dfrac{s - \bar{d}t_l}{\sigma_d\sqrt{t_l}}\right) = \Phi\left(\dfrac{s - \bar{d}t_l}{\sigma_d\sqrt{t_l}}\right) \geq \alpha$

where $\Phi(x)$ is the Standard Normal cumulative distribution function of $\tilde{x} \sim N(0,1)$. Then

$$x = \frac{s - \bar{d}t_l}{\sigma_d\sqrt{t_l}} \quad \Leftrightarrow \quad s - \bar{d}t_l = x\sigma_d\sqrt{t_l} \quad \Leftrightarrow \quad s = \bar{d}t_l + x\sigma_d\sqrt{t_l} \qquad (11.16)$$

So the reorder point s is the expected demand during lead time plus the expression $x\sigma_d\sqrt{t_l} = I_S$, which is called the *safety stock*. The safety stock is the expected inventory level when the order quantity Q^* arrives. Notice that if lead time t_l increases, then the safety stock I_S and the reorder point s also increase.

Example 11.8. (*s*, *Q*) system.

A supermarket uses the (s, Q) system to control the inventory of its packs of coffee. The demand for packs of coffee has a Normal distribution with an average demand $\bar{d} = 1500$ items per week, and a standard deviation $\sigma_d = 200$. Lead time $t_l = 1$ week and is assumed to be constant. The fixed reorder cost k = € 20, the unit value of one pack of coffee is c = € 3, the carrying cost r = 0.08 €/€/year. The supermarket wants to maintain a service level of 99%. What is the order quantity, the reorder point and the safety stock?

Given is $h = r \cdot c = 0.08 \cdot 3 = 0.24$ €/year = 0.0046 €/week

The order quantity $Q^* = \sqrt{\dfrac{2k\bar{d}}{h}} = \sqrt{\dfrac{2 \cdot 20 \cdot 1500}{0.0046}} = 3611.6 \approx 3612$ items

To maintain the service level: $P(\tilde{d}t_l \leq s) \geq \alpha = 0.99 \Rightarrow \Phi(x) = 0.99$. The table for the Standard Normal distribution shows $\Phi^{-1}(0.99) = 2.33$

Equation (11.16) finally gives the reorder point

$s = \bar{d}t_l + x\sigma_d\sqrt{t_l} = 1500 \cdot 1 + 2.33 \cdot 200 \cdot 1 = 1500 + 466 = 1966$ items

and the safety stock $I_S = x\sigma_d\sqrt{t_l} = 466$ items

(s, S) order point, order-up-to-level system

In the (s, S) order point, order-up-to-level system, there is also a continuous review of the inventory level (R = 0). A variable order quantity up-to-level S is ordered whenever the inventory level drops to or below the reorder point s. If the demand transactions are of size 1, then (s, S) = (s, Q) with $S = s + Q$. When demand transactions are >1, then (s, S) becomes variable. This system is called a *min-max system*, because the inventory position (the on-hand inventory + the on-order inventory) is always between a minimum value of s and a maximum value of S.

It can be shown that the total costs of replenishment of the best (s, S) system are smaller than or equal to the costs of the best (s, Q) system. Unfortunately, the computational effort to find the best values for the (s, S) system is substantially more than for the (s, Q) system. Therefore, for suppliers the variable order quantities can be a disadvantage, because errors are made more easily.

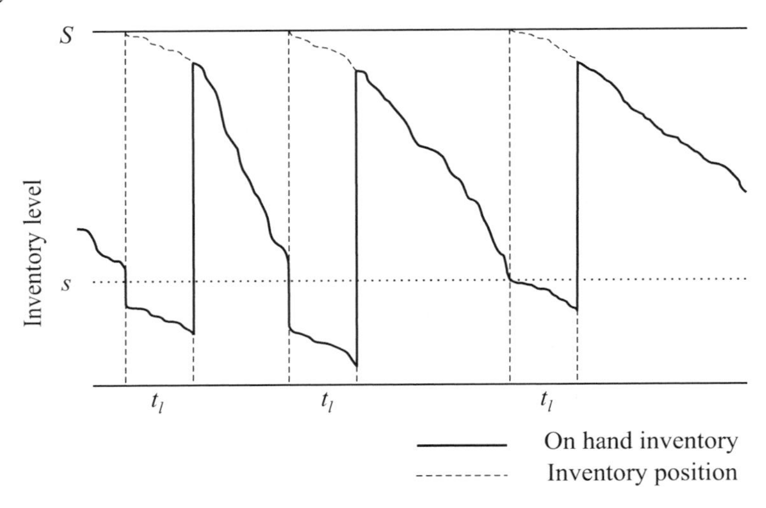

Figure 11.9. Continuous Review: (s, S) system: order-up-to-level S.

(R, S) periodic review, order-up-to-level system

In this control system every R units of time, an order-up-to-level S is placed. The inventory level S should be enough to cover the demand during the review period R and during the lead time t_l, until another replenishment arrives. See Figure 11.10. On average the on-hand-inventory level in this system is higher than in the continuous review systems. Therefore, the inventory holding cost tends to be higher. However, it does reflect practice where stores can only order at specific times.

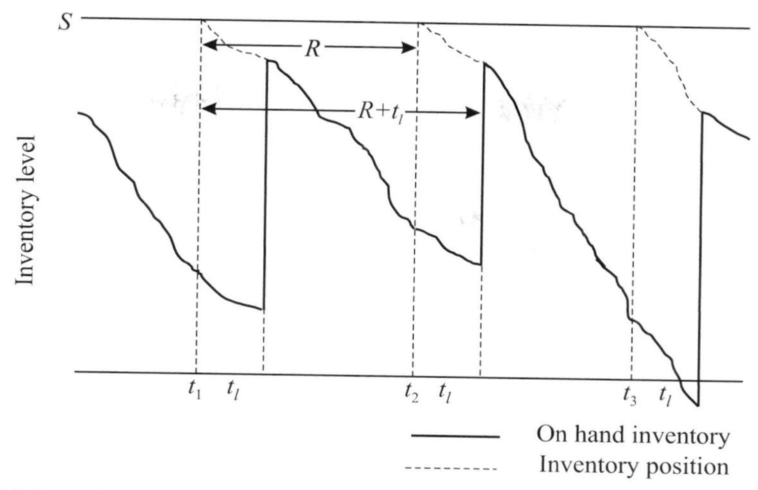

Figure 11.10. Periodic Review: (R, S) system: order-up-to-level S.

So how to calculate the optimal review period R, the order-up-to-level S and the safety stock I_S?

Suppose the same assumptions as for the (s, Q) system hold:

- the demand rate $\tilde{d}$ has a Normal distribution with as expected value the average demand rate $\bar{d}$ and the standard deviation σ_d, so the demand during the review period and the lead time $\tilde{d}(R+t_l)$ has the following distribution: $\tilde{d}(R+t_l) \sim N\left(\bar{d}(R+t_l),\ \sigma_d^2(R+t_l)\right)$.
- $\bar{d}$ and σ_d are constant in time;
- the lead time t_l is deterministic and constant;
- the demand rate and the lead time are statistically independent;
- when a shortage occurs, demand will be backlogged.

Furthermore $t_{i+1} = t_i + R,\ \ R > 0$ and $Q_i = S - I(t_i)$: the order quantity at time t_i is the order-up-to-level S decreased by the inventory position at time t_i.

Now the review period R, the order-up-to-level S and the safety stock I_S can be approached using the classical Economic Order Quantity model derived for the deterministic case.

The review period R is approximated by the optimal time lapse T^* between two consecutive orders according to Equation (11.3):

$$R = \sqrt{\frac{2k}{h\bar{d}}} \tag{11.17}$$

When the required service level is set at α, the probability of not running out of stock during the review period and the lead time should be greater than or equal to α. Or to put it differently: the demand during the review period and the lead time should be smaller than or equal to the order-up-to-level S with probability α.

In formula: $P\left(\tilde{d}(R+t_l) \leq S\right) \geq \alpha$. This can be rewritten as

$$P\left(\frac{\tilde{d}(R+t_l)-\bar{d}(R+t_l)}{\sigma_d\sqrt{R+t_l}} \leq \frac{S-\bar{d}(R+t_l)}{\sigma_d\sqrt{R+t_l}}\right) = \Phi\left(\frac{S-\bar{d}(R+t_l)}{\sigma_d\sqrt{R+t_l}}\right) \geq \alpha$$

where $\Phi(x)$ is the cumulative distribution function of $\tilde{x} \sim N(0,1)$.

So if $\dfrac{S-\bar{d}(R+t_l)}{\sigma_d\sqrt{R+t_l}}$ is represented by x, the following equations can be written:

$$x = \frac{S-\bar{d}(R+t_l)}{\sigma_d\sqrt{R+t_l}} \quad \Leftrightarrow \quad S-\bar{d}(R+t_l) = x\sigma_d\sqrt{R+t_l}$$

The right-hand side of this last equation is also the expression for the safety stock I_S .

$$I_S = x\sigma_d\sqrt{R+t_l} \tag{11.18}$$

The order-up-to-level S can be calculated via:

$$S = \bar{d}(R+t_l) + x\sigma_d\sqrt{R+t_l} \tag{11.19}$$

Notice that in this case the safety stock I_S is greater than in the (s, Q) system, therefore the inventory costs in the (R, S) system are higher than in a continuous review system.

Example 11.9. (*R*, *S*) system.

A supermarket uses the (R, S) system to control the inventory of its packs of coffee. The demand for packs of coffee has a Normal distribution with an average demand $\bar{d} = 1500$ items per week, and a standard deviation $\sigma_d = 200$. Lead time $t_l = 1$ week and is assumed to be constant. The fixed reorder cost k = € 20, the unit value of one pack of coffee is c = € 3, the carrying cost r = 0.08 €/€/year. The supermarket wants to maintain a service level of 99%. What is the review period R, the order-up-to-level S, and the safety stock?

Given $h = r \cdot c = 0.08 \cdot 3 = 0.24$ €/year = 0.0046 €/week.

The review period $R = \sqrt{\dfrac{2k}{h\bar{d}}} = \sqrt{\dfrac{2 \cdot 20}{0.0046 \cdot 1500}} = 2.4$ weeks.

To maintain the service level: $P\left(\tilde{d}(R+t_l) \leq S\right) \geq \alpha = 0.99 \;\Rightarrow\; \Phi(x) = 0.99$, such that $\Phi^{-1}(0.99) = 2.33$.

Equation (11.19) gives the optimal order-up-to-level

$$S = \bar{d}(R+t_l) + x\sigma_d\sqrt{R+t_l} = 1500 \cdot (2.4+1) + 2.33 \cdot 200\sqrt{2.4+1} =$$
$$= 5100 + 859 = 5959 \text{ items.}$$

Equation (11.18) results in a safety stock $I_S = x\sigma_d\sqrt{R+t_l} = 859$ items.

In Example 11.8 and 11.9 the same data have been used, except for the order policy. Comparing Example 11.8 (s, Q) system and Example 11.9 (R, S) system shows in the first system a safety stock of 466 and in the second system a safety stock of 859 items. The (s, Q) system has a maximum inventory level of $s + Q = 1966 + 3612 = 5578$ items. The (R, S) system has a maximum inventory level of order-up-to-level $S = 5959$ items. The difference in maximum inventory level is about the same as the difference in safety stock. This shows that the inventory costs in the periodic review (R, S) system are higher than in the continuous review (s, Q) system.

(R, s, S) periodic review, order point, order-up-to-level system

The (R, s, S) control system is a combination of the (s, S) and the (R, S) system. Every R units of time the inventory level is reviewed. If the level is at or below s, an order-up-to-level S is placed, otherwise, nothing is done. When the optimal values for R, s and S are found, this is the system with the lowest costs, however, the computational effort to find these values is high.

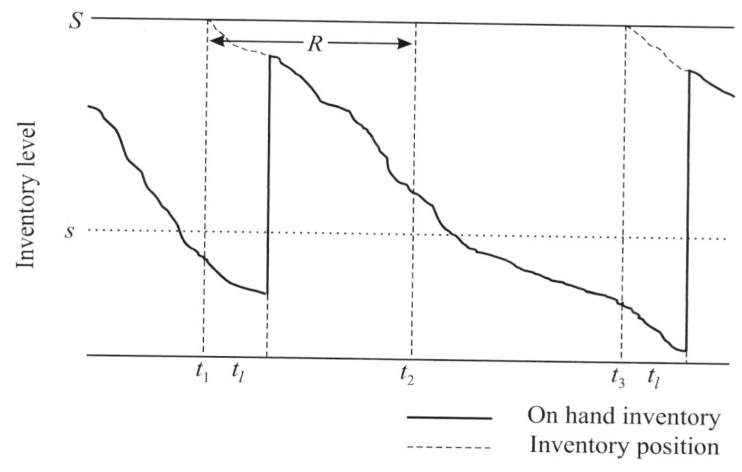

Figure 11.11. Periodic Review: (R, s, S) system.

Comparing continuous versus periodic review systems: continuous review systems need less safety stock, therefore the inventory holding cost are less, but they need more coordination than the periodic review systems.

11.7. Concluding remarks

In the introduction to this chapter the basic questions of inventory management are posed:

- When should an order be released?
- How large should each order be?
- How to control the inventory?

This chapter provides a set of analytical models and algorithms to manage inventory when different demand patterns are in order. In case demand is almost certain (deterministic) and constant the Economic Order Quantity is derived, a simple but robust and therefore often useful approach. Some extensions to the EOQ were also discussed: the Economic Production Quantity where replenishment arrives gradually when production proceeds, and a situation with quantity discounts on greater order quantities. In case demand is deterministic but variable the Silver – Meal heuristic is presented. This heuristic performs well in tests and can be used with a rolling-time horizon, therefore the Silver – Meal heuristic is often a good approach for determining the order quantities. In case demand is uncertain but approximately constant a single-period model (Newsboy Problem) and multiple-period order policies are discussed.

For more extensive information with an operations research point of view (Silver *et al.*, 1998) is a useful book. When demand is uncertain it is necessary to obtain estimates of future demand to be able to manage the inventory. In (Chopra and Meindl, 2007) theory on forecasting is discussed. Finally, this chapter, as part of a decision science book, is written with an operations research view, with emphasis on models and algorithms. The interested reader can find the management science point of view on inventory management in (Slack *et al.*, 2006).

12. Nonlinear programming

E.M.T. (Eligius) Hendrix

12.1. Introduction

In Nonlinear Programming (NLP) one would like to find an extreme (maximum or minimum) of an objective value of a model by varying several parameters or variables. The main difference with Linear Programming (LP) is that the functions are allowed to be nonlinear. The modelling conventions are the same. Usually quantities describing the decisions are given by a vector $\underline{x} \in \mathbb{R}^n$. The property (output) of the model that is optimised (costs, CO_2 emission etc.) is put in a so-called objective function $f(\underline{x})$. Other relevant output properties are indicated by functions $g_i(x)$ and are put in constraints $g_i(\underline{x}) \leq 0$ or $g_i(\underline{x}) = 0$. The so-called feasible area that is determined by the constraints is often summarised by $\underline{x} \in X$. Without loss of generality the general NLP problem can be written as:

$$\begin{aligned} &\min f(\underline{x}) \\ &g_i(\underline{x}) \leq 0 \quad \text{for some properties } i\text{, inequality constraints} \\ &g_i(\underline{x}) = 0 \quad \text{for some properties } i\text{, equality constraints} \end{aligned} \tag{12.1}$$

The general principle of NLP is that the values of the variables can be varied in a continuous way within the feasible set; this is in contrast to integer programming or combinatorial optimisation. To find and characterise the best plan (suggestion for the values of the decision variables), we should define what is an optimum, i.e. maximum or minimum. We distinguish between a local and global optimum, as illustrated in Figure 12.1. In words: a plan is called locally optimal, when the plan is the best in its neighbourhood. The plan is called globally optimal, when there is no better plan in the rest of the feasible area.

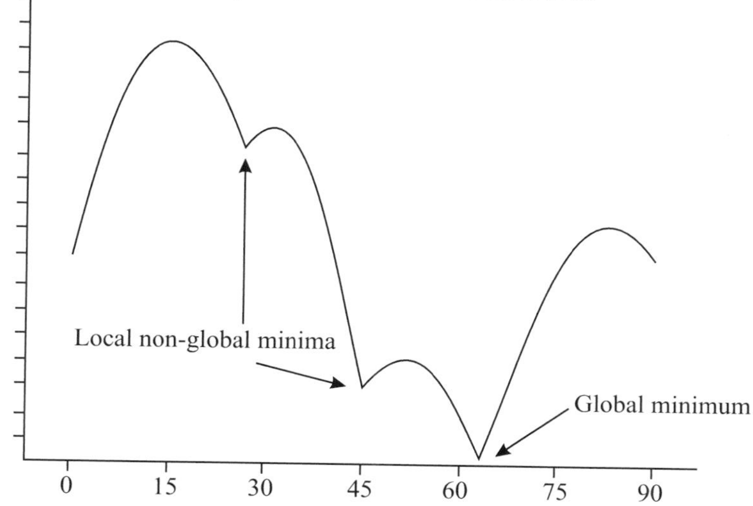

Figure 12.1. Global optimum and local optima.

Three different aspects with respect to NLP are discussed in this chapter.

- How do we formulate a model for a practical decision situation? We discuss this topic with the help of several exercises in Section 12.2.
- How do we recognise an optimal plan? A plan is optimal when it fulfils so-called optimality conditions. Understanding of these conditions is useful for a translation to the practical decision situation. Therefore, it is necessary to go into a mathematical analysis of the underlying model. The topic of optimality conditions is explained in Section 12.3.
- *Effectiveness* of algorithms is defined by their ability to reach the target of the user. *Efficiency* is the effort it costs to reach the target. Traditionally *mathematical programming* is the field of science that studies the behaviour of optimisation algorithms with respect to those criteria depending on the structure of the underlying optimisation problem. A separate section deals with the question of investigating optimisation algorithms in a systematic way. This will be dealt with in Section 12.4.

12.2. Mathematical modelling, cases

This section focuses on the modelling of an optimisation problem where objective and constraint functions are typically nonlinear. Several example cases are given. The reader can practice by trying to formulate the exercise examples based on them.

12.2.1. Enclosing a set of points

We start with a rather abstract application that can be found in data analysis and parameter identification. The problem is sketched and the mathematical formulation is given. The reader is then asked to give the formulation of an easier problem of location theory in an exercise.

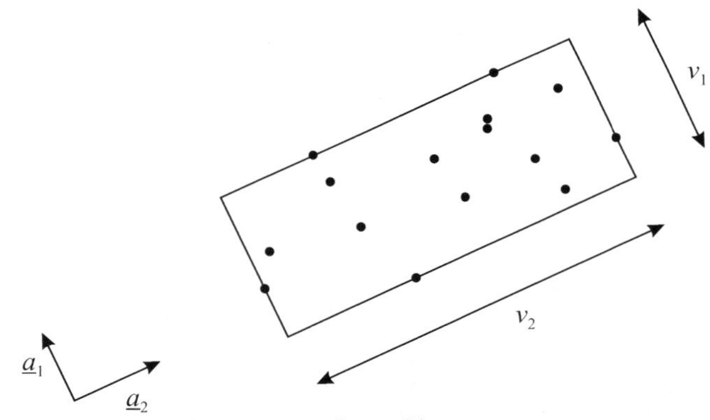

Figure 12.2. Minimum volume hyperrectangle problem.

The problem is to enclose a set of points with a predefined set with a size or volume as small as possible. Depending on the enclosure one is looking for, it can be an easy or very difficult problem to solve. Figure 12.2 the problem of finding a minimum volume hyperrectangle around a set of points. The general problem is defined as: given the set of points $X = \{\underline{x}_1, \ldots, \underline{x}_K\} \in \mathbb{R}^n$, find an enclosing hyperrectangle with minimum volume around the points of which the axes are free to be chosen, see (Keesman, 1992). Mathematically this can be translated into finding axes represented by columns of a so-called orthonormal matrix $A = (\underline{a}_1, \ldots, \underline{a}_n)$ minimising the objective

$$f(A) = \prod_i \{v_i = (\max_j \underline{a}_i^T \underline{x}_j - \min_j \underline{a}_i^T \underline{x}_j)\} \qquad (12.2)$$

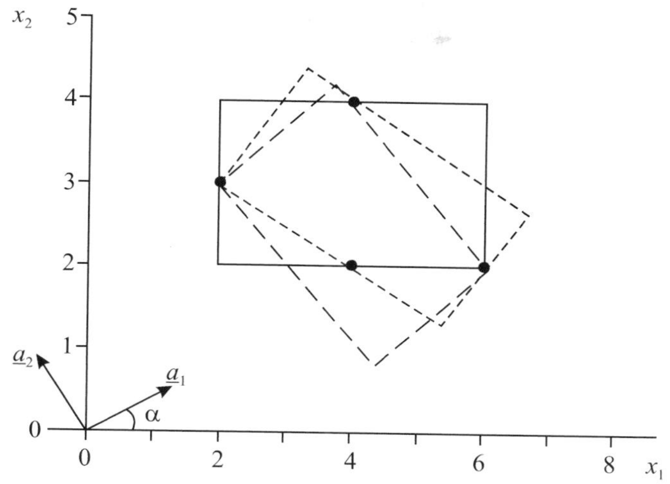

Figure 12.3. Rectangles around 4 points.

Here the axes $\underline{a}$ are seen as decision variables and the final objective function f consists of a multiplication of the lengths v that appear after checking all points. The problem as such has many optima, which is further illustrated in Figure12.4.

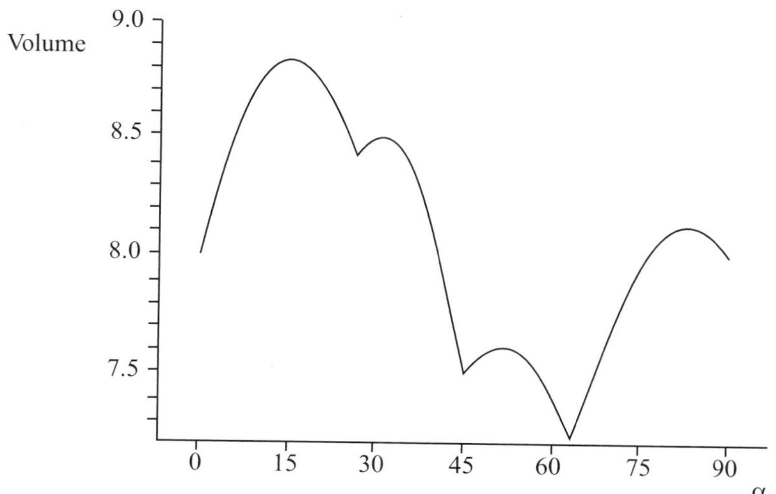

Figure 12.4. Objective function value as function of angle.

It represents the case of Figure12.3 where the set $X = \{(2,3), (4,4), (4,2), (6,2)\}$ is enclosed by rectangles defined by the angle α of the first axis $\underline{a}_1$, such that vector $\underline{a}_1 = (\cos \alpha, \sin \alpha)$. Note that case $\alpha = 0$ represents the same situation as $\alpha = 90$, because the position of the two axes switches. The general problem is not easy to formulate explicitly due to the orthonormality requirements. The requirement of the orthonormality of the matrix of axes of the hyperrectangle, implies the degree of freedom in choosing the matrix to be $n(n-1)/2$. In two dimensions this can be illustrated by using as one parameter the angle of the first vector. In higher dimensions this is not so easy. The number of optima as such is enormous, because it depends on the number of points, or more precisely, on the number of points in the convex hull of X.
A problem similar to the minimum volume hyperrectangle problem, is to find an enclosing or an inscribed ellipsoid, as discussed for example by (Khachiyan and Todd, 1993). The enclosing minimum volume ellipsoid problem can be formulated as finding a positive definite matrix and a centre of the ellipsoid, such that it contains a given set of points or a polytope. This problem is fairly well analysed in literature, but goes too far to be formulated as an example. Instead, as an exercise we describe here the problem of finding the best location for a facility such that its farthest client is at minimum distance.

Exercise Centroid facility

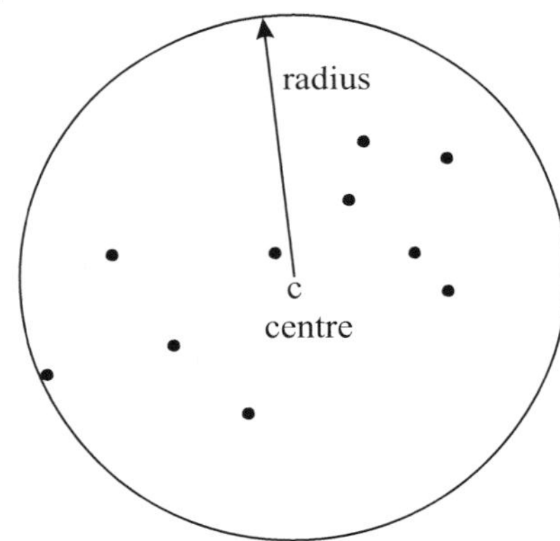

Figure 12.5. Centroid location problem.

Given a set of 10 demand points $\{x_1, \ldots, x_{10}\} \in \mathbb{R}^n$. Find the best location $\underline{c}$ for a facility and distance (radius) r such that the maximum distance over the 10 demand points to the facility $\underline{c}$ is at its minimum value. This means, find a sphere around the set of points with a radius as small as possible. In $\mathbb{R}^2$ this problem is not very hard to solve. In figure12.5, an enclosing sphere is given that does not have the minimum radius.

- Formulate the problem in vector notation (min-max problem).
- Generate with a spreadsheet (or other program) 10 points at random in $\mathbb{R}^2$.
- Make a spreadsheet, that calculates the max distance given location $\underline{c}$.
- Intuitively, does this problem have only one (local) optimum?

12.2.2. Dynamic decision strategies.

The problems in this section involve sequential decision making. The performance, objective function, not only depends on the sequence of decisions, but also on fluctuating data over a given time period, often considered as a stochastic variable. The calculation of the objective typically requires the simulation of the behaviour of a system over a long period.
The first example is derived from an engineering consulting experience dealing with operating rules for pumping water in a higher situated lake in the Netherlands. In general the rainfall exceeds the evaporation and the seepage. In summer however, water has to be pumped from lower areas and is treated to maintain a water level above the minimum with a good water quality. Not only the pumping, but certainly also the treatment to remove phosphate costs money. The treatment installation performs better when the stream is constant, so the pumps should not be switched off and on too frequently. The behaviour of the system is given by Equation (12.3)

$$I_t = \min\{I_{t-1} + \xi_t + x_t, \text{Max}\} \qquad (12.3)$$

with I_t : water level of the lake, period t

ξ_t : natural inflow i.e. rainfall - seepage - evaporation

x_t : amount of water pumped into the lake

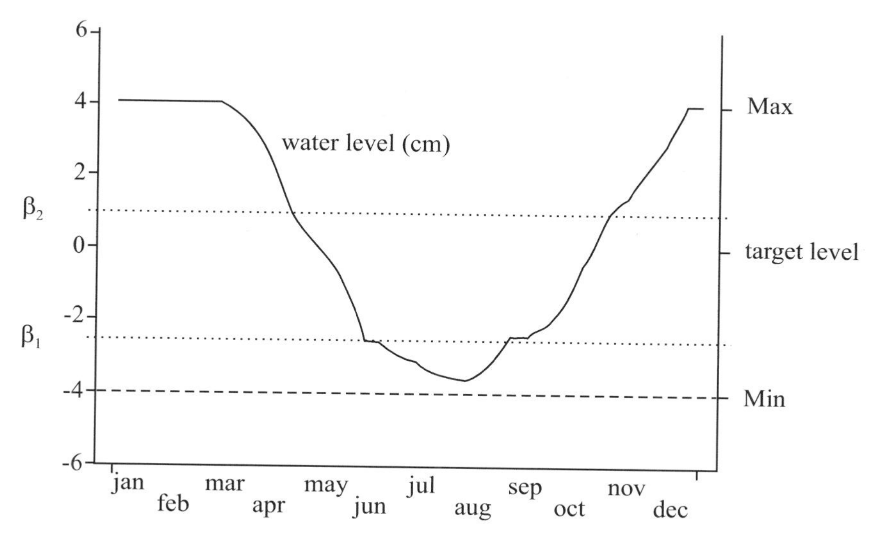

Figure 12.6. Strategy to rule the pumping.

When the water level reaches its maximum (Max), the superfluous water streams downwards through a canal system towards the sea. For the studied case, two pumps were planned to be installed, so that x_t only takes values in $\{0, B, 2B\}$, where B is the capacity of one pump. Decisions are taken on a daily basis. In water management, it is common practice to derive so-called operating rules, decision strategies including parameters. A decision rule instructs on what decision to make in which situation. An example is rule (12.4) with parameters β_1 and β_2.

$$\text{For} \quad \begin{array}{ll} I_t < \beta_1 & x_t = 2B \\ \beta_1 \le I_t \le \beta_2 & x_t = B \\ I_t > \beta_2 & x_t = 0. \end{array} \tag{12.4}$$

Given weather data of a certain period, now the resulting behaviour of a sequence of decisions x_t can be evaluated by measuring performance indicators such as the amount of water pumped Σx_t and the number of switches of the pumps $\Sigma|x_t - x_{t-1}|/B$. Assessment of appropriate values for the parameters β_1 can be considered as a black-box optimisation problem.
For every parameter set, the model (12.3) with strategy (12.4) can be simulated with weather data (rainfall and evaporation) of a certain time period. Some 20 years of data on open water evaporation and rainfall were available. The performance can be measured leading to one (multi-) objective function value. At every iteration an optimisation algorithm delivers a proposal for the parameter vector $\underline{\beta}$, the model simulates the performance and after some time returns an objective function value $f(\underline{\beta})$. One possibility is to create a stochastic model of the weather data and resulting ξ_t and to use the model to 'generate' more years by Monte Carlo simulation, i.e. simulation using (pseudo) random numbers; see Chapter 13. In this way, it is possible to extend the simulation run over many years. The model run can be made arbitrarily long. Notice that in our context it is useful for every parameter proposal to use the same set of random numbers (seed), otherwise the objective function $f(\underline{\beta})$ becomes a random variable. The problem sketched here is an example of so-called *parameterised decision strategies*.

Exercise inventory control (s,Q)

In Management Science and Logistics a well-known application is that of Stochastic Inventory Control. The variables in Equation (12.3) now read as follows, see (Hax and Candea, 1984) and Chapter 11.

I_t	:	level of inventory
x_t	:	amount produced or ordered
$\tilde{\xi}_t$	:	(negative) demand, considered stochastic.

Criteria that play a role are inventory holding costs, ordering costs and backordering or out of stock costs. In this simple case a so-called (s, Q)-policy is followed. Data: Inventory holding costs: 0.3 per unit per day. Ordering costs 750.

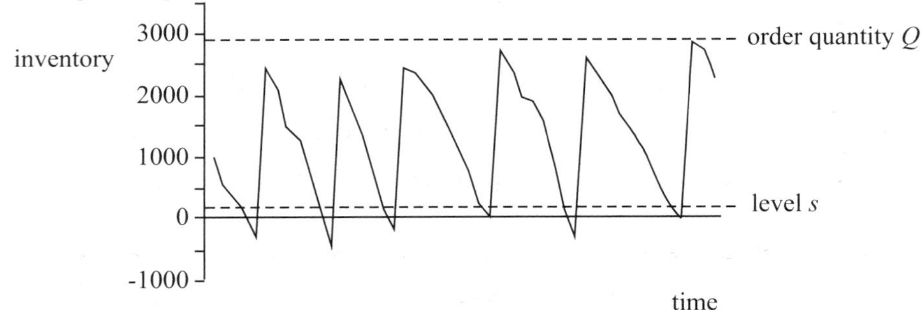

Figure 12.7. Inventory level given parameter values.

An order becomes available at the end of the next day. As soon as the inventory is below level s, an order is placed of size Q. If there is not sufficient stock (inventory), the client is supplied next day (backordering) at an additional cost of 3 per unit. The stochastic daily demand in this case follows a triangular distribution with values between 0 and 800. This is the same as the addition $\tilde{\xi} = \tilde{u}_1 + \tilde{u}_2$ of two uniformly distributed random variables $\tilde{u}_1$ and $\tilde{u}_2$ between 0 and 400.

- Generate with a spreadsheet (or other program) 200 daily demand figures.
- Make a program or spreadsheet, that determines the total costs given values for Q and s and the generated figures.
- Determine good values for Q and s.
- Is the objective value (costs) very sensitive to the values of s ?

Usually in this type of problem an analysis is performed, based on integrating over the probability density function of the uncertain demand and delivery time, see Chapter 11. Let us for the exercise, consider what is the optimal order quantity Q if the demand is not uncertain, but fixed at 400 every day, the so-called *deterministic* situation. The usual approach is to minimise the average daily costs. The length of a cycle in the saw-tooth is defined by $Q/400$, such that the order cost per day is $750/(Q/400) = 300000/Q$. The average inventory costs are derived from the observation that the average inventory over one cycle is $Q/2$. The total relevant cost per day, TRC(Q) is given by:

$$\text{TRC}(Q) = 300000/Q + 0.3 \cdot Q/2$$

- What is the order quantity Q that minimises the total daily costs for this deterministic situation?

One could use the optimal quantity of this exercise as a starting value to derive the optimal value for the stochastic case.

12.2.3. Design and factorial or quadratic regression

Regression analysis is a technique which is very popular in scientific research and in design. Very often it is a starting point for the identification of relations between inputs and outputs of a system. In a first attempt one tries to verify a linear relation between output y, called regressand or dependent variable, and the input vector x, called regressor, factor or independent variable. A so-called linear regression function is used:

$$y = \beta_0 + \beta_1 x_1 + \beta_2 x_2 + \ldots \beta_n x_n$$

For the estimation of the coefficients β_j and to check how good the function 'fits reality', either data from the past can be used or experiments can be designed to create new data for the output and input variables. The data for the regression can be based on a design of a computer experiment which uses a simulation model to generate the data on input and output. The generation of regression relations out of experiments of a relatively large simulation model is called metamodelling and is discussed in (Kleijnen and Van Groenendaal, 1992). The regression model is called a metamodel, because it models the input-output behaviour of the underlying simulation model. In theory about design, the word response surface methodology is more popular and promoted by Taguchi among others, see (Taguchi *et al.* 1989).

The regression functions based on either historical data, special field experiments or computer experiments can be used in an optimisation context. As long as the regression function is linear in the input variables x_j, linear programming can be applied. The optimisation becomes more complicated when interaction between the input variables is introduced in the regression function. Interaction means that the effect of an input variable depends on the values of another input variable. This is usually introduced by allowing so-called two-factor interaction, i.e. multiplications of two input variables in the regression function. An example of such a factorial regression model is

$$y = \beta_0 + \beta_1 x_1 + \beta_2 x_2 + \beta_{12} x_1 x_2$$

The introduction of multiplications implies the possibility to have several optima in an optimisation context. Consider the minimisation of $y = 2 - 2x_1 - x_2 + x_1 x_2$ with $0 \le x_1 \le 4$ and $0 \le x_2 \le 3$. This problem has two minima: $y = -1$ for $\underline{x} = (0,3)$ and $y = -6$ for $\underline{x} = (4,0)$.

Example 12.1.

Given problem *IQP:*

$$\min_{x \in X} \{ f(\underline{x}) = (x_1 - 1)^2 - (x_2 - 1)^2 \}$$

X is given by

$$\begin{aligned} x_1 - x_2 &\le 1 \\ 4x_1 - x_2 &\ge -2 \\ 0 \le x_1 &\le 3 \\ 0 \le x_2 &\le 4 \end{aligned}$$

Contour lines and the feasible set of (*IQP*) are given in Figure 12.8. The problem has two local minimum points, namely (1, 0) and (1, 4) (the global one).

A further extension in regression analysis is the inclusion of quadratic terms which results in a complete second order Taylor series approximation of the relation one intends to find. In two dimensions the quadratic regression function is:

$$y = \beta_0 + \beta_1 x_1 + \beta_2 x_2 + \beta_{12} x_1 x_2 + \beta_{11} x_1^1 + \beta_{22} x_2^2 .$$

Notice that in regression terms this is called linear regression, as the function is linear in the parameters β_j. When these functions are used in an optimisation context, it depends on the second order derivatives β_{ij} whether the function is convex and consequently whether it may have only one or multiple optima such as in Example 12.1.

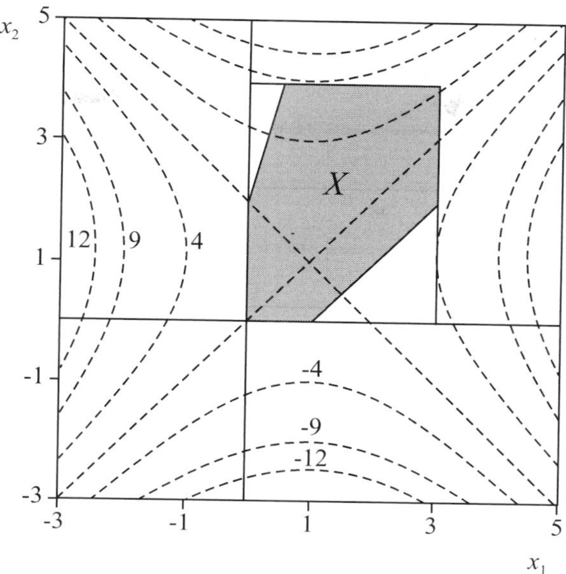

Figure 12.8. Problem IQP.

The use in a design case is illustrated here with the mixture design problem, which can be found in (Hendrix and Pintér, 1991). In particular the so-called rum-coke example, where one tries to find a mix of rum, coke and ice cubes, such that the properties $y_i(\underline{x})$ fulfil some requirements as follows:

$$y_1(\underline{x}) = -2 + 8x_1 + 8x_2 - 32x_1x_2 \leq -1$$
$$y_2(\underline{x}) = 4 - 12x_1 - 4x_3 + 4x_1x_3 + 10x_1^2 + 2x_3^2 \leq 0.4$$

The possible mixtures are described by the so-called unit simplex S, where $x_1 + x_2 + x_3 = 1$. A projection (triangle) of the simplex S on the x_1, x_2 plane can be generated. In Figure 12.9 vertex $\underline{x}_p$ represents a product consisting for 100% of component p, $p = 1,2,3$ (rum, coke and ice cubes). The area in which the feasible products are situated is given by F. One could try to find a feasible design for a design problem defined by inequalities $y_i(\underline{x}) \leq b_i$, by minimising an objective function

$$f(\underline{x}) = \max_i \{y_i(\underline{x}) - b_i\} \tag{12.5}$$

or by minimising

$$f(\underline{x}) = \sum_i \max\{y_i(\underline{x}) - b_i, 0\}. \tag{12.6}$$

The problem of minimising (12.6) over S has a local optimum in $\underline{x}_{loc} = (0.125, 0, 0.875)$, $f(\underline{x}_{loc}) = -0.725$, and of course a global optimum (= 0) for all elements of $F \cap S$.

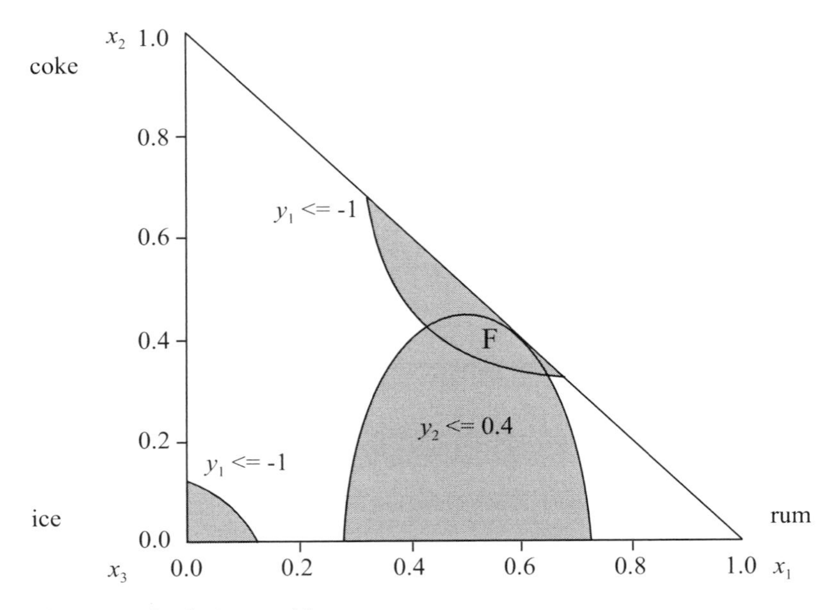

Figure 12.9. Rum-coke design problem.

Exercise quadratic

Given property $y(\underline{x})$ as a result of a quadratic regression:
$y(\underline{x}) = -1 - 2x_1 - x_2 + x_1x_2 + x_1^2$. We would like to minimise $y(\underline{x})$ on a design space defined by $0 \leq x_1 \leq 3$ and $1 \leq x_2 \leq 4$.

- Determine whether $y(\underline{x})$ has several local optima on the design space.

12.3. NLP optimality conditions

12.3.1. Intuition with some examples

After an optimisation problem has been formulated (or during the formulation), methods can be used to determine an optimal plan $\underline{x}^*$. In reality, software only finds an approximation of an optimal plan. In the application of NLP algorithms, the user should normally indicate how close an optimum should be approximated before stopping. We will discuss this in Section 12.4. Software could be based among others on modelling languages such as GAMS and GINO. Nowadays also a so-called solver add-in on Excel, a popular spreadsheet program, exists. To get a feeling for the theories and examples here, the reader could use one of these programs. In the appendices an example can be found of the output of these programs.

The result of a method gives in practice an approximation of an optimal solution that fulfils the optimality conditions and that moreover gives information on sensitivity with respect to the data. First an example is introduced before going into abstraction and exactness.

Example 12.2.

A classical problem in economics is the so-called utility maximisation. In the two goods case, x_1 and x_2 represent the amount of goods of type 1 and 2 and a utility function $U(\underline{x})$ is maximised. Given a budget (here 6 units) and prices for goods 1 and 2, with a value of 1 and 2 respectively the optimisation NLPU appears:

$$\max\{U(\underline{x}) = x_1 x_2\}$$
$$x_1 + 2x_2 \leq 6$$
$$x_1 \geq 0, x_2 \geq 0$$

To describe NLPU in the terminology of general NLP problem (12.1) one can define, $f(\underline{x}) = -U(\underline{x})$, and the feasible area X is described by three inequalities $g_i(\underline{x}) \leq 0$; $g_1(\underline{x}) = x_1 + 2x_2 - 6$, $g_2(\underline{x}) = -x_1$ and $g_3(\underline{x}) = -x_2$.

In order to find the best plan, we should first define what an optimum, i.e. maximum or minimum, is. In Figure12.1, the concept of a global and local optimum has been sketched. In words: a plan is called locally optimal, when it is the best plan in its close environment. A plan is called globally optimal when it is the best plan in the total feasible area. In order to formalise, it is necessary to define the concept of 'close environment' in a mathematical way. The mathematical environment of $\underline{x}^*$ is given as a sphere (ball) with radius ε around $\underline{x}^*$.

Definition 12.1.

Let $\underline{x}^* \in \mathbb{R}^n$, $\varepsilon > 0$. Set $\{\underline{x} \in \mathbb{R}^n \mid \|\underline{x} - \underline{x}^*\| < \varepsilon\}$ is called an ε-environment of $\underline{x}^*$.

For $\|\cdot\|$ usually the Euclidean norm is taken: $\|\underline{x} - \underline{y}\| = \sqrt{\sum (x_j - y_j)^2}$.

Definition 12.2.

Function f has a minimum (or local minimum) over set X at $\underline{x}^*$ if there exists an ε-environment W of $\underline{x}^*$, such that: $f(\underline{x}) \geq f(\underline{x}^*)$ for *all* $\underline{x} \in W \cap X$. In this case, vector $\underline{x}^*$ is called a minimum point (or local minimum point) of f. Function f has a global minimum in $\underline{x}^*$ if $f(\underline{x}) \geq f(\underline{x}^*)$ for all $\underline{x} \in X$. In this case, vector $\underline{x}^*$ is called a *global minimum point*. The terminology strict minimum point is used when above $f(\underline{x}) \geq f(\underline{x}^*)$ is replaced by $f(\underline{x}) > f(\underline{x}^*)$ for $\underline{x} \neq \underline{x}^*$.

Note: For a maximisation problem, in Definition 12.2 'minimum' is replaced by 'maximum', the '≥' sign by the '≤' sign and '>' by '<'.

To determine the optimal plan for the example, the *contour* is introduced.

Definition 12.3.

A contour of $f: \mathbb{R}^n \rightarrow \mathbb{R}$ of altitude h is defined as the set $\{\underline{x} \in \mathbb{R}^n \mid f(\underline{x}) = h\}$.

A contour can be drawn in a figure like lines of altitude on a map as long as $\underline{x} \in \mathbb{R}^2$. Specifically for LP, a contour $c_1x_1 + c_2x_2 = h$ is a line perpendicular to vector $\underline{c}$. The contours of problem NLPU are given in Figure 12.10.

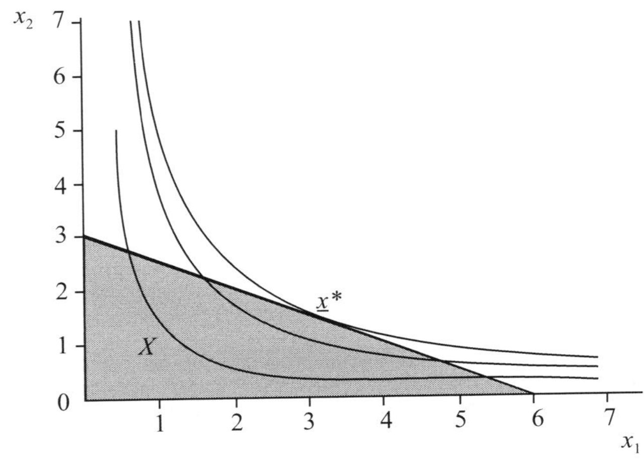

Figure 12.10. Maximising utility function $U(\underline{x}) = x_1x_2$.

Example 12.3.

Similar to graphically solving LP, we can try to find the highest contour of $U(\underline{x})$, that has a point in common with feasible area X. The maximum point is $\underline{x}^* = (3, 3/2)^T$, as indicated in Figure 12.10. The contours are so-called hyperbolas defined by $x_2 = h/x_1$.

It is not a coïncidence that the optimum can be found where the contour of $U(\underline{x})$ touches the binding budget constraint $x_1 + 2x_2 \leq 6$. As with many LP problems, the feasible area here consists of a so-called polytope. However, the optimal point $\underline{x}^*$ can not be found in a vertex. For NLP problems, the optimum can be found on a line, a plane and even in the interior of the feasible area, where no constraint is binding. So the number of binding constraints in the optimal point is not known in advance. We will come back to this phenomenon, but proceed now with the next example.

Example 12.4.

In NLPU, the utility function is changed to $U(\underline{x}) = x_1^2 + x_2^2$. The contours are now circles around the origin. In $\underline{x}_1^* = (6,0)^T$ the global maximum point can be found; $f(\underline{x}_1^*) = 36 \geq f(\underline{x})\ \forall \underline{x} \in X$ corresponds to the highest feasible contour. The point $\underline{x}_2^* = (0,3)^T$ is a local, non-global maximum point. I.e., there exists an ε-environment of $\underline{x}_2^*$ (for instance for $\varepsilon = 0.01$), such that all points situated in this environment as well as X, have a lower objective function value. The point $\underline{x}_3^* = \frac{1}{5}(6,12)^T$ is a point where a contour touches a binding constraint. In contrast to Example 12.3, such a point is not a maximum point.

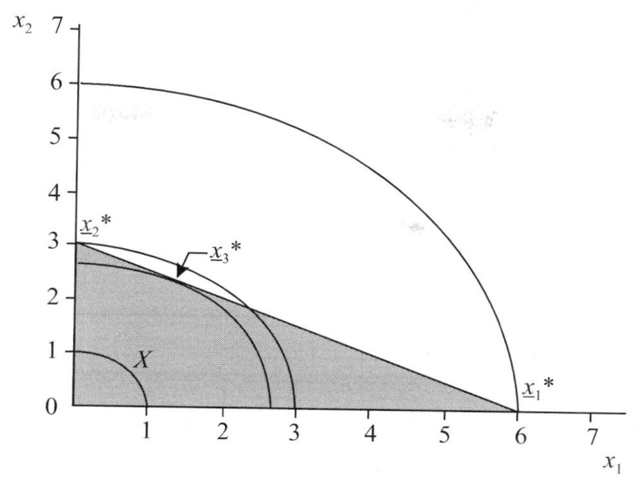

Figure 12.11. Several optima other utility function.

What can be derived from Example 12.4? An optimal point possibly can be found in a vertex of the feasible area. In contrast to Linear Programming, there can exist points that are local non-global optima. A last example in this section is discussed.

Example 12.5.

The concept of investing money in a number of goods can also be found in the decisions of investing in funds, so-called portfolio selection. Traditionally a trade-off has to be made between return on investment and risk. The classical E, V (Expected value, Variance) model introduced by Markowitz (1959) is discussed. A number of funds are given, so-called portfolios. Every portfolio has an expected return μ_j and variance σ^2_j , $j = 1, \ldots , n$. A fixed amount, let's say €100 should be invested in funds, such that the total expected return is as high as possible and the risk (variance) as low as possible. Let x_j be the amount invested in portfolio j, then:

The feasible area $\sum x_j = 100, x_j \geq 0 \quad j = 1,\ldots,n$

Expected return $\sum x_j \mu_j$

Variance of return $V = \sum x_j^2 \sigma_j^2 + 2\sum_{i=1}^{n} \sum_{j+1}^{n} \sigma_{ij} x_i x_j$.

The term σ_{ij} is the so-called covariance between funds i and j. This can be positive (sunglasses and sun-screen), negative (sunglasses and umbrellas) and zero; the funds have no relation. The investor wants to maximise E and at the same time minimise V. This is a so-called multi-objective problem. In this case one can generate a trade-off curve between maximising E and minimising V.

The simplest case, similar to the examples discussed, is when there are two funds. Substituting the binding budget constraint $x_2 + x_1 = 100 \rightarrow x_2 = 100 - x_1$, a problem in one decision variable appears that can be analysed relatively easily. Given two funds with $\mu_1 = 1$, $\sigma_1 = 1$, $\mu_2 = 2$, $\sigma_2 = 2$ and $\sigma_{12} = 0$. The variance $V = x_1^2 + 4_2^2$ by substituting $x_2 = 100 - x_1$ becomes $V = (x_1) = 5x_1^2 - 800x_1 + 40{,}000$.

This describes a parabola with a minimum at $x_1^* = 80$. An investor who wants to minimise the risk has an optimal plan $\underline{x}^* = (80, 20)$ where $E = 120$ and $V = 8000$. This optimal plan is not a vertex of the feasible area. Maximising the expected return E gives the optimal solution $\underline{x}^* = (0, 100)$ where $E = 200$ and $V = 40{,}000$. This is a vertex of the feasible area.

In practice an investor will make a trade-off between expected return and acceptable risk. In this specific case of two funds, if one makes a graph with E and V on the axes as a function of x_1, the trade-off line of Figure 12.12 appears. In the general Markowitz model, points on this curve are derived by maximising an objective function $f(\underline{x}) = E - \beta V$, where β is called a risk aversion parameter. For every value of β, a so-called efficient point or Pareto-point appears.

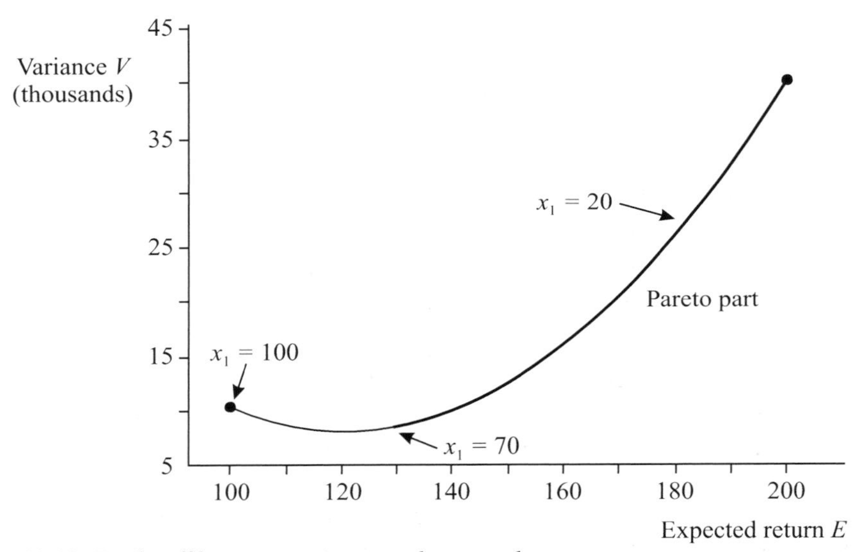

Figure 12.12. Trade-off between variance and expected return.

Determination of the optimal plan for varying parameter values is called *parametric programming*. In Linear Programming this leads to piecewise linear curves as shown in Sections 2.5 and 4.5. As can be observed from the small example, in Nonlinear Programming this can be a smooth nonlinear curve.

Exercising with simple examples based on graphical analysis has produced several insights:

- An optimal point can not always be found in a vertex of the feasible area.
- The point where a contour touches a constraint is a special point.
- Local, non-global optimal points may exist.
- Changes in parameter values (parametric programming) may lead to nonlinear curves.

The rest of the section concerns the formalisation of these insights and the question as to which phenomena appear in which situations.

12.3.2. Derivative information

Some terminology is introduced for the formalisation. An objective function $f: \mathbb{R}^n \to \mathbb{R}$ is analysed. In the first section the contour, $\{\underline{x} \in \mathbb{R}^n \mid f(\underline{x}) = h\}$ has already been introduced. A wider perspective are the concepts of a *graph* $\{(\underline{x}, y) \in \mathbb{R}^{n+1} \mid f(\underline{x}) = y\}$ and a *level set* $\{\underline{x} \in \mathbb{R}^n \mid f(\underline{x}) \leq h\}$. The graphical perception is limited to $\mathbb{R}^3$. A graph as such can be seen from the point of view of a landscape. In the development of algorithms as well as theory, many concepts can be derived by cross-cutting a function $f: \mathbb{R}^n \to \mathbb{R}$ considering a function φ_r of one variable starting in a fixed point $\underline{x}$ and looking in the direction $\underline{r}$:

$$\varphi_r(\lambda) = f(\underline{x} + \lambda \underline{r}) \tag{12.7}$$

Derivatives

The notion of a derivative is first considered for a function in one variable and then via (12.7) extended to functions of several variables. The derivative of $f: \mathbb{R} \to \mathbb{R}$ in the point x is defined as:

$$f'(x) = \lim_{h \to 0} \frac{f(x+h) - f(x)}{h} \tag{12.8}$$

whenever this limit exists. For instance for $f(x) = \sqrt{|x|}$ this limit does not exist for $x = 0$. The use of absolute value and the max-min structure in an optimisation model as illustrated in Section 12.2, in practice causes an objective function not to be differentiable everywhere. Another important concept in mathematical literature is that of *continuously differentiable*, i.e. the derivative function is a continuous function. This is mainly a theoretical concept. It requires some imagination to come up with a function that is differentiable, but not continuously differentiable. In textbooks one often finds the example: $f(x) = x^2 \sin \frac{1}{x}$ when $x \neq 0$ and $f(x) = 0$ when $x = 0$. Exercising with limits shows that this function is differentiable for $x = 0$, but that the derivative is not continuous.
Algorithms often make use of the value of the derivative. Unless computer programs can manipulate formulas (so-called automatic differentiation), the user has to feed the program explicitly with the formulas of the derivatives. Usually, however, a computer package makes use of so-called *numerical differentiation.* In many cases as outlined in Section 12.2, the calculation of a function value is the result of a long calculation process and an expression for the derivative is not available. A numerical approximation of the derivative is determined for instance by the *progressive* or *forward difference* approximation:

$$f'(x) \approx \frac{f(x+h) - f(x)}{h} \tag{12.9}$$

by taking a small step (e.g. $h = 10^{-5}$) forward. Numerical errors can occur that, without going into detail, are smaller in the *central difference approximation*:

$$f'(x) \approx \frac{f(x+h) - f(x-h)}{2h} \tag{12.10}$$

Directional derivative

By considering functions of several variables from a one-dimensional perspective via (12.7), the notion of *directional derivative* $\varphi_r'(0)$ appears, depending on point $\underline{x}$ and direction $\underline{r}$:

$$\varphi'_r(0) = \lim_{h \to 0} \frac{f(\underline{x} + h\underline{r}) - f(\underline{x})}{h} \tag{12.11}$$

This notion is relevant for the design of search algorithms as well as for the test on optimality. When an algorithm has generated a point $\underline{x}$, a direction $\underline{r}$ can be classified according to:

$\underline{r}$ with $\varphi'_r(0) < 0$: descent directions
$\underline{r}$ with $\varphi'_r(0) > 0$: ascent directions
$\underline{r}$ with $\varphi'_r(0) = 0$: directions in which *f* does not increase nor decrease, it is situated in the tangent plane of the contour.

For algorithms looking for the minimum of a differentiable function, in every generated point, the descent directions are of interest. For the test of whether a certain point $\underline{x}$ is a minimum point of a differentiable function, the following reasoning holds. In a minimum point $\underline{x}^*$ there exists no search direction $\underline{r}$ that points into the feasible area and is also a descent direction, $\varphi'_r(0) < 0$. Derivative information is of importance for testing optimality. The test of a set of possible search directions requires the notion of *gradient*.

Gradient

By using in (12.11), $\underline{e}_j$ for the direction $\underline{r}$, the so-called partial derivative appears:

$$\frac{\partial f}{\partial x_j}(\underline{x}) = \lim_{h \to 0} \frac{f(\underline{x} + h\underline{e}_j) - f(\underline{x})}{h} \tag{12.12}$$

The vector of partial derivatives is called the gradient:

$$\nabla f(\underline{x}) = \left(\frac{\partial f}{\partial x_1}(\underline{x}), \frac{\partial f}{\partial x_2}(\underline{x}), \ldots, \frac{\partial f}{\partial x_n}(\underline{x}) \right)^T \tag{12.13}$$

Example 12.6.

Let $f: \mathbb{R}^n \to \mathbb{R}$ be linear function $f(\underline{x}) = \underline{c}^T\underline{x}$. Partial derivatives of f in $\underline{x}$ with respect to $\underline{x}_j$:

$$\frac{\partial f}{\partial x_j}(\underline{x}) = \lim_{\lambda \to 0} \frac{f(\underline{x} + \lambda \underline{e}_j) - f(\underline{x})}{\lambda} =$$

$$\lim_{\lambda \to 0} \frac{\underline{c}^T(\underline{x} + \lambda \underline{e}_j) - \underline{c}^T \underline{x}}{\lambda} = \lim_{\lambda \to 0} \frac{\lambda \underline{c}^T \underline{e}_j}{\lambda} = \underline{c}^T \underline{e}_j = c_j$$

The gradient $\nabla f(\underline{x}) = (c_1, \ldots, c_n)^T = \underline{c}$ for linear functions does not depend on $\underline{x}$.

Example 12.7.

We again consider the utility optimisation problem NLPU. The utility function $U(\underline{x}) = x_1x_2$ has a gradient $\nabla U(\underline{x}) = (x_2, x_1)^T$. The gradient ∇U is depicted for several plans $\underline{x}$ in Figure 12.13(a). The arrow $\nabla U(\underline{x})$ is orthogonal to the contour; this is not a coincidence.

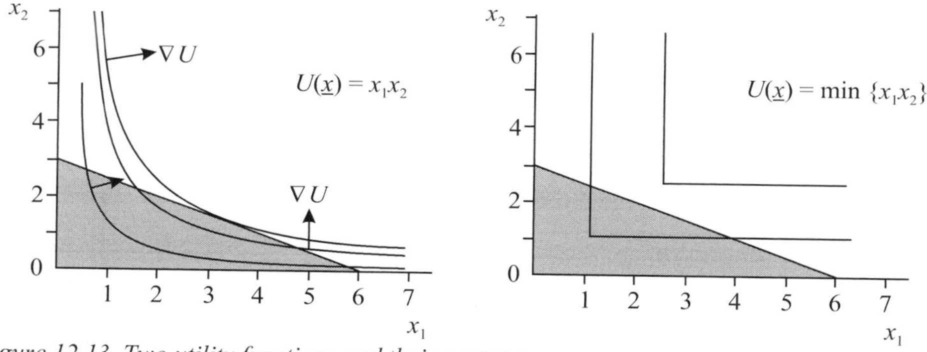

Figure 12.13. Two utility functions and their contours.

In Figure 12.13(b), contours of another function can be found from the theory of utility maximisation. It concerns the notion of complementary goods. A classical example of complementary goods is where x_1: number of right shoes, x_2: number of left shoes and $U(\underline{x}) = \min\{x_1, x_2\}$ the number of pairs of shoes, that seems to be maximised by some individuals. This utility function is not differentiable everywhere. Consider cross-cutting the graph over line $\underline{x} + \lambda\underline{r}$ with $\underline{x} = (0,1)^T$ and with $\underline{r} = (2,1)^T$. The function over this line is

$$\varphi(\lambda) = U\left(\begin{pmatrix} x_1 \\ x_2 \end{pmatrix} = \begin{pmatrix} 0 \\ 1 \end{pmatrix} + \lambda \begin{pmatrix} 2 \\ 1 \end{pmatrix}\right).$$

Note that $\varphi(\lambda)$ is a parabola when $U(\underline{x}) = x_1x_2$ and a piecewise linear curve when $U(\underline{x}) = \min\{x_1, x_2\}$.

If f is continuously differentiable in $\underline{x}$, the directional derivative is:

$$\varphi'_r(0) = \underline{r}^T \nabla f(\underline{x}) \qquad (12.14)$$

This follows from the chain rule for differentiating a composite function with respect to λ :

$$\varphi'_r(\lambda) = \frac{d}{d\lambda} f(\underline{x}+\lambda\underline{r}) = r_1 \frac{\partial}{\partial x_1} f(\underline{x}+\lambda\underline{r}) + r_2 \frac{\partial}{\partial x_2} f(\underline{x}+\lambda\underline{r}) + \ldots + r_n \frac{\partial}{\partial x_n} f(\underline{x}+\lambda\underline{r}) = .$$
$$\underline{r}^T \nabla f(\underline{x}+\lambda\underline{r})$$

Using (12.14), the classification of search directions towards descent and ascent directions becomes relatively easy. For a descent direction $\underline{r}$ holds $\underline{r}^T \nabla f(\underline{x}) = \varphi'_r(0) < 0$, such that $\underline{r}$ has a negative angle with $\nabla f(\underline{x})$. Directions that make a positive angle, $\underline{r}^T \nabla f(\underline{x}) > 0$, are directions where f increases.
The second order derivative in the direction $\underline{r}$ is defined similarly. However, the notation gets more complicated, as every partial derivative $\frac{\partial f}{\partial x_j}$ has derivatives with respect to $x_1, x_2, \ldots,$ x_n. All derivatives can be summarised in the so-called Hesse matrix $H(\underline{x})$ with elements $h_{ij} = \frac{\partial^2 f}{\partial x_i x_j}(\underline{x})$. The matrix is named after the German mathematician Ludwig Otto Hesse (1811-1874). In honour of this man, in this text the matrix is not called Hessian (as usual), but Hessean:

$$H_f(\underline{x}) = \begin{pmatrix} \frac{\partial^2 f}{\partial x_1 \partial x_1}(\underline{x}) & . & . & . & . & \frac{\partial^2 f}{\partial x_1 \partial x_n}(\underline{x}) \\ . & . & . & . & . & . \\ . & . & . & . & . & . \\ . & . & . & . & . & . \\ \frac{\partial^2 f}{\partial x_n \partial x_1}(\underline{x}) & . & . & . & . & \frac{\partial^2 f}{\partial x_n \partial x_n}(\underline{x}) \end{pmatrix}$$

Example 12.8.
Let $f: \mathbb{R}^n \to \mathbb{R}$ be defined by $f(\underline{x}) = x_1^3 x_2 + 2x_1 x_2 + x_1$

$$\frac{\partial f}{\partial x_1}(\underline{x}) = 3x_1^2 x_2 + 2x_2 + 1 \qquad \frac{\partial f}{\partial x_2}(\underline{x}) = x_1^3 + 2x_1$$

$$\frac{\partial^2 f}{\partial x_1 \partial x_1}(\underline{x}) = 6x_1 x_2 \qquad \frac{\partial^2 f}{\partial x_1 \partial x_2}(\underline{x}) = 3x_1^2 + 2$$

$$\frac{\partial^2 f}{\partial x_2 \partial x_1}(\underline{x}) = 3x_1^2 + 2 \qquad \frac{\partial^2 f}{\partial x_2 \partial x_2}(\underline{x}) = 0$$

$$\nabla f(\underline{x}) = \begin{pmatrix} 3x_1^2 x_2 + 2x_2 + 1 \\ x_1^3 + 2x_1 \end{pmatrix} \qquad H_f(\underline{x}) = \begin{pmatrix} 6x_1 x_2 & 3x_1^2 + 2 \\ 3x_1^2 + 2 & 0 \end{pmatrix}$$

In the rest of this text, the suffix f of the Hessean will only be used when it is not clear from the context which function is meant. The Hessean in this example is a symmetric matrix. It can be shown that the Hessean is symmetric if f is *twice continuously differentiable*. For the optimality conditions we are interested in the second order derivative $\varphi''_r(\lambda)$ in the direction $\underline{r}$. Proceeding with the chain rule on (12.14) results in

$$\varphi''_r(\lambda) = \underline{r}^T H(\underline{x} + \lambda \underline{r})\underline{r} \tag{12.15}$$

if f is twice continuously differentiable.

Example 12.9.

The function $U(\underline{x}) = x_1 x_2$ implies

$$\nabla U(\underline{x}) = \begin{pmatrix} x_2 \\ x_1 \end{pmatrix} \text{ and } H(\underline{x}) = \begin{pmatrix} 0 & 1 \\ 1 & 0 \end{pmatrix}$$

The Hessean is independent of the point $\underline{x}$. Consider the one-dimensional function $\varphi_r(\lambda) = U(\underline{x} + \lambda \underline{r})$ with $\underline{x} = (0, 0)^T$. In the direction $\underline{r} = (1,1)^T$ is $\varphi_r(\lambda) = \lambda^2$ (parabola),

$$\varphi'_r(\lambda) = \underline{r}^T \nabla U(\underline{x} + \lambda \underline{r}) = (1,1)\begin{pmatrix} \lambda \\ \lambda \end{pmatrix} = 2\lambda \text{ and}$$

$$\varphi''_r(\lambda) = \underline{r}^T H(\underline{x} + \lambda \underline{r})\underline{r} = (1,1)\begin{pmatrix} 0 & 1 \\ 1 & 0 \end{pmatrix}\begin{pmatrix} 1 \\ 1 \end{pmatrix} = 2.$$

In the direction $\underline{r} = (1, -1)^T$ is $\varphi_r(\lambda) = -\lambda^2$ (parabola with maximum) such that

$$\varphi'_r(\lambda) = (1,-1)\begin{pmatrix} -\lambda \\ \lambda \end{pmatrix} = -2\lambda \text{ and}$$

$$\varphi''_r(\lambda) = (1,-1)\begin{pmatrix} 0 & 1 \\ 1 & 0 \end{pmatrix}\begin{pmatrix} 1 \\ -1 \end{pmatrix} = -2$$

In $\underline{x} = (0,0)^T$ there are directions in which $\underline{x}$ is a minimum point and there are directions in which $\underline{x}$ is a maximum point. The point $\underline{x}$ is called a *saddle point*.

Taylor

The first and second order derivatives play a role in the so-called mean value theorem and Taylor's theorem. Higher order derivatives that are usually postulated in the theorem of Taylor are left out here.

The mean value theorem says that for a differentiable function between two points a and b a point x_i exists where the derivative has the same value as the slope between a and b.

Theorem 12.1.
Mean value theorem
Let $f: \mathbb{R} \rightarrow \mathbb{R}$ be continuous on the interval $(a,b]$ and differentiable on (a,b) then $\exists\xi$, $a \leq \xi \leq b$ such that

$$f'(\xi) = \frac{f(b) - f(a)}{b - a} \tag{12.16}$$

As a consequence, considered from a point x_1, the function value is

$$f(x) = f(x_1) + f'(\xi)(x - x_1). \tag{12.17}$$

So $f(x)$ equals $f(x_1)$ plus a residual term that depends on the derivative in a point in between x and x_1 and the distance between x and x_1. The residual or error idea can also be found in Taylor's theorem. For a twice differentiable function, (12.17) can be extended to

$$f(x) = f(x_1) + f'(x_1)(x - x_1) + \frac{1}{2} f''(\xi)(x - x_1)^2. \tag{12.18}$$

It tells us that $f(x)$ can be approximated by the tangent line through x_1 and that the error term is determined by the second order derivative in a point ξ in between x and x_1. The tangent line $f(x_1) + f'(x_1)(x - x_1)$ is called the first order Taylor approximation. The equivalent terminology for functions of several variables can be derived from the one-dimensional cross-cut function φ_r given in (12.7). We consider vector $\underline{x}_1$ as a fixed point and do a step into direction $\underline{r}$, such that $\underline{x} = \underline{x}_1 + \lambda \underline{r}$; given $\varphi_r(\lambda) = f(\underline{x}_1 + \lambda \underline{r})$ consider $\varphi_r(1) = f(\underline{x})$. The mean value theorem gives

$$f(\underline{x}) = \varphi_r(1) = \varphi_r(0) + \varphi'_r(\xi) = f(\underline{x}_1) + \underline{r}^T \nabla f(\underline{\theta}) = f(\underline{x}_1) + (\underline{x} - \underline{x}_1)^T \nabla f(\underline{\theta})$$

where $\underline{\theta}$ is a vector in between $\underline{x}_1$ and $\underline{x}$. The *first order Taylor approximation* becomes

$$f(\underline{x}) \approx f(\underline{x}_1) + (\underline{x} - \underline{x}_1)^T \nabla(\underline{x}_1) \tag{12.19}$$

This line of reasoning via (12.15) results in Taylor's theorem (2nd order)

$$\begin{aligned} f(\underline{x}) &= \varphi_r(1) = \varphi_r(0) + \varphi'_r(0) + \tfrac{1}{2}\varphi''_r(\xi) = \\ & f(\underline{x}_1) + (\underline{x} - \underline{x}_1)^T \nabla f(\underline{x}_1) + \tfrac{1}{2}(\underline{x} - \underline{x}_1)^T H(\underline{\theta})(\underline{x} - \underline{x}_1) \end{aligned} \tag{12.20}$$

where $\underline{\theta}$ is a vector in between $\underline{x}_1$ and $\underline{x}$. The second order Taylor approximation appears when in (12.20) $\underline{\theta}$ is replaced by $\underline{x}_1$. The function of Equation (12.20) is a so-called quadratic function. In the following section we will first focus on this type of function.

Example 12.10.

Let $f(\underline{x}) = x_1^3 x_2 + 2x_1x_2 + x_1$ like in Example 12.8. The first order Taylor approximation of $f(\underline{x})$ around $\underline{0}$ is

$$f(\underline{x}) \approx f(\underline{0}) + \underline{x}^T \nabla f(\underline{0})$$
$$\nabla f(\underline{0}) = \begin{pmatrix} 1 \\ 0 \end{pmatrix}$$
$$f(\underline{x}) \approx 0 + (x_1, x_2)\begin{pmatrix} 1 \\ 0 \end{pmatrix} = x_1$$

In Section 12.3.4, the consequence of first and second order derivatives with respect to optimality conditions is considered. First the focus will be on the specific shape of quadratic functions at which we arrived in Equation (12.20).

12.3.3. Quadratic functions

In this section, we focus on a special class of functions and their optimality conditions. In the following sections we expand on this towards general smooth functions. At least for any smooth function the second order Taylor Equation (12.20) is valid, which is a quadratic function. In general a quadratic function $f: \mathbb{R}^n \to \mathbb{R}$ can be written as:

$$f(\underline{x}) = \underline{x}^T A\underline{x} + \underline{b}^T \underline{x} + c \tag{12.21}$$

where A is a symmetric $n \times n$-matrix and $\underline{b}$ an n vector. Besides constant c and linear term $\underline{b}^T\underline{x}$ (12.21) has a so-called quadratic form $\underline{x}^T A\underline{x}$. Let us first consider this quadratic form, as has already been exemplified in Example 12.9:

$$\underline{x}^T A\underline{x} = \sum_{i=1}^{n}\sum_{j=1}^{n} a_{ij}x_i x_j \tag{12.22}$$

or alternatively as was written in the portfolio example:

$$\underline{x}^T A\underline{x} = \sum_{i=1}^{n} a_{ii}x_i^2 + 2\sum_{i=1}^{n}\sum_{j=i+1}^{n} a_{ij}x_i x_j \tag{12.23}$$

Example 12.11.

Let $A = \begin{pmatrix} 2 & 1 \\ 1 & 1 \end{pmatrix}$ then $\underline{x}^T A\underline{x} = 2x_1^2 + x_2^2 + 2x_1x_2$

The quadratic form $\underline{x}^T A\underline{x}$ determines whether the quadratic function has a maximum, minimum or neither of them. The quadratic form has a value of 0 in the origin. This would be a minimum of $\underline{x}^T A\underline{x}$, if $\underline{x}^T A\underline{x} \geq 0$ or similarly in the line of the cross cut function $\varphi_r(\lambda) = f(\underline{0} + \lambda \underline{r})$, walking in any direction $\underline{r}$ would give a nonnegative value:

$$(\underline{0}+\lambda \underline{r})^T A(\underline{0}+\lambda \underline{r}) = \lambda^2 \underline{r}^T A \underline{r} \geq 0, \forall r \tag{12.24}$$

We will continue this line of thinking in the following section. For quadratic functions, it brings us to introduce a useful concept.

Definition 12.4.

Let A be a symmetric $n \times n$-matrix. A is called positive definite if $\underline{x}^T A \underline{x} > 0$ for all $\underline{x} \in \mathbb{R}^n$, $\underline{x} \neq 0$. Matrix A is called positive semi-definite if $\underline{x}^T A \underline{x} \geq 0$ for all $\underline{x} \in \mathbb{R}^n$. The notion of negative (semi-) definite is defined analogously. Matrix A is called indefinite if vectors $\underline{x}_1$ and $\underline{x}_2$ exist such that $\underline{x}_1^T A \underline{x}_1 > 0$ and $\underline{x}_2^T A \underline{x}_2 < 0$.

The status of matrix A with respect to positive, negative definiteness or indefiniteness determines whether quadratic function $f(\underline{x})$ has a minimum, maximum or neither of them. The question is of course, how to check the status of A. One can use a mathematical theorem on the eigenvalues of a matrix. It can be shown that for the quadratic form

$$\mu_1 \|\underline{x}\|^2 \leq \underline{x}^T A \underline{x} \leq \mu_n \|\underline{x}\|^2$$

where μ_1 the smallest and μ_2 the highest eigenvalue of A and $\|\underline{x}\|^2 = \underline{x}^T \underline{x}$. This means that for a positive definite matrix A all eigenvalues are positive and for a negative definite matrix all eigenvalues are negative.

Theorem 12.2.

Let A be a symmetric $n \times n$-matrix. A is positive definite $\Leftrightarrow$ all eigenvalues of A are positive.

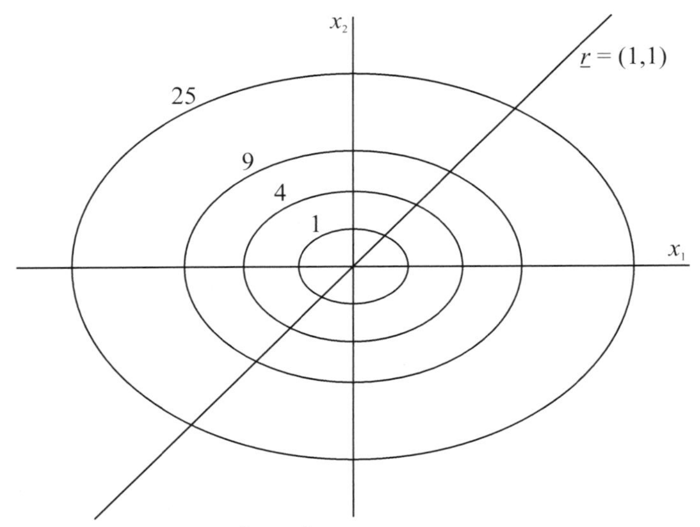

Figure 12.14. Contours of $f(\underline{x}) = x_1^2 + 2x_2^2$.

Moreover, the corresponding eigenvectors are orthogonal to the contours of $f(\underline{x})$. The eigenvalues of A can be determined by finding those values of μ for which $A\underline{x} = \mu\underline{x} \rightarrow$ $(A - \mu E)\ \underline{x} = \underline{0}$ with E the unit matrix, such that the determinant $|A - \mu E| = 0$. Let us look at some examples.

Example 12.12.

Consider $A = \begin{pmatrix} 1 & 0 \\ 0 & 2 \end{pmatrix}$ such that $f(\underline{x}) = x_1^2 + 2x_2^2$. The corresponding contours as sketched in Figure 12.14 are ellipsoids. The eigenvalues can be found on the diagonal of A and are 1 and 2 with corresponding eigenvectors $\underline{r}_1 = (1, 0)^T$ and $\underline{r}_2 = (0, 1)^T$. Following cross-cut functions from the origin according to (12.24) gives positive parabolas $\varphi_{r1}(\lambda) = \lambda^2$ and $\varphi_{r2}(\lambda) = 2\lambda^2$. Walking into the direction $\underline{r} = (1, 1)^T$ also results in a positive parabola, but as depicted, the corresponding line is not orthogonal to the contours of $f(\underline{x})$.

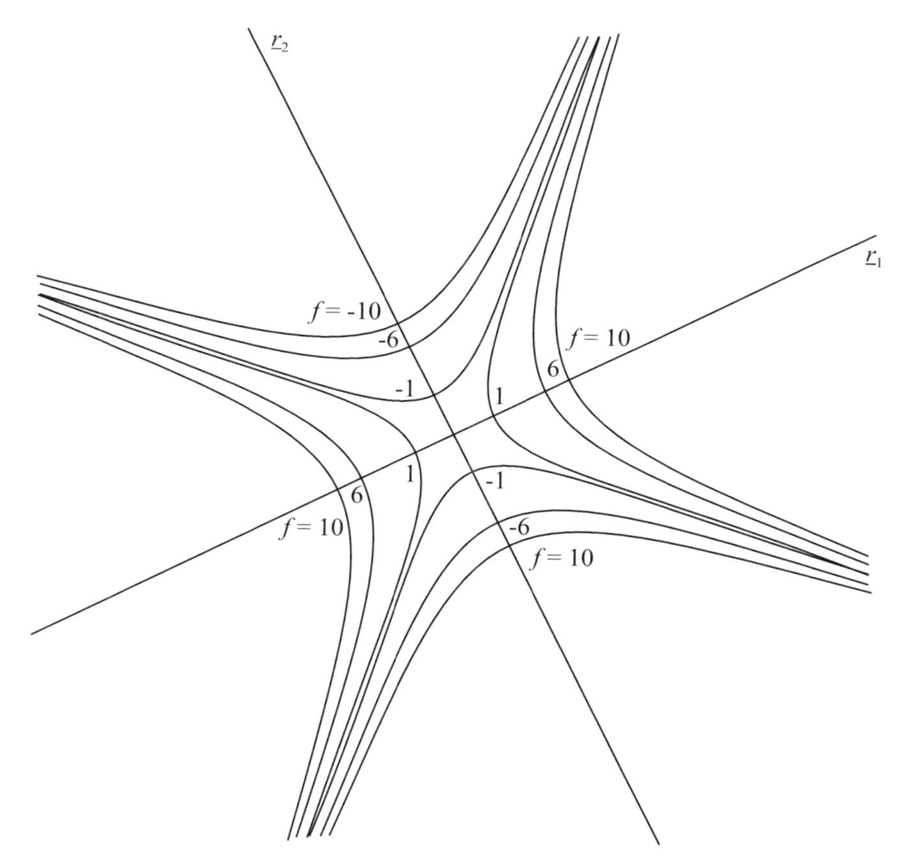

Figure 12.15. Contours of $f(\underline{x}) = 3x_1^2 - 3x_2^2 + 8x_1x_2$.

We have seen already in Example 12.9 a case of an indefinite quadratic form. In some directions the parabola curves downward and in some directions it curves upward. We consider here one example function for which this is less obvious.

Example 12.13.

Consider $A=\begin{pmatrix}3 & 4\\ 4 & -3\end{pmatrix}$ such that $f(\underline{x})=3x_1^2-3x_2^2+8x_1x_2$. The corresponding contours are sketched in Figure 12.15. The eigenvalues of A can be determined by finding those values of μ for which:

$$A=\begin{vmatrix}3-\mu & 4\\ 4 & -3-\mu\end{vmatrix}=0\rightarrow\mu^2-25=0 \tag{12.25}$$

such that the eigenvalues are $\mu_1 = 5$ and $\mu_2 = -5$; A is indefinite. The eigenvectors can be found from $A\underline{r} = \mu\underline{r} \rightarrow (A - \mu E)\underline{r} = \underline{0}$. In this example, they are any multiple of $\underline{r}_1 = (2, 1)^T$ and $\underline{r}_2 = (1, -2)^T$. The corresponding lines, also called axes, are given in Figure 12.15. In the direction of $\underline{r}_1$, $\varphi_{r1}(\lambda) = 5\lambda^2$ is a positive parabola. In the direction of $\underline{r}_2$ we have a negative parabola. Specifically in the direction $\underline{r} = (1, 3)^T$, $f(\underline{x})$ is constant.

When the linear term $\underline{b}^T\underline{x}$ is added to the quadratic form, the centre of the contours is shifted towards

$$\underline{x}^*=-\frac{1}{2}A^{-1}\underline{b} \tag{12.26}$$

where $\underline{x}^*$ can only be determined if the columns of A are linearly independent. This means that in that case (12.21) can be written as

$$f(\underline{x})=\underline{x}^TA\underline{x}+\underline{b}^T\underline{x}+c=(\underline{x}-\underline{x}^*)^TA(\underline{x}-\underline{x}^*)+constant \tag{12.27}$$

where $constant$ = $c-\frac{1}{4}A^{-1}\underline{b}$. Combining Definition 12.4 with Equation (12.27) gives that apparently $\underline{x}^*$ is a minimum point if A is positive semi-definite and a maximum point if A is negative semi-definite. The derivative information of quadratic functions is typically linear. The gradient of quadratic (12.27) is given by

$$\nabla f(\underline{x})=2A\underline{x}+\underline{b} \tag{12.28}$$

Note that point $\underline{x}^*$ is a point where the gradient is the zero vector, a so-called stationary point. The Hessean of a quadratic function is constant:

$$H(\underline{x})=2A \tag{12.29}$$

Example 12.14

Consider $A=\begin{pmatrix}1 & 0\\ 0 & 2\end{pmatrix}, b=\begin{pmatrix}2\\ 4\end{pmatrix}, f(\underline{x})=x_1^2+2x_2^2+2x_1+4x_2$.

The centre of Figure 12.14 is now determined by

$$\underline{x}^*=-\frac{1}{2}A^{-1}\underline{b}=-\frac{1}{2}\begin{pmatrix}1 & 0\\ 0 & \frac{1}{2}\end{pmatrix}\begin{pmatrix}2\\ 4\end{pmatrix}=-\begin{pmatrix}1\\ 1\end{pmatrix} \tag{12.30}$$

and constant $= -\frac{1}{4}\underline{b}^T A^{-1}\underline{b}=-\frac{1}{4}(2,4)^T\begin{pmatrix}1 & 0\\ 0 & \frac{1}{2}\end{pmatrix}\begin{pmatrix}2\\ 4\end{pmatrix}=-3$ such that $f(\underline{x})$ can be written as $f(\underline{x})=x_1^2+2x_2^2+2x_1+4x_2=(x_1+1)^2+2(x_2+1)^2-3$

12.3.4. Optimality conditions

An optimal point is determined by the behaviour of the objective function in all feasible directions. If $f(\underline{x})$ is increasing from $\underline{x}^*$ in all feasible directions $\underline{r}$, then $\underline{x}^*$ is a minimum point. The feasibility of directions is determined by the constraints that are binding in $\underline{x}^*$. Traditionally, two situations are distinguished:

1. There are no binding constraints in $\underline{x}^*$, $\underline{x}^*$ is an interior point of X.
2. There are binding constraints, $\underline{x}^*$ is situated at the boundary of X.

12.3.5. No binding constraints

The same line is followed as in Section 12.3.2 starting with one-dimensional functions via the cross-cut functions $\varphi_r(\lambda)$ to functions of several variables. Mathematical background and education often give the principle of putting derivatives to zero, popularly called 'finding an analytical solution'. The mathematical background of this principle is sketched here and commented.

First order conditions (F.O.C.)

First of all one should assume f is a continuously differentiable function. Considering a minimum point x^* of a one-dimensional function, gives via the definition of derivative

$$f'(x^*)=\lim_{x\to x^*}\frac{f(x)-f(x^*)}{x-x^*} \tag{12.31}$$

The nominator of the quotient is nonnegative (x^* is a minimum point) and the denominator is either negative or positive depending on x approaching x^* from below or from above. So the limit in (12.31) can only exist if $f'(x^*)=0$. More dimensional functions follow the same property with the additional complication that the directional derivative

$$\varphi_r'(0) = \lim_{h \to 0} \frac{f(\underline{x}^* + h\underline{r}) - f(\underline{x}^*)}{h} = \underline{r}^T \nabla f(\underline{x}^*) \tag{12.32}$$

depends on direction $\underline{r}$. The directional derivative being zero $\underline{r}^T \nabla f(\underline{x}^*) = 0$ for all possible directions $\underline{r}$, implies $\nabla f(\underline{x}^*) = 0$. A point $\underline{x}$ with $\nabla f(\underline{x}) = 0$ is called a *stationary point*. Finding one (or all) stationary points results in a set of n equalities and n unknowns and in general cannot be easily solved. Moreover, a stationary point can be:

- A minimum point; $f(x) = x^2$ and $x = 0$
- A maximum point; $f(x) = -x^2$ and $x = 0$
- A point of inflection; $f(x) = x^3$ and $x = 0$
- A saddle point i.e. in some directions a maximum point and in others a minimum point (Example 12.9).
- Combination of point of inflection, minimum or maximum point.

The variety is illustrated by Example 12.15

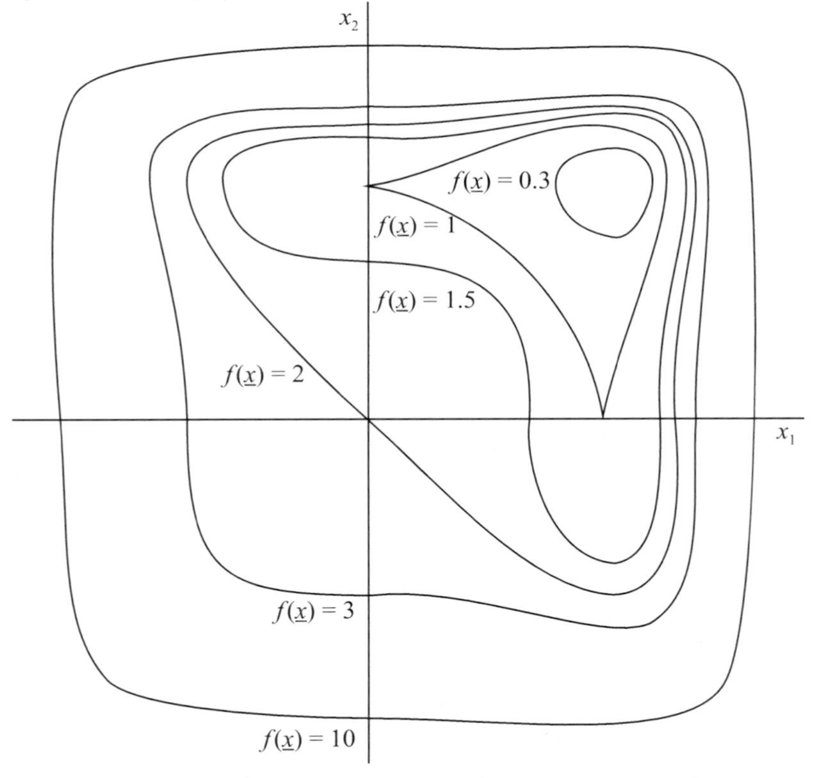

Figure 12.16. Contours of $f(\underline{x}) = (x_1^3 - 1)^2 + (x_2^3 - 1)^2$.

Example 12.15.

Let $f(\underline{x}) = (x_1^3 - 1)^2 + (x_2^3 - 1)^2$. The contours of f are depicted in Figure 12.16. The gradient of f is

$$\nabla f(\underline{x}) = \begin{pmatrix} 6x_1^2(x_1^3 - 1) \\ 6x_2^2(x_2^3 - 1) \end{pmatrix}$$

The stationary points can easily be found for this case: $\nabla f(\underline{x}) = \underline{0}$ gives $6x_1^2(x_1^3 - 1) = 0$ and $6x_2^2(x_2^3 - 1) = 0$. The stationary points are (0, 0); (1, 1); (1, 0) and (0, 1). The function value f in (1, 1) equals zero and it is easy to see that $f(\underline{x}) > 0$ for all other points $\underline{x}$. So point (1, 1) is a global minimum point. The other stationary points (0,0); (1, 0) and (0, 1) are situated on a contour such that in their direct environment there are points with a higher function value as well as points with a lower function value; they are neither minimum nor maximum points.

Second order conditions

The assumption is required that f is twice continuously differentiable. Now Taylor's theorem can be used. Given a point x^* with $f'(x^*) = 0$, then (12.18) tells us that

$$f(x) = f(x^*) + \frac{1}{2} f''(\xi)(x - x^*)^2. \qquad (12.33)$$

Whether x^* is a minimum point is determined by the sign of $f''(\xi)$ in the environment of x^*. If $f'(x^*) = 0$ and $f''(\xi) \geq 0$ for all ξ in an environment, then x^* is a minimum point. When f'' is a continuous function and $f''(x^*) > 0$ then there exists an environment of x^* such that for all points in that environment $f''(x) > 0$, so x^* is a minimum point. However, if $f''(x) = 0$, as for $f(x) = x^3$ and $f(x) = x^4$ in $x^* = 0$, then higher order derivatives should be considered to determine the status of x^*.

Theorem 12.3.

Let $f: \mathbb{R} \to \mathbb{R}$ be twice continuously differentiable in x^*. If $f'(x^*) = 0$ and $f''(x^*) > 0$ then x^* is a minimum point. If x^* is a minimum point then $f'(x^*) = 0$ and $f''(x^*) > 0$.

Extending Theorem 12.3 towards functions of several variables requires studying $\varphi''_r(0)$ in a stationary point x^* where $\varphi_r(\lambda) = f(\underline{x}^* + \lambda \underline{r})$. According to (12.15) we should know the sign of

$$\varphi''_r(0) = \underline{r}^T H(\underline{x}^*)\underline{r} \qquad (12.34)$$

in all directions $\underline{r}$. Expression (12.34) is a quadratic form. The derivation of Theorem 12.3 via (12.18) also applies for functions in several variables via (12.20).

Theorem 12.4.

Let $f: \mathbb{R}^n \to \mathbb{R}$ be twice continuously differentiable in $\underline{x}^*$. If $\nabla f(\underline{x}^*) = \underline{0}$ and $H(\underline{x}^*)$ is positive definite, then $\underline{x}^*$ is a minimum point. If $\underline{x}^*$ is a minimum point then $\nabla f(\underline{x}^*) = \underline{0}$ and $H(\underline{x}^*)$ is positive semi-definite.

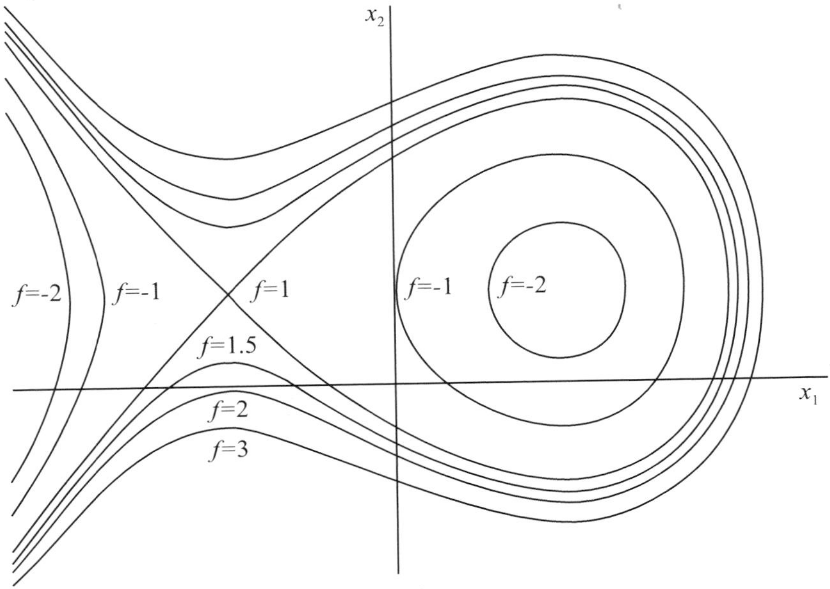

Figure 12.17. Contours of $f(\underline{x}) = x_1^3 - 3x_2 + x_2^2 - 2x_2$.

Example 12.16.

Consider the contours of $f(\underline{x}) = x_1^3 - 3x_2 + x_2^2 - 2x_2$ in Figure 12.17. A minimum point and a saddle point can be recognised. Both are stationary points, but the Hessean has a different character. The gradient is $\nabla f(\underline{x}) = \begin{pmatrix} 3x_1^2 - 3 \\ 2x_2 - 2 \end{pmatrix}$ and the Hessean $H(\underline{x}) = \begin{pmatrix} 6x_1 & 0 \\ 0 & 2 \end{pmatrix}$. The eigenvalues of the Hessean are $6x_1$ and 2. The stationary points are determined by $\nabla f(\underline{x}) = \underline{0}$, such that $\underline{x}_1^* = (1,1)^T$ and $\underline{x}_2^* = (-1,1)^T$.

$H(\underline{x}_1^*) = \begin{pmatrix} 6 & 0 \\ 0 & 2 \end{pmatrix}$ is positive definite and $H_f(\underline{x}_2^*) = \begin{pmatrix} -6 & 0 \\ 0 & 2 \end{pmatrix}$ is indefinite. This means that $\underline{x}_1^*$ is a minimum point and $\underline{x}_2^*$ is not a minimum point.

2.3.6. Binding constraints

To check the optimality of a given $\underline{x}^*$, it should be verified that $f(\underline{x})$ is optimal in $\underline{x}^*$ for all feasible directions. Mathematical theorems have been formulated to help the verification. If

one doesn't carefully consider the underlying assumptions, the applications of such theorems may lead to incorrect conclusions about the status of $\underline{x}^*$. We try to avoid all mathematical details and with the aid of illustrative examples make the reader aware of possible mistakes and the value of the assumptions.
Many names are connected to the mathematical statements with respect to optimality when there are binding constraints, in contrast to the theorems mentioned before. Well-known conditions are the so-called Karush-Kuhn-Tucker conditions (KKT conditions). We will first have a look at the historic perspective, see (Kuhn, 1991).
J.L. Lagrange studied the questions of optimisation subject to equality constraints back in 1813. In 1939, W. Karush presented in his student thesis, conditions that should be valid for an optimal point with equality constraints. Independently F. John presented some optimality conditions for a specific problem in 1948. Finally, the first order conditions really became known after a presentation given by H.W. Kuhn and A.W. Tucker to a mathematical audience at a symposium in 1950. Their names were connected to the conditions that are nowadays known as the *KKT conditions*.

Lagrange multiplier method

The KKT conditions are often explained with the so-called Lagrange function or *Lagrangean*. It was developed for equality constraints $g_i(\underline{x}) = 0$, but can also be applied to inequality constraints $g_i(\underline{x}) \leq 0$.

$$L(\underline{x},\underline{u}) = f(\underline{x}) + \Sigma u_i g_i(\underline{x}) \tag{12.35}$$

where $f(\underline{x})$ is the objective function that should be *minimised*. The constraints with respect to $g_i(\underline{x})$ are added to the objective function with so-called Lagrange multipliers u_i that can be interpreted as dual variables. The most important property of this function is that under some conditions it can be shown that for any minimum point $\underline{x}^*$ of (12.1), there exists a dual solution $\underline{u}^*$ such that $(\underline{x}^*,\underline{u}^*)$ is a saddle point of $L(\underline{x},\underline{u})$ via

$$\underline{x}^*,\underline{u}^* \text{ is a solution of } \min_x \max_u L(\underline{x},\underline{u}) \tag{12.36}$$

So $\underline{x}^*$ is a minimum point of L ($\underline{u}^*$ constant) and $\underline{u}^*$ a maximum point. We are going to experiment with this idea. Why is it important to get some feeling for (12.36)? Often implicit use is made of (12.36) following the concept of the 'Lagrange multiplier method'. In this concept one uses the idea of the saddle point for putting the derivatives to $\underline{u}$ and $\underline{x}$ to zero and trying to find analytical solutions $\underline{x}^*$, $\underline{u}^*$ of

$$\nabla L(\underline{x},\underline{u}) = 0 \tag{12.37}$$

Example 12.17.

In Example 12.3 (see Figure 12.10), the optimum can be determined graphically. In the optimal point $\underline{x}^*$ = (3,3/2) only $g_1(\underline{x}) = x_1 + 2x_2 - 6 \leq 0$ is a binding constraint. Given point $\underline{x}^*$ in the Lagrangean $L(\underline{x},\underline{u}) = -U(\underline{x}) + \Sigma u_i g_i(\underline{x})$ one can put $u_2^* = u_3^*$, because the 2nd and

3rd constraint $x_1 \geq 0$ and $x_2 \geq 0$ are non-binding. The optimal point is the same if the 2nd and 3rd constraint are left out of the problem. This illustrates the notion of *complementarity*, that is also valid in Linear Programming (Chapter 4); $u_i^* g_i(\underline{x}^*) = 0$. If $u_2 = u_3 = 0$, then $L(\underline{x}, u_1) = -x_1 x_2 + u_1(x_1 + 2x_2 - 6)$. So (12.37) leads to $\nabla L(\underline{x}, \underline{u}) = \underline{0}$:

$$\partial L / \partial x_1 = 0 \Rightarrow -x_2 + u_1 = 0$$
$$\partial L / \partial x_2 = 0 \Rightarrow -x_1 + 2u_1 = 0$$
$$\partial L / \partial u_1 = 0 \Rightarrow x_1 + 2x_2 - 6 = 0$$
$$x_1^* = 3, x_2^* = 3/2, u_1^* = 3/2, U(\underline{x}^*) = 4/5$$

is a unique solution and $\underline{x}^* = (3,3/2)$, $\underline{u}^* = (3/2,0,0)$ is a stationary point of the Lagrangean corresponding to the optimum. The value of $u_1^* = 3/2$ has the interpretation of shadow price; an additional unit of budget results into 3/2 units of additional utility.

The Lagrange multiplier method is slightly tricky:

1. Finding a stationary point analytically may not be easy
2. An optimal solution may be just one of (infinitely) many solutions of (12.37).
3. Due to some additional constraints the saddle point (12.36) of L may not coincide with a solution of (12.37).
4. For the inequality constraints one should know in advance which $g_i(x) \leq 0$ are binding. Given a specific point $\underline{x}^*$ this is of course known.

Karush-Kuhn-Tucker conditions

The Lagrange multiplier method may not always be appropriate for finding an optimum. On the other hand, an optimal point $\underline{x}^*$ (under regularity conditions and differentiability) should correspond to a stationary point of the Lagrangean (12.37) via the Karush-Kuhn-Tucker conditions, in which the notion of complementarity is more explicit.

Theorem 12.5.

Karush-Kuhn-Tucker conditions

If $\underline{x}^*$ is a minimum point of (12.1), then there exist numbers $\underline{u}^*$ such that

$$-\nabla f(\underline{x}^*) = \sum u_i \nabla g_i(\underline{x}^*)$$

$$u_i^* g_i(\underline{x}^*) = 0 \qquad \text{complementarity}$$

$$u_i^* \geq 0 \qquad \text{for constraints } g_i(\underline{x}) \leq 0$$

In mathematical terms this theorem shows us that the direction of optimisation ($-\nabla f(\underline{x}^*)$ in a minimisation problem and $\nabla f(\underline{x}^*)$ in a maximisation problem) in the optimum is a combination of the gradients of the active constraints.

Example 12.18.

Problem NLPU (Example 12.4) with $U(\underline{x}) = x_1^2 + x_2^2$ can be formulated as

$$\begin{aligned}
&\min\{ f(\underline{x}) = -x_1^2 - x_2^2 \} \\
&\quad g_1(\underline{x}) = x_1 + 2x_2 - 6 \le 0 \\
&\quad g_2(\underline{x}) = -x_1 \le 0 \\
&\quad g_3(\underline{x}) = -x_2 \le 0
\end{aligned}$$

so

$$\nabla f(\underline{x}) = \begin{pmatrix} -2x_1 \\ -2x_2 \end{pmatrix}, \nabla g_1(\underline{x}) = \begin{pmatrix} 1 \\ 2 \end{pmatrix}, \nabla g_2(\underline{x}) = \begin{pmatrix} -1 \\ 0 \end{pmatrix}, \nabla g_3(\underline{x}) = \begin{pmatrix} 0 \\ -1 \end{pmatrix}$$

In the (local) minimum point $\underline{x}_2^* = (0,3)^T$, g_1 and g_2 are binding and $g_3(\underline{x}_2^*) = -1 < 0$ is non-binding, so that $u_3^* = 0$.

$$-\nabla f(\underline{x}_2^*) = \begin{pmatrix} 0 \\ 6 \end{pmatrix} = u_1^* \nabla g_1(\underline{x}_2^*) + u_2^* \nabla g_2(\underline{x}_2^*) + 0 \nabla g_3(\underline{x}_2^*) \Rightarrow$$

$$\begin{pmatrix} 0 \\ 6 \end{pmatrix} = u_1^* \begin{pmatrix} 1 \\ 2 \end{pmatrix} + u_2^* \begin{pmatrix} -1 \\ 0 \end{pmatrix} \Rightarrow u_1^* = 3, = u_2^* = 3, = u_3^* = 0$$

For the global minimum point $\underline{x}_1^* = (6,0)^T$, can be derived analogously

$$-\nabla f \begin{pmatrix} 6 \\ 0 \end{pmatrix} = \begin{pmatrix} 12 \\ 0 \end{pmatrix} = u_1^* \begin{pmatrix} 1 \\ 2 \end{pmatrix} + 0 \begin{pmatrix} -1 \\ 0 \end{pmatrix} + u_3^* \begin{pmatrix} 0 \\ -1 \end{pmatrix} \Rightarrow u_1^* = 12, u_2^* = 0, u_3^* = 24$$

Note that $\underline{x}_3^* = \frac{1}{5}\begin{pmatrix} 6 \\ 12 \end{pmatrix}$ is a KKT point (not optimal) according to

$$-\nabla f\left(\frac{1}{5}\begin{pmatrix} 6 \\ 12 \end{pmatrix} \right) = \frac{1}{5}\begin{pmatrix} 12 \\ 24 \end{pmatrix} = \frac{12}{5} \nabla g_1 + 0 \nabla g_2 + 0 \nabla g_3$$

Under regularity conditions, the KKT conditions are necessary for a point $\underline{x}^*$ to be optimal. The KKT conditions are not sufficient, as has been shown by Example 10.18. Similarly to the case without binding constraints *second order conditions* exist based on the Hessean. Those conditions are far more complicated, because the sign of the second order derivatives should be determined in the tangent planes of the binding constraints. We refer to the literature on the topic, such as (Scales, 1985), (Gill *et al.*, 1981) and (Bazaraa *et al.*, 1993). In the following section the notion of convexity will be discussed and its relation to the second order conditions.

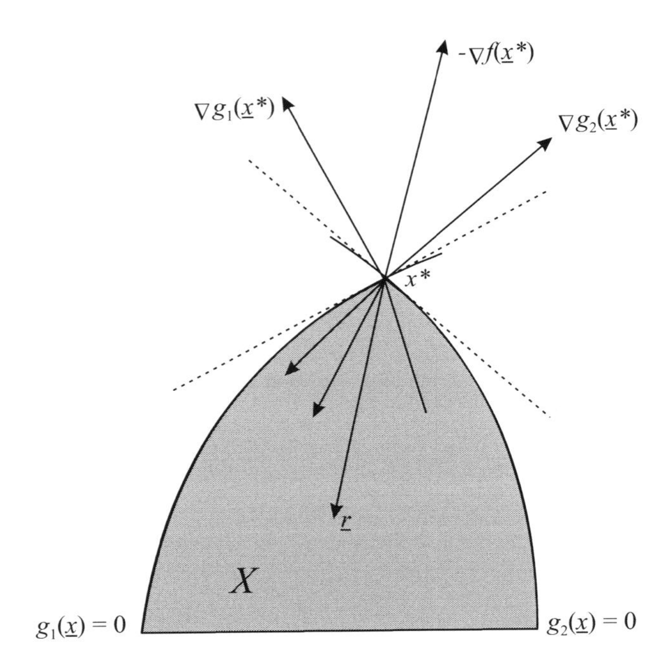

Figure 12.18. Feasible directions.

We end with one of the starting observations of this chapter: $\underline{x}^*$ is a minimum point, if in any feasible direction it is a minimum point. A small positive step into a feasible direction cannot generate a lower objective function value. Graphically seen, the directions that point into the feasible area are related to the gradients of the active constraints (see Figure12.18). Mathematically this can be seen as follows. If constraint $g_i\underline{x}^* \leq 0$, is binding (active) in $\underline{x}^*, g_i(\underline{x}^*) = 0$ a direction $\underline{r}$ fulfilling $\underline{r}^T\nabla g_i(\underline{x}^*) < 0$ is pointing into the feasible area and a direction such that $\underline{r}^T\nabla g_i(\underline{x}^*) > 0$ points out of the area. In a minimum point $\underline{x}^*$ every feasible direction $\underline{r}$ should lead to an increase in the objective function value i.e. $\underline{r}^T\nabla f(\underline{x}^*) \geq 0$. If a direction $\underline{r}$ fulfils $\underline{r}^T\nabla g_i(\underline{x}^*) < 0$ for every binding constraint, then it should also fulfil $\underline{r}^T\nabla f(\underline{x}^*) \geq 0$. Because of the KKT conditions $-\nabla f(\underline{x}^*) = \Sigma u_i \nabla g_i(\underline{x}^*)$ with $u_i \geq 0$ feasible direction for $\underline{r}$ holds

$$-\underline{r}^T\nabla f(\underline{x}^*) = \Sigma u_i \underline{r}^T \nabla g_i(\underline{x}^*) \geq 0 \qquad (12.38)$$

So the KKT conditions are necessary to imply that $\underline{x}^*$ is a minimum point in all feasible directions. Graphically this means that arrow $-\nabla f(\underline{x}^*)$ is situated in between the gradients $\nabla g_i(\underline{x}^*)$ for all binding inequalities.

12.3.7. Consequences of convexity

The relevance of convexity for a general NLP problem is mainly due to three properties. For a so-called convex optimisation problem (12.1) applies:

1. If f and g_i are differentiable functions, a KKT point (and a stationary point) is also a minimum point. The KKT conditions are sufficient for optimality.
2. If a minimum point is found, it is also a global minimum point.
3. A maximum point can be found at the boundary of the feasible region. It is even a so-called extreme point.

Note that the notion of convexity is not directly related to differentiability. Differentiability is relevant for property 1. The second and third property are also valid for non-differentiable cases. How can one test the convexity of a specific problem? That is a difficult point. For many black box applications and formulations in Section 12.2 where the calculation of the function is the result of a long calculation process, analysis of the formulas is not possible. The utility maximisation examples in this chapter reveal their expressions and one can check the convexity. In economics literature where NLP is applied, other more weak assumptions can often be found; the functions g_i are quasi convex. For a more detailed overview we refer to (Bazaraa *et al.*, 1993).

A stationary point is a minimum point

We show that for a convex function f, a stationary point is a minimum point (property 1). This can be seen from the observation that for a convex function a tangent line (plane) is below the graph of f.

Theorem 12.6.

Let f be a convex and continuously differentiable function on X. For any two points $\underline{x}, \underline{x}_1 \in X$

$$f(\underline{x}) \geq f(\underline{x}_1) + \nabla f(\underline{x}_1)^T(\underline{x} - \underline{x}_1) \qquad (12.39)$$

This can be seen as follows. For a convex function f

$$f(\lambda \underline{x} + (1-\lambda)\underline{x}_1) \leq \lambda f(\underline{x}) + (1-\lambda) f(\underline{x}_1)$$

Rewriting gives

$$f(\underline{x}_1 + \lambda(\underline{x} - \underline{x}_1)) \leq f(\underline{x}_1) + \lambda(f(\underline{x}) - f(\underline{x}_1))$$

such that

$$f(\underline{x}) - f(\underline{x}_1) \geq \frac{f(\underline{x}_1 + \lambda(\underline{x} - \underline{x}_1)) - f(\underline{x}_1)}{\lambda}$$

Limit $\lambda \to 0$ at the right hand side gives directional derivative f in $\underline{x}_1$ in the direction $(\underline{x} - \underline{x}_1)$, so that $f(\underline{x}) - f(\underline{x}_1) \geq \nabla f(\underline{x}) \geq f(\underline{x})^T(\underline{x} - \underline{x}_1)$.

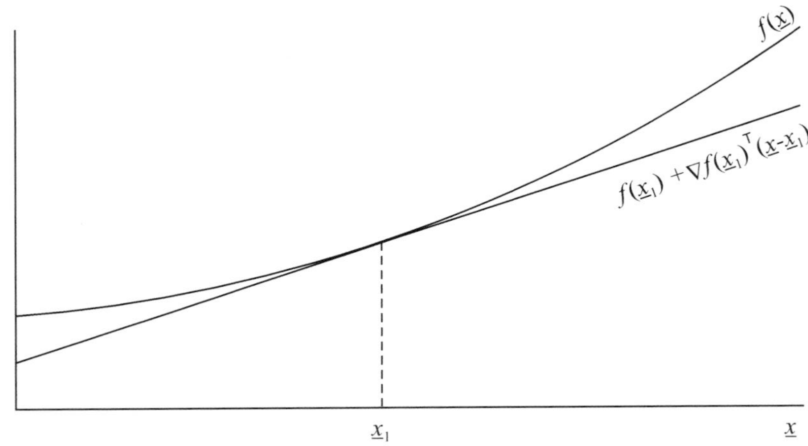

Figure 12.19. Tangent plane below graph.

Property 1 follows directly from (12.39) because in a stationary point $\underline{x}^*, \nabla f(\underline{x}^*) = \underline{0}$.

Theorem 12.7.

If *f* is convex in an ε-environment of stationary point $\underline{x}^*$, then $\underline{x}^*$ is a minimum point of *f*.

Convexity and the Hessean is positive semi-definite

Combining Theorem 12.7 with the second order conditions of Theorem 12.4 shows a relationship between convexity and the Hessean for twice differentiable functions.

Theorem 12.8.

Let $f: X \rightarrow \mathbb{R}$ be twice continuously differentiable on open set *X*:
f is convex $\Leftrightarrow$ H_f is positive semi-definite on *X*.

Theorem 12.8 follows from combining (12.31) and (12.39). The theorem shows that in some cases convexity can be checked.

Example 12.19.

The function $f(\underline{x}) = x_1^2 + 2x_2^2$ is convex. The Hessean is $H_f = \begin{pmatrix} 2 & 0 \\ 0 & 4 \end{pmatrix}$. The eigenvalues of the Hessean are 2 and 4, so H_f is positive definite. Theorem 12.8 tells us that *f* is convex.

Convex optimisation: a local minimum point is a global minimum point

For the other properties, the notion of convex set is required. A *convex optimisation problem* is defined as a problem where the objective function *f* is convex in case of minimisation (concave in case of maximisation) and feasible set *X* is a convex set.
When is the feasible area *X* convex? In problem (12.1), *X* is defined by inequalities $g_i(\underline{x}) \leq 0$ and equalities $g_i(\underline{x}) = 0$. Linear equalities (LP) lead to a convex area, but if an equality

$g_i(\underline{x}) = 0$ is nonlinear e.g. $x_1^2 + x_2^2 - 4 = 0$, a nonconvex area appears. In contrast to the mentioned equality, the inequality $x_1^2 + x_2^2 - 4 \le 0$ describes a circle with its interior and this is a convex set. Considering the inequality $g_i(\underline{x}) \le 0$ more abstractly; it is a level set of the function $g_i(\underline{x})$. The relation with convex functions is given in Theorem 12.9.

Theorem 12.9.

Let $g : X \rightarrow \mathbb{R}$ be a convex function on a convex set X and $h \in \mathbb{R}$.
Level set $S_h = \{x \in X \mid g(\underline{x}) \le h\}$ is a convex set.

The proof proceeds as follows. Given two points $\underline{x}_1, \underline{x}_2 \in S_h$ so $g(\underline{x}_1) \le h$ and $g(\underline{x}_2) \le h$. The convexity of g shows that point $x = \lambda\underline{x}_1 + (1 - \lambda)x_2$ in between $\underline{x}_1$ and $\underline{x}_2$ is also in S_h:

$$g(\underline{x}) = g(\lambda\underline{x}_1 + (1-\lambda)\underline{x}_2) \le \lambda g(\underline{x}_1) + (1-\lambda) g(\underline{x}_2) \le \lambda h + (1-\lambda)h = h \qquad (12.40)$$

A last property often mentioned in literature is that the functions g_i are quasi convex. This is a weaker assumption than convexity for which Theorem 12.9 also applies. The notion of a convex optimisation problem (12.1) is important for the three properties we started with. The KKT conditions are sufficient to determine the optimality of a stationary point in a convex optimisation problem, see (Bazaraa *et al.*, 1993). Property 2 (local is global) can now be derived.

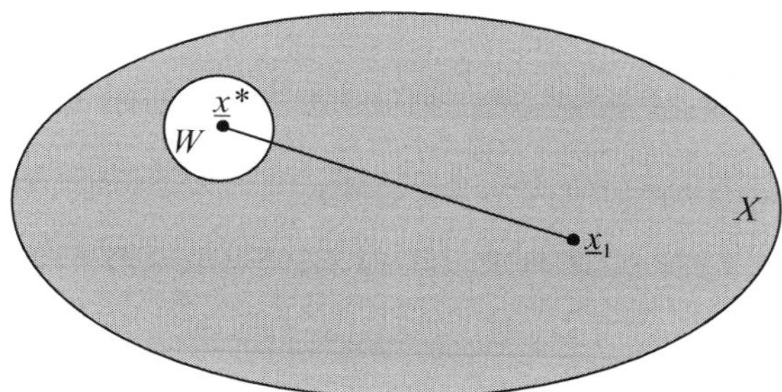

Figure 12.20. Local, non-global doesn't exist.

Theorem 12.10.

Let f be convex on a convex set X. Every local minimum point is a global minimum point.

Showing the validity of Theorem 12.10 is usually done in the typical mathematical way of demonstrating that assuming non-validity will lead to a contradiction. For a local minimum point $\underline{x}^*$, an ε-environment W of $\underline{x}^*$ exists where $\underline{x}^*$ is minimum; $f(\underline{x}) \ge f(\underline{x}^*), x \in X \cap W$. Suppose that Theorem 12.10 is not true. Then a point $\underline{x}_1 \in W$ should exist such that $f(\underline{x}_1) < f(\underline{x}^*)$. By logical steps and the convexity of f and X can be shown that the existence of $\underline{x}_1$ leads to a contradiction. Points on the line between $\underline{x}_1$ and $\underline{x}^*$ are situated in X, $x \in X$ and can be described by $\underline{x} = \lambda\underline{x}_1 + (1-\lambda)\underline{x}^*, 0 \le \lambda \le 1$. Convexity of f implies

$$f(\underline{x})=f(\lambda \underline{x}_1+(1-\lambda)\underline{x}^*)\le \lambda f(\underline{x})+(1-\lambda)f(\underline{x}^*)< \\ \lambda f(\underline{x}^*)+(1-\lambda)f(\underline{x}^*)=f(\underline{x}^*) \quad (12.41)$$

So the convexity of f and the assumption $f(\underline{x}_1)<f(\underline{x}^*)$ implies that all points on the cord between $\underline{x}_1$ and $\underline{x}^*$ have an objective value lower than $f(\underline{x}^*)$. For λ small, the point x is situated in W in contradiction to $\underline{x}^*$ being a local minimum. So the assumption that a point $\underline{x}_1$ exists with $f(\underline{x}_1)<f(\underline{x}^*)$ cannot be true.

The practical importance of Theorem 12.10 is that software like GINO, GAMS/MINOS and the Excel solver return a local minimum point depending on the starting value. If one wants to be certain it is a global minimum point, then the optimisation problem should be analysed further on convexity. We already have seen that this may not be easy.

A maximum point at the boundary of the feasible area

The last mentioned property is: for a convex function, a maximum point (if it exists) can be found at the boundary of the feasible area. A special case of this property is LP.

Theorem 12.11.

Let $f: X \to \mathbb{R}$ be a convex function on a closed set X. If f has a maximum on X, then there exists a maximum point $\underline{x}^*$ that is an extreme point.

Mathematically, extreme means that $\underline{x}^*$ cannot be written as a convex combination of two other points in X. A typical extreme point is a vertex (corner point). At the boundary of a circle, all points are extreme points. The proof of Theorem 12.11 also uses contradiction. The proof is constructed by assuming that there is an interior maximum point $\underline{x}^*$; more exactly, a maximum point $\underline{x}^*$ with a higher function value than the points at the boundary. Point $\underline{x}^*$ can be written as a convex combination of two points $\underline{x}_1$ and $\underline{x}_2$ at the boundary:

$$f(\underline{x}^*)>f(\underline{x}_1), f(\underline{x}^*)>f(\underline{x}_2) \text{ and } \underline{x}^*=\lambda \underline{x}_1+(1-\lambda)\underline{x}_2.$$

Just like in (12.41) this leads to a contradiction.

$$f(\underline{x}^*)\le \lambda f(\underline{x}_1)+(1-\lambda)f(\underline{x}_2)<\lambda f(\underline{x}^*)+(1-\lambda)f(\underline{x}^*)=f(\underline{x}^*).$$

The consequence in this case is that if the feasible area is a polytope, one can limit the search for a maximum point to the vertices of the feasible area. Life doesn't necessarily become very easy with this observation. In linear programming we have already seen that the number of vertices can explode in the number of decision variables. A traditional example to illustrate the relationship between NLP and combinatorial optimisation is the following

Example 12.20.

$$\max \{f(\underline{x})=\Sigma(x_i-\varepsilon)^2\}$$
$$-1\le x_i\le 1,\ i=1,\dots,n$$

where ε is a small number e.g. 0.01. The problem describes the maximisation of the distance to a point that is nearly in the middle of a square/cube. With increasing dimension n, the number of vertices explodes, increases with 2^n. Every vertex is a local maximum point. Moreover, this problem has a multiple of KKT points that are no maximum points. For the case of a cube ($n = 3$) for instance 18.

12.4. Nonlinear and global optimisation algorithms

In this section, several criteria are discussed to measure effectiveness and efficiency of algorithms. Moreover, examples are given of basic algorithms that are analysed. GO concepts such as *region of attraction, level set, probability of success* and *Performance Graph* are introduced. To investigate optimisation algorithms, we start with a definition. An algorithm is a description of steps, preferably implemented into a computer program, which finds an approximation of an optimum point. The aims can be several: reach a local optimum point, reach a global optimum point, find all global optimum points, reach all global and local optimum points. In general, an algorithm generates a series of points $\underline{x}_k$ that approximate an (or the or all) optimum point. According to the generic description of (Törn and Žilinskas, 1989):

$$\underline{x}_{k+1} = \text{Alg}(\underline{x}_k, \underline{x}_{k-1}, \ldots, \underline{x}_0, \tilde{\xi}) \tag{12.42}$$

where $\tilde{\xi}$ is a random variable and index k is the iteration counter. This represents the idea that a next point $\underline{x}_{k+1}$ is generated based on the information in all former points $\underline{x}_k, \underline{x}_{k-1}, \ldots, \underline{x}_0$ ($\underline{x}_0$ usually being the starting point) and possibly some random effect. This leads to three classes of algorithms discussed here:

- Nonlinear optimisation algorithms, that from a starting point try to capture the 'nearest' local minimum point.
- Deterministic GO methods which guarantee to approach the global optimum and require a certain mathematical structure.
- Stochastic GO methods based on the random generation of feasible trial points and nonlinear local optimisation procedures.

12.4.1. Effectiveness and efficiency of algorithms

We will consider several examples illustrating two questions to be addressed to investigate the quality of algorithms see (Baritompa and Hendrix, 2005).

- Effectiveness: does the algorithm find what we want?
- Efficiency: what are the computational costs?

Several measurable performance indicators can be defined for these criteria.

Consider minimisation algorithms. Focusing on *Effectiveness*, there are several targets a user may have:

1. To discover all global minimum points. This of course can only be realised when the number of global minimum points is finite.
2. To detect at least one global optimal point.
3. To find a solution with a function value as low as possible.

The first and second targets are typical satisfaction targets; was the search successful or not? What are good *measures of success*? In the literature, convergence is often used i.e. $\underline{x}_k \to \underline{x}^*$, where $\underline{x}^*$ is one of the minimum points. Alternatively one observes $f(\underline{x}_k) \to f(\underline{x}^*)$. In tests and analyses, to make results comparable, one should be explicit in the definitions of success. We need not only specify ε and/or δ such that

$$\left\|\underline{x}_k - \underline{x}^*\right\| < \varepsilon \text{ and/or } f(\underline{x}_k) < f(\underline{x}^*) + \delta \tag{12.43}$$

but also specify whether success means that there is an index K such that (12.43) is true for all $k > K$. Alternatively, success may mean that a record $\min_k f(\underline{x}_k)$ has reached level $f(\underline{x}^*) + \delta$. Whether the algorithm is effective also depends on its stochastic nature. When we are dealing with stochastic algorithms, effectiveness can be expressed as the probability that a success has been reached. In analysis, this probability can be derived from sufficient assumptions on the behaviour of the algorithm. In numerical experiments, it can be estimated by counting repeated runs how many times the algorithm converges. We will give some examples of such analysis. In Section 12.4.2 we return to the topic of efficiency and effectiveness considered simultaneously.

Efficiency

Globally efficiency is defined as the effort the algorithm needs to be successful. A usual indicator for algorithms is the (expected) number of *function evaluations* necessary to reach the optimum. This indicator depends on many factors such as the shape of the test function and the termination criteria used. The indicator more or less suggests that the calculation of function evaluations dominates the other computations of the algorithm.
Several other indicators appear in literature.
In nonlinear programming e.g. (Scales, 1985) and (Gill *et al.*, 1981) the concept of *convergence speed* is common. It deals with the convergence limit of the series $\underline{x}_k$. Let $\underline{x}_0$, $\underline{x}_1$, . . . , $\underline{x}_k$ converge to point $\underline{x}^*$. The largest number α for which

$$\lim_{k\to\infty} \frac{\left\|\underline{x}_{k+1} - \underline{x}^*\right\|}{\left\|\underline{x}_k - \underline{x}^*\right\|^{\alpha}} = \beta < \infty \tag{12.44}$$

gives the order of convergence, whereas β is called the convergence factor. In this terminology, the special instances are

- linear convergence with $\alpha = 1$ and $\beta < 1$
- quadratic convergence with $\alpha = 2$ and $0 < \beta < 1$
- super linear convergence: $1 < \alpha < 2$ and $\beta < 1$, i.e. $\beta = 0$ when using $\alpha = 1$ in (12.44).

Mainly in deterministic GO algorithms, information on past evaluations is stored in the computer memory. This requires efficient data handling for looking up necessary information during the iterations. Furthermore, *memory requirements* become a part of the computational burden as retrieving actions cannot be neglected compared to the computational effort due to function evaluations.

We do not go deeper into theoretical aspects of performance indicators here. Instead some basic algorithms are introduced and analysed. In Section 12.4.3, systematic investigation of algorithms is expanded upon.

12.4.2. Some basic algorithms analysed

In this section, several classes of algorithms are analysed for effectiveness and efficiency. Two test cases are introduced first for which the performance of the algorithms are investigated. We consider the minimisation of the following functions.

$$g(x) = \sin(x) + \sin(3x) + \ln(x), x \in [3,7] \tag{12.45}$$

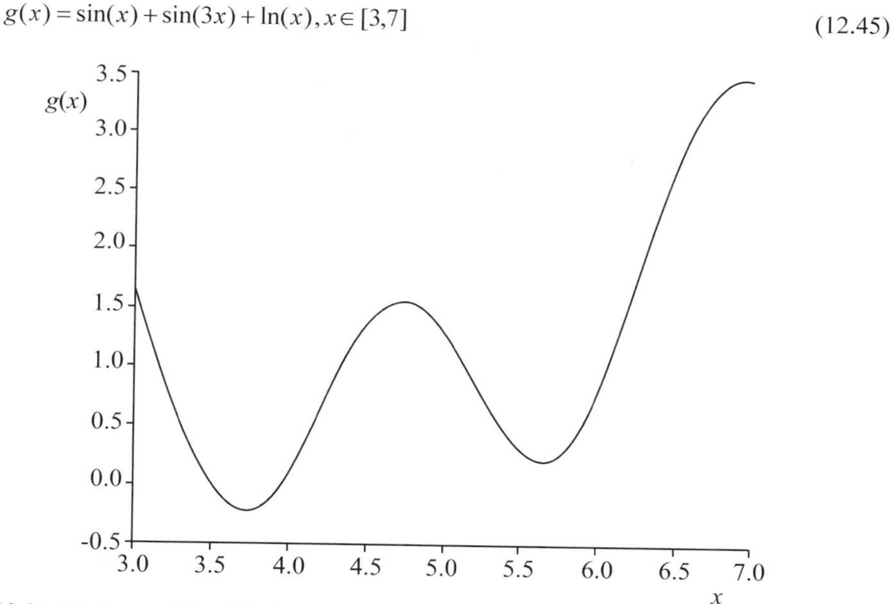

Figure 12.21. Test case g(x) with three optima.

Function g is depicted in Figure 12.21 and has three minimum points on the interval. The global minimum is attained at about $x^* = 3.73$, where $g(x^*) = -0.220$. The derivative is

$$g'(x) = \cos(x) + 3\cos(3x) + \frac{1}{x} \tag{12.46}$$

on the interval [3,7]. Alternatively to function g, we introduce a function h with more local minimum points by adding to function g a bubble function based on frac(x) = $x - \lfloor x \rfloor$ where $\lfloor x \rfloor$ rounds x down to the greatest integer smaller than or equal to x. Now the second case is defined as

$$h(x) = g(x) + 1.5\text{frac}^2(4x) \tag{12.47}$$

In Figure 12.22, the graph of function h is shown. It has 17 local minimum points on the interval [3,7]. Although neither g nor h are convex on the interval, at least function h is piecewise convex on the intervals in between the points of $S = \{x = \frac{1}{4}k + \frac{1}{8}, k \in Z\}$. At these points, h is not differentiable. For the rest of the interval one can define the derivative

$$h'(x) = g'(x) + 12\text{frac}(4x) \quad \text{for} \quad x \notin S \tag{12.48}$$

The global minimum point of h on [3,7] is shifted slightly compared to g towards $x^* = 3.75$, where $h(x^*) = -0.217$.

In the following Sections, we will test algorithms on their ability to find minima of these two functions. One should set a target for what is considered an acceptable or successful result. For instance, one can aim at detecting a local minimum or detecting the global minimum. For the neighbourhood we will take an acceptance of $\varepsilon = 0.01$. For determining an acceptable low value of the objective function we take $\delta = 0.01$. Notice that ε represents 0.25% of the argument range [3,7] and δ is about 0.25% of the function values range.

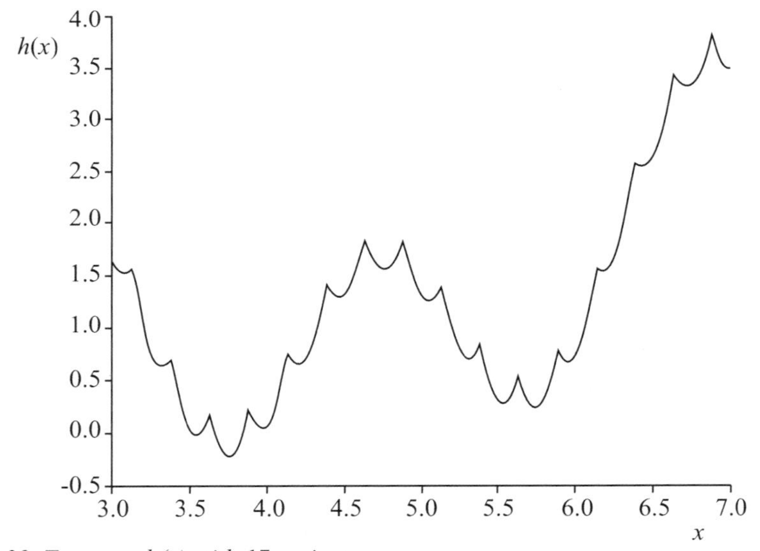

Figure 12.22. Test case h(x) with 17 optima.

NLP local optimisation: Bisection and Newton

Two nonlinear programming algorithms are sketched and their performance measured for the two test cases. First the bisection algorithm is considered.

Algorithm 12.1. Bisection **($|l, r,|, f, \varepsilon$).**

Set $k = 0$, $l_0 = l$ and $r_0 = r$
while

$(r_k - l_k > \varepsilon)$

$x_k = \frac{l_k + r_k}{2}$

if $f'(x_k) < 0$

$l_{k+1} = x_k$ and $r_{k+1} = r_k$

else

$l_{k+1} = l_k$ and $r_{k+1} = x_k$

End while

The algorithm departs from a starting interval $[l, r]$ that is halved iteratively based on the sign of the derivative in the midpoint. This means that the method is only applicable when the derivative is available at the generated midpoints. The point x_k converges to a minimum point within the interval $[l, r]$. If the interval contains only one minimum point, it converges to that. In our test cases, several minima exist and one can observe the convergence to one of them.

Table 12.1. Bisection for functions g and h, first 6 iterations.

	function g					function h				
k	l_k	r_k	x_k	$g'(x_k)$	$g(x_k)$	l_k	r_k	x_k	$h'(x_k)$	$h(x_k)$
0	3.00	7.00	5.00	−1.80	1.30	3.00	7.00	5.00	−1.80	1.30
1	5.00	7.00	6.00	3.11	0.76	5.00	7.00	6.00	3.11	0.76
2	5.00	6.00	5.50	−1.22	0.29	5.00	6.00	5.50	−1.22	0.29
3	5.50	6.00	5.75	0.95	0.24	5.50	6.00	5.75	0.95	0.24
4	5.50	5.75	5.63	−0.21	0.20	5.50	5.75	5.63	−6.21	0.57
5	5.63	5.75	5.69	0.36	0.20	5.63	5.75	5.69	−2.64	0.29
6	5.63	5.69	5.66	0.07	0.19	5.69	5.75	5.72	−0.85	0.24

The algorithm is effective in the sense of converging to a local (nonglobal) minimum point for both cases. Another starting interval could have lead to another minimum point. In the end, we are certain that the current iterate x_k is not further away than ε from a minimum point. Many other stopping criteria like convergence of function values or derivatives going to zero could be used. The current stopping criterion is easy for analysis of efficiency. One question could be: how many iterations (i.e. corresponding derivative function evaluations) are necessary to come closer than ε to a minimum point. The bisection algorithm is a typical case of linear convergence with a convergence factor of $\frac{1}{2}$, $\frac{|r_{k+1} - l_{k+1}|}{|r_k - l_k|} = \frac{1}{2}$. This means one can determine the number of iterations necessary for reaching ε-convergence:

$$|r_k - l_k| = (\tfrac{1}{2})^k \times |r_0 - l_0| < \varepsilon \Rightarrow$$

$$(\tfrac{1}{2})^k < \frac{\varepsilon}{|r_0 - l_0|} \Rightarrow k > \frac{\ln \varepsilon - \ln|r_0 - l_0|}{\ln \frac{1}{2}}$$

The example case requires at least 9 iterations to reach an accuracy of $\varepsilon = 0.01$.

An alternative for finding the zero point of an equation, in our case the derivative, is the so-called method of Newton. The idea is that its efficiency is known to be super linear (e.g. (Scales, 1985)), so it should be faster than bisection. We analyse its efficiency and effectiveness for the two test cases.
In general, the aim of the Newton algorithm is to converge to a point where the derivative is zero. Depending on the starting point x_0, the method may converge to a minimum or maximum. Also, it may not converge at all, for instance when a minimum point does not exist. Specifically in the version of Algorithm 2, a safeguard is built in to ensure the iterates remain in the interval; it can converge to a boundary point. If x_0 is in the neighbourhood of a minimum point where f is convex, then convergence is guaranteed and the algorithm is effective in the sense of reaching a minimum point. Let us consider what happens in the two test cases.

Algorithm 12.2. Newton (|l, r,|, x_0, f, α).

Set $k = 0$,
while $(|f'(x_k)| > \alpha$

$$x_{k+1} = x_k - \frac{f'(x_k)}{f''(x_k)}$$

! safeguard for staying in interval

if $x_{k+1} < l$, $x_{k+1} = l$

if $x_{k+1} > r$, $x_{k+1} = r$

if $x_{k+1} = x_k$, STOP

$k = k + 1$

End while

Table 12.2. Newton for functions g and h, $\alpha = 0.001$.

	function g				function h			
k	x_k	$g'(x_k)$	$g''(x_k)$	$g(x_k)$	x_k	$h'(x_k)$	$h''(x_k)$	$h(x_k)$
0	5.000	−1.795	−4.934	1.301	5.000	−1.795	43.066	1.301
1	4.636	0.820	−7.815	1.511	5.042	0.018	43.953	1.264
2	4.741	−0.018	−8.012	1.553	5.041	0.000	43.944	1.264
3	4.739	0.000	−8.017	1.553	5.041	0.000	43.944	1.264

When choosing the starting point x_0 in the middle of the interval [3,7], the algorithm converges to the closest minimum point for function h and to a maximum point for the function g, i.e. it fails for this starting point. This gives rise to introducing the concept of a *region of attraction* of a minimum point x^*. A region of attraction of point x^*, is the region of starting points x_0 where the local search procedure converges to point x^*. We elaborate this concept further in Section 12.4.2.

One can observe here when experimenting further, that when x_0 is close to a minimum point of g, the algorithm converges to one of the minimum points. Moreover, notice now the effect of the safeguard to keep the iterates in the interval [3,7]. If for instance $x_{k+1} < 3$, it is forced to a value of 3. In this way, the left point $l = 3$ is also an attraction point of the algorithm. Function h is piecewise convex, such that the algorithm always converges to the closest minimum point.

Deterministic GO: Grid search, Piyavskii-Shubert

The aim of deterministic GO algorithms is to approach the optimum with a given certainty. We sketch two algorithms for the analysis of effectiveness and efficiency. The idea of reaching the optimum with an accuracy of ε can be done by so-called 'everywhere dense sampling', as introduced in literature on Global Optimisation see e.g. (Törn and Žilinskas, 1989). In a rectangular domain this can be done by constructing a grid with a mesh of ε. By evaluating all points on the grid, the best point found is a nice approximation of the global minimum point. The difficulty of GO is, that even this best point found may be far away from the global minimum point, as the function may have a needle shape in another region in between the grid points. As shown in literature, one can always construct a polynomial of sufficiently high degree, which fits all the evaluated points and has a minimum point more than ε away from the best point found.

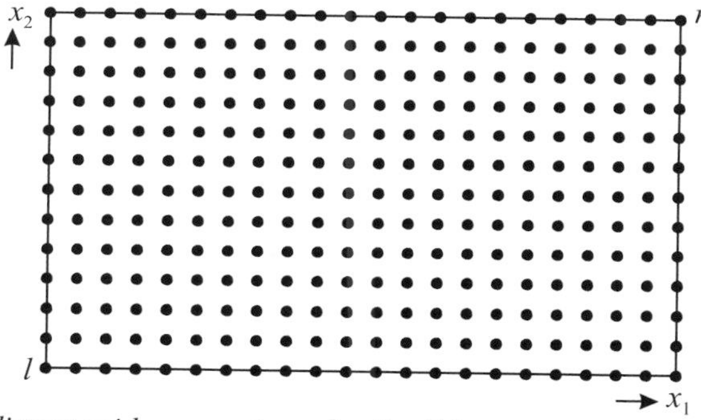

Figure 12.23. Equidistant grid over rectangular feasible set.

Actually, grid search is theoretically not effective if no further assumptions are posed on the optimisation problem to be solved.

Let us have a look at the behaviour of the algorithm for our two cases. For ease of formulation, we write down the grid algorithm for one dimensional functions. The algorithm starts with the domain $[l, r]$ written as an interval and generates $M = \lceil (r - l / \varepsilon \rceil + 1$ grid points, where $\lceil x \rceil$ is the lowest integer greater than or equal to x. The best function value found f^U is an upper bound for the minimum over the feasible set. We denote by x^U the corresponding best point found. Experimenting with test functions g and h gives reasonable results for $\varepsilon = 0.01$, ($M = 401$) and $\varepsilon = 0.1$, ($M = 41$). In both cases one finds an approximation x^U less than ε from the global minimum point. One knows exactly how many function evaluations are required to reach this result in advance.

Algorithm 12.3. Grid ([l, r], f, ε).

$M = \lceil (r - l / \varepsilon \rceil + 1, f^U = \infty$

for (k = 1 to M) do

$$x_k = l + \frac{(k-1)\times(r-l)}{M-1}$$

if $f(x_k) < f^U$

$f^U = f(x_k)$ and $x^U = x_k$

End for

The efficiency of the algorithm in higher dimensions is also easy to establish. Given the lower left vector $\underline{l}$ and upper right vector $\underline{r}$ of a rectangular domain, one can easily determine how many grid co-ordinates M_j , j = 1, =1 , . . . , n in each direction should be taken and the total number of grid points is $\Pi_j M_j$. This number is growing exponentially in the dimension n. As mentioned before, the effectiveness is not guaranteed in the sense of being closer than ε from a global minimum point, unless we make an assumption about the behaviour of the function. A common assumption in the literature is Lipschitz continuity.

Definition 12.5.

L is called a Lipschitz constant of f on X if:

$$|f(x) - f(y)| \le L\|x - y\|, \forall x, y \in X$$

In a practical sense it means that big jumps do not appear in the function value; slopes are bounded. With such an assumption, the δ-accuracy in the function space translates into an ε-accuracy in the x-space. Choosing $\varepsilon = \delta/L$ implies that the best point x^U is in function value finally close to minimum point x^*:

$$|f^U - f^*| \le L\|x^U - x^*\| \le L\varepsilon = \delta \qquad (12.49)$$

In higher dimension, one should be more exact in the choice of the distance norm $\|\cdot\|$. Here, for the one-dimensional examples we can focus on deriving the accuracy for our cases in a simple way. For a one-dimensional differentiable function f, L can be taken as

$$L = \max_{x \in X} |f'(x)| \qquad (12.50)$$

Using Equation (12.50), one can now derive valid estimates for the example functions h and g. One can derive an over-estimate L_g for the Lipschitz constant of g on [3,7] as

$$\begin{aligned}
&\max_{x\in[3,7]}|g'(x)| = \max_{x\in[3,7]}\left|\cos(x) + 3\cos(3x) + \tfrac{1}{x}\right| \le \\
&\max_{x\in[3,7]}\{|\cos(x)| + |3\cos(3x)| + |\tfrac{1}{x}|\} \le \\
&\max_{x\in[3,7]}|\cos(x)| + \max_{x\in[3,7]}|3\cos(3x)| + \max_{x\in[3,7]}|\tfrac{1}{x}| = \\
&1 + 3 + \tfrac{1}{3} = L_g
\end{aligned} \tag{12.51}$$

The estimate of L_h based on (12.48) is done by adding the maximum derivative of the bubble function $12 \times \frac{1}{2}$ to L_g for illustrative purposes rounded down to $L_h = 10$. We can now use (12.49) to derive a guarantee for the accuracy. One certainly arrives closer than $\delta = 0.01$ to the minimum in function value by taking a mesh size of $\varepsilon = \frac{0.01}{4.33} = 0.0023$ for function g and taking $\varepsilon = 0.001$ for function h. For the efficiency of grid search this means that reaching the δ-guarantee requires the evaluation of $M = 1733$ points for function g and $M = 4001$ points for function h. Note, that due to the one-dimensional nature of the cases, ε can be taken twice as big, as the optimum point x^* is not further than half the mesh size from an evaluated point.

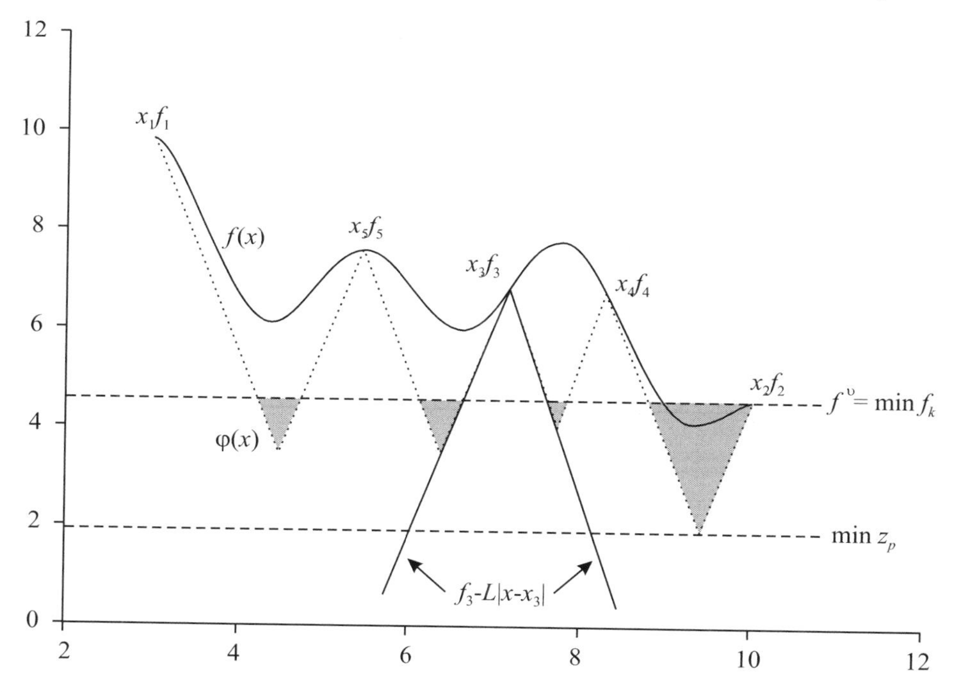

Figure 12.24. Piyavskii-Shubert algorithm.

The main idea of most deterministic algorithms is not to generate and evaluate points everywhere dense, but to throw out those regions where the optimum cannot be situated. Giving a Lipschitz constant, Piyavskii and Shubert independently constructed similar algorithms, see (Shubert, 1972) and (Danilin and Piyavskii, 1967). From the point of view of the graph of the function f to be minimised and an evaluated point (x_k, f_k), one can say that the

region described by x, $f < f_k - L|x - x_k|$ cannot contain the optimum; the graph is above the function $f_k - L|x - x_k|$. Given a set of evaluated points $\{x_k\}$, one can construct a lower bounding function, a so-called saw-tooth underestimator that is given by
$\varphi(x) = \max_k (f_k - L|x - x_k|)$ as illustrated by Figure 12.24. Given that we also have an upper bound f^U on the minimum of f being the best function value found thus far, one can say that the minimum point has to be in one of the shaded areas.

Algorithm 12.4. PiyavShub ([l, r], f, L, δ).

Set $p = 1$, $l_1 = l$ and r_1, $\Lambda = \{[l_1, r_1]\}$

$z_1 = \frac{f(l_1) + f(r_1)}{2} - \frac{L(r_1 - l_1)}{2}, f^U = \min\{f(l), f(r)\}, x^U = \arg\min\{f(l), f(r)\}$

while ($\Lambda \neq \varnothing$)

  remove an interval $[l_k, r_k]$ from Λ with $z_k = \min_p z_p$

  evaluate $f(m_k) = f\left(\frac{f(l_k) - f(r_k)}{2L} + \frac{r_k + l_k}{2}\right)$

  if $f(m_k) < f^U$

    $f^U = f(m_k), x^U = m_k$ and remove all C_p from Λ with $z_p > f^U - \delta$

  split $[l_k, r_k]$ into 2 new intervals $C_{p+1} = [l_k, m_k]$ and $C_{p+2} = [m_k, r_k]$
  with corresponding lower bounds z_{p+1} and z_{p+2}

  if $z_{p+1} < f^U - \delta$ store C_{p+1} in Λ

  if $z_{p+2} < f^U - \delta$ store C_{p+2} in Λ

  $p = p + 2$

End while

We will describe here the algorithm from a Branch-and-Bound point of view, where the subsets are defined by intervals $[l_p, r_p]$ and the end points are given by evaluated points. The index p is used to represent the intervals in λ. For each interval, a lower bound is given by

$$z_p = \frac{f(l_p) + f(r_p)}{2} - \frac{L(r_p - l_p)}{2} \tag{12.52}$$

The gain with respect to grid search is that an interval can be thrown out as soon as $z_p > f^U$. Moreover, δ works as a stopping criterion as the algorithm implicitly (by not storing) compares the gap between f^U and $\min_p z_p$; stop if ($f^U - \min_p z_p$) < δ. The algorithm proceeds by selecting the interval corresponding to $\min_p z_p$ (most promising) and splitting it over the minimum point of the saw-tooth cover $\varphi(x)$ defined by:

$$m_p = \frac{f(l_p) + f(r_p)}{2L} - \frac{L(r_p - l_p)}{2} \tag{12.53}$$

being the next point to be evaluated. By continually evaluating, splitting and throwing out intervals where the optimum cannot be, the stopping criterion is finally reached and we are certain to be closer than δ from f^* and therefore closer than $\varepsilon = \delta/L$ from one of the global minimum points. The consequence of using such an algorithm, in contrast to the other algorithms, is that we now have to store information in a computer consisting of a list Λ of intervals.

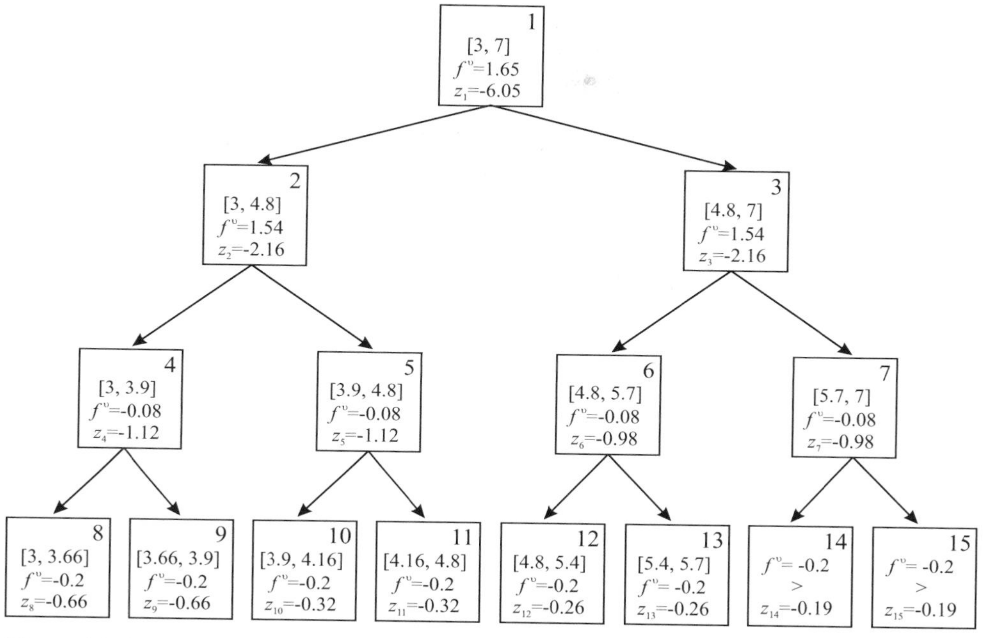

Figure 12.25. Branch-and-Bound tree of Piyavskii-Shubert algorithm for function g.

This computational effort is now added to that of evaluating sample points and doing intermediate calculations. This concept becomes more clear when running the algorithm on the test function g using an accuracy of $\delta = 0.01$. The Lipschitz constant $L_g = 4.2$ is used for illustrative purposes. As can be seen from Table 12.3, the algorithm is slowly converging. After some iterations, 15 intervals have been generated of which 6 are stored and 2 can be discarded due to the bounding; it has been proven that the minimum cannot be in the interval [5.67, 7]. The current estimate of the optimum is $x^U = 3.66$, $f^U = -0.19$ and the current lower bound is given by $\min_p z_p = -0.66$. Figure 12.25 illustrates the appearing binary structure of the search tree.

The maximum computational effort with respect to storing intervals is reached when the branching proceeds and no parts can be thrown out; 2^K intervals appear at the bottom of the tree, where K is the depth of the tree. This mainly happens when the used Lipschitz parameter L drastically overestimates the maximum slope, or seen from another angle, the function is very flat compared to the used constant L. In that case, the function is evaluated in more than the M points of the regular grid. With a correct constant L, the number of evaluated points is less, as part of the domain can be discarded as illustrated here.

Table 12.3. Piyavskii-Shubert for function g, $\delta = 0.01$.

p	l_p	rl_p	$f(l_p)$	$f(r_p)$	m_p	z_p	f^U	x^U	
1	3.00	7.00	1.65	3.44	4.79	−5.85	1.65	3.00	split
2	3.00	4.79	1.65	1.54	3.91	−2.16	1.54	4.79	split
3	4.79	7.00	1.54	3.44	5.67	−2.16	1.54	4.79	split
4	3.00	3.91	1.65	−0.08	3.66	−1.12	−0.08	3.91	split
5	3.91	4.79	−0.08	1.54	4.15	−1.12	−0.08	3.91	split
6	4.79	5.67	1.54	0.20	5.39	−0.98	−0.08	3.91	split
7	5.67	7.00	0.20	3.44	5.95	−0.98	−0.08	3.91	split
8	3.00	3.66	1.65	−0.20	3.55	−0.66	−0.20	3.66	
9	3.66	3.91	−0.20	−0.08	3.77	−0.66	−0.20	3.66	
10	3.91	4.15	−0.08	0.47	3.96	−0.32	−0.20	3.66	
11	4.15	4.79	0.47	1.54	4.34	−0.32	−0.20	3.66	
12	4.79	5.39	1.54	0.46	5.22	−0.26	−0.20	3.66	
13	5.39	5.67	0.46	0.20	5.56	−0.26	−0.20	3.66	
14	5.67	5.95	0.20	0.61	5.76	−0.19	−0.20	3.66	discarded
15	5.95	7.00	0.61	3.44	6.14	−0.19	−0.20	3.66	discarded

Stochastic GO: PRS, Multistart, Simulated Annealing

Stochastic methods are understood to contain some stochastic elements. Either the outcome of the method is a random variable or the objective function itself is considered a realisation of a stochastic process. For an overview of stochastic methods we refer to (Boender and Romeijn, 1995), (Törn and Žilinskas, 1989) and (Törn *et al.* 1999). Two classical approaches from Global Optimisation, Pure Random Search (PRS) and Multistart are analysed for the test cases. This is followed by a classical variant of Simulated Annealing, a so-called heuristic.

Pure Random Search (PRS) generates points uniformly over the domain and stores the point corresponding to the best value as the approximation of the global minimum point.

Algorithm 12.5. PRS (X, f, N).

$f^U = \infty$
for ($k = 1$ to N) do
  Generate x_k uniformly over X
  if $f(x_k) < f^U$
  $f^U = f(x_k)$ and $x^U = x_k$
End for

The algorithm is popular as a reference algorithm since it can easily be analysed. The question can now be, how does it behave for our test cases g and h. The domain is clearly the interval [3, 7], but what can be defined as the success region now? Let a success be defined as the case that one of the generated points is closer than $\varepsilon = 0.01$ to the global minimum point. The probability that we do NOT hit this region after $N = 50$ trials is $(3.98/4)^{50} \approx 0.78$. In the specific case, the size of the success region is namely $2 \times \varepsilon$ and the size of the feasible area is 4. The probability of NOT hitting is $(1-\frac{0.02}{4})$ and of NOT hitting 50 times is $(1-\frac{0.02}{4})^{50}$. This means that the probability of success as efficiency indicator has a value of about 0.22 for both cases h and g.

A similar analysis can be done for determining the probability that the function value of PRS after $N = 50$ iterations is less than $f^* + \delta$ for $\delta = 0.01$. The usual tool in the analysis on the function space is to introduce $\tilde{y} = f(\tilde{x})$ as a random variate representing the function value, where $\tilde{x}$ is uniformly distributed over X. Value $\tilde{y}$ has distribution function $F(y) = P(f(\tilde{x}) \le y)$. Keeping this in mind, analysis with so-called extreme order statistics has shown that the outcome of PRS as record value of N points can be easily derived from $F(y)$. For a complete introduction to extreme order statistics in optimisation, we refer to (Zhigljavsky, 1991). Under mild assumptions it can be shown that $\tilde{y}_{(1)} = \min\{f(\tilde{x}_1),..,f(\tilde{x}_N)\}$, has the distribution function $F_{(1)}(y) = 1-(1-F(y))^N$. This means that for the question about the probability that $\tilde{y}_{(1)} \le f^* + \delta$, we don't have to know the complete distribution function F, but only the probability mass $F(f^* + \delta)$ of the success level set where $f(x) \le f^* + \delta$, i.e. the probability that one sample point hits this low level set. Here the 2 test cases differ considerably. One can verify that the level set of the more smooth test function g is about 0.09 wide, whereas that of function h is only 0.04 wide for a δ of 0.01. This means that the probability of PRS to reach a level below $f^* + \delta$ after 50 evaluations for function g is $(1 - \frac{0.09}{4})^{50} = 0.68$, whereas the same probability for function h is $(1 - \frac{0.04}{4})^{50} = 0.40$.

In conclusion, stochastic algorithms show something which in the literature is called the *infinite effort property*. This means that if one proceeds long enough (read $N \to \infty$) in the end the global optimum is found. The problem with such a concept is that infinity can be pretty far away. Moreover, we have seen in the earlier analyses that the probability of reaching what one wants, depends considerably on the size of the success region. One classical way of increasing the probability of reaching an optimum is to use (nonlinear optimisation) local searches. This method is called *multistart*.
Define a local optimisation routine $LS(x): X \to X$ as a procedure which given a starting point returns a point in the domain that approximates a local minimum point. As an example, one can consider the Newton method of Section 12.4.2. Multistart generates convergence points of a local optimisation routine from randomly generated starting points.

Algorithm 12.6. Multistart (X, f, LS, N).

$f^U = \infty$
for ($k = 1$ to N) do
    Generate x uniformly over X
    $x_k = LS(x)$
    if $f(x_k) < f^U$
    $f^U = f(x_k)$ and $x^U = x_k$
End for

Note that the number of iterations N is not comparable with that in PRS, as every local search requires several function evaluations. Let us for the example cases assume that the Newton algorithm requires 5 function evaluations to detect an attraction point, as is also implied by Table 12.2. As we were using $N = 50$ function evaluations to assess the success of PRS on the test cases, we will use $N = 10$ iterations for Multistart. In order to determine a similar probability of success, one should find the relative size of the region of attraction of the global minimum point. Note again, that the Newton algorithm does not always converge to the nearest optimum; it only converges to a minimum point in a convex region around it.

For function g, the region of attraction of the global minimum is not easy to determine. It consists of a range of about 0.8 on the feasible area of size 4, such that the probability of one random starting point leading to success is $(0.8)/4 = 0.2$. For function h, the good region of attraction is simply the bubble of size 0.25 around the global minimum point, such that the probability of finding the global minimum in one iteration is about 0.06. Reaching the optimum after $N = 10$ restarts is $1 - 0.8^{10} \approx 0.89$ for g and $1 - 0.94^{10} \approx 0.48$ for h. In both examples, the probability of success is larger than that of PRS.

As sketched so far, the algorithms of Pure Random Search and Multistart have been analysed widely in the literature of GO. Algorithms that are far less easy to analyse, but very popular in applications, are the collection of so-called metaheuristics. This term was introduced by (Fred Glover in Glover, 1986) and includes Simulated Annealing, Evolutionary algorithms, Genetic algorithms, tabu search and all the fantasy names derived from crossovers of the other names. Several are discussed in Chapter 10.

Originally these algorithms were not only aimed at continuous optimisation problems, see (Aarts and Lenstra, 1997). An interesting research question is whether they are really better than combining classical ideas of random search and nonlinear optimisation local searches. We discuss here a variant of *Simulated annealing*, a concept that also got attention in the GO literature, see (Romeijn, 1992).

Simulated annealing describes a sampling process in the decision space where new sample points are generated from a so-called neighbourhood of the current iterate. The new sample point is always accepted when it is better and with a certain probability when it is worse. The probability depends on the so-called temperature that is decreasing (cooling) during the iterations.

The algorithm contains the parameter CR representing the *Cooling rate* with which the temperature variable decreases. A fixed value of 1000 was taken for the initial temperature to avoid creating another algorithm parameter. The algorithm accepts a worse point depending on how much worse it is and the development of the algorithm. This is a generic concept in Simulated Annealing.

Algorithm 12.7. SA (X, f, CR, N).

$f^U = \infty . T_1 = 1000$
Generate x_1 uniformly over X
for ($k = 1$ to N) do
    Generate x from a neighbourhood of x_k
    if $f(x) < f(x_k)$
        $x_{k+1} = x$
        if $f(x) < f^U$
            $f^U = f(x)$ and $x^U = x$
    else with probability $\exp\frac{f(x_k) - f(x)}{T_k}$ let $x_{k+1} = x$
    $T_{k+1} = CR \times T_k$
End for

There are several ways to implement the concept of 'sample from neighbourhood'. In one dimension one would perceive intuitively a neighbourhood of x_k in real space $[x_k - \varepsilon, x_k + \varepsilon]$, which can be found in many algorithms, e.g. see (Baritompa *et al.*, 2005). As such heuristics were not originally aiming aimed at continuous optimisation problems, but at integer problems, one of the first approaches was the coding of continuous variables in bit strings. For the illustrations, we elaborate this idea for the test case. Each point $x \in [3,7]$ is represented by a bit string $(B_1, \ldots, B_9) \in \{0, 1\}^9$ where,

$$x = 3 + 4\frac{\sum_{i=1}^{9} B_i 2^{i-1}}{511} \qquad (12.54)$$

Formula (12.54) describes a regular grid over the interval, where each of the $M = 512$ bit strings is one of the grid points, such that the mesh size is $\frac{4}{511}$. The sampling from a neighbourhood of a point x is done by flipping at random one of its bit variables B_i from a value of 0 to 1, or the other way around. Notice that by doing so, the generated point is not necessarily in what one would perceive as a neighbourhood in continuous space. The question is therefore, whether the described SA variant will perform better than an algorithm where the new sample point does not depend on the current iterate, PRS.

To test this, a figure is introduced that is quite common in experimenting with meta-heuristics. It is a graph with the effort on the x-axis and the reached success on the y-axis, usually the *record value*, i.e. the best objective function value found. The GO literature often looks at the two criteria effectiveness and efficiency separately. Figure 12.26 being a special case of what in (Baritompa and Hendrix, 2005) was called the *performance graph*, gives a trade-off between the two main criteria. One can also consider the x-axis to give a budget with which one has to reach a level as low as possible, see (Hendrix and Roosma, 1996). In this way one can change the search strategy depending on the amount of available running time. The figure suggests, for instance, that a high cooling rate (the process looks like PRS) does better for a lower number of function values (iterations) and worse for a higher number of function values.

Figure 10.26 gives an estimation of the expected level one can reach by running SA on function *g*. Implicitly it says the user wants to reach a low function value; not necessarily a global minimum point. The figure is common in metaheuristic approaches. The reader will not be surprised that the figure looks similar for function *h*, as the number of local optima is not relevant for the bit-string perspective and the function value distribution is similar. Theoretically, one can also derive the expected value of the minimum function value reached by PRS. It is easier to consider the theoretical behaviour from the perspective where success is defined Boolean, as has been done so far.

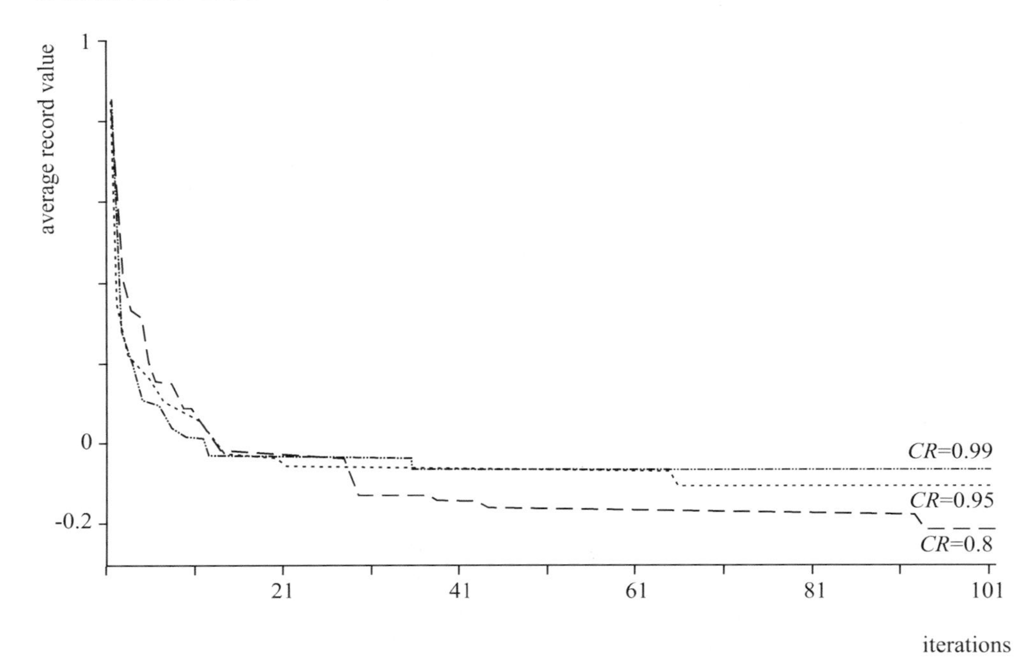

Figure 12.26. Average record value over 10 runs of SA, 3 different values of CR for a given amount of function evaluations, test case g.

Let us consider again the situation of the algorithm reaching the global optimum as a success. For stochastic algorithms we are interested in the probability of success. Define reaching the optimum again as finding a point with function value closer than $\delta = 0.01$ to the minimum. For function *g* this is about 2.2% (11 out of the 512 bit strings) of the domain. For PRS, one can determine the probability of success as $P_{PRS}(N) = 1 - 0.978^N$. For SA this is much harder to determine, but one can estimate the probability of success empirically. The result is the performance graph in Figure 12.27.

Let us have a look at the figure critically. In fact it suggests that PRS is doing as well as the SA algorithms. As this is verifying a hypothesis (not falsifying), this is a reason to be suspicious. The following critical remarks can be made.

- the 10 runs are enough to illustrate how the performance can be estimated, but is too low to discriminate between methods. Perhaps the author has even selected a set of runs which fits the hypothesis nicely.
- one can choose the scale of the axes to focus on an effect. In this case, one can observe that up to 40 iterations, PRS does not look better than the SA variants. By choosing the x-axis to run to 100 iterations, it looks much better.
- the graph has been depicted for function g, but not for function h, where the size of the success region is twice as small. One can verify that, in the given range, the SA variants nearly always do better.

This brings us to the general scientific remark, that all results should be described in such a way that they can be *reproduced*, i.e. one should be able to repeat the experiment. For the exercises reported in this section, this is relatively simple. Spreadsheet calculations and, for instance, Matlab implementations can easily be made.

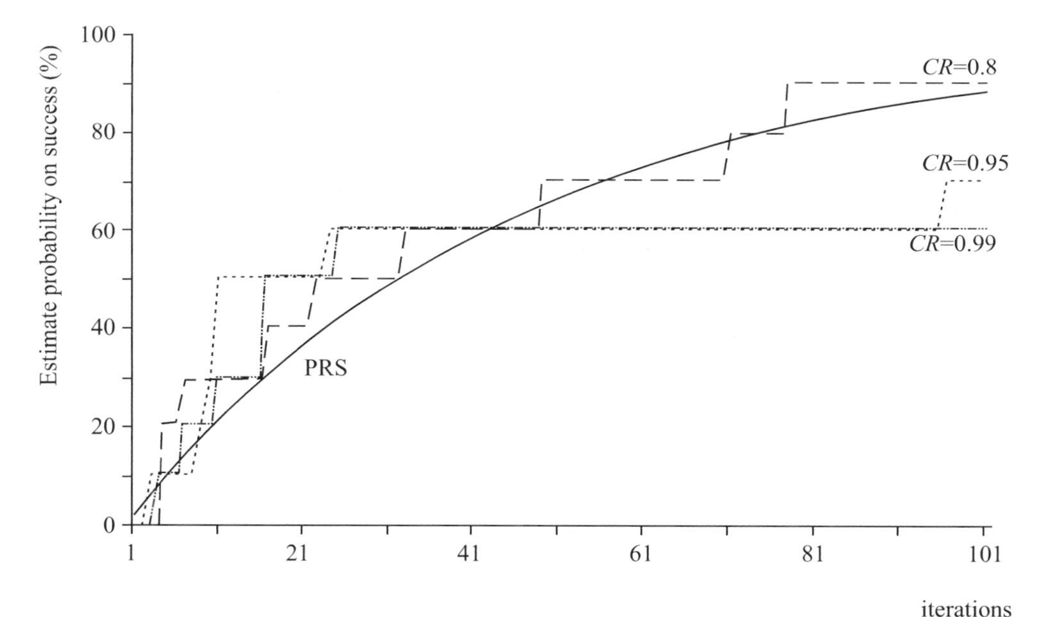

Figure 12.27. Estimate of probability of reaching the minimum for PRS and SA (average over 10 runs) on function g given a number of function evaluations

12.4.3. Investigating algorithms

In Section 12.4.2, we have seen how several algorithms behave on two test cases. What did we learn from that? How do we carry out the investigation systematically? Figure 12.28 depicts some relevant aspects. All aspects should be considered together.

The following steps are described in (Baritompa and Hendrix, 2005).

1. Formulation of performance criteria.
2. Description of the algorithm(s) under investigation.
3. Selection of appropriate algorithm parameters.
4. Production of test functions (instances, special cases) corresponding to certain landscape structures or characteristics.
5. Analysis of its theoretical performance, or empirical testing.

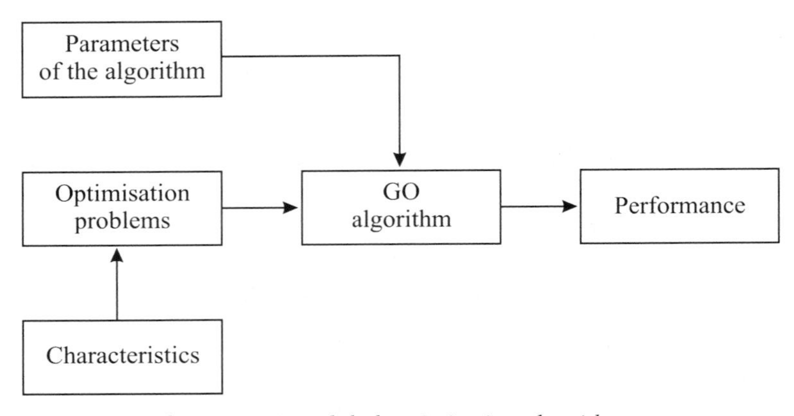

Figure 12.28. Aspects of investigating global optimisation algorithms.

Several criteria and performance indicators have been sketched: to get a low value, to reach a local minimum point, a high probability to hit an ε neighbourhood of a global minimum point, obtain a guarantee to be less than δ from the minimum function value, etc. Several classes of algorithms have been outlined. The number of parameters has been kept low, a Lipschitz constant L, the number of iterations N, cooling rate CR, stopping accuracy α. Many modern heuristics contain so many tuning parameters, that it is hard to determine the effect of their value on the performance of the algorithm.
Only two test functions were introduced to experiment with. The main difference between them is the number of optima, which in literature is seen as an important characteristic. However, in the illustrations, the number of optima was only important for the performance of multistart. The piecewise convexity appeared important for the local search Newton algorithm and further the difference in size of the δ-level set was of especial importance for the probability of success of stochastic algorithms. It teaches us that, in a research setting, one should think carefully, make hypotheses and design corresponding experiments, to determine which characteristics of test functions are relevant for the algorithm under investigation.

12.4.4. Comparison of algorithms

When comparing algorithms, a specific algorithm is *dominated*, if there is another algorithm which performs better (e.g. has a higher probability performance graph) in all possible cases under consideration. Usually however, one algorithm runs better on some cases and another on other cases.
So basically, the performance of algorithms can be compared on the same test function, or preferably for many test functions with the same characteristic, where that characteristic is the

only parameter that matters for the performance of the compared algorithms. As we have seen, it may be very hard to find out such characteristics. The following principles can be useful:

- Comparability: when comparing several algorithms, they should all make use of the same type of (structural) information (same stopping criteria, accuracies, etc.).
- Simple references: it is wise to include in the comparison simple benchmark algorithms such as Pure Random Search, Multistart and Grid Search in order not to let analysis of the outcomes get lost in parameter tuning and complicated schemes.
- Reproducibility: in principle, the description of the method that is used to generate the results, has to be so complete that someone else can repeat the exercise obtaining similar results (not necessarily the same).

12.5. Summary and discussion points

The topics discussed in this chapter focus on modelling, on analysis and optimality conditions and finally on the investigation of algorithms. The following points have been highlighted.

- In modelling optimisation problems, one should clearly distinguish decision variables, given parameters and data, criteria and model structure.
- First order conditions for differentiable functions are based on stationary points and Karush-Kuhn-Tucker conditions where complementarity plays a big role.
- Secondary order conditions are based on the eigenvalues of the Hessean.
- A convex optimisation problem has at most one minimum.
- Results of investigating algorithms consist of a description of the performance of algorithms (parameter settings) depending on characteristics of test functions.
- To obtain good performance criteria, one needs to identify the target of an assumed user and to define what is considered a success.
- The performance graph is a useful instrument for comparing performance.
- The relevant characteristics of the test cases depend on the type of algorithm and performance criterion under consideration.

13. Discrete-event simulation

D.L. (Dik) Kettenis and J.G.A.J. (Jack) van der Vorst

13.1. Introduction

The previous chapters in this book have focussed on analytical methods to model and solve decision problems and find the optimal solution. Most of the time, managers are already satisfied with suitable and acceptable outcomes that improve (but do not optimise) performance; computer simulation can provide this. Because of this, simulation is one of the most widely used operations research and management science techniques. This chapter will introduce simulation in general and discrete-event simulation in particular. However, you do not learn how to perform a simulation study by reading books; a lot of practice is required. In this text not all important subjects can be discussed. We have made a selection of the issues we thought to be of importance as an introduction. The books by Law and Kelton (2000) and Banks *et al.*, (2001) are highly recommended for further study. While writing this text, we have used both books mentioned as a guideline and the discussion of some issues is along the lines of the discussions in the books mentioned.

The next section discusses the background to simulation and present definitions. Section 13.3.1 compares Operations Research techniques with simulation. Section 13.4 provides the layout of the remainder of this chapter; it presents the steps to be taken in a simulation study, which are successively discussed in more detail in the subsequent sections, and discusses the principles of modelling. Section 13.5 deals with random numbers and data collection, Section 13.6 with experiments and Section 13.7 with verification and validation. We conclude this chapter with some final remarks.

13.2. Simulation, system, model

A general definition of *simulation* is 'the act of imitating the behaviour of some situation or some process by means of something suitably analogous (especially for the purpose of study or personnel training)'. When we focus on computer simulation we refer to the technique of representing part of the real world by a computer program. Many definitions for computer simulation can be found in literature, our favourite is:

Simulation is the process of designing a model of a system and conducting experiments with the model for the purpose either of understanding the behaviour of the system or of evaluating various strategies for the operation of the system (Shannon, 1975).

Let us explain the underlined words. A *system* is a collection of components, for example people, machines, or both like a factory or warehouse that act and interact together toward the accomplishment of some purpose. A system is part of reality identified and investigated by the researcher. It is rarely feasible to experiment with the actual system, because such an experiment would often be too costly or too disruptive to the system, or because the required system might not even exist. For these reasons, it is usually necessary to build a model as a representation of the real system and to study it as a surrogate for the real system (see Figure

13.1). To do this we, often have to make a set of assumptions about how the system works. These assumptions, which usually take the form of mathematical or logical relations, constitute a model. Pidd (1999) defines a *model* as follows:

A model is an external and explicit representation of part of reality (a system) as seen by the people who wish to use that model to understand, to change, to manage, and to control that part of reality in some way or another.

This definition indicates that a specific system can be represented by different models, depending on the assumptions made and components of the system that the researcher believes should be included in the model. Wilson (1993) refers to the *World view*, i.e. that view of the world which enables each observer to attribute meaning to what is observed. As a result a large part of scientific literature on simulation is devoted to the validity and applicability of the developed model. We will come back to this in later sections.

Law and Kelton (2000) distinguish between physical and mathematical models (Figure 13.1). Physical models refer, for example, to cockpit simulators or miniature super tankers in a pool. Mathematical models represent a system in terms of logical and quantitative relationships that are manipulated and changed to see how the model reacts, and thus how the actual system would react – if the mathematical model is valid. If the relationships that compose the model are simple enough, it may be possible to use analytical methods (such as discussed in the previous chapters or differential equations) to obtain exact information on questions of interest. In analytical models the relationships between the elements of the system are expressed through mathematical equations. However, most real-world systems are too complex to allow for analytical modelling, and these systems are preferably studied by means of simulation.

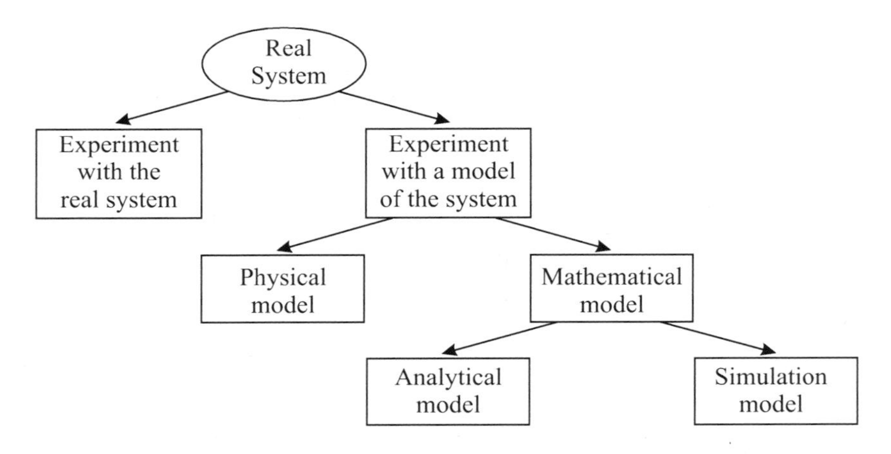

Figure 13.1. Ways to study a system (Law and Kelton, 2000).

Simulation can be applied in various contexts, such as development of control systems, distribution systems, mechanical systems, information systems or communication networks, weather forecasts, forecast of crop growth, organization of a production process, scheduling of machine maintenance, and training of personnel. Simulation is applied for several reasons. For example, experimenting with the real-world system is too costly, too dangerous, impossible, or not ethical. The goal of a simulation study is, among other things, to solve a problem or to optimize the performance of a real-world system. Computer simulation has proved to be a flexible tool. Therefore, it has replaced mathematical analysis and physical prototyping as the preferred method in many applications.

13.2.1. Types of simulation models

A model consists of variables and instructions that specify the changes of the state of the system. A *system's state* can be regarded as a snapshot of the system, showing all relevant details. The instructions may have the following type:

- mathematical equations such as algebraic and differential equations. Models using these instructions are called *continuous models*, since the variables in the model are (at least conceptually) continuous, i.e. state changes happen continuously in time;
- description of events in the system (an event signifies the transition from one state to another at a specific point in time). Models using these instructions are called discrete or *discrete-event simulation models*.

In discrete-event simulation models, the variables change in steps. If the model is partly defined by differential equations and other parts are defined by descriptions of events one calls this a combined continuous discrete model or mixed continuous discrete model. In this text we will not discuss continuous or mixed continuous discrete models.

Models may also be categorized in another way: deterministic versus stochastic models. *Deterministic models* contain no probabilistic components; the output is determined once the set of input quantities and relationships in the model have been specified. In a *stochastic model*, random effects are present. These random effects may be caused by a simplification applied while building the model. For example, the arrival process of customers at a supermarket or gas station can be simplified by using a stochastic variable. In general, continuous models are deterministic and discrete-event models are stochastic.

Simulation may be categorized as dynamic versus static simulation. In a dynamic simulation, models represent systems as they change over time. In a static simulation, however, time plays no role. Monte Carlo simulations are static, see the following example. This text is devoted to dynamic simulation, so we will not discuss Monte Carlo simulation in more detail.

Example 13.1.

Monte Carlo simulation to determine the value of π (=3.141592653589793).

Suppose you have a square piece of paper and on that paper you have drawn the largest possible circle. You fix the piece of paper on a wall and you randomly throw a dart many (100 or more) times at the piece of paper. Afterwards you count the total number of hits on the paper (including hits inside the circle) (N) and the total of hits inside the circle (C). Since we throw the dart in a random fashion, we may expect N and C to be in proportion to the area of the square and circle respectively. Suppose the length of the square is d. The area of the square and circle then are d^2 respectively $\pi\left(\frac{d}{2}\right)^2$. So $\frac{C}{N}=\frac{\pi d^2/4}{d^2}=\frac{\pi}{4}$. If N is large enough then the following relation holds

$$\frac{C}{N}=\frac{\pi}{4}$$

With this relation it is simple to compute the value for π. We may do this through simulation. In that case we let the computer throw the dart. How is it possible for a computer to throw a dart? First you let the computer generate a random number (see Section 13.5.1) for the x-position of the dart and then another number for the y-position of the dart. The computer checks whether the position (x, y) will be inside the circle or not. In this way, the computer may determine after, say 1000, 'throws' C and N and may compute the value of π. The accuracy of the approximation of π will increase when the number of throws increases.

13.2.2. Discrete event simulation models

Let us now go into the specific type of models that are central in this chapter, namely Discrete Event Simulation Models (DESM). These models are characterised by the fact that the modelled system has a vector of possible states and there is a sequence of events that moves the system from one state to another. DESM can suppose the presence of one or more probabilistic components. For DESM it is assumed that state changes are connected to a certain moment in time and result from the initiation or completion of activities.

The set of all states that a system may have is called a *state space*. An *event* signifies the transition from one state to another at a specific point in time. *Entities* are the dynamic objects in a simulation; they move around, change status, affect and are affected by other entities and the state of the system, and they affect the output performance measures. When the system state is depicted in a graph for some variable during a certain time period, we see a state-time diagram. It represents the path of the variable in the system, i.e. a sequence of events. The time domain is a set of so-called time points; the state of the system is only defined at these points. Which paths a certain system can have is determined by a function called the transition law. It represents a list of possible next events. For example, an order can be accepted or not, resulting in different activities and thus system states.

The event list keeps track of the events that are supposed to happen in the (simulated) future. When the logic of the simulation calls for it, a record of information for a future event is placed on the event list. This event record contains identification of the entity involved, the event time, and the kind of event it will be. Newly scheduled events are placed on the list so that the next event is always at the top of the list (the events are sorted). When it is time to execute the next event, the top record is removed from the list and the information on this record is used to execute the appropriate logic.

Table 13.1 presents the main semantic concepts of discrete-event simulation models. Note that different simulation languages use different terminology for the same or similar concepts.

We will now first present a case example that will be used in the remainder of this chapter and which illustrates the main semantic concepts.

Table 13.1. Semantic concepts in discrete-event simulation.

Concepts	Definition
System	A group of objects in reality that are joined together in some regular interaction or interdependence toward the accomplishment of some purpose.
Model	An abstract representation of a system, usually containing structural, logical, or mathematical relationships which describe a system in terms of state, entities and their attributes, events, activities, and delays.
System state	The collection of variables necessary to describe a system at any time (entities) relative to the objectives of a study.
Entity	Any object or component in the system that requires explicit representation in the model.
Attribute	A property of an entity.
Entity list	A collection of associated entities, ordered in some logical fashion.
Event	An instantaneous occurrence that may change the state of the system.
Event notice	A record of an event scheduled to occur at the current or some future time, along with any associated data necessary to execute the event.
Event list	A list of event notices for future events, ordered by time or occurrence.
Activity	Represents a time period of specified length (e.g. a service time or inter arrival time) which is known when it begins.
Delay	Duration of time of unspecified indefinite length, not known until it ends.
Clock	A variable representing simulated time.

13.2.3. Illustrative case example: post office

The case deals with a post office where customers arrive to be served by one of the clerks. One can easily generalise this case to car drivers arriving at a gas station, customers arriving at a check-out counter in a supermarket or customer orders coming in at a production plant to be produced on one of the production lines.

Suppose our position is at the entrance of a post office. We observe that people arrive and move into the post office and we see that people leave the post office. If we are not interested in what happens inside the post office we may see the post office as a system with an input, the incoming people, and an output, the people that leave the post office. In a way, we may see the post office as a system that transforms the input to an output (see Figure 13.2).

Usually a real-life system has several inputs, called *input variables*, and several outputs that are called *output variables*. In the post office example, the input is the stream of customers. The input variable then is a distribution function for the inter-arrival time between customers arriving at the post office. In a manufacturing system, the inputs will be a stream of raw material and orders for products and the output may be a stream of end products. In general terminology, we may consider a model as something that transforms input variables into output variables.

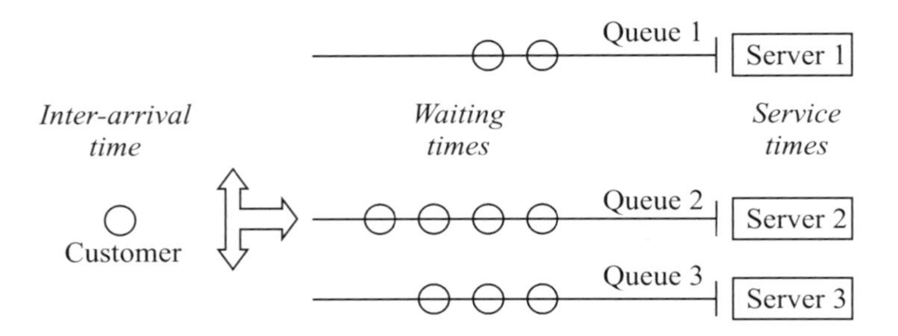

Figure 13.2. Illustration of the post office.

In the description of the model, we need other types of variables, such as *state variables*, *decision variables*, and *parameters*. A parameter is a variable that will normally be constant, but may vary in different experiments. A decision variable is, for example, the number of servers available in the post office. We will simulate the system with, for example, 3, 4, or 5 servers to decide on the optimal number of servers. Optimal can be in relation to the cost, customer satisfaction, or both.

In simulation, the state of a model is determined by the values of the so-called state variables. The state variables are the variables needed to describe a system at a particular time. It is not so that in a discrete-event model all variables are state variables. In the post office simulation, variables are, for example, number of busy servers, waiting time of a customer, service time of a customer, and the total time a customer stays in the post office. There is a relation between these variables: the time a customer stays in the post office is the sum of the waiting time and service time. Because of the relation we are able to compute the value of one of the variables when the values of the other two variables are known. Therefore, only two of these three variables suffice as state variables. In general, the values of all variables of a model can be computed when the values of the state variables are known.

In a discrete-event model, the values of the state variables change instantaneously at separate points in time. These points in time are the ones at which an event occurs. Suppose we study the post office with three servers and queues. In this post office simulation, the state of the model changes when (i.e. the following events occur):

1. a customer arrives at the post office;
2. a customer joins one of the queues in front of a server;
3. a customer leaves the queue to go to the server;
4. a server starts serving a customer;
5. a server finishes serving a customer; and
6. a customer leaves the post office.

Often it is possible to reduce the number of events and in this way simplify the model. For example, if the time a customer needs to travel from the entrance of the post office to the tail of the queue is negligible we may combine event 1 and 2. Alternatively, we may think that the customer enters the post office at the moment the customer enters the queue. We may combine event 3 and 4 when the time needed for a customer to travel from the head of the queue to the server is small. If the time is not negligible, we may increase the service time with the time needed for the transfer to the server. Events 5 and 6 may be seen as one event, since the time needed to travel from the server to the door of the post office is negligible or is not important for the study. A further simplification is to combine event 4 and 5 and rename the combined event as 'end service or start service.' We may do so when the time the server needs to switch from one customer to the next customer is negligible or, if the period is not negligible, we increase the service time with the time needed to switch to the next customer. The result is that the simplified model of the post office has the following events:

- arrival of customer;
- start service or end service.

A simulation practitioner must determine what aspects of the real-world system actually need to be incorporated in the simulation model and at what level of detail, and what aspects or events can be safely ignored. It is rarely necessary to have a one-to-one correspondence between each element of the system and each element of the model. Modelling each aspect of the system will seldom be required to make effective decisions based on the simulation and will also be infeasible due to time, money, or computer constraints. However, it is essential to include in the model the elements that are important for the study.

13.2.4. World views

Underlying the simulation package, but usually hidden from a modeller's view, events are being scheduled causing one process to temporarily suspend its execution while other processes proceed. It is important that the modeller has a basic understanding of the concepts and the simulation language being used, and a detailed understanding of the built-in but hidden rules of operation (Van der Zee and Van der Vorst, 2005).

In a discrete-event model we need to describe the events that occur in the system. For each event we must describe the time or the condition in which it occurs and we must describe what happens to the system when the event occurs. In simulation methodology, there are several ways to do that. Such a method is called a '*world view*'. The most important world views are: event scheduling, activity scanning, and process interaction. In the event-scheduling world view, the modeller concentrates on events and their effect on the system state. In the process-interaction world view, the modeller thinks in terms of processes. The modeller defines the simulation model in terms of entities and objects and their life cycle. In the simulation program, all events of one element are grouped in one software component. With the activity-scanning approach, a modeller concentrates on the activities of a model, and those conditions that allow an activity to begin.

Most of the present day simulation tools, like Enterprise Dynamics (ED) or Arena, adopt a special type of process interaction. In this world view the model consists of permanent elements (so-called 'building blocks' or 'atoms') like servers, trucks, or operators. These permanent elements are connected by channels and form a so-called network. Temporary elements, like customers or orders, flow through the network of the permanent elements. This software allows for the building of relatively complex models with a minimum of effort.

The events that occur in the model are hidden in this world view. Nevertheless, the events are still present in the model stored in the event list. For each event on the event list, the simulation tool stores the future time at which the event occurs and the type of the event (arrival of a customer, start serving, and so on). In a discrete-event simulation, the time advances from the time an event occurs to the next event. Time, so to speak, jumps from one event time to the next. That means the simulation tool has to sort the events according to the time they occur, smaller event time before larger event time.

In Figure 13.3 you see a simplified snapshot of the so-called time-axis of the post office simulation. In contrast to Figure 13.2, in this post office one server is available. The first event depicted in this figure is the arrival of customer 5. Because the (single) server starts serving this customer later, we may conclude that the server is busy. Service of this customer starts at the time of event e_8. Pay attention to event e_{10}. Here customer 7 arrives when the server is busy. This illustrates that arrival of customers and serving of customers run concurrently. In this model, the arrival process and the serving process run concurrently, in more complex models more processes may run in parallel. The snapshot of the time axis is of a simulation with the model built according to the simplifications discussed before. Without the applied simplifications more events would occur in the model. Examples are 'customer joins the queue' and 'end service of a customer'.

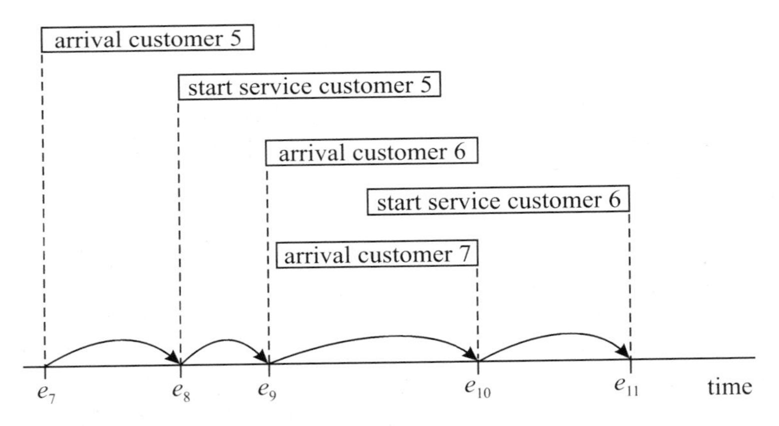

Figure 13.3. Snapshot of time-axis of the post office simulation model.

13.2.5. Queuing theory

In some instances, the queuing system is simple enough to study the system analytically. In such a study one computes measures of performance developed in *queuing theory*. For more information about queuing theory see, for instance, Chapter 6 in (Banks *et al.* 2001). Many systems, however, are so complex that it is virtually impossible to generate responses analytically. In these instances the computer will be used to generate data representing the system responses. The simulation-generated data is used to estimate the measures of performance of the system.

13.3. When is simulation appropriate?

Simulation is a powerful tool that is applied frequently. The popularity of simulation has increased with the increase in computer power, development of sophisticated software and the decrease in computer costs. This does not mean that simulation is the most appropriate tool in each situation. Simulation is not appropriate when:

- System behaviour is too complex or cannot be defined.
- It is easier, cheaper, or both to perform experiments with the real-world system.
- The measures of performance can be computed analytically, see for instance the calculation of waiting times and capacity requirements using queuing theory.
- Other techniques, such as operational research techniques or just using common sense, answer the questions at a lower cost. Note that sometimes it looks as though common sense can eliminate the need for simulation. Assume you want to know the minimum number of servers needed in a post office to minimize waiting times. The customers arrive randomly at an average rate of 75 per hour and are served at a mean rate of 20 per hour. The minimum number of servers is 75/20= 3.75. So, at first glance the minimum number of servers is four. In practice, however, four servers will not be sufficient. Since the customers arrive randomly, at some times the arrival rate will be too large to serve the customers, so some clients have to wait. At other times the arrival rate will be such that some servers will be idle. Simulation may be needed to find more adequate information.
- The costs exceed the possible savings. The resources or time to develop a simulation model and to collect the data for the input variables are not available.
- It is not possible to verify and validate the model (see Section 13.7).
- The problem itself can never be solved. In an ideal production system all products are delivered in time, there is no stock, and capacity utilization is 100%. Such a system does not exist due to uncertainty in customer demand, machine breakdowns, and so on. Simulation will not help to build such a perfect system; it may be of help to design the best system such that it will operate at the lowest possible cost or the highest client satisfaction.

An advantage of simulation related to experimenting with the real-world system is that the speed of the simulation may be faster than real time. For example, to perform a one-day simulation of a post office will take only a few seconds. One run, however, will not be sufficient to base conclusions on about the real-world system. Because of the stochastic effects in a model, several runs need to be performed for each set of input variables and model characteristics. Even if the number of the runs needed is large, because of the speed of the present generation of computers, a simulation will be generally faster than the real-world system. Another advantage of simulation is that the real-world system will not be disturbed by collecting data, and it is easier to change the way of operation in a simulation model than in reality. A disadvantage of simulation is that simulation modelling and analysis can be time consuming and expensive. Table 13.2 provides an overview of the advantages and disadvantages of simulation.

Table 13.2. Advantages and disadvantages of simulation, adapted from (Law and Kelton, 2000).

Advantages	Disadvantages
– Most systems with stochastic elements are too complex for analytical evaluation. Thus simulation is the only possibility. – Simulation allows one to estimate the performance of an existing system under some projected set of operating conditions. – Alternative proposed system designs can be compared to see which best meets a specified requirement. – Better control over the experimental conditions than when experimenting with the system itself. – Allows one to study a system with a long time-frame, in compressed time, or even in expanded time.	– Each run of a stochastic model produces only estimates of a model's true characteristics for a particular set of input parameters. Thus several independent runs of the model are required. An analytical model, if appropriate, can produce the exact true characteristics. – Expensive and time consuming to develop. – The large volume of numbers produced or the persuasive impact of a realistic animation often creates a tendency to place too much confidence in a study's results. – If a model is not a valid representation of a system under study, the results are of little use.

13.3.1. Operations research techniques versus simulation

To study a system of interest, we often have to make a set of assumptions about how it works. These assumptions are used to construct a model that is used to try and gain insight into the behaviour of the corresponding system. As stated in the introduction to this chapter, if the relationships that compose the model are simple enough, it may be possible to use analytical methods (such as algebra, probability theory, or linear programming) to obtain exact information on questions of interest. In analytical models the relationships between the elements of the system are expressed through mathematical equations. However, most real-world systems are too complex to allow for analytical modelling, and these models are preferably studied by means of simulation (Law and Kelton, 2000).

Hoover and Perry (1989) identify some advantages of analytical models: conciseness in problem description, closed form solutions, ease of evaluating the impact of changes in inputs on output measures, and in some cases the ability to produce an optimum outcome. On the other hand, they recognise some disadvantages, such as assumptions regarding the system description which may be unrealistic, and complex mathematical formulations which defy solutions. For example, many systems can be modelled as queuing networks, but either the assumptions required for analytic solution are somewhat unrealistic (e.g. exponential inter-arrival and service times), or the mathematical formulation necessary to reflect the desired degree of realism is intractable. Silver *et al.* (1998) state that if mathematical models are to be more useful as aids for managerial decision making, they must be more realistic representations of the problem; in particular, they must permit some of the usual 'givens' to be treated as decision variables. Moreover, such models must ultimately be in an operational form such that the user can understand the inherent assumptions, the associated required input data can be realistically obtained, and the recommended course of action can be provided within a relatively short period of time. Simulation models can compensate for the disadvantages of analytical models, but not without sacrificing some of the advantages of the analytical models.

Assume you are asked to find the optimum number of servers needed in a post office. You have a valid criterion to detect what the optimum is. You may solve this problem with operations research techniques or with simulation. Both will give you a result and hopefully the results will be comparable. Operations research models give analytical results, that means, the optimal number of servers will be the result of a mathematical computation. Simulation models, however, provide outcomes for a specific set of input variables and model characteristics. For each setting, the model is run and the simulated behaviour is observed. Next the input, model characteristics, or both are changed and the model is run again and so on. In doing so, one collects a set of scenarios that are evaluated. The best scenario, either in the analysis of an existing system or the design of a new system, is then recommended for implementation. In the post office case this means that we will run the simulation with say 4, 5, 6, and 7 servers. For each situation we will compute the criterion value and choose the number of servers that has the optimal value for the criterion. From the values of the criterion we may decide whether it is needed to perform simulation runs for one or more other numbers of servers.

So-called optimization packages can be integrated with the simulation tool to automate the optimum seeking process. OptQuest (Glover *et al.*, 1999) is an example that can cooperate with both Enterprise Dynamics (ED) and Arena. We conclude this section by stating that simulation is preferred when the system that needs to be analysed is complex, subject to uncertainties, time-dependent and has multiple performance variables as desired output.

13.4. Steps in the simulation study

Some authors consider building a simulation model an art rather than a science. Most probably, this is based on the observation that it is not possible to provide a set of instructions that will lead to the building of successful and appropriate models for every system. Another observation that supports this idea is that different (groups of) people produce different models for the same real-world system. One reason for that is that different people have different world views and abilities to abstract the essential features of a system and select basic assumptions that characterize the system. Furthermore, there are different ways to implement the essential features. In this section, we will present some guidelines for building appropriate discrete-event models.

A simulation study is not a simple sequential process. As one proceeds with a study and obtains a better understanding of the system of interest, it is often desirable to go back to a previous step. Figure 13.4 shows the steps that will compose a typical, sound simulation study and the relationships among them. This section presents an overview of the steps; the main issues are discussed in the following sections. We distinguish the following steps:

1. Problem formulation and preparation of the overall project plan.
Before starting a simulation study the questions to be answered by the study (objectives of the study) should be clear. If the study is performed for another person or organization, it is important that that person or organization agrees with the problem formulation. The objectives should be validated, meaning it is possible to answer the questions by the simulation study. If applicable to the study, a project plan is defined that contains the alternatives to be considered and how the alternatives will be evaluated. Furthermore, the plan contains the number of people involved in various stages of the study, the cost of the study, and the dates on which each phase of the study is supposed to be ready.

2. Development of the conceptual model.
To be able to build a model of a real-world system, it is necessary to observe the real-world system to detect the various components and the interactions among these components. The conceptual model consists, among others, of a textual description of the components involved in the model and their interactions, and the input data needed. The decision on how much of the real system should be included in the conceptual model to bring about a valid representation of the real system must be jointly agreed upon by the simulation analyst and the decision-makers. See Section 13.4.1 for more details about this topic.

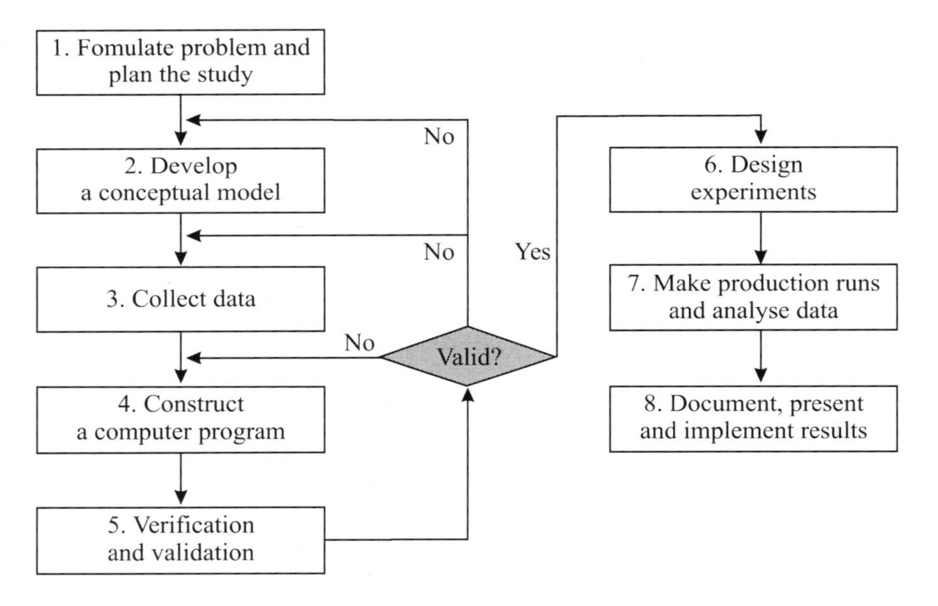

Figure 13.4. Steps in a simulation study.

3. Data collection
The model needs input data, like arrival time and service time of clients. Furthermore, data will be needed in the model validation phase. These data should be collected. Usually, for input data a distribution function is fitted and validated. Since this step is rather laborious, one should not delay the start of the data collection. During the model development phase it may frequently become clear that more data is needed. Section 13.5 goes deeper into this subject.

4. Construct a computer program
The conceptual model should be translated to a computer program in a specified simulation language, which will execute the logic described. In general, with the current simulation tools this translation is not too complicated and laborious (see Section 13.4.2).

5. Verification and validation of the model.
Conclusions based on a study with an invalid model are of no value to the real-world system. Section 13.7 will discuss this subject. Note that experimenting with the model is also needed to validate the model – therefore, we decided to discuss step 6 first in this chapter.

6. Experimental design.
For a study where different designs need to be evaluated, we decide here the alternatives that are to be simulated. For each system design that is simulated, we have to decide the length of the simulation runs and the number of replications to get statistical sound performance measures. For some systems it is necessary to decide about the length of the initialization period. Section 13.6 discusses this subject.

7. Make production runs and analyze the results.
In this phase we estimate measures of performance for the system designs that are being simulated.

8. Documentation and reporting.
The model and the simulation program should be documented. Furthermore, it is wise to communicate the progress of the study to the customer. Implementation of the best strategy is usually no part of the simulation study. However, it may be part of the project.

Example 13.2. A wrong problem formulation.

The management of a distribution centre observed that it is not possible to load all the goods into trucks. Based on these observations they decide to start a simulation study to see whether it is possible to realise the target with an increase in terminals. However, the real problem is not the number of terminals. If the management had studied the real situation better, they would have found that several terminals were not in use during long periods. The problem in reality is the number of trucks available. In this particular situation the problem could never be solved by the suggested solution and, therefore, the study could not be successful.

One of the first phases in a simulation study is to make a model. To be able to model a system quantitatively, knowledge of the scientific discipline the system is based on should be available. In addition to that, the variables in the system have to be quantifiable. The models employed in the technological field are often based on physical laws and the variables can be quantified easily. Consequently, technological systems in general meet the requirements for quantitative modelling. In contrast, so-called soft systems correspond to models that are based on intuition, Such systems have properties that make a quantitative approach cumbersome or even inadequate. Soft systems are found, for example, in biology, ecology, and sociology. Such systems are sometimes classified as ill-defined. Validation of models of technological systems is relatively easy. In contrast, validation of ill-defined systems is difficult. Nevertheless, models of ill-defined systems are useful, since experimenting with those models provides more insight into the field. However, one has to be careful with extrapolations based on experiments with intuitive models.

13.4.1. From reality to conceptual model

The ultimate success of simulation is determined by a combination of the analyst's skills, participants' involvement and the quality of the simulation tool (Van der Zee and Van der Vorst, 2005). Together they should lay the basis for a realistic simulation model, which is both transparent and complete.

Starting with the real system, we first form a *conceptual model* of the system that contains the elements of the real system that we believe should be included in the model. That is, one should identify all facilities, equipment, events, operating rules, and descriptions of behaviour, state variables, decision variables, measures of performance, and so on, that will be part of the model. Consecutively, we must identify the relationships between the elements identified.

The model is an abstraction of the real-world system and will not duplicate the real-world system in all its aspects. The questions to be answered about the real-world system determine the components to be included and the level of detail of these components in the model. Suppose you are studying a distribution centre. If you are interested in overall performance of the distribution centre, it is not necessary to include a detailed order-picking process in the model. In such a model the time needed to pick an order may be modelled by a random variable. If the simulation study is meant to solve a problem that may be caused by the order picker, the order-picker process needs to be implemented in detail. To prepare a model to determine the number of beds for a new regional hospital, to give another example, it is not necessary to include a detailed model of the surgery. Frequently, some processes will be implemented as random variables. The arrival process of clients in a post office is an example. In fact, modelling a process by a random variable is a simplification. The details of such a process will not be modelled; only the overall behaviour, normally the duration of a part of the process, will be represented in the model.

From the conceptualization of the system a logical model (or flow chart model) is formed that contains the classification of and logical relationships among the elements of the system, as well as the exogenous variables that affect the system.

13.4.2. Building the operational model

Simulation models may be developed with the help of simulation tools or general programming languages, such as C++ or Java. The effort needed to implement a model in a general programming language is greater than implementing a model in object-oriented simulation tools like ED and Arena. The majority of the simulation tools available today, are oriented towards manufacturing systems, ED and Arena are no exception. However, the tools are applied in all kinds of settings. The software contains several elements (building blocks) and the idea is that you select the appropriate elements to represent the real-world components. ED contains elements like:

- queue to represent a storage area or waiting line;
- server to represent a machine or a clerk at a window; in general a server will be used in case we have to model an operation that takes some time;
- conveyor to transport goods.

It goes beyond the objective of this chapter to discuss the actual building of an operational model.

13.4.3. Principles and pitfalls of modelling

Law and Kelton (2000) present some guidelines for determining the level of model detail. The most important are to understand the decision maker's needs and specify the measures of interests; to concentrate on the most important factors by using 'experts' and sensitivity analysis (see Section 13.7.3) of the model; and, to ensure that the level of model detail is consistent with the type of data that are available. Once developed and validated, a model can be used to investigate a wide variety of 'what if' questions about the real system. Potential changes to the system, scenarios, can first be simulated in order to predict their impact on system performance. Pidd (1999) gives some general principles of modelling:

- Model simple; think complicated. Models should be transparent and should be easy to manipulate and control. Transparency is desirable because successful Decision Science practice depends on trust between consultant and client.
- Start small and add. Start with simple assumptions and add complications only if necessary.
- Avoid Megamodels. Use object-oriented methodology in which the idea is to divide the model into components (building blocks) that can be replaced if they are too simple, without the need to redevelop the entire model. But be aware of system effects: the whole is more than the sum of its parts, and the part is more than a fraction of the whole. Each component model should be developed with exactly the same set of modelling assumptions to prevent problems caused by the interaction of the components.
- Do not fall in love with data. Only after thinking about the model, one can determine the type of data to be collected. Also remember that information equals data plus interpretation and be careful when extrapolating results into the future. The future may simply differ from the past and the population from which the data sample comes may behave differently in the future.
- Modelling is not a linear process. Develop models not in one burst, but over an extended period of time marked by heavy client interactions.

The literature identifies a number of generic pitfalls that can prevent successful completion of a simulation study:

- A set of well-defined objectives and performance measures should be defined at the beginning of the simulation study that suit all involved parties. Furthermore, key persons should be involved in the project on a regular basis.
- Some degree of abstraction is usually necessary. Some parts may be left out of the model completely; others may be aggregated. The summarised characteristics of the aggregated parts must be checked against expert opinion to see if they represent the situation fairly.
- Data must often come from different disparate locations: to ensure the success of the modelling effort, it is necessary to obtain sufficient commitment of resources to ensure accurate, useful data.
- Accuracy of data: often the collected data paints a false image of an operation due to data entry errors, inconsistent collection procedures, and system incompatibilities.
- Commercial simulation software may contain errors or complex macro statements that may not be well documented and may not implement the modelling logic desired.

13.5. Input modelling

One of the standard expressions when modelling is: '*garbage in = garbage out*'. In other words, even though the model fits its purpose, the output of the model study can be unsatisfactory because wrong data are used. Detection of the input variables and collecting data on its behaviour is thus essential to test the validity of the model (for instance, the arrival time of customers, average waiting times, and the service times in the post office example). Observing the system, however, is rarely sufficient. For complex systems in particular, it is useful to discuss the system with people such as operators, maintenance personnel, and managers. In case the system under study does not exist, you may study the design of the system and collect as much data as possible. Most probably the majority of the components of the system are operating in real life, so you can observe these components and collect data of its behaviour.

As indicated before, in discrete-event models often stochastic variables appear as input data. For stochastic variables we need to be able to generate so-called random numbers. In this section, we will successively discuss methods to produce random numbers by a computer, properties of random-number generators relevant for simulation applications, fitting data to distribution functions and software for selection of distribution functions.

13.5.1. Random-number generators

You are no doubt familiar with the concept of throwing a dice to introduce random effects in games. Parts of discrete-event models are often modelled by stochastic variables. Therefore, In simulation applications, one frequently needs random effects. To mention a few:

- arrival times of customers;
- service time of a clerk serving a customer or the time a machine needs to produce a product;
- the time needed to repair a broken machine.

For the computer throwing a dice is impossible. Furthermore, for simulation purposes, throwing a dice will not be acceptable. This is because:
- the computer needs to generate a large number of random values;
- frequently we need numbers in a range of zero to a number larger than six;
- sometimes it is necessary to generate exactly the same numbers during (a part of) the simulation. In other words, the numbers should be replicable. This is essential when we test the implementation of the model or we are searching for the reason why the model does not operate as expected. Evaluation of alternative system designs is another example.

Therefore, computers generate random numbers in a different way. Many generators have been designed and are described in, among others, (Kleijnen and Groenendaal, 1992; Banks *et al.*, 2001; Law and Kelton, 2000). One popular generator is the so-called *linear congruential generator*. The discussion about random-number generators in this chapter is limited to this method. We discuss this method to learn about the properties of random-number generators in general. The linear congruential generator is based on the fixed recursive formula

$$X_{n+1} = (aX_n + c) \bmod m,\ n = 0,1,2,\ldots \tag{13.1}$$

In (13.1), a, c and m are positive constant integer values and $x \bmod m$ is the remainder on dividing x by m. When we repeatedly apply this formula, we produce a sequence of integers $X_1, X_2, \ldots$ between zero and $m-1$. To be able to start the computation, this relation needs the value X_0; this is the start value and is called the *seed*. To illustrate some properties of this generator we will show some random numbers generated by this method. We take X_0 = 7, a = 19, c = 41, and m = 100. Because the result of recursion (13.1) is the remainder on dividing by m, here 100, we may expect values between 0 and 99. This range is not sufficient for a simulation, so we may expect that a simulation tool for discrete-event simulation will use a larger value for m. This generator generates the following values

7, 74, 47, 34, 87, 94, 27, 54, 67, 14, 7, 74, 47, 34, …

To compute, for example, X_1 we compute (19·7+41) mod 100 = 174 mod 100, so X_1 = 74. X_2 = (19·74 + 41) mod 100 = 1447 mod 100 = 47, and so on.

Some observations are:
- At first sight the numbers seem to be random. However, only 4 and 7 appear as last digits of the numbers generated.
- Only 10 different values appear. These values are 7, 74, 47, 34, 87, 94, 27, 54, 67, and 14. The values are repeating themselves in the same order, and this cycle will repeat forever. The number of random values created is called the cycle length or period. In this case the cycle length is 10.
- The seed, here 7, determines the number of and the order in which the values appear. This is true for other random-number generators as well.
- When we start the procedure a second time with 7, the same values appear in exactly the same order.

The seed may affect the cycle length (period). For example, with $X_0 = 3$ the generator with $a = 19$, $c = 41$, and $m = 100$ generates the following data 3, 98, 3, 98,.... So, the same generator produces with the seed 3 only two different values (3 and 98) and has a cycle length of 2. The behaviour of the linear congruential generator is determined by the value for a, c, and m in relation (13.1). When we choose other values for these constants, the behaviour of the linear congruential generator will be different. For example, for $a = 17$ and all other values the same (so $X_0 = 7$, $c = 41$, and $m = 100$) we find

7, 60, 61, 78, 67, 80, 1, 58, 27, 0, 41, 38, 87, 20, 81, 18, 47, 40, 21, 98, 7, 60,...

The cycle length (period) now is 20.

In practice we need a random-number generator to have the following behaviour:

- Maximum density of the numbers, which means that the values X_i, $i = 1, 2, \ldots$ leave no large gaps on the interval $[0, m - 1]$. Or in other words, almost all values appear in the cycle, in the ideal case all values show up.
- In practical applications we want to avoid cycling. That means, the cycle length (period) should be large. Since, the maximum cycle length (period) of the linear congruential generator is m, we need a large m.

From this we may conclude that the linear congruential generators with the presented values for a, c, and m are not good enough to be used in a real simulation study. Other values for the constants, however, will produce a better performance.

The linear congruential generators available in simulation tools have fixed values for a, c, and m. The values of the constants mentioned, normally, are chosen such that the generator functions optimally. The seed will be the only variable that may be chosen to generate independent sequences of random numbers. This is true for other random-number generators as well. Some simulation tools choose the seeds for the user, other packages leave that to the user. Important to remember is the following:

- The seed determines the numbers generated and the order in which the numbers appear.
- The length of a simulation run is limited by the cycle length (period) of the random-number generator.

Since the seed determines the numbers generated and the order in which they appear, we may conclude this section by stating that the random numbers drawn by the computer are not random at all. Despite that, at a practical level, we are able to use these random-number generators in simulation experiments. Because the random-number generators used by computers are not fully random, the numbers drawn are sometimes called *pseudo-random numbers*. Modern and carefully constructed random-number generators generally succeed in producing a sequence of numbers that appear to be truly random. These numbers pass various statistical tests for both uniformity and independence. If you do not use professional software, it might be useful to test the available random-number generator. For these tests we refer to (Law and Kelton, 2000) and (Banks *et al.*, 2001).

13.5.2. Random variables

In simulation applications one often wants to model some events with the help of stochastic or random variables. There are two types of random variables: discrete and continuous. *Discrete random variables* are used to describe random phenomena in which only integer values occur. Discrete random variables may be used to model, for example, demands for inventory items. *Continuous random variables* may be used to describe random phenomena in which the variable of interest can take on any value in some interval. Inter-arrival time at a post office is an example. One normally collects data from the real-world system. When these data are known, the next task is to identify the distribution and the parameter(s) of the distribution function with which the random variable can be modelled. In this modelling process the family (shape) of the distribution plays an important part. Appendix B presents some well-known and frequently used distribution functions. It will be sufficient to memorize the names of the distributions mentioned, their characteristics, and the shape of the density functions. You will need the shape to compare it with a histogram derived from experimental data in order to select a proper distribution function for that data and determine the parameters of the distribution function. This is discussed in the next section.

13.5.3. Fitting data to a distribution function

In general, three factors characterize a distribution function: the mean, the variance, and the family. Hillen (1993) states that the most important factor is the mean, then the variance, and finally the family of the distribution function. That means an error of say 10 percent in the mean produces a larger error in the results than an error of 10 percent in the variance. In other words, in practice it is more important to compute the correct value for the mean and the correct value for the variance than it is to select the correct family of the distribution function. This does not mean that it is wise to replace a distribution function in a model by its mean alone. As an illustration, Law and Kelton (2000) give on page 260 the following example.

Example 13.3.

In a manufacturing system consisting of a single machine, raw parts arrive at the machine with exponential inter-arrival times having a mean of 1 minute. Processing times at the machine are exponentially distributed with a mean of 0.99 minute. With the help of queuing theory we can compute the average delay of a part in the queue to be 98.01 minutes in the long run. From this we may conclude that at times there will be parts waiting in the queue. If in a simulation of this system we replace the distribution functions by their mean then no part is ever delayed in the queue. This is so, because then the arrival time (1 minute) is larger than the service time (0.99 minutes). It is the variance that makes parts queue up sometimes. In general, the variance as well as the mean of input distributions affect the output measures for queuing-type systems.

In search for the type of random variable we may distinguish four steps that will be discussed in the subsequent sub-sections:

1. Collect the data and compute the mean and the variance.
2. Make a histogram of the collected data and select the family of the distribution function based on a visible comparison with the known distribution functions.

3. Determine the parameter(s) of the selected distribution.
4. Test whether the selected distribution and parameter value(s) are acceptable.

Step 1. Data Collection

We will illustrate the process of fitting the data to a distribution function with an example. Suppose we need a distribution for the inter-arrival times at a bank terminal. We have to collect the inter-arrival times first. That means somebody, most probably you, has to stand at the bank terminal and record the times clients arrive. The inter-arrival times can be computed from these data. There are several issues related to data collection that need some elaboration.

Homogeneity of the data

One important issue is whether the arrival of clients is homogeneous. It may happen, for example, that the bank terminal will be popular at specific hours (for example, during lunch time) and less popular during other time periods. You must study the arrival process carefully and during a period that is long enough to detect inhomogeneity. To detect inhomogeneity it is necessary to check data in successive time periods and during the same time period on successive days. When checking for homogeneity, an initial test is to see if the average inter-arrival times are the same during such a period for all days.

When the arrival process is not homogeneous enough to be able to describe the process with one distribution, a solution might be to distinguish several periods. For example, you may distinguish a period with high, medium, or low arrival rate. Then we may choose to select a different distribution, different parameters, or both for each period. This procedure, however, might be acceptable for non-arrival processes (see Law and Kelton (2000; 337) for how to handle in those situations), however for arrival processes that is not a good idea. Kelton *et al.* (1998) illustrates this with the following example.

Example 13.4.

Assume we apply an exponential distribution function with for every interval a different value for the mean. If the mean in one interval is large (this means arrival rate is low) and in the next interval the mean is small (meaning arrival rate is high), this method may give misleading and wrong results. Assume we have only two intervals, each 30 minutes long. The mean for the first interval is 20 minutes and the mean for the second interval is 1 minute. Furthermore, assume that the last arrival in the first period occurred at time 29 minutes. The next arrival time will be generated with the distribution function for the first period (mean is 20 minutes). The distribution function with the mean of 20 minutes can easily return a value larger than 31 minutes. This means, during the second period no arrivals could occur. In other words, during the second, more busy, period no arrival could occur. For more information about so-called non-stationary Poisson Processes, see (Kelton *et al.* 1998; 244-245) and (Law and Kelton, 2000; 485-489).

Rounding applied to the data

Another point that needs some attention is rounding of the collected data. If the data, for example, will be expressed in minutes, it is not wise to round these data to the nearest 5 minutes. This may create problems when fitting the data to a distribution function.

Bias of the supplier of the data
If you use data provided by others, be aware that these data may be biased. If you ask, for example, the manufacturer of a machine to provide the intervals between the successive breakdowns of the machine, you may expect the delivered data to be too optimistic.

Independence of the data
Many of the statistical techniques require that the observations $X_1, X_2, \ldots, X_n$ are independent. The *chi-square test* is an example. If the observations are dependent or correlated, these statistical techniques may not be valid. Data collected in reality or in a simulation are, often, dependent. For example, there is a correlation between the arrival times of customers at a bank terminal and the time a client has to wait for service. If a number of clients arrive at a faster rate than the time needed for service, the clients will queue up. The time these clients have to wait will be significant and, therefore, the observations of these successive clients will be correlated. If at a certain moment the temperature in a room is high, the temperature will probably be high during the next hour too. Therefore, hourly temperatures close together are correlated. Correlation within a series in time is called *autocorrelation* There are several techniques available to detect dependence of the collected data. We will illustrate this with the bank terminal example.

Suppose, over a period of 200 minutes we found the values of the inter-arrival times in minutes at the bank terminal given in Table 13.3. Then the mean of sample of n data X_i is

$$\overline{X}(n) = \frac{\sum_{i=1}^{n} X_i}{n} \tag{13.2}$$

The sample variance follows from

$$S^2(n) = \frac{\sum (X_i - \overline{X}(n))^2}{n-1} \tag{13.3}$$

For the data given in Table 13.3, the mean of the sample is 3.33 and the variance is 9.157.

Table 13.3. Inter-arrival times in minutes at a bank terminal measured during 200 minutes.

0.5	1.6	5.0	0.9	3.2	0.8	7.5	2.4	0.8	1.7	0.5	1.6	5.0
1.3	2.6	0.4	3.4	0.9	0.3	5.6	0.4	1.1	3.7	1.3	2.6	0.4
8.2	1.5	3.2	0.5	1.8	1.9	2.3	11.2	6.0	1.2	8.2	1.5	3.2
0.6	0.1	0.9	2.7	3.8	6.2	7.1	13.1	6.7	4.7	0.6	0.1	0.9
4.3	2.5	1.1	0.7	5.3	5.4	7.8	1.0	0.7	6.8	4.3	2.5	1.1
9.5	9.9	0.6	1.5	4.6	2.9	2.3	0.9	1.9	2.3	9.5	9.9	0.6

A *scatter diagram* of the observations is a means to detect dependence of the collected data. The scatter diagram of the observations $X_1, X_2, \ldots, X_n$ is a plot of the pairs (X_i, X_{i+1}) for $i = 1, 2, \ldots, n-1$. In Figure 13.5 the scatter diagram is given for the data collected at the bank terminal, see Table 13.3. If the X_i's are independent, the points are scattered randomly throughout the (X_i, X_{i+1}) plane. This is the case in Figure 13.5 showing that the data are independent.

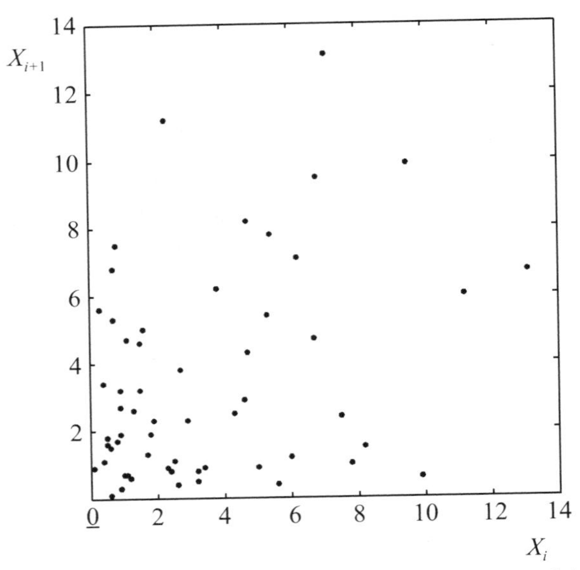

Figure 13.5. Scatter diagram indicating independence of the collected data.

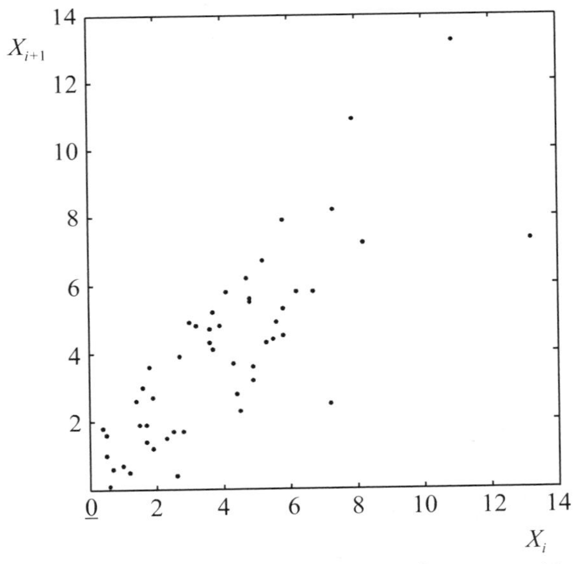

Figure 13.6. Scatter diagram indicating autocorrelation between waiting times.

Normally, there is a clustering of points in the lower left corner of a scatter diagram of independent data, as in Figure 13.5. This, however, does not mean that the data are dependent. Compare Figure 13.5 and 13.6 to see the difference between a scatter diagram of dependent and independent data. If the X_i's are (auto)correlated, the points will lie on a line, as in Figure

13.6. When the slope of the line is positive, as in Figure 13.6, there is a positive correlation, whereas a negative correlation is indicated by a negative slope.

Another way to detect dependencies in the collected data is the so-called *correlation plot* or *autocorrelation function*. The correlation plot is a graph of the sample autocorrelation $\hat{\rho}_j$ for j = 1, 2, ..., l, where l is a positive integer, called *lag*. The sample autocorrelation $\hat{\rho}_j$ is an estimate of the autocorrelation ρ_j between two observations that are j observations separated in time. For sample X_1, X_2, ..., X_n, the estimate of the sample autocorrelation $\hat{\rho}_j$ is computed as follows (Law and Kelton, 2000)

$$\hat{\rho}_j = \frac{\hat{C}_j}{S^2(n)} \tag{13.4}$$

where

$$\hat{C}_j = \frac{\sum_{i=1}^{n-j} \{X_i - \overline{X}(n)\}\{X_{i+j} - \overline{X}(n)\}}{n-j} \tag{13.5}$$

$S^2(n)$ and $\overline{X}(n)$ are determined via (13.3) and (13.2) respectively.

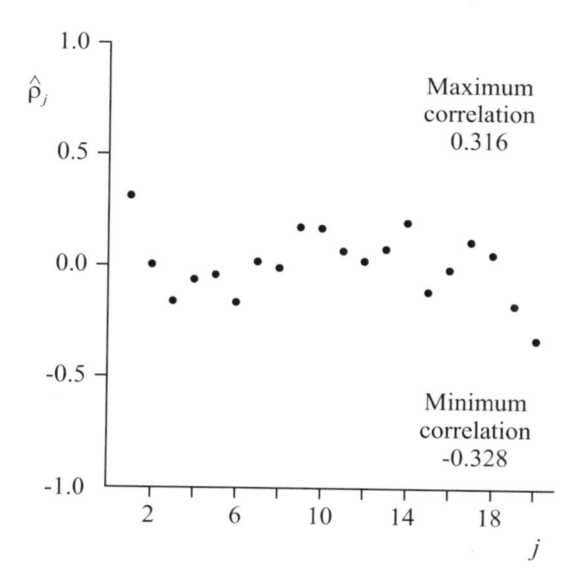

Figure 13.7. Autocorrelation plot for inter-arrival times at the bank terminal.

If the observations X_1, X_2,...,X_n are independent, then $\rho_j = 0$ for j = 1, 2, ..., n – 1. The $\hat{\rho}_j$, however, will not be exactly zero even when the X_i's are independent. If the $\hat{\rho}_j$'s differ from 0 by a significant amount, then this is strong evidence that the X_i's are not independent. For more information, see Law and Kelton (2000). The autocorrelation plot of the data given in

Table 13.3 is presented in Figure 13.7. Note that the (auto)correlation is $-1 \leq \rho_j \leq 1$. For the data collected at the bank terminal, most elements of the vector $\hat{\rho}$ are close to zero. Therefore, we expect these data to be independent. The correlation plot for the data of Figure 13.6 is given in Figure 13.8. In Figure 13.8, the minimum and maximum correlation is larger than in Figure 13.7. Therefore, we conclude that the data used to generate Figure 13.6 are dependent.

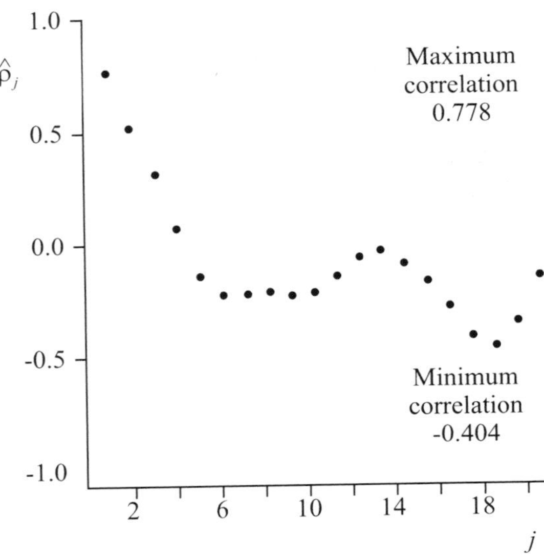

Figure 13.8. Autocorrelation plot for dependent data.

Step 2. Shape of distribution function

A histogram is a good approximation for the underlying continuous distribution function of the sampled data. Therefore, it is a good idea to make a histogram of the collected data and compare the histogram with the known distribution functions. Normally, one or a couple of distribution functions may be selected to model the data. This technique may be applied even when the collected data are dependent. To illustrate the procedure, we will prepare a histogram for the data presented in Table 13.3. We prepare a histogram by arranging the data in class intervals. A rule of thumb is to take the number of class intervals approximately equal to the square root of the sample size. The square root of the sample size, 60, is 7.7, which gives 7 or 8 classes. Law and Kelton (2000) recommend drawing a number of histograms with a different number of class intervals and selecting the histogram with the largest number of classes that gives a smooth histogram. Because the histogram with 8 classes was not smooth, we tried some other numbers of classes and decided to use 14 classes with an interval width of 1. In Table 13.4 the intervals and the frequencies of the data in the class intervals are presented. The histogram is given in Figure 13.9.

Table 13.4. Frequency of arrival times at the bank terminal.

Inter-arrival time (min)	Frequency	Fraction
$0 \leq X_i < 1.0$	16	0.267
$1.0 \leq X_i < 2.0$	12	0.200
$2.0 \leq X_i < 3.0$	8	0.133
$3.0 \leq X_i < 4.0$	5	0.083
$4.0 \leq X_i < 5.0$	3	0.050
$5.0 \leq X_i < 6.0$	4	0.067
$6.0 \leq X_i < 7.0$	4	0.067
$7.0 \leq X_i < 8.0$	3	0.050
$8.0 \leq X_i < 9.0$	1	0.017
$9.0 \leq X_i < 10.0$	2	0.033
$10.0 \leq X_i < 11.0$	0	0.000
$11.0 \leq X_i < 12.0$	1	0.017
$12.0 \leq X_i < 13.0$	0	0.000
$13.0 \leq X_i < \infty$	1	0.017

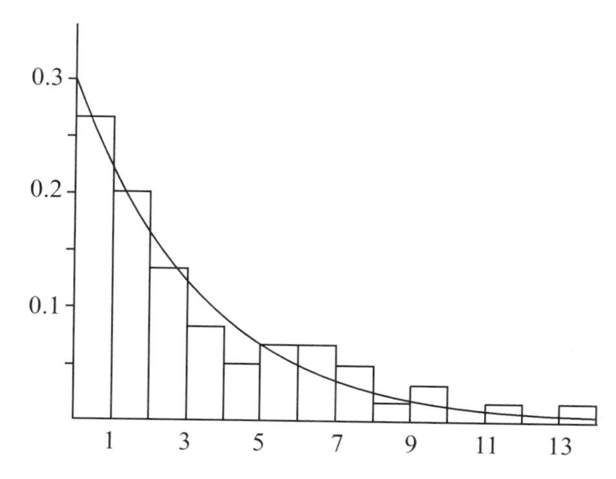

Figure 13.9. Comparing histogram with exponential distribution (λ = 0.3). The left box represents the frequency in class 1, next one the frequency in class 2, etcetera.

To be able to compare the histogram with the density function of a distribution, one needs to depict the frequencies as a fraction of the total number of data. In this case we estimate the distribution function to be exponential. This corresponds to the theory that inter-arrival times will often be exponentially distributed. In the foregoing, the selection for the family of the distribution function was based on the histogram of the data. In addition, it is important to take into account the physical characteristics of the process when selecting a distribution. A guideline that may be used when the physical characteristics are not known is presented in Figure 13.10 taken from Hillen (1993). In this guideline there is a reference to an empirical distribution. So far, we have met so-called theoretical distributions like the normal or exponential distributions. An empirical distribution is a distribution directly based on collected data. The empirical distributions simply produce a function based on either the collected data or based on data represented in a histogram. For more details, see Law and Kelton (2000).

Events independent?
Yes: Select exponential distribution
No: Histogram symmetrical?
  Yes: Lower and upper value are bounded?
    Yes: Select uniform distribution
    No: Select normal distribution
  No: Histogram has several peaks?
    Yes: Select an empirical distribution
    No: Standard deviation is larger than mean?
      Yes: Select lognormal distribution
      No: Select Erlang or lognormal distribution

Figure 13.10. Selecting the family of a distribution function (Hillen, 1993).

Step 3. Determining the parameters

When the family of the distribution has been selected, we have to select the parameters of the distribution. Banks *et al.* (2001) presents the estimators given in Table 13.5 and some additional estimators.

Table 13.5. Suggested estimators for distributions.

Distribution	Parameter(s)	Estimator(s)
Exponential	λ	$\hat{\lambda} = \frac{1}{\overline{X}}$
Normal	μ, σ^2	$\hat{\mu} = \overline{X}$ $\hat{\sigma}^2 = S^2(n)$
Lognormal	μ, σ^2	$\hat{\mu} = \overline{X}$ after taking the ln of data $\hat{\sigma}^2 = S^2(n)$ after taking the ln of data

Based on the physical characteristics and the histogram we decide that the stochastic variable for the inter-arrival time has the exponential distribution. The exponential distribution has only one parameter, previously called λ. For the exponential distribution an estimate for the parameter is

$$\hat{\lambda} = \frac{1}{\overline{X}} = \frac{1}{3.33} = 0.3 \tag{13.6}$$

The variance of an exponential distribution is

$$V(\tilde{x}) = \frac{1}{\lambda^2} \tag{13.7}$$

For the exponential distribution with $\lambda = 0.3$ the variance is $V(x) = 11.089$. In the ideal situation, $V(\tilde{x})$ is equal to $S^2(n) = 9.157$. Here, $V(\tilde{x})$ and $S^2(n)$ differ. Since the variance is less important than the mean, we accept the difference in the variance and, therefore, accept $\lambda = 0.3$. See for a graph of this distribution function Figure 13.9.

Step 4. Goodness-of-fit tests

In this section we will discuss two methods to examine how well the selected distribution represents the true underlying distribution for our data. The family of the distribution function and its parameter(s), in fact, are a hypothesis.

Goodness-of-fit: Chi-square Test

The *chi-square test* is one of the procedures for testing the hypothesis that a random sample of a random variable follows a specific distribution form. This test compares the histogram of the data to the shape of the candidate density function. We will illustrate the chi-square test with the bank terminal case presented before. The collected data is given in Table 13.3. The distribution function we will use in this case, exponential, is a continuous distribution function. For discrete distribution functions the test is similar; see Law and Kelton (2000).

The chi-square test is valid for large sample sizes; in literature one finds a minimum of 50 samples. The test procedure begins by preparing a *frequency table.* Because of that we arrange the observations into a set of k class intervals. In some chi-square tests one applies class intervals equal in width. Law and Kelton (2000) recommend using class intervals for the goodness-of-fit test that are equal in probability rather than equal in width of interval. Therefore, the endpoints of each class interval will be different and must be computed.

First we need to know the number k of class intervals. If using equal probabilities, then the probability that a value is in interval i is $p_i = 1/k$. Banks *et al.* (2001) recommend selecting $np_i \geq 5$ with n the number of observations. Substituting for p_i yields

$$\frac{n}{k} \geq 5 \tag{13.8}$$

and solving for k yields

$$k \leq \frac{n}{5} \tag{13.9}$$

Since the number of data $n = 60$, the condition for the number of intervals is $k \leq 12$. In this example we select $k = 10$. A consequence is that each interval will have probability $p = 0.1$. The endpoints for each interval, a_i; $i = 1, 2, \ldots k$ are computed from the cumulative distribution function for the exponential distribution

$$F(a_i) = 1 - e^{-\lambda a_i} \quad (13.10)$$

To be able to compute a_i we need the inverse function $F^{-1}(x)$. This function is

$$a_i = F^{-1}(x) = -\frac{1}{\lambda}\ln(1-x) \quad (13.11)$$

Since $F(a_i)$ is the cumulative area from zero to a_i, $x = F(a_i) = ip$, therefore $F(a_i) = 1 - e^{-\lambda a_i}$. $F(a_i) = ip$, so $ip = 1 - e^{-\lambda a_i}$ or $e^{-\lambda a_i} = 1 - ip$. Taking the natural logarithm of both sides and solving for a_i gives the relation

$$a_i = -\frac{1}{\lambda}\ln(1-ip), i = 0,1,\ldots,k-1 \quad (13.12)$$

With $\lambda = 0.3$ and $k = 10$ we find the endpoints of the class intervals given in the first column of Table 13.6. For the exponential distribution the first interval starts at 0. The last interval ends at ∞.

Table 13.6. Chi-square test for goodness-of-fit.

Interval	actual frequency f_i	theoretical frequency e_i	difference err_i	$\frac{err_i^2}{e_i}$
$0 \le X_i < 0.351$	2	6	-4	2.667
$0.351 \le X_i < 0.743$	8	6	2	0.667
$0.743 \le X_i < 1.188$	9	6	3	1.5
$1.188 \le X_i < 1.701$	6	6	0	0
$1.701 \le X_i < 2.308$	6	6	0	0
$2.308 \le X_i < 3.051$	5	6	-1	0.167
$3.051 \le X_i < 4.009$	5	6	-1	0.167
$4.009 \le X_i < 5.359$	5	6	-1	0.167
$5.359 \le X_i < 7.668$	8	6	2	0.667
$7.668 \le X_i < \infty$	6	6	0	0
Total	60	60		$X^2 = 6.0$

For the exponential distribution it is rather simple to compute the endpoints of the class intervals. This is because the inverse function of the cumulative distribution function $F^{-1}(x)$ is known. For other distributions this is more complicated. For more details, see (Law and Kelton, 2000) or (Banks *et al.*, 2001).

The test statistic of the chi-square test is

$$X^2 = \sum_{i=1}^{k} \frac{(f_i - e_i)^2}{e_i} \quad (13.13)$$

Here f_i is the observed frequency in the ith class interval and e_i is the expected (theoretical) frequency in that class interval. The expected frequency is $np_i = 60 \times \frac{1}{10} = 6$.

The computations for the chi-square test are given in Table 13.6. The critical value for X^2 can be found in a table for critical values for the chi-square distribution. To be able to find a value in such a table, we need to know the *degrees of freedom* and to specify a *significance level*. At the significance level $\alpha = 0.05$, there is a probability of 0.05 of rejecting a true hypothesis. Usually, α is set to 0.01 or 0.05. The degrees of freedom are given by $k - s - 1$, where k is the number of interval classes used in the test and s is the number of parameters estimated from the data. For the exponential distribution we have estimated one parameter, λ. So, in this test we have $k = 10$ and $s = 1$. Therefore the degree of freedom is 8. Taking $\alpha = 0.05$ for the significance level, the critical value is $\chi^2_{8,0.05} = 15.5$. Since $X^2(= 6.0) < \chi^2_{8,0.05}(= 15.5)$ we do not reject the hypothesis that the selected distribution fits the data well.

Table 13.7. Sorted arrival times at the bank terminal measured during a period of 200 minutes.

0.1	0.3	0.4	0.4	0.5	0.5	0.6	0.6	0.7	0.7
0.8	0.8	0.9	0.9	0.9	0.9	1.0	1.1	1.1	1.2
1.3	1.5	1.5	1.6	1.7	1.8	1.9	1.9	2.3	2.3
2.3	2.4	2.5	2.6	2.7	2.9	3.2	3.2	3.4	3.7
3.8	4.3	4.6	4.7	5.0	5.3	5.4	5.6	6.0	6.2
6.7	6.8	7.1	7.5	7.8	8.2	9.5	9.9	11.2	13.1

Goodness-of-fit: Kolmogorov-Smirnov test

The hypothesis we test here is that the distribution of the inter-arrival times is exponential. The correctness of the parameter of the distribution (in the example λ) is not part of the test. It can be shown that if the underlying distribution function of the inter-arrival times collected over the interval $(0, T)$ is exponential, the arrival times are uniformly distributed on that interval. The test statistic of the Kolmogorov-Smirnov test is the largest vertical distance between the selected distribution and the distribution of the collected data.

Table 13.8. Normalized arrival times based on Table 13.7.

0.0005	0.002	0.004	0.006	0.0085	0.011	0.014	0.017
0.0205	0.024	0.028	0.032	0.0365	0.041	0.0455	0.05
0.055	0.0605	0.066	0.072	0.0785	0.086	0.0935	0.1015
0.11	0.119	0.1285	0.138	0.1495	0.161	0.1725	0.1845
0.197	0.21	0.2235	0.238	0.254	0.27	0.287	0.3055
0.3245	0.346	0.369	0.3925	0.4175	0.444	0.471	0.499
0.529	0.56	0.5935	0.6275	0.663	0.7005	0.7395	0.7805
0.828	0.8775	0.9335	0.999				

The distribution of the arrival times can be computed as follows. The inter-arrival times are given in Table 13.7 and consist of 60 arrivals over a period of 200 minutes. We use the sorted data set for this purpose. Suppose the inter-arrival times are t_1, t_2, t_3, ..., t_{60} and the arrival times are T_1, T_2, T_3, …, T_{60}. Then the arrival times are $T_1 = t_1$, $T_2 = T_1 + t_2$, $T_3 = T_2 + t_3$, …, $T_{60} = T_{59} + t_{60}$. The next step is to normalize the computed arrival times to a [0,1) interval. The result of this operation is presented in Table 13.8.

To understand the : Kolmogorov-Smirnov test, we illustrate the test with an example of five arrivals, such that we can draw a clear picture. Assume $F(x)$ is the (cumulative) distribution function of the uniform distribution function on the interval $0 \leq x \leq 1$. Furthermore, assume R_1, R_2, …, R_N is the sample of arrival times. The (cumulative) distribution function of this sample is

$$S_N(x) = \frac{number\ of\ R_1, R_2, \ldots, R_N\ which\ are \leq x}{N} \quad (13.14)$$

$S_N(x)$ is a discrete function with jumps at each observed value, see, for example, Figure 13.11. $F(x)$ is a continuous function. The Kolmogorov-Smirnov test compares $S_N(x)$ with $F(x)$. Suppose the arrival times over a period of 1 hour ranked from smallest to largest and normalized (R_i) are 0.06, 0.24, 0.44, 0.54, and 0.74. Note that the number of data and the values have been selected for the purpose of illustration. Since we have five numbers, the probability of one on each number is $i/N = 1/5 = 0.2$. The distribution function $S_N(x)$ is depicted in Figure 13.11 and is also presented in the second row of Table 13.9. The test statistic of the Kolmogorov-Smirnov test is the largest absolute deviation between $F(x)$ and $S_N(x)$ over the range of the random variable. Because $F(x)$ is a linear increasing function and the graph of $S_N(x)$ consists of a couple of lines parallel to the x-axis with discrete steps, the maximum deviation will be at one of the steps in $S_N(x)$. The differences at these steps are

$$\frac{i}{N} - R_i,\ i \leq i \leq N \quad (13.15)$$

and

$$R_i - \frac{i-1}{N},\ 1 \leq i \leq N. \quad (13.16)$$

Table 13.9. Kolmogorov-Smirnov test.

	$i = 1$	$i = 2$	$i = 3$	$i = 4$	$i = 5$
R_i	0.06	0.24	0.44	0.54	0.74
i/N	0.20	0.40	0.60	0.80	1.00
$i/N - R_i$	0.14	0.16	0.16	0.26	0.26
$R_i - (i-1)/N$	0.06	0.04	0.04	-0.06	-0.06

See Table 13.9 for the differences in our example. Some of the differences are indicated in Figure 13.11. The difference might be positive and negative, see for a negative value the fourth and fifth sample in Table 13.9 and the fourth point in Figure 13.11. The maximum absolute value of the difference, the test statistic, is 0.26.

Now we return to the collected arrival times and test whether they are uniformly distributed. With R_i a number from Table 13.8 and N the total number of collected values, the differences with the uniform distribution follow from relations (13.15) and (13.16). These numbers have to be computed and the maximum over all these numbers will be the test statistic. For the given data the test statistic is computed to be 0.363. This number must be compared to the critical value taken from the table of Kolmogorov-Smirnov critical values, see (Banks *et al.*, 2001). Again, we select a significance level of 0.05. The critical value (with $N = 60$) then is

$$\frac{1.36}{\sqrt{N}} = \frac{1.36}{\sqrt{60}} = 0.176$$

The conclusion is that we must reject the hypothesis since $0.363 \geq 0.176$.

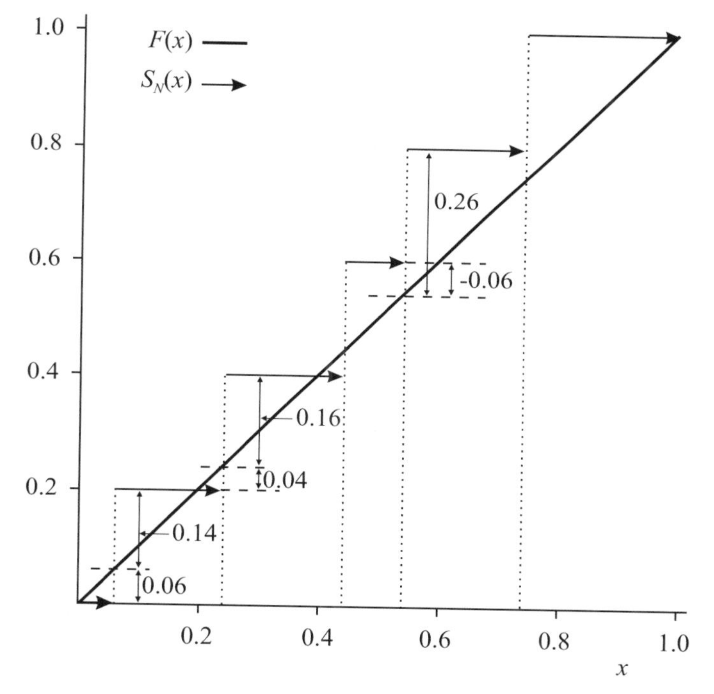

Figure 13.11. Kolmogorov-Smirnov test.

From a theoretical point of view the Kolmogorov-Smirnov test is useful when no parameters have been estimated from the data. When parameter estimates have been made, the critical values found in the Kolmogorov-Smirnov critical values table tend to be too large. In the example, we do not use the estimated parameter in the test, therefore we expect that the critical values may be used as presented in the table. Modified Kolmogorov-Smirnov tests have been developed so that they can be used in situations where parameters are estimated from the data. The Kolmogorov-Smirnov test can also be used to test for distributions other than the exponential distribution. For more information about these subjects, see (Law and Kelton, 2000) and (Banks *et al.*, 2001).

Discussion
Goodness-of-fit tests are useful in the process of selecting a distribution function. Goodness-of-fit tests, however, are just a part of the story. Frequently, several distribution functions may fit the data well. Furthermore, the sample size affects the result of the goodness-of-fit tests: if the sample size is small then many distribution functions may be accepted by the goodness-of-fit tests and if the sample size is large then no distribution function will be acceptable. In general, we may say that the chi-square test is valid for a sample of at least 50 while the Kolmogorov-Smirnov test is valid for a smaller sample size. For the chi-square test we should take into account that the selection of the class intervals affects the result of the test. Concluding we may say that a test only provides us with pieces of evidence in favour of or against the candidates.

For fitting the data of Table 13.7 to the exponential distribution, we have the following situation. The physical characteristics of the arrival process and the histogram give us evidence for the exponential distribution. Furthermore, the mean of the sample is 3.33 and we have selected λ in correspondence with this value. The chi-square test confirms our choice. The Kolmogorov-Smirnov test, however, rejects the hypothesis. Since the Kolmogorov-Smirnov test is the only negative indication, we decide to model the arrival process in accordance with our proposal. The guidelines presented in Figure 13.10 supports this choice.

If the tests show that the selected distribution is not a good approximation for the data, then the analyst selects a different family of distributions and repeats the process of computing estimates for the parameters of the distribution and testing this distribution with the tests presented in this chapter. For example, an exponential distribution function was assumed, but it was found not to fit the data. If the histogram of the collected data shows that the lack of fit was in one of the tails of the distribution function, a gamma or Weibull distribution would more adequately fit the data (Banks *et al.*, 2001). If several iterations of this procedure fail to yield a fit between an assumed distribution family and the collected data, an empirical distribution might solve the problem. Another solution may be to apply a *shifted distribution* or to split up the simulation period into several periods and apply a different distribution function for each period. An alternative option is to use the collected data as such. For example, we could use the data in Table 13.3 to model the arrival process at the bank terminal. If we do so, the simulation can only reproduce what has happened historically. As we will see in Section 13.6 we need several, 100 or even more, simulation runs to be able to study a system in a statistically reliable manner. We seldom have enough data available to make all the required simulation runs. With a theoretical distribution it is easy to generate another arrival pattern and it is possible to perform a large number of so-called independent runs. To make the simulation runs independent, it is sufficient to choose another seed for each distribution before we start a new run. The conclusion so far is that if a theoretical distribution can be found that fits the observed data reasonably well, then this will generally be preferable.

13.5.4. Software for selection of a distribution function

As illustrated above, a lot of work has to be done to fit the correct distribution function to the data. Some simulation tools have software available to support you. In Arena, for example, you can use the so-called 'Input Analyzer' for this task. You can select the distribution yourself and use the Input Analyzer to provide numerical estimates of the appropriate parameters, or you can fit a number of distribution functions to the data and select the most appropriate one. In either case, the Input Analyzer provides you with estimates for the parameter values based on the data you supply (Kelton *et al*, 1998). Enterprise Dynamics offers the possibility to fit a number of distribution functions to the data and presents them in order of the best fit to the worse. @Risk's BestFit and ExpertFit (Law and Vincent, 1999) offer the same functionality.

13.6. Experiments

As we have seen before, in a discrete-event simulation model some effects are modelled by random variables. The consequence is that the observations we collect during a simulation run are just particular realizations of random variables that may have large variances. As a result, the estimates based on a particular simulation run could differ greatly from the corresponding true characteristics for the model. At the very least, we want to know a mean and a range in which the result will most likely be. Therefore, we need to repeat a simulation run several times. One run is called a *replication*. If the results of a simulation study are to have any meaning, appropriate statistical techniques must be used to design and analyze the simulation experiments. Some of the most common statistical techniques will be discussed in this chapter. We need enough independent data to apply the classical statistical techniques. The data obtained through virtually all simulations, however, are dependent. Therefore, we may only apply the classical statistical techniques with precautions.

13.6.1. Terminating and Non-terminating simulations

Design and analysis of simulation experiments depend on the type of simulation at hand: terminating or non-terminating. A *terminating simulation* is one for which two prerequisites hold: 1) there is a natural event that specifies the end of each run, and 2) the model starts each run in the same state. A simulation of a supermarket from 8 a.m. to 9 p.m. will be terminating. Each day the supermarket starts in the empty state, which means no customers in the shop. The simulation run ends when the last customer who entered before the doors closed at 9 p.m. has been served. On the other hand consider, for example, a manufacturing company that operates 16 hours a day and work in progress at closing time will be resumed the next day. Here a natural event stops the operation. The initial state of the simulation, however, is not the same each day. Therefore, the simulation is not terminating. For a terminating simulation run the terminating event is specified before any runs are made. The time of occurrence of the terminating event for a particular run might be a random variable. This is the case in the simulation of the supermarket that cannot close exactly at 9 p.m. because it is not known in advance how long it will take to serve the last customers for that day.

In a *non-terminating simulation* we are interested in the behaviour of the system in the long run when it is operating normally. Or in other words, we want to study the system in the so-called steady state. In such simulations we have a problem deciding when to start and when to stop a simulation run. In the sequel we will discuss these issues. An example of a non-terminating simulation is the manufacturing company discussed before. Because the work in progress after 16 hours of operation will be resumed the next day, we may paste the 16-hours days together, thus ignoring the period the system is not in operation. Other examples of non-terminating simulations are simulations of communications systems such as telephone systems and hospital emergency rooms.

We may raise the question whether we should apply the adjective terminating to the system under study or to the simulation. Each day the supermarket opens in the empty state and closes at closing time. In simulation, such a system is treated as a terminating system when we study the system to answer questions about, for example, the time customers have to wait at the checkout or how many customers visit the supermarket each day. If the purpose of the simulation, however, is to study the flow of goods, this system is non-terminating as well since, for example, staff will continue to fill shelves where they stopped the previous day. Therefore, whether a simulation is considered to be terminating or not depends on both the objectives of the simulation study and the nature of the system.

13.6.2. Dependence of observations

As we have mentioned before, the observations collected in one simulation run in virtually all simulations are correlated. The next example will demonstrate this.

Example 13.5.
Suppose we study a simple queuing model. In such a model customers arrive from time to time and join a queue (or waiting line), are served, and finally leave the system. Note that the term 'customer' may refer to any type of entity that may be viewed as requesting 'service' from a system (think of production systems, hospitals, and transport and material handling systems). In this study we collect data about the length of the queue and the time a customer spent in the system. These data are not independent. We may expect the time a particular customer has to wait is long when the length of the queue is large at the moment the customer arrives. Therefore, we may expect that the time this particular customer spends in the system, i.e. the sum of the time to wait and the time to be served, to be large as well. Thus different random variables related to a single customer will be dependent. But also the waiting time of different customers are dependent. This is so because when a customer has to wait a long time then the probability is high that the next customer must also wait a long time.

To be able to generate independent data we will perform several replications and make the replications independent of each other. We accomplish independence of different runs as follows. Each random variable in the model needs a *random-number generator*. You may remember that such a generator needs a number to start (*seed*), see Section 13.5.1. You may also remember that the seed determines the values generated for the random variable. If we select another seed for a random variable, different values will be generated for that random variable. If we repeat a run without specifying other seeds for each random variable, we will just duplicate the run. The result is that no new information is collected (note that this is useful when we wish to compare different systems with the same random numbers). Runs become independent if we specify per replication different seeds for each random variable. Make sure that the simulation tool selects new seeds before a replication starts. If the simulation tool does not, select other seeds yourself. As we have mentioned before, the observations from a particular replication are not independent. The observations across runs, however, are independent if we select for each run another seed for each random variable. This independence across runs is the key to the relatively simple output-data analysis methods described in the coming sections of this chapter. In the sequel we will discuss methods for statistical analysis of terminating and non-terminating simulations.

13.6.3. Statistical analysis for terminating simulations

Example 13.6.

Consider a post office with one desk and one server. Suppose the number of people living in the neighbourhood increases such that more people visit the post office and customers have to wait longer before they are served. The people complain about this and these complaints arrive at the management of the post offices of the district. The management of the district wants to know whether it is possible to solve the problem by opening an additional desk and hiring an additional server. Moreover, the management wants to know whether it is profitable to do so.

In this situation, we will first simulate the system with one server and we want to know what the waiting time of the customers is. Since the post office opens each morning without any customers present and closes at 5 p.m., this is a terminating simulation. A simulation run may start at opening of the post office and finish at closing time. During one run,0 we collect the waiting time for each customer that arrives during that day. Because the customers do not arrive uniformly distributed over the day, it would only be possible to simulate the system during rush hours. We will postpone this idea for the moment.

At the end of each run we compute the mean of the waiting times we have collected. Since we will repeat the simulation for several runs we will use index *j* to indicate that the information comes from replication *j*. For replication *j* the mean waiting time is

$$X_j = \frac{\sum_{i=1}^{N_j} W_{i,j}}{N_j} \qquad (13.17)$$

Here $W_{i,j}$ represents the waiting time of customer i in replication j, and N_j the number of customers that visited the post office during replication j. Note that N_j will be a random variable, because the number of collected data may differ over the replications.

In this study we want to find the mean μ of the waiting time of the customers. We consider the average waiting time $\tilde{x}$ as a random variable with an unknown distribution. One simulation run provides a single sample observation X_j from the population of all possible observations on $\tilde{x}$. In fact, we would want to know the distribution function for $\tilde{x}$. However, that is in general not possible. We will be able to find a point estimate for μ and we can find an indication of how accurate this estimate is, the so-called *confidence interval*.

If we repeat the runs n times, we collect $X_1, X_2, X_3, ..., Xn$. An unbiased point estimator of the mean (μ) is

$$\overline{X}(n) = \frac{\sum_{j=1}^{n} X_j}{n} \tag{13.18}$$

And the sample variance

$$S^2(n) = \frac{\sum_{j=1}^{n} (X_j - \overline{X}(n))^2}{n-1} \tag{13.19}$$

is an unbiased estimator of the variance σ^2 of $\widetilde{x}$.

According to the well-known *Central Limit Theorem* we may assume for large n that X_j is normally distributed. We could compute the variance based on that. However, a question is how large n should be. Therefore, simulation tools use another strategy and compute the so-called confidence interval to indicate the accuracy of the estimate of the mean. The $100(1-\alpha)$ percent confidence interval for μ is given by

$$\overline{X}(n) \pm t_{n-1,1-\frac{\alpha}{2}} \sqrt{\frac{S^2(n)}{n}} \tag{13.20}$$

where $t_{n-1,1-\frac{\alpha}{2}}$ is the upper $1-\dfrac{\alpha}{2}$ critical point for the t distribution function with $n-1$ degrees of freedom. The critical points can be found in tables in statistical literature, for example Table T.1 in the Appendix of (Law and Kelton, 2000) or Table A.5 in (Banks *et al.*, 2001).

The quantity $t_{n-1,1-\frac{\alpha}{2}} \sqrt{\dfrac{S^2}{n}}$, see relation (13.20), is called the *half-length of the confidence interval.* The confidence interval is a measure of how precisely we know μ. For example, suppose we have the following 10 observations

1.30, 1.45, 1.78, 0.97, 0.67, 1.13, 0.88, 1.07, 0.58, and 1.39.

From these data we get $\overline{X}(10) = 1.12$ and $S^2(10) = 0.136$. Here we have $n-1 = 9$ degrees of freedom. The 90 percent confidence interval (so $\alpha = 0.1$ and thus $\dfrac{\alpha}{2} = 0.05$) is

$$\overline{X}(10) \pm t_{9,0.95} \sqrt{\frac{S^2(10)}{10}} = 1.12 \pm 1.83 \sqrt{\frac{0.136}{10}} = 1.12 \pm 0.21$$

From this we conclude with 90 percent confidence that μ is in the interval [0.91, 1.33].

If we increase the sample size from n to $4n$ in relation (13.20), then we expect that the confidence interval decreases by a factor 2, because there is a $\sqrt{n}$ in the denominator of the expression for the confidence-interval half-length. However, since $S^2(n)$ changes also with n, the factor is not exactly 2. An important conclusion of the so-called 'strong law of large numbers' is that if we take n sufficiently large, the $\overline{X}(n)$ will be arbitrary close to μ for almost all experiments. Therefore, in an experiment it is wise to have a large number of runs.

The procedure described above computes the accuracy of the experimental results. An interesting question is how many replications are necessary to obtain a given accuracy. Several methods are available to answer this question, see (Law and Kelton, 2000). A pragmatic approach is to perform experiments with n, $2n$, ... replications and stop when the required precision has been reached or does not change anymore. It is also possible to watch the so-called *running mean* of the performance measure over these runs. Increasing the number of replications does not make sense when the running mean no longer changes.

Initial conditions

Assume we want to study the post office during the busiest period, say from 12 noon to 1 p.m.. In reality, the state of the post office at noon differs from day to day. In fact, this is not a terminating simulation, since the initial state is different in every run. The easiest way to handle this is to start the simulation in the empty state (in the empty state there are no customers in the post office). What will happen if we start our simulation at noon without any customers present? At that time the post office will be quite congested, so the estimate for the mean of the waiting time of the customers will be too low. We may solve this problem in the following way. We start the simulation at 9 a.m. in the empty state and run it until 1 p.m.. To compute the performance measures, such as average waiting time, we only include the data collected from noon to 1 p.m.. The purpose of the simulation between 9 a.m. and noon is to make sure that the model is in the correct state at noon. The period between 9 a.m. and noon is called *warm-up period.* To save some computer time we may take a shorter warm-up period. For example, we might start the simulation at 11 a.m. in the empty state. If we do so, however, there is no guarantee that the state of the model at noon will be representative of the actual conditions in the post office at noon.

Law and Kelton (2000) suggest two other approaches. One is to initialize the model in a state representative for the situation in the post office at noon. This is rather laborious and, depending on the simulation tool you use, impossible. The other one is to select the initial state from a distribution function.

Finishing the study

The purpose in Example 13.6 is to determine whether appointing a second server for the post office solves the problem of long waiting times and is economically feasible. Until now we have suggested performing a number of replications of the situation with one server. To complete the study, it is necessary to perform a number of replications with a model of the post office with two servers. The results of these two studies may be presented to the management of the district and this management may make a decision based on the results.

13.6.4. Statistical analysis for non-terminating simulations

For non-terminating systems we are interested in how the system operates in the long run. In other words we are interested in the so-called *steady state*. To compute a mean and a confidence interval, we may follow the same strategy as for terminating systems and make several replications. The main problem with this strategy is that the computed mean will be biased by the initial conditions. This is so because the initial transient is not representative for the steady state. Suppose you study a hospital in the steady state. If the simulation starts in the empty state, so no patients occupy a bed, the transient period is relative long. In the transient period we may expect, for example, that many patients, possibly the majority of the patients, will be discharged at the same time, which is not realistic.

One method is to initialize the model in a state that is representative for the simulation. This approach was discussed in the previous section. In general, it is laborious to initialize the model in such a state. Suppose you study a hospital in the steady state. To initialize the hospital, for each patient, for example, you have to specify the date of discharge from the hospital. If there are 500 patients, this is a rather laborious task.

Most simulation tools do not support this idea of so-called *intelligent initialization*, and it is difficult to find a state representative for the long run. In this section we will discuss how to solve the problem of the initial transient in a different way. Furthermore, there are several approaches for computing the mean, of which we will discuss the replication/deletion approach and batching. We refer to (Law and Kelton, 2000) for a description of other approaches. As mentioned before, if we start a simulation with the model in a state not representative for the steady state, the computed mean will be biased by the initial transient. One solution is to follow the strategy of the *warm-up period* we discussed in Section 13.6.3 (Initial conditions). In that strategy we run the model from T_0 to T_E and forget data collected during the period T_0 to T_S. T_S is the time at which the model is in the steady state. It is difficult to find T_S because of the variability of the observations. Welch developed a graphical procedure to determine T_S. For a discussion of this method, see (Law and Kelton, 2000).

Replication/deletion approach for the mean

The *Replication/Deletion approach* is similar to the procedure for terminating simulations we discussed in Section 13.6.3. The difference is that now only those observations beyond the warm-up period in each replication are used to compute the estimates. Therefore, we need to make some replications to determine the warm-up period. Assume there are l observations in the warm-up period. In general, one specifies the time needed to warm up the model, so l will vary from run to run. Then we make n replications of the simulation. Assume in replication j we have m observations. In general, m is much larger than the number of observations in the warm-up period l and like l, m varies from run to run. Let $Y_{j,i}$ be the observations of replication j. Then the mean of the replication j is

$$X_j = \frac{\sum_{i=l+1}^{m} Y_{j,i}}{m-l}, \; for \; j = 1, 2, \ldots, n \tag{13.21}$$

Then we may estimate the mean (μ) by $\overline{X}(n)$ in (13.18) and construct an approximate $100(1-\alpha)$ percent confidence interval via (13.20), where $S^2(n)$ is computed by (13.19).

This approach gives reasonable good statistical performance if the warm-up period is long enough to bring the model into the steady state. Increasing the number of replications will generate a smaller estimate of the confidence interval. If the warm-up period is not long enough (a so-called *initialization bias* remains), however, the confidence interval will be around the wrong point (Banks *et al.*, 2001; 430). It is impossible to correct the initialization bias by increasing the number of replications. The bias may be corrected by deleting more data, i.e. increasing the warm-up period, or extending the length of each run. The best approach, however, is to do a thorough job of investigating the required warm-up period.

One objection that might be raised to the replication/deletion approach is that it uses one set of replications to determine the warm-up period, and then uses only part of the observations from a different set of replications to perform the actual analysis. In the past this was a problem, however, nowadays this is seldom a problem due to the relatively low cost of computer time.

Batch means approach

The main objection to the replication/deletion approach is that for every run it has to go through the transient period. In this respect, the batch means approach saves some computer time. The *batch means approach*, also known as subruns approach, is based on one long run, so it only has to go through the warm-up period once. The approach works as follows.

Suppose we make one long simulation run with the observations Y_1, Y_2, ..., Y_m. In general, we will delete the first l observations measured during the warm-up period. The other $m-l$ observations will be divided in n batches of length k. So $m-l=nk$. Thus batch 1 contains the observations $Y_{l+1}, Y_{l+2}, \ldots, Y_{l+k}$, batch 2 contains the observations $Y_{l+k+1}, Y_{l+k+2}, \ldots, Y_{l+2k}$, and so on. The mean of replication j is

$$X_j = \frac{\sum_{i=1}^{k} Y_{l+(j-1)k+i}}{k}. \qquad (3.22)$$

The estimates for the mean, variance, and confidence interval may be computed with (13.18), (13.19), and (13.20) respectively. Independence of the observations is required for application of the estimates for the mean, variance, and confidence interval. The observations of succeeding batches, however, are not independent. In fact the number of products waiting for processing at the end of the previous batch is equal to the number of products at the beginning of the succeeding batch. Since computers nowadays are relatively cheap and fast, and observations of succeeding batches are not independent, this approach is hardly ever used these days. The only reason for mentioning this approach is that most of the simulation tools provide for this approach. Our advice is not to use this approach in your experiments but to rely on the replication/deletion approach because it generates reliable results at an acceptable price.

13.6.5. Antithetic variates

Stochastic simulations require repetition of measurements to be able to attain an acceptable accuracy of results (computed by the confidence interval). A replication of a large and complex model may require so much effort that it requires too much time and money to attain an acceptable accuracy. In such a situation a so-called *variance-reduction technique* may solve the problem. The *antithetic variates technique* is one of the variance-reduction techniques applicable in simulation studies. In the antithetic variates technique we make pairs of runs with the model. We organize the pair of runs such that one run produces a small observation, for example service time, and the other run produces a large observation. The average of the two observations in the pair will be used as a basic data point for the analysis. So, we make two runs to produce one observation. This seems to be a waste; however, the average is closer to the common expectation μ of the observations. The result is that a smaller confidence interval may be computed based on a relatively smaller number of runs. Most simulation tools, like ED and Arena, apply antithetic variates on request. Antithetic variates are generated as follows. In Section 13.5.1 we have seen that random variates are generated based on numbers R_i in the interval [0, 1). Suppose, for each random variable in the model several R_i's will be generated in the first run of the pair. During the second run of the pair, the variate will be computed based on $1 - R_i$ which is the complement of R_i. That means that when R_i is large, $1 - R_i$ will be small and the reverse. The generation of the random numbers in both runs need to be synchronized. For more information about variance-reduction techniques in general and antithetic variates see the references mentioned in the next section.

13.6.6. Some concluding remarks and pitfalls

The discussion of statistical techniques has not been extensive. We have limited ourselves to some important techniques. More advanced topics are worth studying. For that purpose we have presented pointers to literature. Issues we have not discussed are, among others, the design of experiments where many factors may differ and other variance-reduction techniques, such as the common random numbers technique (these are applicable when different designs are compared), control variates and indirect estimation. Once again, the books of (Law and Kelton, 2000) and (Banks *et al.*, 2001) are valuable sources for studying these advanced issues. The remainder of this section aims to present some instructive examples collected from various literature. The collection shows some pitfalls in designing experiments.

Example 13.7.

This example is taken from (Law and Kelton, 2000). It shows that when comparing two or more systems by some sort of mean system response, misleading conclusions may occur. Consider a bank with several tellers. In this example we compare the policy of having one queue for each teller (with jockeying, i.e. customers may move from one queue to another queue) on the one hand, and the policy of having one queue feeding all tellers on the other hand. Table 13.10 gives the results of making one simulation run for each policy. These simulation runs were performed so that the time of arrival of the *i*th customer was identical for both policies and so that the service time of the *i*th customer to begin service was the same for both policies. In principle this is a wise strategy to compare two or more policies.

Table 13.10. Simulation results for two bank policies: estimated means (Law and Kelton, 2000).

Measure of performance	Five queues	One queue
Expected operating time, hours	8.14	8.14
Expected average delay, minutes	5.57	5.57
Expected average number in queue	5.52	5.52

Based on the average system response, see Table 13.10, we may conclude that the two policies are equivalent. However, are the systems comparable? Since customers need not be served in the order of their arrival with the multi-queue policy, we would expect this policy to result in greater variability of a customer's delay. If we measure the delays of the customers, the systems are not equivalent at all.

Table 13.11. Simulation results for two bank policies: estimates of expected proportions of delays in interval (Law and Kelton, 2000).

Interval (minutes)	Five queues	One queue
[0, 5)	0.626	0.597
[5, 10)	0.182	0.188
[10, 15)	0.076	0.107
[15, 20)	0.047	0.095
[20, 25)	0.031	0.013
[25, 30)	0.020	0
[30, 35)	0.015	0
[35, 40)	0.003	0
[40, 45)	0	0

Consider Table 13.11 that presents a proportion of customers with a delay in various intervals for both strategies. These data have been computed from the same two simulation runs used to generate the data in Table 13.10, so the averages are the same. Observe from Table 13.11 that a customer is more likely to have a large delay with the multi-queue policy than with the single-queue policy which is in accordance with our expectations. For example, if 480 customers arrive in a day, then 33 customers of them are expected to have delays greater than or equal to 20 minutes for the five-queue policy ($480 \times (0.031 + 0.020 + 0.015 + 0.03) = 480 \times 0.069 = 33.12$.). For the one-queue policy this will be 6 customers ($0.013 \times 480 = 6.24$).

Example 13.8.

The second example is taken from (Kleijnen, 1987). A manager at a bank is interested in the waiting times of clients. In the simulation of this queuing problem, the analyst studied the average waiting time in the long run or steady state. The steady state was reached after approximately 36 hours. The analyst overlooked the fact that the bank is open only from 9 a.m. to 4 p.m., that is 7 hours a day. The steady state will never be reached in this case. The analysis of peak hours would have been more appropriate.

13.7. Verification and validation

It is clear that experimenting with an incorrect model results in incorrect conclusions. Therefore, it is important to make as sure as possible that the model behaves closely enough to be used as a substitute for the actual system. Two words play a central role in model validation: verification and validation:

- *Verification* is determining that a simulation computer program performs as intended, i.e. debugging the computer program. Verification checks the translation of the conceptual simulation model (the process models and assumptions) into a correctly working program.
- *Validation* is concerned with determining whether the conceptual model on which the simulation model is based (see Section 13.4.1) is an accurate representation of the system under study.

Since end users in general are sceptical about models, it is wise to involve the end users in the validation process. In Section 13.4 a number of steps in a simulation study have been listed. The list suggests that the steps will be performed in the sequential order mentioned. This will seldom be true. In fact, modelling is an iterative process including, the following steps:

- Formulation of the model.
- Comparison of the behaviour of the model to the actual system behaviour.
- Studying of the discrepancies between the behaviour of the model and the actual system.
- Using the insights gained to improve the model.

This process is repeated until model accuracy is judged to be acceptable. In general, however, we will not be able to develop a model that is accurate under all circumstances. Therefore, we validate the model under conditions representative for the study. In the remainder of this chapter we will discuss verification and validation of models.

13.7.1. Verification

The *verification* process is to assure that the conceptual model has been transformed accurately into the simulation program. For the implementation of a model, one may use a general-purpose language, for example C++ or Java, or a simulation tool, like Enterprise Dynamics or Arena. The effort required to implement a conceptual model, in general, is considerable larger in a general-purpose language than in a simulation tool. Moreover, implementing a model in a general-purpose language is more error prone. That means, verification requires more effort if the model is implemented in a general-purpose language. A model represented in a simulation tool is self-documenting. In contrast, one must carefully document the program in a general-purpose language. We will now limit the discussion to points that are important if the model is implemented in a simulation tool. To verify a model one may do, among other things, the following:

- Make sure that the important elements of the system are present in the model. Here it is important that the elements of the system are clearly represented in the model, such that it can be checked by someone other than the developer. Preferably the model will be checked by people that know the real-world system and the end users well.

- Closely examine the model output for reasonableness under a variety of settings of the input parameters. For example, examine output statistics such as utilization of servers and time-average number of customers in various subsystems. If the utilization of a server is unreasonably low or high this may indicate that there is an error in the model logic or the mean service time is wrong.
- Current simulation tools are able to animate the model. Carefully watch the flow of (temporary) elements like customers and see whether these elements disappear (unintentionally) during the simulation. Compare the flow with the flow of elements in the real-world system.
- For queuing models, it is possible to compute certain long-run measures of performance analytically. Compare such measures of performance with the measures of performance of the model. If the queuing model is too complex to compute the measures of performance analytically, often it is valuable to use measures of performance of a somewhat simplified system.
- Modern simulation tools have options available to observe the state of the simulation model as it changes over time. Through this it is possible to watch the changes in state and values of variables over a simulation period.
- Another approach is to change all the stochastic variables in your model to constant values. Then the results of the model are predictable and you can more easily check the results of your model. In a complex model it might be necessary to apply some additional simplifications in your model such as releasing a limited number of products or types of products in a production process.

13.7.2. Validation

Validation is the process of checking whether the behaviour of the model and the real-world system are similar. In literature, verification and validation are considered distinct phases in the modelling process. In practice, however, they are often conducted simultaneously by the modeller. Some of the points mentioned in the previous section about verification are valuable in the validation phase as well.

Models should be based on solid knowledge of the underlying disciplines. Models, however, are frequently based on assumptions and often apply simplifications. The results of the model apply only if those assumptions and simplifications hold. Therefore, validation of the assumptions and simplifications is important. Often, assumptions and simplifications hold only under certain limited experimental conditions. Therefore, it is important to have documentation of the so-called *experimental frame* (Zeigler *et al.*, 2000) stating under which circumstances the model is valid. In other words, no model is ever 100 percent representative of the system under study under every experimental condition.

Law and Kelton (2000) present two basic thoughts in model validation that we adhered to. First, it is extremely important for the modeller to interact with the problem owner(s) on a regular basis throughout the course of the simulation study. The model is more credible when the manager understands and accepts the model's assumptions. Another important idea for validity/credibility enhancement is for the modellers to perform a structured walk-through of the conceptual model (prior to the beginning of coding) before an audience of all key people.

This meeting helps ensure that the model's assumptions are correct, complete and consistent (i.e. that 'local' information obtained from difference people is not contradictory). Law and Kelton specified these thoughts in a three-step approach for developing valid and credible simulation models based on (Naylor and Finger, 1967):

1. Develop a model with high face validity, i.e. a model that, on the surface, seems reasonable to people who are knowledgeable about the system under study. Use conversations with system 'experts' in multiple layers of the organisation, observations of the system and collections of empirical data, existing theory, relevant results from similar simulation models and your own experience/intuition.
2. Test the assumptions of the model by sensitivity analysis.
3. Determine how representative the simulation output data are.

A model, in fact, is a hypothesis and, therefore, it is impossible to prove that the model is a correct representation of the real-world system. One of the major techniques applied in validation is comparing the results of a simulation run with the responses of the real-world system under similar conditions. If these results are close enough, however, this is not a guarantee that the model will be a correct representation of the real-world system under other circumstances. The problem now is that we, in general, will apply the model under conditions where we do not know the behaviour of the real-world system. If we knew the behaviour of the real-world system under such conditions, there would be no reason to perform a simulation study. Despite this problem, there are some techniques which may be applied to increase confidence in the developed model. These techniques will be discussed in the sequel.

Comparing with the real-world system

The ultimate test is to see whether the behaviour of the model and the real-world system match when they experience similar circumstances. We may compare the two systems in two ways:

1. use comparable input for the model and the real-world system, or
2. use historical data.

Both situations will be discussed in the sequel.

Before we do so an important remark. When comparing model behaviour with behaviour of the real-world system, it is wise to make sure that the available real-world data represent the real-world system. Kelton *et al.* (1998) report about their experience with the validation of a model of a production line. In their study data is used about the real-world system stored in a database. The differences between the model and real-world system behaviour were too large to accept the model as valid. A closer look at the system showed that the differences were caused by the data stored in the database which in no way represented the real-world processing times.

Using comparable input

In this test one compares performance measures of the real-world system with performance measures collected with the model one wants to validate. If the decisions to be made with a simulation model are of particularly great importance, field tests can be used to obtain system output data from a version of the proposed system (or a subsystem) for validation purposes. According to (Law and Kelton, 2000) the most definitive test of a simulation model's validity is establishing that its output data closely resemble the output data that would be expected from the actual (proposed) system. If the two sets of data compare 'favourably', then the model is considered 'valid'. Unfortunately, how one defines correspondence and determines if it is sufficient is not universally agreed upon (Law and Kelton, 2000).

Figure 13.12 presents an approach to model validation. After the model is developed we observe the system for a period of time, collecting data for all exogenous variables and performance measures. The exogenous variables are then used as model inputs, which yield performance measures from the model. A decision on model validity is based on the degree to which the performance measures produced by the model and those observed in the system are similar.

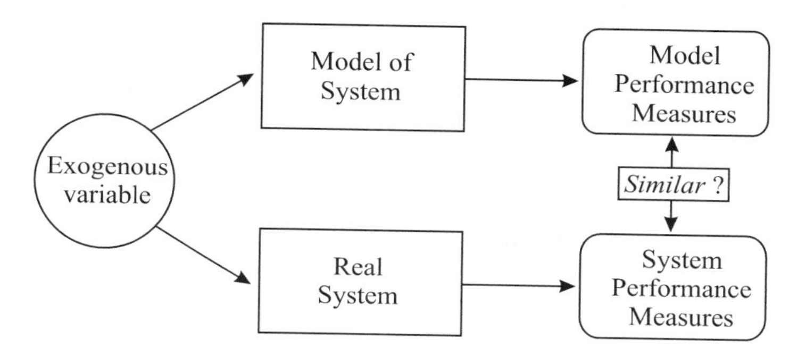

Figure 13.12. An approach to model validation (Hoover and Perry, 1989).

Example 13.9.

Assume one has observations about service time and waiting time of customers in a service oriented system during a period of three hours. Based on these data one computes the mean service time and the mean waiting time. To be able to compare the performance measures of the real-world system and the model, one performs an experiment consisting of several independent replications with the model. The run length of every replication will be three hours (the same period as the real-world system). One should make sure that the state of the model and the real-world system at the beginning of the observation period are identical or, if that is impossible, at least comparable. Furthermore, for the arrival pattern of customers one uses a distribution function that is representative for the real-world system during the three-hour period. In the experiment one computes the same performance measures, so in this example one computes the mean service time and waiting time and the confidence intervals of the two. If the model is valid, one expects that these performance measures are comparable with the performance measures of the real-world system. Because of the variability of both the real-world system and the model, however, the results will not be exactly the same. Therefore, it is necessary to apply an appropriate statistical test. According to (Banks *et al.*, 2001), the appropriate statistical test here is the t-test.

Using historical data

In the technique described in the previous section we used data generated by random-number generators for the inter-arrival time and service time. If we do so, we may expect the model to produce event patterns that are comparable with, but not identical to, event patterns that occurred in the real-world system during the period of data collection. A closer duplication is possible if the model experiences the same input as the real-world system. In this example system this means that we use the inter-arrival times of clients in the real-world system during the period of data collection. Moreover, the service times collected during this period will be used in the model as well. In such a simulation, no random numbers will be generated for the arrival and service times. If the model does not contain other random variables then the model is deterministic. In such a model we may expect that the performance measures of the model and the real-world system are comparable and the times at which certain critical events occur are similar.

In general, we do not have many historical data sets available. If we have only one historical data set available, it is advisable to check whether the critical events occur at the same times. This is because there is no statistical test available based on one data set. If we have several historical input data sets available an objective statistical test is available, see Banks *et al.* (2001). In this situation, however, it is a good idea to check the times critical events occur in one or more of the runs anyway.

Turing test

A Turing test may also be used to compare the output data from the model to those from the real-world system. Turing (1950) developed this test to validate artificial intelligence computer programs. The test runs as follows. One presents a mixture of output of simulation runs and real-world system output to one or more people, managers or engineers, knowledgeable about the behaviour of the system. The output should all be in exactly the same format and should contain information of the type the managers and engineers have previously seen on the real-world system. One challenges them to identify the data that was generated by the simulation. Of course, they may correctly identify some of the data by mere chance; this, however, can be tested statistically. When a substantial number of reports based on experiments with the model are identified as such by someone, this is evidence that the model is not valid. Then the model builder questions such a person and uses the information gained to improve the model. If no one can distinguish between reports based on simulation and the real-world system, the modeller may conclude that this test provides no evidence of model inadequacy.

What to do if the results are not comparable?

If the statistical tests discussed in the two previous sections reject the hypothesis that the results of the model are comparable to the real-world measurements, the modeller should investigate the real-world system and the implicit or explicit assumptions applied in the model. Then adapt the model and repeat the verification and validation.

It is usually very hard to perform a statistical validation between model output data and the corresponding system output data (if it exists), due to the nature of these data. The output processes of almost all real-world systems and simulations are non-stationary (the distributions of the successive observations change over time) and auto-correlated (the observations in the process are correlated with each other). Law and Kelton (2000) believe that it is most useful to ask whether or not the differences between the system and the model are considerable enough to affect any conclusions derived from the model.

13.7.3. Sensitivity analysis

In general, a so-called *sensitivity analysis* may be used to check a model's validity. In a sensitivity analysis, one changes a parameter or input variable of the model and observes whether the model operates in the expected way. Furthermore, if a moderate change in a parameter of input variable has a large impact on the system, it is wise to invest time and money in determining the correct value or the correct distribution function. If there are many parameters or input variables in the model, one should ideally execute all possible sensitivity tests. That is, change a specific parameter or input variable and leave the others constant, and change combinations of two or more parameters or input variables and evaluate the consequences. If that is not feasible, the modeller must attempt to choose the most critical input variables or parameters. People knowledgeable about the real-world system may be of help and may be a source for the expected changes in the behaviour of the system. If real-world system data are available for at least two settings of the input variables or parameters, objective scientific sensitivity tests may be conducted. See (Law and Kelton, 2000) for more information about these tests.

Example 13.10.

A simple example of a sensitivity test is the following. Suppose we are simulating the checkouts in a supermarket. The arrival rate of customers at the checkout is such that the customers form a waiting line. Now we reduce the number of open checkouts, for example, by one and run the simulation again. When the arrival of customers and other parameters in the model are not changed you expect that in this run the time customers have to wait will increase. In fact, a decrease in the mean of the waiting times indicates that the model is not correct. If the modeller wants to know what the effect is of a change in a parameter or input variable, it is not normally necessary to conduct many replications with the model. Often, two runs will be sufficient: one run for the original value of the parameter or input variable and one run for the changed value. To be sufficient it is necessary 1) to use the same seed for each random variable in the two runs, and 2) to start the model in both runs in the same initial state.

13.7.4. Animation

A model designed with a simulation tool like ED or Arena contains permanent and temporary elements. In 2D-animation, elements of the system are represented on the screen by icons. On the screen, icons of temporary elements move from one permanent element (building blocks) to another. For example, in a supermarket simulation an icon representing a customer will change position in the shop and will join the queue at the checkout. On the screen, the state of the cashier (a permanent element) may be indicated to be busy or idle. Furthermore, the number of clients waiting for service by the checkout is indicated. Moreover, present day simulation tools provide for 3D-animation where a realistic three-dimensional movie will be shown.

For discrete-event simulation models, animation is often used to partly document the model. To the domain expert, animation shows the operation of the model in a way similar to reality. Therefore, animation is a good technique to show the validity of a model to domain experts who are non-simulation experts. An animation may be an effective way to find invalid model assumptions and to enhance the credibility of a simulation model. According to (Kleijnen, 2001), however, animation may be misleading, since it uses very short runs such that the events that are responsible for the incorrect behaviour of the system may not occur during the animation, and animation may hide critical details.

13.7.5. No real-world system

In the validation methods described before, the modeller in one way or another compares the real-world system with the simulation model. The question now is what the modeller should do when there is no real-world system. This occurs, for example, when the purpose of the model is to develop a completely new system. In such a case one should draw conclusions with particular caution. Perhaps parts of the system exist and one might validate these parts. Sensitivity analysis and animation may be helpful in increasing the credibility of such models.

13.8. Conclusion

This chapter presented a basic introduction to discrete-event simulation. We defined simulation as the process of designing a model of a system and conducting experiments with the model for the purpose either of understanding the behaviour of the system or of evaluating various strategies for the operation of the system. We gave definitions of (types of) models and discussed the main steps in a simulation study. Specific attention is given to data collection and the fitting of data to a distribution function, because random variables play an important role in discrete-event simulation studies. Next, attention is given to appropriate statistical techniques for designing and analyzing simulation experiments. Simulation models have to be run several times (replicated) since the use of random variables results in different outcomes for a particular simulation run. We ended this chapter with a discussion on model verification and validation. It is clear that experimenting with an incorrect model results in incorrect conclusions. Therefore, it is important to make as sure as possible that the model behaves closely enough to the actual system to be used as a substitute for that system.

Exercises

Chapter 1

1.1. Farmers problem

Consider a farmer who cultivates grain, potatoes and sugar beets on 10 hectares of land. During the winter he wants to decide how much land to devote to each crop. The following data are available:

	Labour requirement in hours ha^{-1}						
crop	May	June	July	Aug	Sept	Oct	Profit €/ha
grain	2			45			4500
potatoes	24	17	5		198	120	5000
sugar beets	158	98	11				6000
Available Labour (hours)	350	300	250	250	400	400	

Some additional constraints are:

- grain can be grown on at most $\frac{1}{2}$ of the total area
- potatoes can be grown on at most $\frac{1}{3}$ of the total area
- sugar beets can be grown on at most $\frac{1}{4}$ of the total area.

a. The farmer wants to maximise the profit for the entire farm. Formulate the problem as a linear programming problem.

b. Suppose the farmer uses for each hectare of grain, potatoes and sugar beets 200, 600 and 500 kg fertilizer respectively. The agricultural information service pleads for lower levels of (artificial) fertilizer. A substantial loss of nitrogen and pollution of the ground water would be inevitable. Now, the farmer wants to calculate a new plan in which the total amount of fertilizer is at a minimum level. However, the yearly profit should be at least €36,000. Reformulate the LP model.

1.2. Assignment problem

A coach wants to make a team for the 4 × 100 medley race. Five swimmers are available. Their personal records (in seconds) are:

	Swimmers				
Type of stroke	A	B	C	D	E
Back stroke	58	55	56	59	60
Breast stroke	67	64	68	65	63
Butterfly	56	54	58	57	56
Freestyle	50	49	51	51	52

The coach wants to assign a swimmer to each type of stroke in such a way that the total time for the medley race is minimised. Note that a swimmer cannot swim more than one type of stroke. Formulate an (integer) linear programming model for this problem.

1.3. Skating problem

The coach of the Dutch skating team struggles with the strategy for the pursuit races during the Olympic Games. Drawing up the optimal team for each race turns out to be quite complicated.

The main characteristics of the pursuit race are:

- For each race three skaters are selected. The three selected skaters are a team and skate (closely) together as a group.
- The leading skater of the group determines the speed of skating i.e. the time needed for each lap (round).
- A pursuit race consists of 8 laps (rounds) for the complete group.

For each race the coach has to choose 3 skaters out of 5 (i.e. skater A, B, C, D and E). It is determined that in every race two skaters will take the lead for three complete laps and the third skater will lead for only two complete laps. Taking the lead takes a lot of energy. The next table shows for all skaters the time needed for each lap in case the skater will take the lead for two or three laps.

	No. of rounds to lead	
Skater	2	3
A	27	28
B	27	27.5
C	27.5	29
D	28	28
E	28	28.5

Example: in case skater A will take the lead for two laps the time for each lap will be 27 seconds. If skater A takes the lead for three laps the racing time will be 28 seconds for each lap.

The coach is convinced that his team will pass without difficulties the races preliminarily to the last two races i.e. the semi final and the final. The decision problem will only focus on the semi final and the final race. The coach knows that the total time of 225 seconds for 8 laps in the semi final race is sufficient to reach the final race. In order to avoid exhaustion none of the selected skaters may take the lead for more than 5 laps in the semi- and final race together. The coach wants to minimise the total time needed in the final race.

The coach wants to know which three skaters he has to select for the semi final and the final race. Moreover, he wants to know which of the selected skaters for the races have to take the lead for two or three laps. Formulate the problem as an (integer) LP-problem.

1.4. Production planning

Suppose the following data are known at the production department of a brewery: Demand for a beer is 10 hectolitre in the first month. The regular production capacity is 25 hectolitres and the production costs for each hectolitre are € 110 in the first month. Management may decide to produce in overtime. The additional costs for production in overtime are 10% of the regular costs. During the first month the brewery can produce as much as the demand of the first month. However, they can also decide to produce more in order to cope with future demand. For each hectolitre in stock the inventory holding costs are € 5 monthly.

Month	Demand	Production capacity	Production costs	Additional costs (overtime)	Inventory holding costs
1	10	25	110	10%	5
2	15	20	120	10%	5
3	35	30	115	10%	5
4	20	15	130	10%	

Management of the brewery wants to create an optimal production plan for the next four months. The total costs (regular production costs, inventory holding costs and additional costs for production in overtime) should be at a minimum level.
Formulate an LP model for this problem.

1.5. A set-covering problem

The municipality wants to create a number of fire stations in the city such that all districts can be reached within five minutes. It is possible to establish a fire station in every district. However, this option will be rather expensive. The table gives the travelling times in minutes between all districts (note the non-symmetrical structure).

From \ To → ↓	1	2	3	4	5	6
1	0	7	3	6	9	6
2	6	0	6	7	6	8
3	3	6	0	4	6	7
4	7	4	4	0	6	6
5	9	4	7	6	0	6
6	6	8	7	6	4	0

The question is: what is the minimum number of fire stations and where should they be located such that every district can be reached within 5 minutes. Formulate a (binary) LP-model for this problem.

1.6. A simple example in food supply chains

Management of a large supermarket chain wants to decrease the transportation costs for the distribution chain of potatoes within the company. In the current situation all local stores buy their potatoes directly at a single (agricultural) cooperative centre

(AC) in the north of the country. According to the management the total transportation costs from the cooperative centre to the (individual) local stores can be reduced by building one or more distribution centres (DC) which will be scattered all over the country. Moreover they want to order the potatoes at more then one cooperative centre. Management has decided that the local stores are no longer allowed to buy their potatoes directly at the cooperative centres. Fixed costs for building the distribution centres are disregarded.

Suppose the new structure is based on the following data (2 cooperative centres, 2 distribution centres and 2 local stores):

	Demand of local store	Distribution capacity distribution centre	Supply capacity cooperative centre
1	10	40	15
2	30	30	30

The old and new distribution structure can be shown schematically (transportation costs per unit are printed next to the arrows):

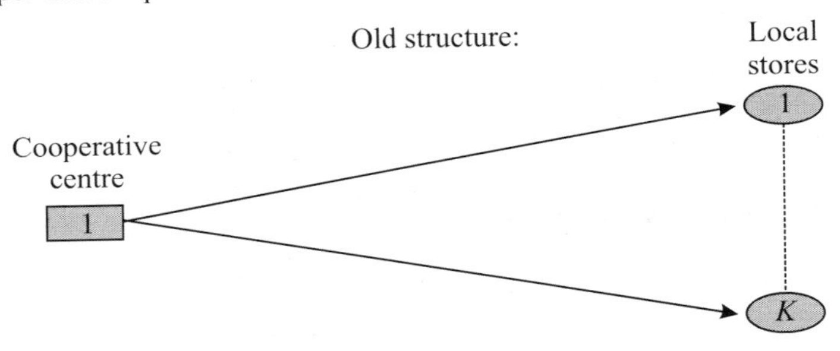

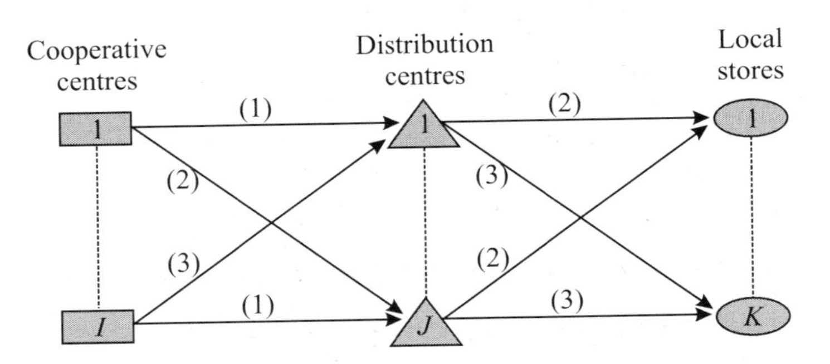

a. The management wants to meet the demand levels in the local stores such that the total transportation costs will be at a minimum level. The available capacity in the distribution centres should be taken into account. Moreover, the supply levels of the cooperative centres are restricted too. Formulate an LP-model for the problem.

b. In a more realistic case the potential number of cooperative centres, distribution centres and (most of all) the number of local stores will be much larger than $I=J=K=2$. Suppose I denotes the potential number of cooperative centres ($i = 1...I$), J denotes the potential number of distribution centres ($j = 1...J$) and K represents the total number of local stores ($k = 1...K$).

Moreover:

s_i	~	the supply of potatoes from cooperative centre i	(10^3 kg)
ac_j	~	the available capacity of distribution centre j	(10^3 kg)
d_k	~	the demand for potatoes at local store k	(10^3 kg)
p_{ij}	~	transportation costs per unit of weight (tons) from cooperative centre i to distribution centre j.	
q_{jk}	~	transportation costs for one unit of weight (tons) from distribution centre j to local store k.	

Formulate a general LP-model for this problem. How many decision variables and how many constraints has the problem in case $I = 20$, $J = 10$ en $K = 100$?

1.7. A blending problem

The production facility of a company produces several final products (concentrates) on a weekly basis. Concentrates are composed of different raw materials. The inventory or stock levels of the available raw materials are restricted to certain quantities. Suppose the company takes in the cargo of raw materials only once a week. During a week the stock levels of all different raw materials together is sufficient for the weekly production of concentrates.
The amount of individual raw materials in a concentrate is restricted to a minimum and/or maximum level. Moreover, the percentages of starch and protein in each concentrate should satisfy certain levels. Different raw materials are denoted by $i = 1,\ldots,I$ and concentrates (or final products) are denoted by $j = 1,\ldots,J$. The following data are given:

pp_i	~	percentage of protein in raw material i	(%)
ps_i	~	percentage of starch in raw material i	(%)
c_i	~	(prime) costs of raw material i	(€ / ton)
sl_i	~	stock level of raw material i	(ton)
d_j	~	(weekly) demand for concentrate j	(ton)
p_j	~	the required percentage of protein in concentrate j	(%)
s_j	~	the required percentage of starch in concentrate j	(%)
min_{ij}	~	the minimum quantity of raw material i in concentrate j	(ton)
max_{ij}	~	the maximum quantity of raw material i in concentrate j	(ton)

Formulate a (general) LP-model in which the total (prime) costs of all concentrates together are at a minimum level.

Chapter 2

2.1. A company manufactures two brands of cereals: Brunchy and Cruesli. Three raw materials are needed: bran, nuts and raisins (see table).

Raw material	kg raw material for one box of		Available quantity in kg
	Brunchy	Cruesli	
bran	1	2	1500
nuts	4	1	2000
raisins	1	1	1000
Profit per box	€ 60	€ 10	

The company wants to maximise the profit. Formulate an LP problem and determine the optimal solution graphically.

2.2. Given the inequalities (restrictions) $x_1 \geq 0$ and $x_2 \geq 0$ and:

(1) $-x_1 + 2x_2 \geq -1$
(2) $3x_1 + 3x_2 \geq 3$
(3) $-x_1 + x_2 \leq 2$
(4) $x_1 + x_2 \leq 4$
(5) $2x_1 + 3x_2 \geq 12$

a. Draw in a figure in $\mathbb{R}^2$ the feasible regions:
V_1: formed by (1), (2) and (3)
V_2: formed by (1), (2), (3) and (4)
V_3: formed by (1), (2), (3), (4) and (5).

b. Which of these regions is unbounded and which empty?

c. Maximise the following objective functions on V_1, V_2 and V_3.
$w_1 = -x_1 + \frac{1}{10}x_2$
$w_2 = -x_1 + x_2 + 100$
$w_3 = -x_1 + 3x_2$

d. On V_1 the function $w = -x_1 + \alpha x_2$ is maximised.
- For which values of α is $(x_1, x_2) = (0,1)$ the optimal solution?
- For which values of α is the optimal solution bounded?

2.3. Max$\{w = \alpha x_1 + x_2\}$
$3x_1 + x_2 \leq 9$
$x_1 + x_2 \leq 5$
$x_1 \geq 0, x_2 \geq 0$
Give the optimal objective function value w as a function of α. Solve this problem graphically.

2.4. Max$\{w = 2x_1 + x_2\}$
$3x_1 + x_2 \leq 9$
$x_1 + x_2 \leq \beta$
$x_1 \geq 0, x_2 \geq 0$
Give the optimal objective function value w as a function of β. Solve this problem graphically.

Chapter 3

3.1. Transform the following LP problem into the standard form (S):

$$\begin{aligned} &\min\{w = 4x_1 + 5x_2 + 6x_3\} \\ &\quad x_1 - 2x_2 + 3x_3 \le 6 \\ &-2x_1 + 3x_2 - x_3 \ge 7 \\ &\quad x_1 + x_2 + x_3 = -3 \\ &\quad x_1 \ge 0,\ x_2 \le 0,\ x_3 \text{ free.} \end{aligned}$$

3.2. Given system $A\underline{x} = \underline{b}$, where

$$A = (\underline{a}_1, \underline{a}_2, \underline{a}_3) = \begin{pmatrix} 1 & 2 & 1 \\ 2 & 1 & 5 \end{pmatrix}; \quad \underline{x}' = (x_1, x_2, x_3); \quad \underline{b}' = (4, 5).$$

a. Write the complete set of equalities $A\underline{x} = \underline{b}$.

b. Show that $\underline{a}_1$ and $\underline{a}_2$ are linearly independent.

c. Give $\underline{b}$ as a linear combination of $\underline{a}_1$ and $\underline{a}_2$.

d. Determine the basic solution $\underline{x}$ corresponding to the basis $\underline{a}_1$ and $\underline{a}_2$.

e. Let $B = (\underline{a}_1, \underline{a}_2)$. Multiply both sides of $A\underline{x} = \underline{b}$ with B^{-1}.
Verify that $\underline{x}' = (\underline{x}_B', \underline{0}')$ with $\underline{x}_B = B^{-1}\underline{b}$.

3.3. Given system $A\underline{x} = \underline{b}$, $\underline{x} \in \mathbb{R}^7$, $\underline{b} \in \mathbb{R}^3$ where

$$A = \begin{pmatrix} 3 & -1 & 1 & 3 & 0 & -1 & 0 \\ -4 & -2 & 0 & -1 & 1 & 2 & 0 \\ 2 & -1 & 0 & 4 & 0 & 1 & 1 \end{pmatrix}, \quad \underline{b} = \begin{pmatrix} 6 \\ 7 \\ 3 \end{pmatrix}.$$

a. Determine the corresponding trivial basic solution.

b. Determine the basic solution where x_2, x_5, x_7 are basic variables.

c. Does a nonnegative basic solution exist where x_2 and two of the original basic variables are basic variables?

d. Determine a nonnegative basic solution, where x_1 and two of the original basic variables are basic variables.

3.4. Given LP problem (P):

$$\begin{aligned} &\max\{w = 3x_1 + x_2\} \\ &\quad x_1 + x_2 \le 5 \\ &\quad x_1 + 2x_2 \le 8 \qquad (P) \\ &\quad x_1 \le 4 \\ &\quad x_1 \ge 0,\ x_2 \ge 0 \end{aligned}$$

a. Solve problem (P) graphically.

b. Construct the simplex tableau. Which two elements a_{ij} can serve as a pivot? Determine the optimal solution starting from both elements. Write down the objective function value in every generated tableau.

c. Indicate which vertices in the figure correspond to the generated basic solutions.

d. Determine in the final tableau the r-row using $r_j = \underline{c}_B'\hat{\underline{a}}_j - c_j = \underline{c}_B' B^{-1}\underline{a}_j - c_j$ and write $w = 3x_1 + x_2$ as a function of the non-basic variables.

3.5. Given the following problem:

$$\min\{w = -4x_1 - 3x_2 - x_3 + 10\}$$
$$x_1 + x_2 + x_3 \leq 3$$
$$-2x_1 - x_2 - 3x_3 \geq -8$$
$$3x_1 + 2x_2 + 2x_3 \leq 10$$
$$x_1 \geq 0,\ x_2 \text{ free},\ x_3 \geq 0$$

a. Construct the corresponding initial simplex tableau.

b. Solve the problem.
- Give $\underline{c}_B$ in every iteration (generated tableau).
- Check in the final tableau the r-row with $r_j = \underline{c}_B^{\,'} B^{-1} \underline{a}_j - c_j$ and show $\underline{x}_B = B^{-1}\underline{b}$ and $w = \underline{c}_B^{\,'} B^{-1}\underline{b}$

3.6. Given the simplex tableau corresponding to a maximisation problem:

$\underline{c}'$:	2	3	−10	1	−8	2	
	x_1	x_2	x_3	x_4	x_5	x_6	
	1	0	-3	1	-2	0	3
	-1	0	-2	0	0	1	2
	1	1	-1	0	-2	0	1

a. Verify that it is a final tableau.

b. Determine the set of alternative optimal solutions that follow directly from the tableau.

3.7. Given the problem:

$$\max\{w = -2x_1 + 5x_2\}$$
$$x_1 + x_2 \leq 6$$
$$x_1 \geq 2$$
$$-x_2 \geq -3$$
$$x_2 \geq 1$$
$$x_1 \geq 0,\ x_2 \geq 0$$

a. Make the right-hand side positive, add slack variables to convert inequalities into equalities, and add artificial slack variables, such that a unit matrix appears. Solve the problem with the two-phase method.

b. Draw in $\mathbb{R}^2$ the feasible area. Solve the problem graphically. Indicate the vertices that correspond to the generated simplex tableaus.

c. Verify that the optimal objective value $w_0 = \underline{c}_B^{\,'} B^{-1}\underline{b}$

3.8. Solve as well graphically as with the simplex method:

a. $\min\{w = x_1 + 2x_2\}$

$$x_1 + x_2 \leq -1$$
$$x_1 - x_2 \geq 0$$
$$x_1 \geq 0,\ x_2 \geq 0$$

b. $\max\{w = x_1 + x_2\}$

$$x_1 + x_2 \geq 1$$
$$x_1 - x_2 \geq 0$$
$$x_1 \geq 0, x_2 \geq 0$$

3.9. Given the simplex tableau corresponding to a maximisation problem:

x_1	x_2	x_3	x_4	x_5	x_6	x_7	
0	1	0	α	1	0	3	β
0	0	1	−2	2	γ	−1	2
1	0	0	0	−1	2	1	3
0	0	0	δ	3	λ	ξ	0

Variables x_2, x_3, x_1 are basic variables with corresponding basis B. So, $\underline{x}_B = (x_2, x_3, x_1)'$. Determine the values for the parameters (α, β, γ, δ, λ, ξ) for which the following statements hold:

a. Basis B is optimal and feasible.
b. B is feasible, but not optimal.
c. B is feasible and the tableau shows an unbounded solution.
d. B is feasible, x_6 is candidate to enter the basis. If x_6 enters the basis, x_3 leaves the basis.
e. B is feasible, x_7 is candidate to enter the basis. If x_7 enters the basis, the objective function value does not change.
f. B is optimal and feasible. A line segment of alternative optimal solutions exists.
g. B is optimal and feasible. A half-line of alternative optimal solutions exists.

Chapter 4

4.1. Apply scheme 4.1 to derive the dual problem of:

$$\min\{-x_1 + 2x_2 - 38x_3\}$$
$$3x_1 - 4x_2 + 5x_3 \leq -6$$
$$-7x_1 + 8x_2 - 9x_3 \geq 10$$
$$11x_1 - 12x_2 + 13x_3 = 14$$
$$x_1 \geq 0, x_2 \leq 0, x_3 \text{ free}$$

4.2. Let (P) be an LP problem and (D) be the corresponding dual problem.

If ($\hat{\text{P}}$) is equivalent to (P) and ($\hat{\text{D}}$) is the dual of ($\hat{\text{P}}$), then ($\hat{\text{D}}$) is equivalent to (D).

Show that this statement is correct for the following equivalent problems:

$$(P)\quad \min\{\underline{c}'\underline{x}\}, \quad A\underline{x} \geq \underline{b}, \quad \underline{x} \geq \underline{0}$$

$$(\hat{\text{P}})\quad \min\{\underline{c}'\underline{x} + \underline{0}'\underline{y}\}, \quad A\underline{x} - E\underline{y} = \underline{b}, \quad \underline{x} \geq \underline{0},\ \underline{y} \geq \underline{0}$$

Why is this statement relevant?

4.3. Given LP problem (P):

$$\min\{w = 2x_1 + 3x_2 + x_3\}$$
$$\begin{array}{rl} x_1 + 2x_2 + x_3 & \geq 27 \\ x_1 + x_2 & \geq 35 \\ x_1 - x_2 + 6x_3 & \geq 8 \\ x_1 - x_3 & \geq 39 \\ x_1 - x_2 & \geq 3 \\ x_1 + x_2 + 6x_3 & \geq 24 \\ 7x_1 + x_2 & \geq 13 \\ x_1 \geq 0, x_2 \geq 0, x_3 \geq & 0 \end{array} \qquad \text{(P)}$$

a. Construct the initial tableau of (P).
b. Derive the dual (D) of (P).
c. Solve the dual (D) and derive the optimal solution of (P).

4.4. Given LP problem:

$$\min\{w = 420x_1 + 350x_2 + 245x_3 + 210x_4 + 189x_5 + 154x_6\}$$
$$\begin{array}{rl} 14x_1 + 14x_2 + 7x_3 + 0x_4 + 7x_5 + 14x_6 & \geq 63 \\ 21x_1 + 7x_2 + 0x_3 + 7x_4 + 21x_5 + 14x_6 & \geq 133 \\ x_i \geq 0, (i = 1, ..., 6). & \end{array}$$

a. Derive the dual.
b. Given feasible solution $\underline{x}' = (0, 0, 0, 0, 5, 2)$. Do not use the simplex method. Use the complementary slackness relations to show that $\underline{x}'$ is optimal, i.e. derive complementary feasible values for the dual main and slack variables.

4.5. Given LP problem (P):

$$\max\{w = 6x_1 + \alpha x_2 - x_3\}$$
$$\begin{array}{rl} 3x_1 + 2x_2 & = \beta \\ 2x_2 + 4x_3 & \leq 6 \\ -x_1 + 2x_3 & \geq -1 \\ x_1 \text{ free}, x_2 \geq 0, x_3 & \leq 0 \end{array}$$

a. Solve (P) with the simplex method where $\alpha = 2$ and $\beta = 4$.
Use the final tableau under a. to answer the following questions:
b. $\beta = 4$. Give the values of α for which the solution found under a. is optimal.
c. $\alpha = 2$. Give the values of β for which the solution found under a. is feasible.
d. Let α be in the range of b. and β be in the range of c. Give the optimal objective function value w as a function of α and β.
e. $\alpha = 2$ and $\beta = 4$. An extra variable $x_4 \geq 0$ is added to problem (P). The column vector corresponding to x_4 is $(-1, -1, 0)'$ and de objective function coefficient is γ. Give the optimal objective function value w as a function of γ.

Chapter 5

5.1. Consider problem (P):

$$
\begin{array}{llll}
\max\{w_1 = -2x_1 + x_2\} & & & \\
\max\{w_2 = x_1 + 3x_2\} & & & \\
x_1 + x_2 \leq 8 & & (P) & \\
x_2 \leq 3 & & & \\
-x_1 + x_2 \leq 1 & & & \\
x_1 \geq 0\,,\ x_2 \geq 0 & & &
\end{array}
$$

a. Draw the feasible area in the decision space X and show whether the solutions (4, 0) and (2, 3) are efficient (Pareto-optimal) solutions?

b. Give the parameter equation(s) of the set of Pareto-optimal points.

Suppose we replace the objective functions w_1 and w_2 by the two goal constraints:
$w_1 \geq 0$ with penalty weight 2
$w_2 = 6$ with penalty weights 3(−) and 4(+).

c. Give the goal programming formulation of this problem.

d. Are there solutions which fulfil both goals (solve with the help of reasoning, use the figure. So, do not use the simplex method).

5.2. In February 2005 the Kyoto global warming agreement became effective. There are different opinions concerning the CO_2-reductions (greenhouse gas concentration). Is it allowed to count investments into woods for CO_2 reduction? How far is trading in emissions allowed and which amount of money should be invested in expensive reduction-alternatives (such as sun and wind energy) in own countries?

Very simplified, the problem can globally be described with the help of the following table (bil. means billion = thousand million).

	reduction in bil. tons of Carbon for every invested bil. Euro			
	expensive energy	emissions trading	woods	reduction in bil. tons Carbon (C) goals
America	0.2	0.7	0.5	7
Europe	0.1	0.4	0.6	10

Explanation of the table:
If America invests 1 bil. Euro in expensive energy then there will be a reduction of 0.2 bil. tons of Carbon in the atmosphere. An investment of 1 bil. Euro in woods gives a reduction of 0.5 bil. tons of Carbon etc.

- America will not invest more than 15 bil. Euro.
- Europe will not invest more than 20 bil. Euro.
- The investment in expensive energy should be at least 80% of the total investments.

Two goals are formulated:

1. America has to reduce at least 7 bil. tons Carbon.
2. Europe has to reduce at least 10 bil. tons Carbon.

Give the formulation of this goal programming problem (priorities are the same).

Chapter 6

6.1. Given LP problem (P):

$$\begin{aligned} &\max\{w = 3x_1 + 5x_2\} \\ &2x_1 + 2x_2 \leq 7 \qquad \text{(P)} \\ &x_2 \leq 2 \\ &x_1 \geq 0,\ x_2 \geq 0 \end{aligned}$$

a. Solve (P) graphically.
b. Additional constraints are added: x_1 and x_2 should take an integer value. Solve the integer LP problem using the Branch-and-Bound method. Give clearly the chosen branches and lower bounds. Start with an initial lower bound of w_b = 12. Solve every generated continuous LP problem graphically.

6.2. Consider the integer problem (IP):

$$\begin{aligned} &\max\{w = \tfrac{1}{2}x_1 + 4x_2\} \\ &2x_1 + 2x_2 \leq 9 \qquad \text{(IP)} \\ &x_2 \leq 3 \\ &x_1 \geq 0,\ x_2 \geq 0 \text{ and integer} \end{aligned}$$

Use the Branch-and-Bound method to solve problem (IP). Apply a search strategy in which the right-hand side problems, i.e. $x_j \geq [\beta_j] + 1$, are systematically examined first. Start with an initial bound of w_b = 8. Solve all generated continuous LP problems graphically.

Chapter 7

7.1. Replacement problem

Many types of machinery tend to need more maintenance when they get older. Therefore they have to be replaced from time to time. The management faces the task of determining the optimal (i.e. cheapest) replacement strategy for a machine. The maintenance and operating cost have to be balanced against the replacement cost.

Suppose management has a planning horizon of 4 years. Assume the following data:

P: the purchase price of the machine (constant),
R_i: the return price of a machine that has been used for i years,
C_j: the maintenance and operating cost in the jth year that a machine is in use (assume $C_{j+1} > C_j$).

This problem can be modelled as a network:

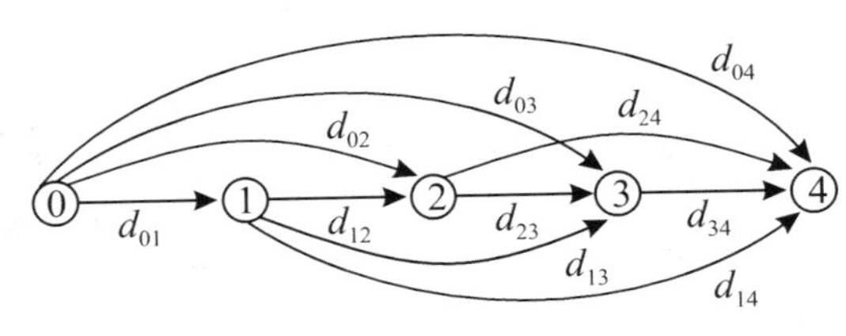

The nodes in the network represent the moments at which it has to be decided whether the machine has to be replaced. As we assume that this decision has to be taken once per year, the nodes represent the years.
The arcs (i,j) in the network represent the cost of going from node i to node j. The length d_{ij} of arc (i,j) is defined as the total cost of a machine that is bought in year i and returned in year j. So,

$$d_{ij} = P - R_{j-i} + \sum_{t=1}^{j-i} C_t$$

Use Dijkstra's algorithm to calculate the optimal replacement strategy if

$P = 100$,
$R_i = 75 - 15i$,
$C_j = 10j$.

7.2. Shortest route

A man living in Wageningen (the Netherlands) wants to visit his girlfriend in Valencia (Spain). He wants to go by motorcycle. There are many ways to drive from Wageningen to Valencia. By close inspection of the map he has reduced the options to those shown in the figure.

Use Dijkstra's algorithm to calculate the shortest route from Wageningen to Valencia.

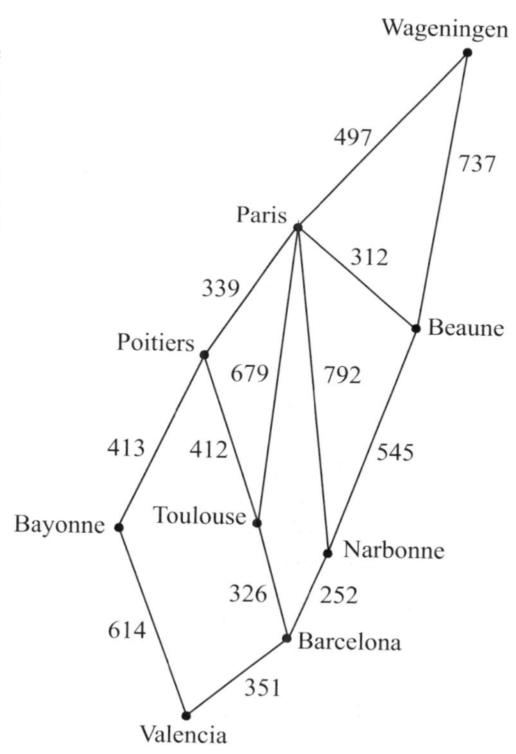

7.3. In a warehouse boxes have to be stored on shelves. There are n kinds of boxes. They have height $H_1, H_2, \ldots, H_n$ with $H_1 < H_2 < \ldots < H_n$. A box with height H_i can be stored on a shelf with height $\geq H_i$. The length of every box is known, so the required length for storing all boxes with height H_i is also known: L_i.

If all boxes are stored on one shelf then the height of this shelf has to be H_n. In that case the total space is $H_n \cdot \sum_{i=1}^{n} L_i$. If every kind of box is assigned its own shelf then the total space is $\sum_{i=1}^{n} H_i \cdot L_i$.

The construction cost of a shelf with height H_i are calculated from:

F = fixed cost (independent of the length of the shelf), $F = 300$ per shelf.

C = cost per unit space (length ·height). $C = 1$ per unit space.

Now the question is: how much shelves should be constructed, which height should they have, and which length, such that the total cost is minimised?

Construct a network representation for this problem.

Use Dijkstra's algorithm to calculate the optimal layout of the shelves for the following data:

	1	2	3	4	5	Box i
H_i	10	12	15	20	30	
L_i	20	40	30	10	50	

7.4. Find the longest st-path in the following network:

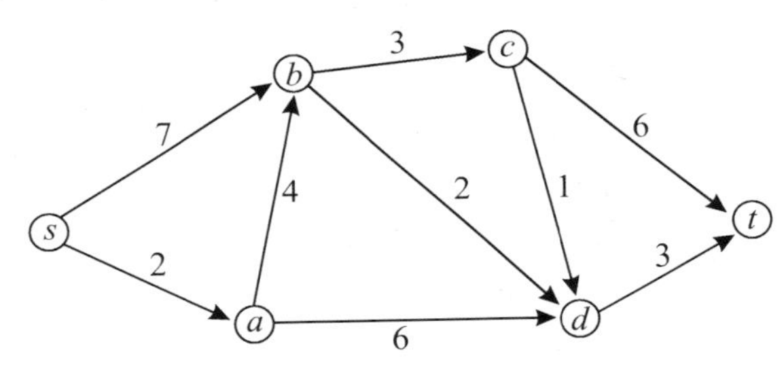

7.5. The figure below shows the road network between the cities s and t, in one direction. Arc st is a highway with capacity 8000 cars per hour. Highway st has to be maintained. This maintenance can be carried out in two ways:

- by closing off so many lanes that the capacity of st is halved (then the maintenance will cost four weeks),
- by closing off so many lanes that the capacity of st is reduced to one quarter (then the maintenance can be carried out in two weeks).

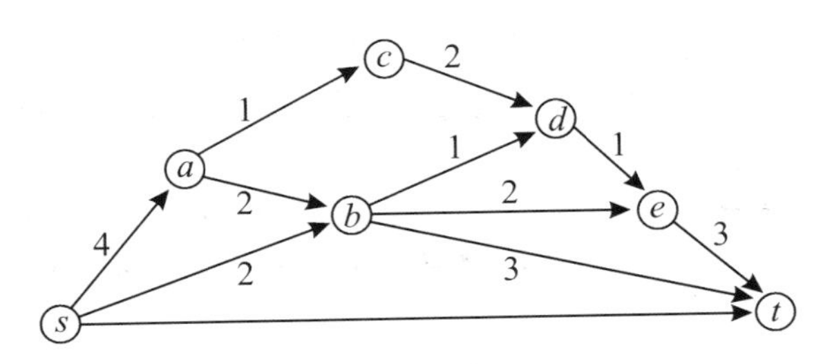

The figure shows that st is not the only road from s to t. Alternative routes via $\{a,b,c,d,e\}$ are possible. It is expected that many car drivers will use these routes during the maintenance period. The numbers along the arcs represent the current slack capacities (in thousands of cars per hour) of the ways. The maintenance has to be carried out in such a way that the residual capacity of st together with the slack capacities of the other ways suffices to facilitate 8000 cars per hour.
Use network theory to formulate an advice for the maintenance of st.

7.6. The figure below shows the road network between cities s and t. The current capacity of this network is not large enough. It is allowed to build a new arc ae. This new arc can either be a small road (capacity 3, cost 2) or a large road (capacity 5, cost 3). No future expansion of the network shall be allowed.

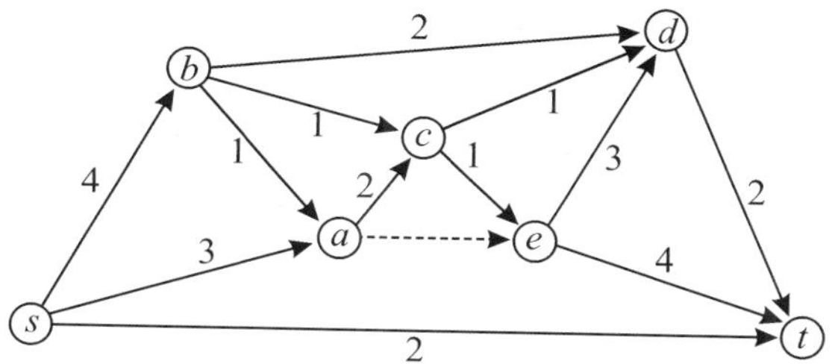

Use network theory to formulate an advice about the size of the new road.

7.7. A university has just reorganised its structure. As a result several chair groups are combined into one new group. Many of the group members are relatively unknown to each other. In order to encourage the contacts within the new group a dinner party shall be given. The seating arrangement must be such that no two members of a chair group eat at the same table. The problem of finding such a seating arrangement can be formulated as a maximum flow problem.
Draw the network for this maximum flow problem for a simplified version of this problem: there are 3 chair groups (with 4, 3, 2 members) and 4 tables (with 3, 3, 2, 3) chairs.

Chapter 8

8.1. A mountaineer can take at most 15 kg in his backpack. Three different items can be chosen (each item at most once). Item i weighs m_i kg (see table). The importance of item i to the mountaineer is denoted by b_i.

a. Apply DP to determine which items the mountaineer should take along.

Item	weight m_i (kg)	importance b_i
1. tent	9	5
2. sleeping bag	5	10
3. cooker	6	4

b. Give the general Bellman equation for $t = 0$ (state $s = 15$) and $t = 1$ for any state s of the problem.

8.2. In wood processing industry, trees are converted into assortments. The operator has to decide how to cut a tree in such a way that the total return is as high as possible. If the tree is straight, the optimal saw pattern can be determined by dynamic programming. Suppose the following data are given:

Tree: length 7 meter (thickness 20 cm, completely straight). Assortments:

- assortment 1: length 1 metre (thickness 20 cm); return € 5 per unit of 1 metre
- assortment 2: length 2 metre (thickness 20 cm); return € 10 per unit of 2 metres
- assortment 3: length 4 metre (thickness 20 cm); return € 25 per unit of 4 metres

a. Solve the problem via dynamic programming.

b. Give the corresponding general Bellman equation for $t = 1$ and a general state s.

8.3. Draw a convenient network for the following knapsack problem and solve it by means of dynamic programming. Each product may be chosen several times. The maximum weight is set on 9 kg and the maximum volume is 10 litres.

product i	weight m_i in kg	volume l_i in litres	importance b_i
1	4	3	3
2	3	7	6
3	5	9	7

8.4. Equest is a successful company in horse trading. Its activities in the Netherlands include visiting the yearly horse market in Noord-Laren. The specific actions (purchasing or selling horses) depend on the price p_t of horses on January 1st of year t. If the price p_t is high, Equest will sell a number of horses; if the price is low, Equest will buy horses. The following data are relevant:

- On January 1st 2007, Equest has a budget of 1 million Euro.
- The policy of the company does not allow to borrow money and to exceed the budget for purchasing horses in any year.
- On January 1st 2007, Equest has 20 horses.
- The capacity of the stables allows a maximum of 30 horses to be kept at any period t.
- On January 2nd 2015, Equest will be liquidated and all horses will be sold; no matter the price p_t of horses at that time.
- The costs for training and maintenance of a horse are € 10,000 yearly.

Equest wants to maximise the final capital on January 2nd in 2015. Give the DP formulation of the problem, i.e. formulate the Bellman equation for any state $\underline{s}_t$ in phase (decision moment) t and give the value function for $t = 2015$.

8.5. Given the continuous non-linear programming problem (DP):

$$\begin{array}{lll} \max\{-x_1^2 + 2x_1 + x_2\} & & \\ \quad x_1 + x_2 & \leq 5 & \quad \text{(DP)} \\ \quad x_1 & \leq 3 & \\ \quad x_1 \geq 0, x_2 \geq 0 & & \end{array}$$

Use dynamic programming to solve problem (DP).

Chapter 9

9.1. A minimax problem

In horse-racing it is very common that observers bet on the winning combination (horse/jockey). Suppose that the quotation for the horse 'Lucky-boy' equals 1 to 3.5. This means that the bookmaker will pay 3.5 Euros for every Euro staked on Lucky-boy in case Lucky-boy wins the race. However, if the horse does not win the race, all stakes on Lucky-boy will be for the bookmaker. Suppose that spectators can bet on the following horses and quotations:

Lucky-boy 3.5	Blacky 4	Lightning 8.5	Speedy 13.5	Giant 16

A student called 'Gamble' disposes of € 1000 and does not know the first thing about horse-racing. Student Gamble wants to bet on one or more horses simultaneously in such a way that the worst outcome will be maximised. In other words: the criterion is concerned with making the worst possible outcome as pleasant as possible. For example: a bet of € 1000 on a single horse will not be optimal. After all, if the horse in view does not win the race 'Gamble' loses all the money. The worst possible outcome is € 1000 with this decision.

Formulate this problem as an LP problem.

9.2. Given the LP relaxation of a (mixed integer) fractional problem (FP).

$$\max\left\{\frac{80x_{1,10}+160x_{1,20}+144x_{2,10}+240x_{2,20}}{8x_{1,10}+8x_{1,20}+16x_{2,10}+16x_{2,20}}\right\}$$

$$\begin{array}{llr} \text{s.t.} & & \text{(FP)} \\ 11x_{1,10}+22x_{2,10}+55x_{1,20}+61x_{2,20} & \leq 55 & \\ x_{1,10}+x_{2,10}+x_{1,20}+x_{2,20} & = 1 & \\ y_{10}+y_{20} & = 1 & \\ x_{1,10}+x_{2,10} & \leq y_{10} & \\ x_{1,20}+x_{2,20} & \leq y_{20} & \\ x_{1,10},x_{2,10},x_{1,20},x_{2,20},y_{10},y_{20}\geq 0 & & \end{array}$$

Transform the non-linear problem (FP) into an equivalent linear programming model.

9.3. Given the non-linear problem (PT)

$$\begin{array}{llr} \max\{x_1 y+5x_2\} & & \\ x_1 & \leq 4y & \\ x_2 & \leq 3 & \text{(PT)} \\ x_1 + x_2 & \leq 6 & \\ x_1,x_2\geq 0 \text{ and } y\in\{0,1\} & & \end{array}$$

Reformulate problem (PT) into a tight mixed integer linear programming model without redundant constraints.

Chapter 10

10.1. An investment problem

A fund of € 1.000.000 is available to support welfare projects. More projects have been submitted than can be supported. Therefore every project is rated on a 1-5 scale (with better projects getting a higher rating).

The fund committee has to decide which projects should be accepted for financial support. The goal is to maximise the sum of the ratings of the projects that are accepted. In the table below the costs and the rating of every project are listed.

Project	Costs (in 10^3 €)	Rating
1	25	1
2	98	5
3	48	2
4	50	1
5	89	3
6	94	3
7	111	4
8	44	2
9	36	2
10	94	4

a. What type of problem is this?
b. Formulate the problem as an Integer Programming problem.
c. Find a solution for the problem with a suitable construction heuristic.
d. Try to improve the solution with a suitable improvement heuristic.

10.2. Assigning tasks to people

A company has several smaller jobs to do. These jobs have to be finished in the next week (40 hours). The company wants to hire students to do the jobs. The following table shows the time t_j (in hours) that it takes to finish job j.

Job j	1	2	3	4	5	6	7	8	9	10	11	12	13	14	15	16	17	18	19	20
t_j	22	5	13	8	28	10	23	7	16	4	4	25	9	2	17	14	6	2	19	3

Jobs can not be shared between students. If a student is hired he has to be paid for the full week. The company wants to hire the minimum possible number of students to do the jobs.

a. What type of problem is this?
b. Formulate the problem as an Integer Programming problem.
c. Find a solution for this problem with a suitable heuristic.

10.3. Ordering books from internet

John needs 5 study books. He wants to order them via internet. He looked up the prices of the five books at three different internet stores (in Euro per book, see table). The table also contains the fixed order cost of the internet stores (in Euro per order; regardless of the number of books). John wants to know which books to order at which internet store in order to minimise his total costs.

Internet bookstore	Fixed costs per order	Price of book 1	Price of book 2	Price of book 3	Price of book 4	Price of book 5
Store 1	3	15	14	24	34	13
Store 2	10	17	14	24	29	14
Store 3	6	20	11	26	32	9

a. What type of problem is this?
b. Formulate the problem as an Integer Programming problem.
c. Find a solution for this problem with a suitable heuristic.

10.4. Drilling holes

The production process of some part requires that 12 holes be drilled in a plate. This is done with a robot arm. It takes time to move the arm from one drilling location to the next. The amount of time is proportional to the distance between the consecutive holes. So it is profitable to use the drilling sequence with lowest total distance.
The robot can only handle one sequence of the holes. Therefore the arm has to return to the location of the first hole after drilling the last hole.

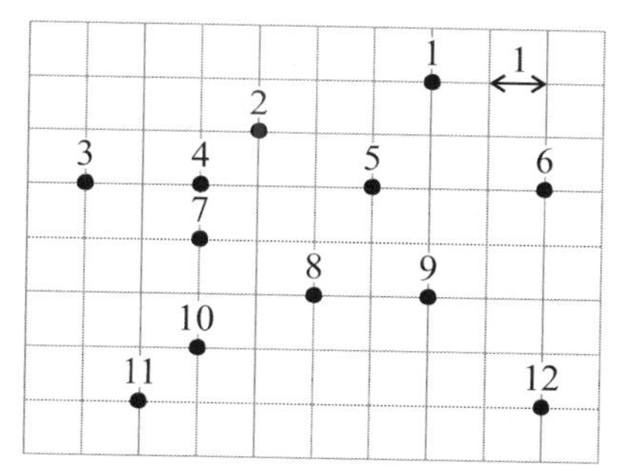

The picture shows the locations of the holes that have to be drilled. The distances between the holes can be calculated with Pythagoras' theorem. Which drilling sequence minimises the production time?

a. What type of problem is this?
b. Formulate the problem as an Integer Programming problem.
c. Find a solution for this problem with a suitable construction heuristic.
d. Try to improve the solution with a suitable improvement heuristic.

Chapter 11

11.1. Demand for a product at a warehouse is at a constant rate of 300 units per week. The product is supplied from the factory to the warehouse. The factory gate price is € 7.50 per unit, while the total cost of a shipment from the factory to the warehouse when the shipment has size B is given by
Shipment cost = € 64 + € 2.50·B

The inventory carrying cost in €/€/week is 0.006. Determine the appropriate shipment size.

11.2. A manufacturing firm located in Apeldoorn produces an item in a two-month time supply. A new logistics manager, attempting to introduce a more logical approach to selecting run quantities, has obtained the following estimates of characteristics of the item:

$d = 6000$ units/yr
$k = €\ 10$
$c = €\ 0.10$ per unit
$r = €\ 0.30$ /€/yr

Assume that the production rate is much larger than d.

a. What is the economic order quantity of the item?
b. What is the time between consecutive replenishments of the item when EOQ is used?
c. The production manager insists that the $k = €\ 10$ figure is only a guess. Therefore, he insists on using his simple two-month supply rule. Indicate how you would find the range of k values for which the EOQ (based on € 10) would be preferable (in terms of a lower total of replenishment and carrying costs) to the two-month supply.

11.3. A company produces cabinet doors. For those doors, doorknobs are needed. The company produces them as well. The level of production $p = 10000$ doorknobs a year. The set-up cost $k = €\ 100$ per production series. The carrying charge $r = 0.4$ €/€/year. The unit variable cost $c = €\ 2.50$ per unit.
The factory considers buying the doorknobs from an outside firm.
In that case the price of the doorknobs is € 2.75 per unit. Ordering costs $k = €\ 25$ per order. Carrying charge $r = 0.4$ €/€/year.
Demand $d = 2000$ doorknobs/year. What should the company do? Make or buy?

11.4. A pharmacy sells 30 bottles of antibiotics per week. The unit variable costs are € 20. Every time the pharmacy orders bottles of antibiotics, they charge an extra € 10. The carrying charge is € 0.2 /€/year. The bottles of antibiotics are very perishable, they cannot be sold anymore if they are more than a week in storage. What is the optimal order quantity?

11.5. A dairy company routinely replaces a specific part on a packaging line. The usage rate is 3000 items per year. The supplier of the part offers the following all-units discount structure.

Range of Q	Unit cost
$0 < Q < 400$ units	€ 16
$Q \geq 400$	€ 15

The fixed cost of a replenishment is estimated to be € 60, and a carrying charge of € 0.25 /€/yr is used by the company.

a. What replenishment size should be used?
b. If the supplier was interested in having the dairy company acquire at least 1000 units at a time, what is the largest unit price they could charge for an order of 1000 units?

11.6. Consider a company facing a demand pattern and costs as following:

t	1	2	3	4	5
d_t	50	40	25	100	70

$K = €\ 50$, $h = €\ 1$ per period
Construct a replenishment schedule and calculate the associated costs using the Silver – Meal Heuristic.

11.7. JoyGarden is a central buying agent and distributor for garden equipment like deckchairs, garden tables and sunshades. For a special collection of sunshades they have one opportunity to order, several months before spring. JoyGarden has to decide how many sunshades to order.
The unit acquisition cost c is € 70 / unit
The selling price re is € 120 / unit

Any units unsold at the end of the winter will be marked down to u = € 60 / unit, ensuring a complete clearance and thus avoiding the prohibitive expense of storing the sunshades until the next season. The probability distribution of regular demand is estimated to be:

demand d	300	400	500	600	700	800
probability $P(\tilde{d} = d)$	0.1	0.1	0.4	0.2	0.1	0.1

a. What is the expected demand?
b. What is the standard deviation of demand?
c. Calculate the expected revenue $\rho(Q)$ (or expected profit) for $Q = 300$ and for $Q = 500$.
d. To maximise expected profit, how many units should JoyGarden (using a discrete demand model) acquire? (Calculate Q^*)
e. What is the expected profit under the strategy of d? (Calculate $\rho(Q^*)$)

11.8. A cook shop for indoor cooking necessities is expanding its range of products with a line of outdoor kitchens. During the winter time the cook shop has to decide on the number of each type of outdoor kitchen to order from its manufacturing supplier for the upcoming summer season. For a particular outdoor kitchen, the cost per kitchen is € 300 and the retail selling price is € 400. The shop estimates an average sale of 50 of these outdoor kitchens but with considerable uncertainty, which the shop can express as a uniform distribution between 25 and 75. Any outdoor kitchen not sold at the end of the summer can be disposed of at a price of € 250 to a kitchen discounter.
a. How many outdoor kitchens should the cook shop order?
b. What is the expected revenue under the strategy of a?

11.9. It is reasonable to assume that the demand of an item, controlled by an (s, Q) system, has a normal distribution over its replenishment lead time ($t_l = 1$) with expected value $\bar{d} = 20$ units and $\sigma = 4.3$ units.
a. Calculate the desired reorder point and the desired safety stock for a service level of 98%.
b. Repeat for a service level of 99%.

Chapter 12

12.1. Solve the following NLP problem graphically

$$\min\{(x_1-3)^2+(x_2-2)^2\}$$
s.t.
$$x_1^2-x_2-3\le 0$$
$$x_2-1\le 0$$
$$-x_1\le 0$$

12.2. Designing a desk with length x and y wide, the following aspects appear:

- The surface has to be as big as possible: max xy
- The costs of the expensive edge should not be too big : $2x+2y\le 8$
- The desk should not be too wide: $y\le b$.

Solve this NLP problem graphically for $b = 1$.

What happens to the optimal surface when b increases?

12.3. In Example 12.5 determine V, x_1^* and the (E,V)-curve when there is negative correlation according to $\sigma_{12}=-\frac{1}{2}$.

12.4. Is the function $f(\underline{x})=\sqrt{(x_1^2+x_2^2)}$ differentiable in $\underline{0}$?

12.5. Determine gradient and Hessean of $f(\underline{x})=x_1^2x_2^2e^{x_3}$.

12.6. Derive the second order Taylor approximation of $f(\underline{x})=x_1e^{x_2}$ around $\underline{0}$.

12.7. Let $f(\underline{x})=4x_1x_2+6x_1^2+3x_2^2$. Write $f(\underline{x})$ as a quadratic function (12.21). Determine the stationary point and eigenvalues and eigenvectors of A.

12.8. Let $f(\underline{x})=-1-2x_1+x_1x_2+x_1^2$. Write $f(\underline{x})$ as a quadratic function (12.21). Determine the stationary point and eigenvalues and eigenvectors of A.

12.9. Given $f(\underline{x})=2x_1^2+x_2^4$. Derive and draw the contours corresponding to a function value of 3, of $f(\underline{x})$ and the first and second order Taylor approximation around $(1, 1)^T$.

12.10. Determine the minima of

a. $f(\underline{x})=x_1^2+x_2^2$

b. $f(\underline{x})=\sqrt{x_1^2+x_2^2}$

c. $f(\underline{x})=x_1x_2$

12.11. Given function $f(\underline{x}) = x_1^3 - x_2^3 - 6x_1 + x_2$

a. Determine gradient and Hessean of $f(\underline{x})$.

b. Determine the stationary points of $f(\underline{x})$.

c. Which point is a minimum point and which are saddle points?

12.12. Given the utility function $U(\underline{x}) = x_1^2 x_2$ and the budget-constraint $x_1 + x_2 = 3$. Determine the stationary point of the Lagrangean maximising the utility function subject to the budget-constraint.

12.13. Given the concave optimisation problem P:

$$\begin{aligned} &\min\{-(x_1 - 1)^2 - x_2^2\} \\ &2x_1 + x_2 \leq 4 \qquad\qquad (P) \\ &x_1, x_2 \geq 0 \end{aligned}$$

a. Determine graphically the local and global minimum points of P.

b. Check the Karush-Kuhn-Tucker (KKT) conditions for the minimum points.

c. Point $(0, 0)^T$ fulfils the KKT conditions. Show via the definition that $(0, 0)^T$ is not a local minimum point.

d. Give another point that fulfils the KKT-conditions, but is not a minimum point.

12.14. Show $f(\underline{x}) = \max\{g_1(\underline{x}), g_2(\underline{x})\}$ is a convex function if $g_1(\underline{x})$ and $g_2(\underline{x})$ are convex.

12.15. Check whether $f(\underline{x}) = 2x_1 + 6x_2 - 2x_1^2 - 3x_2^2 + 4x_1x_2$ is convex.

12.16. Determine the validity of Theorem 12.6 for $f: (0, \infty) \rightarrow \mathbb{R}$ with $f(\underline{x}) = 1/r$.

Chapter 13

13.1. To be able to select a proper distribution function for the inter-arrival times of customers at a kiosk, one observes the inter-arrival times over a period of 100 minutes. We assume that the inter-arrival times have a (negative) exponential distribution. To test this hypothesis one applies the so-called Kolmogorov-Smirnov test.
One observes the following inter-arrival times in minutes:

2.5	0.3	4.2	3.3	3.6	4.3	14.2	5.8	0.7	3.9
7.0	2.6	13.1	3.3	5.5	1.7	3.9	2.4	8.2	9.4

Apply the Kolmogorov-Smirnov test on the given observations. Show all steps of the process. The critical value for $N = 20$ is 0.294.

13.2. A random number generator produces the following numbers successively:
8, 10, 45, 36, 17, 31, 12, 10, 31, 64, and 85. Is it possible that a linear congruential generator produces these numbers? Please explain your answer.

13.3. In a simulation we have selected for the production time of a part a normal distribution function with a mean of 5 minutes. The mean is based on collected data during a period of several days. We make one run of 1 hour with this model and find for the time needed to produce the part a mean of 7 minutes. What is the reason that in the simulation the mean of the production time differs from the selected distribution function?

13.4. Suppose you study with a model the waiting times of customers at the checkout of a supermarket during rush hours. The purpose of the study is to investigate the waiting times of customers during the rush hours from 5 PM to7 PM. Furthermore, assume that you have a validated model available and that the needed distribution functions are known and validated. The arrival process of customers has been selected for a busy period.
Specify the experiment to be performed with the model. The following issues and related topics should be included:

a. What characteristic of every client do you measure during one replication and what performance measure will be used per replication?
b. Do you apply subruns (batch mean approach) or independent replications (replication/deletion approach)?
c. At what time do the replications start?
Explain your answer.

13.5. Successive inter-arrival times of customers observed at a bank terminal are (in minutes): 2.3, 3.1, 2.8, 3.7, 5.3, 6.9, 7.1, 6.2, 5.2, 4.2, 5.6, 4.9, 5.2, and 5.4. Prepare a scatter diagram of these observations and explain how to prepare a scatter diagram. Furthermore, indicate whether the observations are dependent or independent. Please explain your answer.

13.6. To find a proper distribution function for the arrival process of customers at a kiosk we observe 60 inter-arrival times. We assume that the inter-arrival times are (negatively) exponentially distributed with a mean of 0.3. We test this hypothesis with the so-called Chi-square test. The observations are (in minutes):

12.4	5.2	0.3	0.8	13.5	7.3	1.1	2.5	4.3	1.8
0.4	2.2	7.6	4.5	2.9	2.6	7.7	3.2	3.1	0.1
8.5	1.5	3.7	2.7	2.4	0.3	2.2	0.7	0.5	5.2
7.3	0.3	1.3	0.2	7.5	0.5	4.8	4.3	1.6	0.8
0.2	0.5	1.6	7.0	7.7	0.3	1.0	2.2	3.4	3.7
0.1	0.5	1.5	0.2	1.1	2.2	0.3	1.2	5.1	0.3

Apply the Chi-square test and use ten classes with the following end points: 0.351, 0.743, 1.188, 1.701, 2.308, 3.051, 4.009, 5.359, 7.668, and ∞. Show all steps of the test and indicate what you are doing. For this test apply the critical value $\chi^2 = 15.5$. Indicate whether you can reject the hypothesis or not. Explain your answer.

13.7. The following successive inter-arrival times (in minutes) are observed at a barbershop: 8, 6, 9, 3, 5, 10, 11, 12, 13, 15, 16, and 12. Answer the following questions:

a. Compute the autocorrelation coefficients for the observations one observation apart and observations three observations apart of the given data.
b. The next day one observes the following inter-arrival times during the same period: 10, 7, 12, 8, 11, 7, 15, 13, 16, 14, 17, and 14. Indicate whether it is acceptable to use the data of the first day to select a distribution function for the inter-arrival time. Please explain your answer.

Appendix A. Linear algebra

Th.H.B. (Theo) Hendriks

A.1. Introduction

In Chapter 2, it has been discussed how to solve small LP problems with two decision variables, graphically. We also showed, that with the aid of slack variables, solutions of an LP-problem can be represented as solutions of a set of linear equations. In this chapter we briefly review some key concepts of linear algebra, a part of mathematics that is related to solving linear equations. The Simplex Method, explained in Chapter 3, moves along the boundaries of the feasible area from one vertex to the other until it has found an optimal one. Every vertex corresponds to a solution of the set of linear equations. Only the relevant linear algebra required for understanding the concepts of this book is discussed.

A.2. Basic concepts (1)

In this section we discuss some concepts related to matrices, vectors and scalars.

Concept A.1

A *matrix* is a rectangular array of numbers.

The numbers in a matrix can be real numbers or complex numbers. In this book, a matrix is represented by a capital letter and the numbers are real.

Example A.1.

$$M = \begin{pmatrix} 2 & 3 & -4 \\ 0 & 1 & \sqrt{7} \end{pmatrix}; \qquad E = \begin{pmatrix} 1 & 0 & 0 \\ 0 & 1 & 0 \\ 0 & 0 & 1 \end{pmatrix}; \qquad A = \begin{pmatrix} 3 & 4 \\ -1 & \sqrt{5} \\ 0 & \pi \end{pmatrix}$$

M, E and *A* are matrices.

Concept A.2

A number from a matrix is called an *element* or an entry of the matrix. Matrix *M* has six, *E* has nine and *A* has six elements.

Concept A.3

Horizontally the matrix elements constitute a *row*. Vertically the matrix elements constitute a *column*.

The first row of *M* consists of the elements 2, 3 and –4 and the second row of 0, 1 and $\sqrt{7}$.

Matrix *M* has three columns, respectively $\begin{matrix} 2 \\ 0 \end{matrix}$, $\begin{matrix} 3 \\ 1 \end{matrix}$ and $\begin{matrix} -4 \\ \sqrt{7} \end{matrix}$.

Concept A.4

A matrix with m rows and n columns is called an $m \times n$-*matrix* (notation $A^{m \times n}$). Matrix *M* is a 2×3-matrix, represented by $M^{2 \times 3}$.

Concept A.5

A matrix is called *square* if the number of rows equals the number of columns (so $m = n$). Matrix E is square. In Example A.1 indicated as $E^{3\times 3}$.

Concept A.6

De elements from a matrix are indicated with two *indices*. The first index represents the row number, the second index the column number.

$$A^{m\times n} = \begin{pmatrix} a_{11} & a_{12} & . & . & . & a_{1n} \\ a_{21} & a_{22} & . & . & . & a_{2n} \\ . & . & . & . & . & . \\ . & . & . & a_{ij} & . & . \\ . & . & . & . & . & . \\ a_{m1} & a_{m2} & . & . & . & a_{mn} \end{pmatrix}$$

so a_{ij} means: the element in row i and column j. For example in matrix M, element $m_{23} = \sqrt{7}$. Special matrices are the null matrix and the unit matrix.

Concept A.7

The *null matrix* is a matrix with all elements zero.

Concept A.8

The *unit matrix* (notation E) is a matrix with elements e_{11} , e_{22} , … , e_{nn} equal to 1 and the other elements zero. A unit matrix E is always square.

null matrix : $0^{2\times 2} = \begin{pmatrix} 0 & 0 \\ 0 & 0 \end{pmatrix}$ unit matrix : $E^{3\times 3} = E_3 = \begin{pmatrix} 1 & 0 & 0 \\ 0 & 1 & 0 \\ 0 & 0 & 1 \end{pmatrix}$

Concept A.9

Elements a_{11} , a_{22} , … , a_{nn} of a square matrix constitute the *main diagonal* of a matrix.

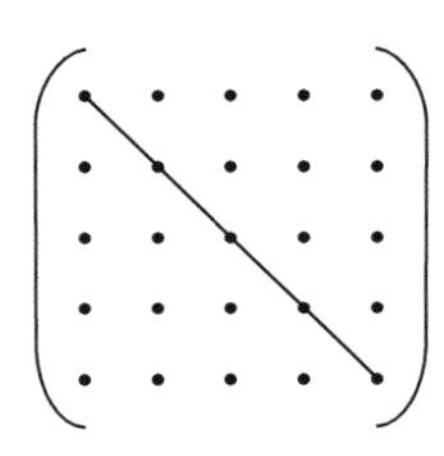

Figure A.1. The main diagonal.

The main diagonal of a matrix starts at the left upper corner and ends at the right lower corner of a matrix (see Figure A.1). The main diagonal of a unit matrix E consists of the elements with value 1, whereas the other elements equal zero. The unit matrix E can be represented by $E^{n\times n}$ or E_n; and is called the nth order unit matrix. Usually the order follows from the context such that E or E_n can be used.

Concept A.10

A *vector* is a matrix with either one row or one column.

Concept A.11

A *column vector* is a matrix with only one column. A *row vector* is a matrix with only one row.

In this book, vectors are represented by italic, underlined lower case letters. For example:

$$\underline{a} = \begin{pmatrix} 1 \\ 3 \\ -5 \end{pmatrix} \text{ is a column vector.}$$

As a vector is a matrix of specific size, in this book its elements are real numbers. One index is sufficient to represent an element of a vector; a_2 represents the second element of vector $\underline{a}$. In the example, $a_2 = 3$.

Concept A.12

A *unit vector* (notation $\underline{e}$) is a vector with one element equal to 1 and all other elements have the value zero. Usually an index is added to indicate which element has the value 1.

$$\underline{e}_1 = \begin{pmatrix} 1 \\ 0 \\ 0 \\ 0 \\ 0 \end{pmatrix}; \quad \underline{e}_3 = \begin{pmatrix} 0 \\ 0 \\ 1 \\ 0 \end{pmatrix}; \quad \underline{e}_3 = \begin{pmatrix} 0 \\ 0 \\ 1 \end{pmatrix} \quad \underline{e}_3 = \begin{pmatrix} 0 \\ 0 \\ 1 \\ 0 \\ 0 \end{pmatrix}$$

Note that the index is not related to the number of elements in the vector; $\underline{e}_3$ is first a vector with four elements, then a vector of three elements and finally it contains five elements.

A vector of four elements is a member of a four dimensional space.
The general notation is:

$$\underline{a} \in \mathbb{R}^n.$$

This notation means vector $\underline{a}$ is a member (notation: $\in$) of a space $\mathbb{R}^n$ of dimension n. Respectively $\underline{e}_3 \in \mathbb{R}^4$, $\underline{e}_3 \in \mathbb{R}^3$ and $\underline{e}_3 \in \mathbb{R}^5$.

Concept A.13

A *null vector* (notation $\underline{0}$) is a vector where all elements equal zero.

$$\underline{0} = \begin{pmatrix} 0 \\ 0 \\ 0 \\ 0 \end{pmatrix} \in \mathbb{R}^4$$

Concept A.14

Partitioning a matrix is dividing it into parts (sub matrices).

The partitioning is represented by dashed lines in the matrix (see Example A.2).

Example A.2.

$$A = \left(\begin{array}{cc|c|cc} 15 & 0 & \pi & 3 & 5 \\ -3 & 4 & 2 & 3 & 7 \\ \hdashline -1 & \sqrt{17} & -2 & 4 & 0 \\ 0 & 1 & 6 & 1 & -7 \end{array}\right)$$

A can be written as:

$$A = \begin{pmatrix} A_{11} & A_{12} & A_{13} \\ A_{21} & A_{22} & A_{23} \end{pmatrix} \text{ where}$$

$$A_{11} = \begin{pmatrix} 15 & 0 \\ -3 & 4 \end{pmatrix}; \quad A_{12} = \begin{pmatrix} \pi \\ 2 \end{pmatrix}; \quad \ldots; \quad A_{23} = \begin{pmatrix} 4 & 0 \\ 1 & -7 \end{pmatrix}$$

A can be partitioned in several ways. Column wise partitioning of A results into:

$$A = \left(\begin{array}{c|c|c|c|c} 15 & 0 & \pi & 3 & 5 \\ -3 & 4 & 2 & 3 & 7 \\ -1 & \sqrt{17} & -2 & 4 & 0 \\ 0 & 1 & 6 & 1 & -7 \end{array}\right)$$

So, A can be written as:

$$A = (\underline{a}_1, \underline{a}_2, \underline{a}_3, \underline{a}_4, \underline{a}_5) \text{ in which } \underline{a}_1 = \begin{pmatrix} 15 \\ -3 \\ -1 \\ 0 \end{pmatrix}, \ldots, \underline{a}_5 = \begin{pmatrix} 5 \\ 7 \\ 0 \\ -7 \end{pmatrix}$$

Partitioning the unit matrix E with respect to columns gives:

$$E_4 = \left(\begin{array}{c|c|c|c} 1 & 0 & 0 & 0 \\ 0 & 1 & 0 & 0 \\ 0 & 0 & 1 & 0 \\ 0 & 0 & 0 & 1 \end{array}\right) = (\underline{e}_1, \underline{e}_2, \underline{e}_3, \underline{e}_4)$$

Concept A.15

The *transpose* of a matrix A, denoted by A', is systematically exchanging rows and columns of a matrix. For the $m \times n$-matrix A, the $n \times m$-matrix A' is called its transpose, in which every row of A corresponds to the same indexed column of A' and vice versa.

Example A.3.

$$A = \begin{pmatrix} 2 & 5 & 7 & -3 \\ 3 & 13 & 1 & -5 \\ 0 & \sqrt{2} & -3 & 4 \end{pmatrix}; \quad A' = \begin{pmatrix} 2 & 3 & 0 \\ 5 & 13 & \sqrt{2} \\ 7 & 1 & -3 \\ -3 & -5 & 4 \end{pmatrix}$$

In Example A.3 the matrix A is a 3×4-matrix and its transpose A' is a 4×3-matrix. Note that the transpose of A', so $(A')'$ is the original matrix A. So $(A')' = A$.

Remark

Vectors can be transposed too. In this book a vector is noted as a column vector, the transpose of a vector is always a row vector.

Example A.4.

$$\underline{a} = \begin{pmatrix} 5 \\ 7 \\ -3 \\ \sqrt{5} \end{pmatrix}; \qquad \underline{a}' = (5, 7, -3, \sqrt{5})$$

Concept A.16

A *symmetric matrix* is a matrix that equals its transpose, so $A' = A$.

Example A.5.

$$B = \begin{pmatrix} 5 & 3 & 1 \\ 3 & 7 & 0 \\ 1 & 0 & 9 \end{pmatrix}; \quad B' = \begin{pmatrix} 5 & 3 & 1 \\ 3 & 7 & 0 \\ 1 & 0 & 9 \end{pmatrix}; \quad B = B'$$

A symmetric matrix is square. Unit matrices are symmetric.

Concept A.17

A relation (greater than, equal to, less than) between matrices or vectors of equal size is valid if it holds for all corresponding elements individually.

Comparison of matrices and vectors is only defined for matrices and vectors of equal size.
$A^{m \times n}$ can be compared to $B^{m \times n}$.
$C^{p \times q}$ can not be compared to $A^{m \times n}$, if $p \neq m$ and/or $q \neq n$.
$\underline{a} \in \mathbb{R}^n$ can be compared to $\underline{b} \in \mathbb{R}^n$, not to $\underline{c} \in \mathbb{R}^m$, $(m \neq n)$.

Example A.6.

$$\underline{a} = \begin{pmatrix} a_1 \\ a_2 \\ a_3 \\ a_4 \end{pmatrix} = \begin{pmatrix} 5 \\ 3 \\ 7 \\ -5 \end{pmatrix}; \qquad \underline{b} = \begin{pmatrix} b_1 \\ b_2 \\ b_3 \\ b_4 \end{pmatrix} = \begin{pmatrix} 4 \\ 3 \\ 2 \\ -8 \end{pmatrix}; \qquad \underline{c} = \begin{pmatrix} c_1 \\ c_2 \\ c_3 \end{pmatrix} = \begin{pmatrix} 6 \\ 4 \\ 8 \end{pmatrix}$$

As $a_1 > b_1$; $a_2 = b_2$; $a_3 > b_3$ and $a_4 > b_4$, vector $\underline{a}$ is greater than or equal to $\underline{b}$. So, $\underline{a} \geq \underline{b}$ or $\underline{b} \leq \underline{a}$. However, vector $\underline{a}$ is not greater than $\underline{b}$, as $a_2 = b_2$. Comparing $\underline{a}$ and $\underline{c}$, or $\underline{b}$ and $\underline{c}$ is not defined as $\underline{a}, \underline{b} \in \mathbb{R}^4$ and $\underline{c} \in \mathbb{R}^3$. If $\underline{x} \in \mathbb{R}^n$ then $\underline{x} \geq \underline{0}$ means that $x_1 \geq 0, \ldots, x_n \geq 0$.

Example A.7.

$$A = \begin{pmatrix} 6 & 3 \\ 1 & 4 \end{pmatrix}; \quad B = \begin{pmatrix} 4 & 2 \\ 0 & -1 \end{pmatrix}; \quad C = \begin{pmatrix} 4 & 7 \\ 0 & -1 \end{pmatrix}$$

The following holds: $A > B$ because $a_{ij} > b_{ij}$ for all i and j; $B \leq C$ as $b_{ij} \leq c_{ij}$ for all i and j.

Concept A.18

The *inner product* of two vectors from the same space $\mathbb{R}^n$ is the summation of the products of corresponding elements in the vectors. An inner product of the vectors $\underline{a}$ and $\underline{b}$ is noted as $\underline{a}'\underline{b}$ and is a real number. If $\underline{a} \in \mathbb{R}^n$ and $\underline{b} \in \mathbb{R}^n$ then $\underline{a}'\underline{b} = \sum_{i=1}^{n} a_i b_i = a_1 b_1 + \ldots + a_n b_n$.

So an inner product $\underline{a}'\underline{b}$ is only defined if the vectors are members of the same space. If $\underline{a} \in \mathbb{R}^n$ then $\underline{a}'\underline{b}$ only exists in case $\underline{b} \in \mathbb{R}^n$.

Example A.8.

$$\underline{a}' = (a_1, a_2, a_3, a_4, a_5) = (5, 3, 7, \sqrt{2}, -2); \qquad \underline{b} = \begin{pmatrix} b_1 \\ b_2 \\ b_3 \\ b_4 \\ b_5 \end{pmatrix} = \begin{pmatrix} -2 \\ 7 \\ 0 \\ \sqrt{8} \\ -2 \end{pmatrix}$$

$$\underline{a}'\underline{b} = a_1b_1 + a_2b_2 + a_3b_3 + a_4b_4 + a_5b_5 = -10 + 21 + 0 + 4 + 4 = 19.$$

Concept A.19

The (*Euclidean*) *length* $|\underline{a}|$ of a vector $\underline{a} \in \mathbb{R}^n$ is defined as:

$$|\underline{a}| = \sqrt{\underline{a}'\underline{a}} = \sqrt{a_1^2 + a_2^2 + \ldots + a_n^2}\,.$$

Example A.9.

For $\underline{a}' = (3, 4)$ holds $|\underline{a}| = \sqrt{3^2 + 4^2} = 5$.

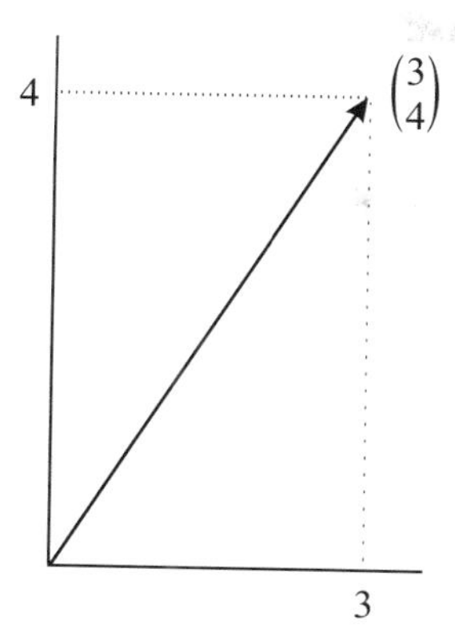

Concept A.20

Multiplication of a matrix by a scalar λ. A *scalar* is a real or complex number. λA is a matrix, in which all elements of the matrix A are λ times the original elements of A.

Example A.10

$$C = \begin{pmatrix} 4 & 3 \\ -2 & -5 \end{pmatrix}; \qquad 3C = \begin{pmatrix} 12 & 9 \\ -6 & -15 \end{pmatrix}$$

Concept A.21

Adding matrices of the same size results into a matrix of the same size where every element equals the sum of the corresponding elements of the matrices that are added. Subtracting matrices is defined analogously. Adding matrices is only defined for matrices of the same size: $A^{m \times n} + B^{m \times n} = S^{m \times n}$

Example A.11

$$\begin{pmatrix} 2 & 3 & -5 \\ 0 & -1 & 7 \end{pmatrix} + \begin{pmatrix} -3 & 0 & 1 \\ 2 & -1 & -4 \end{pmatrix} = \begin{pmatrix} -1 & 3 & -4 \\ 2 & -2 & 3 \end{pmatrix}$$

Concept A.22

Multiplication of matrices

Two matrices A and B can be multiplied if the number of columns of A equals the number of rows of B.

For $A^{m \times n}$ and $B^{n \times p}$ the multiplication $A \times B$, or AB, is defined because the number of columns of A equals the number of rows of B (namely n). For the product matrix $P = AB$, element p_{ij} in row i and column j is the inner product of row i of A and column j of B.

Consequently, matrix P is an $m \times p$-matrix, corresponding to the number of rows of $A^{m \times n}$ and the number of columns of $B^{n \times p}$.

So, $\quad A^{m \times n} \times B^{n \times p} = P^{m \times p}$

Example A.12

$$A^{3 \times 2} \times B^{2 \times 3} = P^{3 \times 3};\; A^{5 \times 1} \times B^{1 \times 6} = P^{5 \times 6}$$
$$A^{2 \times 3} \times B^{2 \times 2} \text{ is not defined.}$$

Notice the importance of the order of A and B.

$$A^{1 \times 5} \times B^{5 \times 1} = P^{1 \times 1}$$
$$B^{5 \times 1} \times A^{1 \times 5} = P^{5 \times 5}$$

$P^{1 \times 1}$ is a scalar and $P^{5 \times 5}$ is a matrix of 5 rows and 5 columns.

The rule $A^{m \times n} \times B^{n \times p} = P^{m \times p}$ facilitates the determination of the size of the product matrix.

$$A^{m\times n} = \begin{pmatrix} a_{11} & a_{12} & \dots & a_{1j} & \dots & a_{1n} \\ a_{21} & a_{22} & \dots & a_{2j} & \dots & a_{2n} \\ . & . & & . & & . \\ . & . & & . & & . \\ a_{i1} & a_{i2} & \dots & a_{ij} & \dots & a_{in} \\ . & . & & . & & . \\ . & . & & . & & . \\ a_{m1} & a_{m2} & \dots & a_{mj} & \dots & a_{mn} \end{pmatrix};\; B^{n\times p} = \begin{pmatrix} b_{11} & b_{12} & \dots & b_{1j} & \dots & b_{1p} \\ b_{21} & b_{22} & \dots & b_{2j} & \dots & b_{2p} \\ . & . & & . & & . \\ . & . & & . & & . \\ b_{i1} & b_{i2} & \dots & b_{ij} & \dots & b_{ip} \\ . & . & & . & & . \\ . & . & & . & & . \\ b_{n1} & b_{n2} & \dots & b_{nj} & \dots & b_{np} \end{pmatrix}$$

$$A = \begin{pmatrix} \underline{a}_1' \\ \underline{a}_2' \\ . \\ . \\ \underline{a}_i' \\ . \\ . \\ \underline{a}_m' \end{pmatrix}; \qquad B = (\underline{b}_1, \underline{b}_2, \ldots, \underline{b}_j, \ldots, \underline{b}_p)$$

where $\underline{a}_1'$ is row vector $(a_{11}, a_{12} \ldots a_{1j}, \ldots, a_{1n})$, and $\underline{a}_i' = (a_{i1}, a_{i2}, \ldots, a_{ij}, \ldots, a_{in})$. In the same way $\underline{b}_1$ is the column vector:

$$\underline{b}_1 = \begin{pmatrix} b_{11} \\ b_{21} \\ \cdot \\ \cdot \\ b_{i1} \\ \cdot \\ \cdot \\ b_{n1} \end{pmatrix}; \qquad \underline{b}_j = \begin{pmatrix} b_{1j} \\ b_{2j} \\ \cdot \\ \cdot \\ b_{ij} \\ \cdot \\ \cdot \\ b_{nj} \end{pmatrix}$$

From this partitioned view, matrix A partitioned row-wise and B column-wise, the product matrix $P = A \times B$ can be written as follows:

$$A \times B = \begin{pmatrix} \underline{a}_1' \underline{b}_1 & \underline{a}_1' \underline{b}_2 & \dots & \underline{a}_1' \underline{b}_j & \dots & \underline{a}_1' \underline{b}_p \\ \underline{a}_2' \underline{b}_1 & \underline{a}_2' \underline{b}_2 & \dots & \underline{a}_2' \underline{b}_j & \dots & \underline{a}_2' \underline{b}_p \\ \cdot & \cdot & \cdot & \cdot & & \cdot \\ \cdot & \cdot & & \cdot & & \cdot \\ \underline{a}_i' \underline{b}_1 & \underline{a}_i' \underline{b}_2 & \dots & \underline{a}_i' \underline{b}_j & \dots & \underline{a}_i' \underline{b}_p \\ \cdot & \cdot & & \cdot & & \cdot \\ \cdot & \cdot & & \cdot & & \cdot \\ \underline{a}_m' \underline{b}_1 & \underline{a}_m' \underline{b}_2 & \dots & \underline{a}_m' \underline{b}_j & \cdot & \underline{a}_m' \underline{b}_p \end{pmatrix} = P$$

Each element of P is an inner product of a row vector of A and a column vector of B. The inner product of row i of A and column j of B results into element p_{ij} of P: $p_{ij} = \underline{a}_i' \underline{b}_j$.

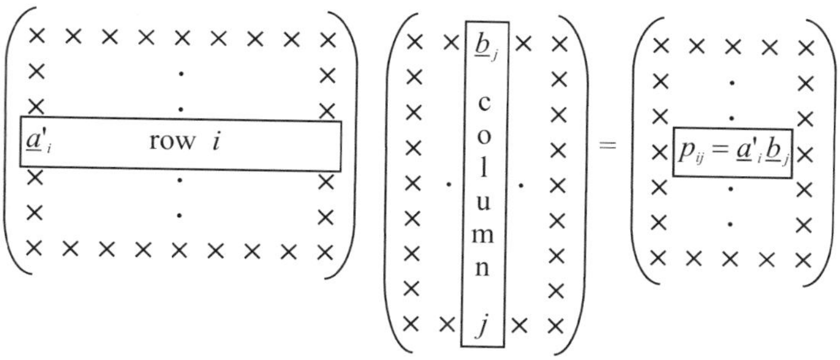

$\underline{a}'_i$ row i $\underline{b}_j$ column j = $p_{ij} = \underline{a}'_i \underline{b}_j$

A B P

Example A.13.

$$A = \begin{pmatrix} 2 & 3 \\ -1 & 4 \\ 0 & 5 \end{pmatrix}; \quad B = \begin{pmatrix} 1 & 0 & 3 \\ 2 & 1 & -1 \end{pmatrix}$$

Multiplication AB gives $A^{3\times2}\times B^{2\times3} = P^{3\times3}$. Product matrix P is of size 3×3. The elements of P are determined as follows:

$$p_{11} = \underline{a}_1'\underline{b}_1 = (2,3)\times\begin{pmatrix}1\\2\end{pmatrix} = 2\times1+3\times2 = 2+6 = 8$$

$$p_{12} = \underline{a}_1'\underline{b}_2 = (2,3)\times\begin{pmatrix}0\\1\end{pmatrix} = 2\times0+3\times1 = 0+3 = 3$$

and so on, up to

$$p_{33} = \underline{a}_3'\underline{b}_3 = (0,5)\times\begin{pmatrix}3\\-1\end{pmatrix} = 0\times3+5\times-1 = 0-5 = -5$$

Such that:

$$P = AB = \begin{pmatrix}2 & 3\\-1 & 4\\0 & 5\end{pmatrix}\begin{pmatrix}1 & 0 & 3\\2 & 1 & -1\end{pmatrix} = \begin{pmatrix}8 & 3 & 3\\7 & 4 & -7\\10 & 5 & -5\end{pmatrix}$$

In this case also the multiplication BA is defined.

$$B^{2\times3}\times A^{3\times2} = Q^{2\times2}$$

$$Q = BA = \begin{pmatrix}1 & 0 & 3\\2 & 1 & -1\end{pmatrix}\begin{pmatrix}2 & 3\\-1 & 4\\0 & 5\end{pmatrix} = \begin{pmatrix}2 & 18\\3 & 5\end{pmatrix}$$

Note that $AB \neq BA$. There are special cases in which $AB = BA$, but in general $AB \neq BA$ holds.

In general holds: $AB \neq BA$.

$$\underline{a} = \begin{pmatrix}1\\2\end{pmatrix}; \qquad \underline{b} = \begin{pmatrix}2\\-1\end{pmatrix}$$

$$\underline{a}'\underline{b} = (1,2)\begin{pmatrix}2\\-1\end{pmatrix} = 0 \qquad (A^{1\times2}\times B^{2\times1} = P^{1\times1};\ \text{a scalar, an inner product})$$

$$\underline{ab}' = \begin{pmatrix}1\\2\end{pmatrix}(2,-1) = \begin{pmatrix}2 & -1\\4 & -2\end{pmatrix} \qquad (A^{2\times1}\times B^{1\times2} = P^{2\times2};\ \text{a matrix}).$$

Matrices cannot be multiplied in an arbitrary sequence. The following calculation rules hold.

Concept A.23

Calculation rules for matrices.

1. $A\times(BC) = (AB)\times C$
2. $A\times(B+C) = AB + AC$
3. $(A+B)\times C = AC + BC$

4. $0A = 0$
5. $AE_n = E_mA = A$

Concept A.24

For the transpose of a product of two matrices holds: $(BA)' = A'B'$.

Example A.14.

$$A = \begin{pmatrix} 4 & 3 \\ 2 & 1 \end{pmatrix}; \qquad B = \begin{pmatrix} 0 & -1 \\ -2 & 3 \end{pmatrix}$$

$$BA = \begin{pmatrix} -2 & -1 \\ -2 & -3 \end{pmatrix} \qquad (BA)' = \begin{pmatrix} -2 & -2 \\ -1 & -3 \end{pmatrix}$$

$$A' = \begin{pmatrix} 4 & 2 \\ 3 & 1 \end{pmatrix} \qquad B' = \begin{pmatrix} 0 & -2 \\ -1 & 3 \end{pmatrix}$$

$$A'B' = \begin{pmatrix} -2 & -2 \\ -1 & -3 \end{pmatrix} \qquad \text{so } (BA)' = A'B'$$

Concept A.25

The *product matrix* of a matrix and its transpose is always a symmetric matrix. Namely, $(BB')' = (B')'B' = B'$; so BB' is symmetric.

A.3. Geometric interpretation of vectors

In $\mathbb{R}^2$ a point $\underline{a}$ with co-ordinates (a_1, a_2) can be represented as the vector $\underline{a} = \begin{pmatrix} a_1 \\ a_2 \end{pmatrix}$, which in turn can be represented geometrically as the arrow starting in the origin $\underline{0}$ and ending in $\underline{a}$.

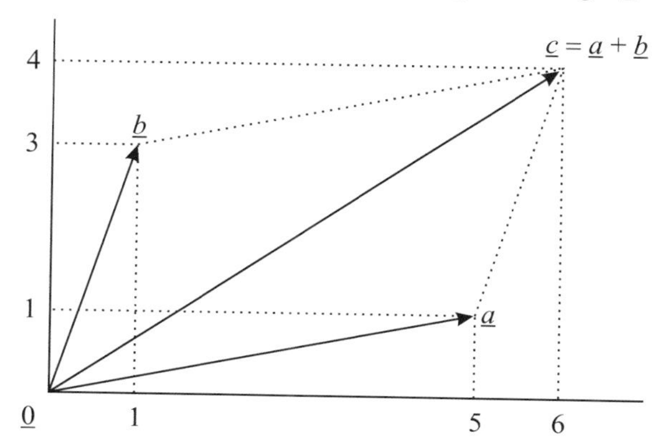

Figure A.2. Sum vector $\underline{c} = \underline{a} + \underline{b}$.

In the same way a vector in $\mathbb{R}^3$ can be represented geometrically. Vectors represented as arrows start in a fixed point, the origin $\underline{0}$. A vector with length zero is the null vector $\underline{0}$. Two vectors $\underline{a}$ and $\underline{b}$ can be added if $\underline{a}$ and $\underline{b}$ belong to the same space (See Figure A.2). In $\mathbb{R}^2$ the *sum vector* $\underline{c} = \underline{a} + \underline{b}$ can be constructed as:

$$\text{If } \underline{a} = \begin{pmatrix} 5 \\ 1 \end{pmatrix} \text{ and } \underline{b} = \begin{pmatrix} 1 \\ 3 \end{pmatrix}, \text{ then } \underline{c} = \underline{a} + \underline{b} = \begin{pmatrix} 6 \\ 4 \end{pmatrix}$$

Multiplication of a vector $\underline{a}$ with a scalar looks is shown in Figure A.3. The presentation shows that the set of all vectors $\lambda\underline{a}$ with $\lambda \in \mathbb{R}$ is a line through the origin.

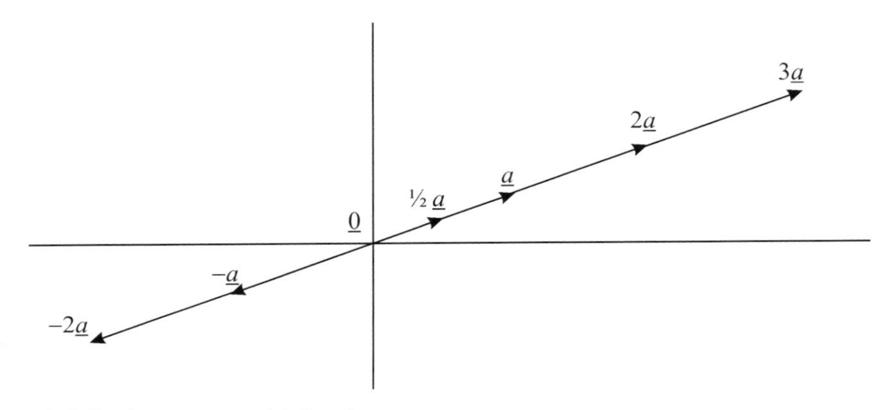

Figure A.3. Scalar vector multiplication.

Concept A.26

The *vector equation* of a line through the origin is $\underline{x} = \lambda\underline{a}$, $\lambda \in \mathbb{R}$.

Drawing the sum vectors $\underline{c} = \underline{a} + \lambda\underline{b}$ for all $\lambda \in \mathbb{R}$ gives a line too (see Figure A.4).

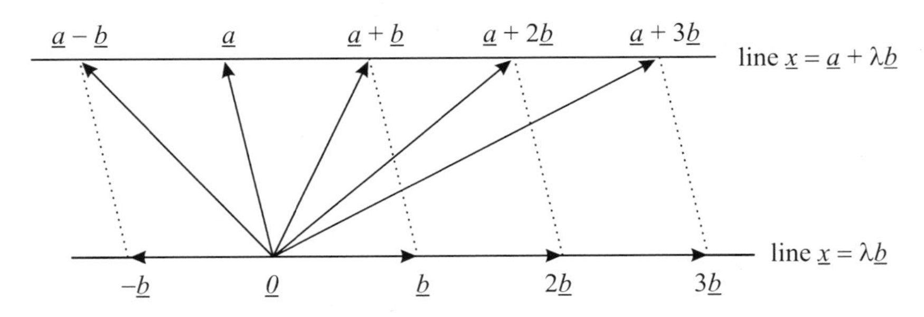

Figure A.4. Line $\underline{x} = \underline{a} + \lambda\underline{b}$ and line $\underline{x} = \lambda\underline{b}$, $\lambda \in \mathbb{R}$.

Line $\underline{x} = \underline{a} + \lambda\underline{b}$ does not go through the origin, but is parallel to line $\lambda\underline{b}$ through the origin.

Concept A.27

The vector equation of line l is $\underline{x} = \underline{a} + \lambda\underline{b}$ where $\underline{b}$ is called the *direction vector*.

Representation of line l as a set is $l := \{\underline{x} \in \mathbb{R}^n \mid \underline{x} = \underline{a} + \lambda\underline{b}, \lambda \in \mathbb{R}\}$. In case only non-negative values of λ are considered, a half-line appears:

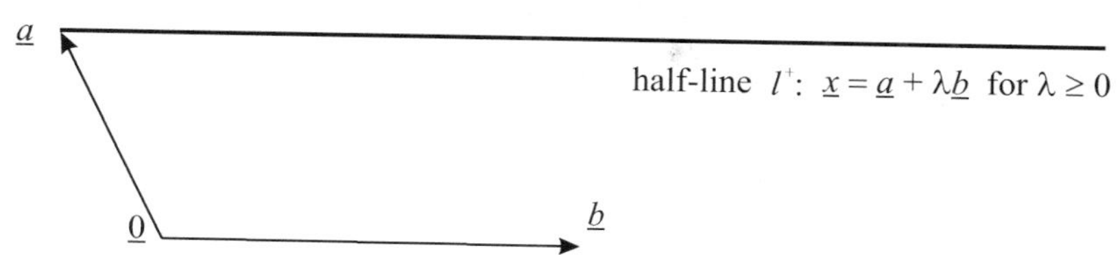

Figure A.5. Half-line $\underline{x} = \underline{a} + \lambda\underline{b}$, $\lambda \geq 0$.

Concept A.28

The vector equation of line l^+ is $\underline{x} = \underline{a} + \lambda\underline{b}$ with $\lambda \geq 0$.
For $\lambda \leq 0$ the half-line $l^- = \{\underline{x} \in \mathbb{R}^n \mid \underline{x} = \underline{a} + \lambda\underline{b}, \lambda \leq 0\}$ appears.

Concept A.29

A *line segment* between points $\underline{a}$ and $\underline{b}$ is given by $\underline{x} = \underline{a} + \lambda(\underline{b} - \underline{a})$, with $0 \leq \lambda \leq 1$ or alternatively $\underline{x} = \lambda\underline{b} + (1 - \lambda)\underline{a}$, $0 \leq \lambda \leq 1$.

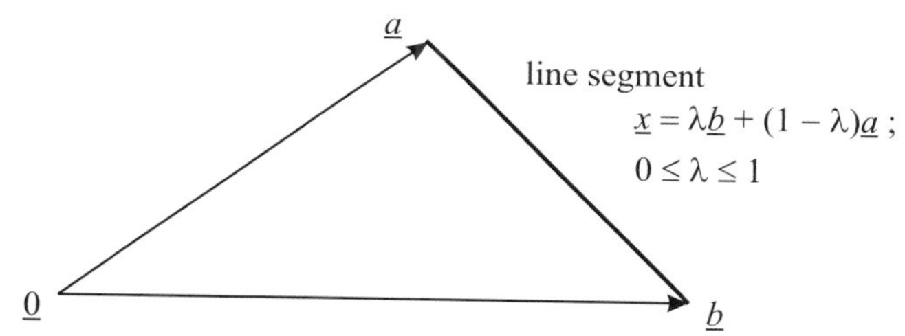

Figure A.6. Line segment $\underline{x} = \lambda\underline{b} + (1 - \lambda)\underline{a}$, $0 \leq \lambda \leq 1$ *between points* $\underline{a}$ *and* $\underline{b}$.

For $\lambda = 0$, $\underline{x} = \underline{a}$; for $\lambda = 1$, $\underline{x} = \underline{b}$ and for $\lambda = \frac{1}{2}$, $\underline{x}$ is the midpoint of the line segment between $\underline{a}$ and $\underline{b}$.

A.4. Basic concepts (2)

If $\underline{a}$ and $\underline{b}$ are situated on one line then $\underline{b}$ can be expressed in $\underline{a}$ and the other way around. If $\underline{b}$ can be written as $\underline{b} = \lambda\underline{a}$, $\lambda \in \mathbb{R}$, then $\underline{b}$ is *linearly dependent* on $\underline{a}$.

Example A.15.

Left in Figure A.7, $\underline{a} = \begin{pmatrix} 2 \\ 1 \end{pmatrix}$ and $\underline{b} = \begin{pmatrix} 4 \\ 2 \end{pmatrix}$ are two linearly dependent vectors in $\mathbb{R}^2$, as $\underline{b} = 2\underline{a}$ or alternatively $2\underline{a} - \underline{b} = \underline{0}$. If $\underline{a}$ and $\underline{b}$ are linearly dependent, not every vector $\underline{x}$ in $\mathbb{R}^2$ can be expressed as a linear combination of $\underline{a}$ and $\underline{b}$. E.g. $\underline{x} = \begin{pmatrix} 2 \\ 5 \end{pmatrix}$ cannot be written as a linear combination of $\underline{a}$ and $\underline{b}$, because $\begin{pmatrix} 2 \\ 5 \end{pmatrix} = \lambda \begin{pmatrix} 2 \\ 1 \end{pmatrix} + \mu \begin{pmatrix} 4 \\ 2 \end{pmatrix}$ leads to an *inconsistent system*. Alternatively, $\left.\begin{matrix} 2 = 2\lambda + 4\mu & \rightarrow & 2 = 2\lambda + 4\mu \\ 5 = \lambda + 2\mu & \rightarrow & 10 = 2\lambda + 4\mu \end{matrix}\right\}$ is inconsistent. The vectors $\underline{a}$ and $\underline{b}$ left in Figure A.7 do not span space $\mathbb{R}^2$ but only line l: $\underline{x} = \lambda\underline{a}$ in $\mathbb{R}^2$.

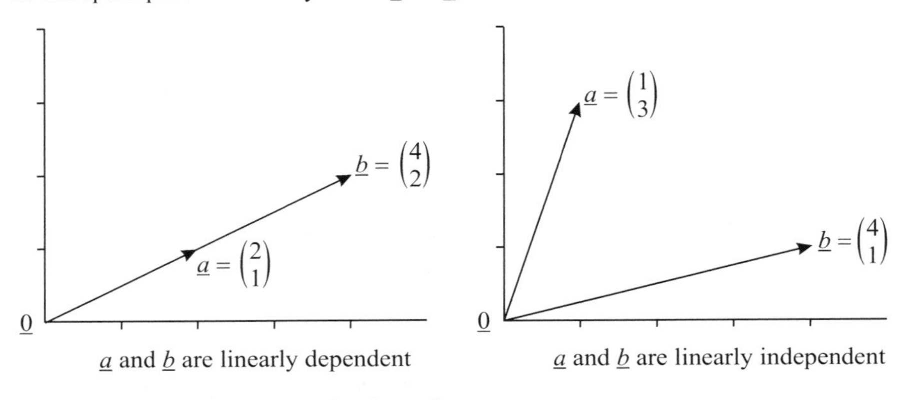

Figure A.7. Linear dependence and independence.

Right in Figure A.7, $\underline{a} = \begin{pmatrix} 1 \\ 3 \end{pmatrix}$ and $\underline{b} = \begin{pmatrix} 4 \\ 1 \end{pmatrix}$ are linearly independent vectors in $\mathbb{R}^2$; $\underline{b} \neq \lambda\underline{a}$.
Every other vector $\underline{x}$ in $\mathbb{R}^2$ can be expressed as a linear combination of $\underline{a}$ and $\underline{b}$. For example, $\underline{x} = \begin{pmatrix} 6 \\ 7 \end{pmatrix}$ can be written as $\underline{x} = 2\underline{a} + \underline{b} = 2\begin{pmatrix} 1 \\ 3 \end{pmatrix} + \begin{pmatrix} 4 \\ 1 \end{pmatrix} = \begin{pmatrix} 6 \\ 7 \end{pmatrix}$.
Generalisation towards $\mathbb{R}^n$ gives the following definitions and concepts:

Concept A.30

A set of vectors $\underline{x}_1, \underline{x}_2, \underline{x}_3, \ldots, \underline{x}_m \in \mathbb{R}^n$ is called *linearly dependent* if scalars $\lambda_1, \lambda_2, \lambda_3, \ldots, \lambda_m$ exist not all equal to zero, such that $\lambda_1\underline{x}_1 + \lambda_2\underline{x}_2 + \lambda_3\underline{x}_3 + \ldots \lambda_m\underline{x}_m = \underline{0}$.

Concept A.31

A set of vectors $\underline{x}_1, \underline{x}_2, \underline{x}_3, \ldots, \underline{x}_m \in \mathbb{R}^n$ is called *linearly independent* if the only solution of system $\lambda_1\underline{x}_1 + \lambda_2\underline{x}_2 + \lambda_3\underline{x}_3 + \ldots \lambda_m\underline{x}_m = \underline{0}$ is $\lambda_1 = \lambda_2 = \lambda_3 = \ldots = \lambda_m = 0$.

Example A.16.

- Vectors $\underline{a} = \begin{pmatrix} 1 \\ 3 \end{pmatrix}$ and $\underline{b} = \begin{pmatrix} 4 \\ 1 \end{pmatrix}$ are linearly independent because solving $\lambda_1\underline{a} + \lambda_2\underline{b} = \underline{0}$

gives $\lambda_1\begin{pmatrix}1\\3\end{pmatrix}+\lambda_2\begin{pmatrix}4\\1\end{pmatrix}=\begin{pmatrix}0\\0\end{pmatrix}$, or $\left.\begin{matrix}\lambda_1+4\lambda_2=0\\3\lambda_1+\lambda_2=0\end{matrix}\right\}$ this system has as just one solution $\lambda_1 = 0$ and $\lambda_2 = 0$.

- The vectors $\begin{pmatrix}2\\1\end{pmatrix}$ and $\begin{pmatrix}4\\2\end{pmatrix}$ are linearly dependent because $\lambda_1\begin{pmatrix}2\\1\end{pmatrix}+\lambda_2\begin{pmatrix}4\\2\end{pmatrix}=\begin{pmatrix}0\\0\end{pmatrix}$ has more solutions than $\lambda_1 = \lambda_2 = 0$. E.g. $\lambda_1 = 2$, $\lambda_2 = -1$ or $\lambda_1 = 100$, $\lambda_2 = -50$ (see Figure A.7).

- Vectors $\underline{a}=\begin{pmatrix}7\\1\end{pmatrix}$; $\underline{b}=\begin{pmatrix}3\\3\end{pmatrix}$; $\underline{c}=\begin{pmatrix}2\\8\end{pmatrix}$ are linearly dependent because $\lambda_1\underline{a} + \lambda_2\underline{b} + \lambda_3\underline{c} = \underline{0}$ has more solutions than $\lambda_1 = \lambda_2 = \lambda_3 = 0$; $\lambda_1\begin{pmatrix}7\\1\end{pmatrix}+\lambda_2\begin{pmatrix}3\\3\end{pmatrix}+\lambda_3\begin{pmatrix}2\\8\end{pmatrix}=\begin{pmatrix}0\\0\end{pmatrix}$ gives system

  $7\lambda_1 + 3\lambda_2 + 2\lambda_3 = 0$
  $\lambda_1 + 3\lambda_2 + 8\lambda_3 = 0$

  A solution of this system is for instance $\lambda_1 = 1$, $\lambda_2 = -3$ and $\lambda_3 = 1$. The system has (infinitely) many solutions, e.g. $\lambda_1 = 10$, $\lambda_2 = -30$ and $\lambda_3 = 10$ or $\lambda_1 = \sqrt{\pi}$, $\lambda_2 = -3\sqrt{\pi}$ and $\lambda_3 = \sqrt{\pi}$. Drawing $\underline{a}$ and $\underline{b}$ in $\mathbb{R}^2$ shows that they do not point in the same direction and therefore these vectors are linearly independent. Vectors $\underline{a}$ and $\underline{b}$ span the space $\mathbb{R}^2$. Vector $\underline{c}$ surely can be written as a linear combination of $\underline{a}$ and $\underline{b}$:

  $\underline{c} = \lambda\underline{a} + \mu\underline{b}$, gives $\begin{pmatrix}2\\8\end{pmatrix}=-\begin{pmatrix}7\\1\end{pmatrix}+3\begin{pmatrix}3\\3\end{pmatrix}$. So, vectors $\underline{a}$, $\underline{b}$ and $\underline{c}$ are linearly dependent.

- The set of unit vectors $\underline{e}_1, \underline{e}_2, \ldots, \underline{e}_n \in \mathbb{R}^n$ is linearly independent.
  The system $\lambda_1\underline{e}_1 + \lambda_2\underline{e}_2 + \ldots + \lambda_n\underline{e}_n = \underline{0}$ has the unique solution $\lambda_1 = \lambda_2 = \ldots = \lambda_2 = 0$.

Concept A.32

A set of linearly independent vectors in $\mathbb{R}^n$ spanning space $\mathbb{R}^n$ is called a *basis* of $\mathbb{R}^n$.

The set of unit vectors $\underline{e}_1$, $\underline{e}_2$ and $\underline{e}_3 \in \mathbb{R}^3$ is linearly independent. Every vector in $\mathbb{R}^3$ can be written in one way as a linear combination of $\underline{e}_1$, $\underline{e}_2$ and $\underline{e}_3$.

$$\underline{x}=\begin{pmatrix}\alpha\\\beta\\\gamma\end{pmatrix}=\begin{pmatrix}\alpha\\0\\0\end{pmatrix}+\begin{pmatrix}0\\\beta\\0\end{pmatrix}+\begin{pmatrix}0\\0\\\gamma\end{pmatrix}=\alpha\begin{pmatrix}1\\0\\0\end{pmatrix}+\beta\begin{pmatrix}0\\1\\0\end{pmatrix}+\gamma\begin{pmatrix}0\\0\\1\end{pmatrix}=\alpha\underline{e}_1+\beta\underline{e}_2+\gamma\underline{e}_3.$$

Vectors $\underline{e}_1$, $\underline{e}_2$ and $\underline{e}_3$ span space $\mathbb{R}^3$ and therefore constitute a basis of $\mathbb{R}^3$. The three vectors:

$$\underline{a}=\begin{pmatrix}2\\1\\1\end{pmatrix};\qquad \underline{b}=\begin{pmatrix}1\\1\\1\end{pmatrix};\qquad \underline{c}=\begin{pmatrix}-1\\1\\0\end{pmatrix}$$

are linearly independent too, so they also constitute a basis of $\mathbb{R}^3$.

Concept A.33

A *basis* of $\mathbb{R}^n$ consists of n linearly independent vectors $\in \mathbb{R}^n$; various sets of n linearly independent vectors $\in \mathbb{R}^n$ can be used as a basis.

If $\underline{a}_1, \dots, \underline{a}_n$ constitute a basis of $\mathbb{R}^n$, then every vector $\underline{x} \in \mathbb{R}^n$ can be written in a unique way as a linear combination of $\underline{a}_1, \dots, \underline{a}_n$.

- Vectors $\underline{a}_1 = \begin{pmatrix} 1 \\ 3 \end{pmatrix}$ and $\underline{a}_2 = \begin{pmatrix} 2 \\ 1 \end{pmatrix}$ are linearly independent and constitute a basis of $\mathbb{R}^2$.
- Vectors $\underline{a}_1 = \begin{pmatrix} 1 \\ 3 \end{pmatrix}, \underline{a}_2 = \begin{pmatrix} 2 \\ 1 \end{pmatrix}, \underline{a}_3 = \begin{pmatrix} 3 \\ 4 \end{pmatrix}$ are linearly dependent, so they do not constitute a basis of $\mathbb{R}^2$.

Concept A.34

A basis of $\mathbb{R}^n$ consisting of n unit vectors $\underline{e}_1$ up to $\underline{e}_n \in \mathbb{R}^n$ is the *standard basis* of $\mathbb{R}^n$.

In an $m \times n$-matrix A appear n columns of length m. Partitioning A column wise gives n column vectors.

Concept A.35

The *rank* of a matrix A is the (maximum) number of linearly independent columns of A. Notation: r(A).

The number of linearly independent columns of matrix A equals the number of linearly independent rows of A.

$$A = \begin{pmatrix} 3 & -5 & \sqrt{2} & 6 & 3/2 & 1 & 0 \\ 7 & \pi & 3 & 0 & 2 & 0 & 1 \end{pmatrix}$$

In this example, the maximum number of independent columns is 2: r(A) = 2. There are also two independent rows: r(A) = 2.

$$A = \begin{pmatrix} 1 & 2 & 0 & -1 & 0 \\ 0 & 3 & 1 & 2 & 0 \\ 0 & 4 & 0 & -3 & 1 \end{pmatrix}$$ at most 3 independent columns exist, so r(A) = 3.

$$A = \begin{pmatrix} 1 & 2 & -3 & 5 \\ 2 & 4 & -6 & 10 \end{pmatrix} \Rightarrow \text{r}(A) = 1$$ (one independent column; one independent row).

A.5. Solving systems of linear equations

In Section A.5.1 a systematic method (pivoting) is discussed to solve systems of equations and to determine whether more than one solution exists or alternatively find out the system is inconsistent. In Section A.5.2, a link is made with the graphical solutions in $\mathbb{R}^2$ and the concepts of linear algebra such as (in)dependent vectors, r(A) and basis.

A.5.1. Solving systems of linear equations by pivoting

Concept A.36

Pivoting is a method to solve systems of linear equations in a systematic way. By pivoting we transform a system, of which the solution can not immediately be seen, into a simpler system of which the solution can immediately be seen.

Two elementary operations are used:
- multiplying an equation with a scalar $\neq 0$;
- add (subtract) equations.

or a combination of these operations.
We illustrate this pivoting procedure by an example:

Example A.17.

Solve system $A\underline{x} = \underline{b}$, alternatively $(\underline{a}_1, \underline{a}_2, \underline{a}_3)\begin{pmatrix} x_1 \\ x_2 \\ x_3 \end{pmatrix} = \underline{b}$ or $x_1\underline{a}_1 + x_2\underline{a}_2 + x_3\underline{a}_3 = \underline{b}$ in which

$$\underline{a}_1 = \begin{pmatrix} 2 \\ -1 \\ 4 \end{pmatrix}; \quad \underline{a}_2 = \begin{pmatrix} -2 \\ 2 \\ -2 \end{pmatrix}; \quad \underline{a}_3 = \begin{pmatrix} -2 \\ 3 \\ 3 \end{pmatrix}; \quad \underline{b} = \begin{pmatrix} 4 \\ -1 \\ 13 \end{pmatrix}$$

System $A\underline{x} = \underline{b}$,

$$\begin{aligned} 2x_1 - 2x_2 - 2x_3 &= 4 \\ -x_1 + 2x_2 + 3x_3 &= -1 \\ 4x_1 - 2x_2 + 3x_3 &= 13 \end{aligned}$$

can be written in so-called tableau form. In the *tableau form* only the coefficients of the variables x_j are shown. The vertical line represents the =–signs.

x_1	x_2	x_3	
2	–2	–2	4
–1	2	3	–1
4	–2	3	13

(A)

A solution does not follow directly from system (A). Therefore, system (A) is transformed in such a way that the equivalent system (S) appears with an obvious solution:

x_1	x_2	x_3	
1	0	0	2
0	1	0	−1
0	0	1	1

(S)

The solution of system (S) is trivial: $x_1 = 2$, $x_2 = -1$ and $x_3 = 1$.Now we transform system (A) into the equivalent system (S) in which the three unit vectors $\underline{e}_1$, $\underline{e}_2$, $\underline{e}_3$ appear. For this transformation we use the two elementary operations (multiplication and addition).

Iteration 1

In the first step, the column vector belonging to x_1 will be transformed into $\underline{e}_1$. Therefore, the element $a_{11} = 2$ is used as a *pivot*, this means:

- multiply equation 1 by $\frac{1}{2}$ (so $a_{11} = 2$ gets the value 1),
- add $\frac{1}{2} \times$ equation 1 to equation 2 (so $a_{21} =$ gets the value $-1 + \frac{1}{2} \times 2 = 0$),
- subtract $2 \times$ equation 1 from equation 3 (so $a_{31} =$ gets the value 0).

This leads to (the pivots are shaded):

x_1	x_2	x_3	
2	−2	−2	4
−1	2	3	−1
4	−2	3	13

(A)

	x_1	x_2	x_3	
½ row 1 →	1	−1	−1	2
row 2 + ½ row 1→	0	1	2	1
row 3 – 2 row 1→	0	2	7	5

(A_1)

Iteration 2

In the second step, pivot $a_{22} = 1$ is used to transform the column vector belonging to variable x_2 in system (A_1) into unit vector $\underline{e}_2$:

	x_1	x_2	x_3	
Row 1 + row 2 →	1	0	1	3
row 2 →	0	1	2	1
row 3 – 2 row 2 →	0	0	3	3

(A_2)

Iteration 3

In the third step, pivot $a_{33} = 3$ is used to pivot the column vector belonging to x_3 in system (A_2) into unit vector $\underline{e}_3$:

	x_1	x_2	x_3	
row 1 – 1/3 row 3 →	1	0	0	2
row 2 – 2/3 row 3 →	0	1	0	−1
1/3 row 3 →	0	0	1	1

(S)

So in three iterations of elementary pivot operations, system (A) has been transformed into the equivalent system (S), which has a trivial solution ($x_1 = 2$, $x_2 = -1$, $x_3 = 1$).

$$2\begin{pmatrix}2\\-1\\4\end{pmatrix}-1\begin{pmatrix}-2\\2\\-2\end{pmatrix}+1\begin{pmatrix}-2\\3\\3\end{pmatrix}=\begin{pmatrix}4\\-1\\13\end{pmatrix}$$

The pivoting procedure is generally applicable, so also for systems with n variables. Solution techniques for linear programming use these pivot operations frequently.

Remarks

1. Given a system like (S):

$$\begin{aligned} x_2 + x_3 &= 5 \\ x_1 + x_2 + x_3 &= 6 \\ -x_1 + x_2 + 2x_3 &= 7 \end{aligned} \qquad \text{(S)}$$

The pivoting procedure seems not to be applicable, as $a_{11} = 0$. There are two methods to solve this system by pivoting:

- Of course the solution for the system is not changed by switching equations. So, switch equations, e.g. exchange equation 1 with equation 2:

x_1	x_2	x_3	
1	1	1	6
0	1	1	5
−1	1	2	7

Solve this system with pivot $a_{11} = 1$ etc.

- Transform the first column of (S) e.g. into $\underline{e}_2$ (choose pivot $a_{21} = 1$); transform the second column into $\underline{e}_1$ and the third column into $\underline{e}_3$. The order of the unit vectors is not relevant to determine the trivial solution.

2. *Inconsistency* of the system is detected during the pivoting procedure.

x_1	x_2	x_3	
0	1	1	2
1	1	2	3
−1	2	1	6

x_1	x_2	x_3	
0	1	1	2
1	1	2	3
0	3	3	9

x_1	x_2	x_3	
0	1	1	2
1	0	1	1
0	0	0	3

The third equation $0x_1 + 0x_2 + 0x_3 = 3$ shows the system has no solutions.

3. A d*ependent* system is also detected.

$$\begin{aligned} x_1 + 2x_2 &= 2 \\ 3x_1 + 6x_2 &= 6 \end{aligned}$$

x_1	x_2	
1	2	2
3	6	6

x_1	x_2	
1	2	2
0	0	0

The second equality $0x_1 + 0x_2 = 0$ is always fulfilled. In this case the set of solutions is described by $x_1 + 2x_2 = 2$.

4. If in a system with m equations and n variables a unit matrix E_m appears, it is easy to give a description of the solution set. In the following system of 2 equations and 4 variables, the unit matrix E_2 appears:

$$\left.\begin{matrix} x_1 \quad -2x_3 + 3x_4 = 5 \\ x_2 + \ x_3 - 4x_4 = 6 \end{matrix}\right\} \text{this implies} \begin{cases} x_1 = 5 + 2x_3 - 3x_4 \\ x_2 = 6 - \ x_3 + 4x_4 \end{cases}$$

The set of solutions can be described as $\underline{x} = \begin{pmatrix} x_1 \\ x_2 \\ x_3 \\ x_4 \end{pmatrix} = \begin{pmatrix} 5 \\ 6 \\ 0 \\ 0 \end{pmatrix} + x_3 \begin{pmatrix} 2 \\ -1 \\ 1 \\ 0 \end{pmatrix} + x_4 \begin{pmatrix} -3 \\ 4 \\ 0 \\ 1 \end{pmatrix}$

5. In general (Sections A.5.2 and 4.4), a system $A\underline{x} = \underline{b}$ has solutions if $r(A) = r(A, \underline{b})$ i.e. the rank of matrix A = rank of matrix A where vector $\underline{b}$ has been added (see examples in the next Section A.5.2). If $r(A) = r(A, \underline{b})$ then vector $\underline{b}$ is situated in the space spanned by the column vectors of A.

A.5.2. Graphical illustration

In this section, solving systems of linear equations is considered from various viewpoints. Pivoting, solving with linear algebra and the graphical background are discussed.

System of equations with one solution

Solve system $A\underline{x} = \underline{b}$, in which $A = \begin{pmatrix} 1 & 1 \\ 1 & -1 \end{pmatrix}$ $\underline{x} = \begin{pmatrix} x_1 \\ x_2 \end{pmatrix}$ $\underline{b} = \begin{pmatrix} 3 \\ 1 \end{pmatrix}$. So,

$$A\underline{x} = \underline{b} \Rightarrow \begin{pmatrix} 1 & 1 \\ 1 & -1 \end{pmatrix}\begin{pmatrix} x_1 \\ x_2 \end{pmatrix} = \begin{pmatrix} 3 \\ 1 \end{pmatrix} \text{ or } \begin{cases} x_1 + x_2 = 3 \\ x_1 - x_2 = 1 \end{cases}$$

Moreover, note that

$$r(A) = r\begin{pmatrix} 1 & 1 \\ 1 & -1 \end{pmatrix} = r(A, \underline{b}) = r\begin{pmatrix} 1 & 1 & 3 \\ 1 & -1 & 1 \end{pmatrix} = 2$$

Is the system solvable

- $r(A) = r(A,\underline{b}) = 2$ so $A\underline{x} = \underline{b}$ can be solved i.e. vector $\underline{b}$ is situated in the space spanned by the columns of A (Figure A.9).

Solving by pivoting

- Solving the system of equations $\begin{cases} x_1 + x_2 = 3 \\ x_1 - x_2 = 1 \end{cases}$ by pivoting gives:

x_1	x_2	
1	1	3
1	–1	1

x_1	x_2	
1	1	3
0	-2	–2

x_1	x_2	
1	0	2
0	1	1

With the corresponding solution $x_1 = 2$, $x_2 = 1$.

Graphical solution

- The solution $(x_1, x_2) = (2, 1)$ follows from the intersection of the lines $\begin{cases} x_1 + x_2 = 3 \\ x_1 - x_2 = 1 \end{cases}$

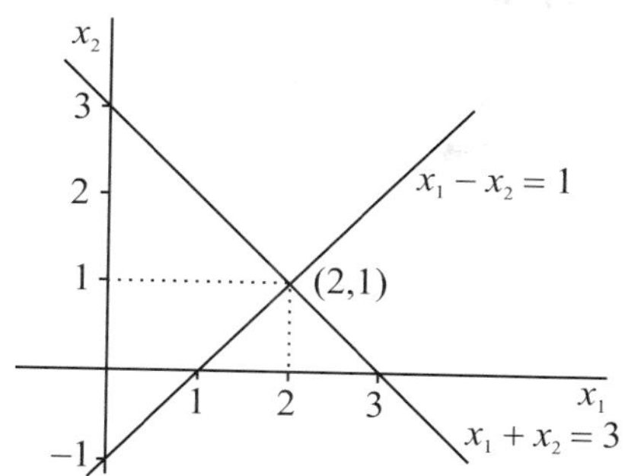

Figure A.8. Solving a system of equations with one solution, graphically.

Solving by vector calculation

- The system $\begin{cases} x_1 + x_2 = 3 \\ x_1 - x_2 = 1 \end{cases}$ can be written as $x_1\begin{pmatrix}1\\1\end{pmatrix} + x_2\begin{pmatrix}1\\-1\end{pmatrix} = \begin{pmatrix}3\\1\end{pmatrix}$. From Figure A.9 follows $2\begin{pmatrix}1\\1\end{pmatrix} + 1\begin{pmatrix}1\\-1\end{pmatrix} = \begin{pmatrix}3\\1\end{pmatrix}$. This corresponds to the solution $(x_1, x_2) = (2, 1)$.

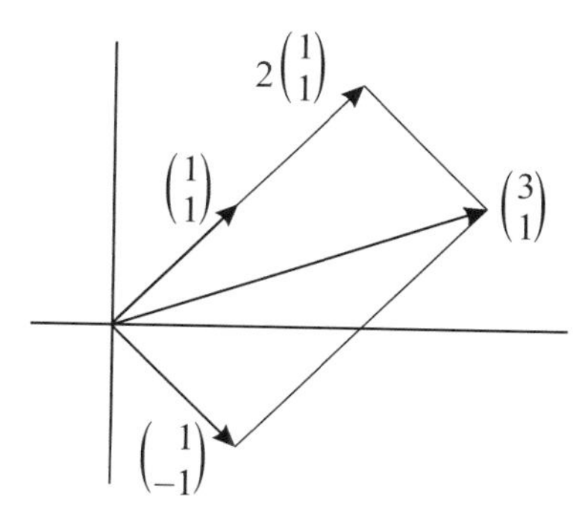

Figure A.9. Solving a system of equations with one solution by vector calculation.

Systems of equations with infinitely many solutions

Consider a system $A\underline{x} = \underline{b}$ in which $A = \begin{pmatrix}2 & 4\\1 & 2\end{pmatrix}$; $\underline{b} = \begin{pmatrix}6\\3\end{pmatrix}$ $\underline{x} = \begin{pmatrix}x_1\\x_2\end{pmatrix}$

Is the system solvable

- $r(A) = r(A, \underline{b}) = 1$ so the system can be solved.

Solving by pivoting

- Solving the system of equations by pivoting gives:

$$A\underline{x} = \underline{b} \quad \Rightarrow \quad \begin{pmatrix} 2 & 4 \\ 1 & 2 \end{pmatrix}\begin{pmatrix} x_1 \\ x_2 \end{pmatrix} = \begin{pmatrix} 6 \\ 3 \end{pmatrix} \quad \Rightarrow \quad \begin{cases} 2x_1 + 4x_2 = 6 \\ x_1 + 2x_2 = 3 \end{cases}$$

x_1	x_2	
2	4	6
1	2	3

x_1	x_2	
1	2	3
0	0	0

The second equality in tableau, $0x_1 + 0x_2 = 0$ always holds. The set of solutions is described by the first equality: $x_1 + 2x_2 = 3$. Apparently, system $A\underline{x} = \underline{b}$ has infinitely many solutions e.g. $x_1 = 3, x_2 = 0$, $x_1 = 1, x_2 = 1$ or $x_1 = 23, x_2 = -10$.

Graphical solution

- $$A\underline{x} = \underline{b} \quad \Rightarrow \quad \begin{cases} 2x_1 + 4x_2 = 6 \\ x_1 + 2x_2 = 3 \end{cases}$$

The two lines coincide. So, all points on line $x_1 + 2x_2 = 3$ fulfil the set of equations.

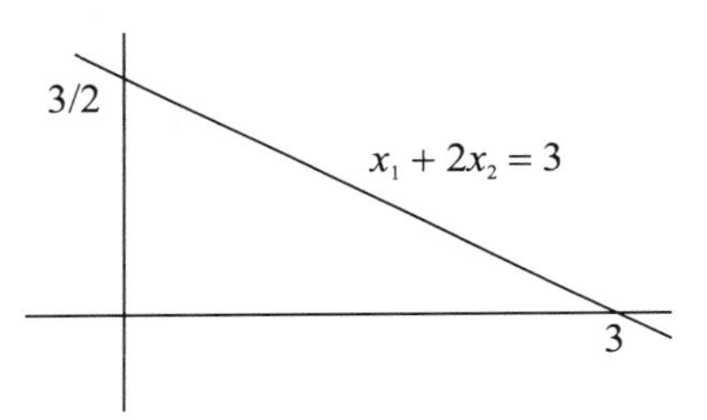

Figure A.10. Solving a system of equations with infinitely many solutions, graphically.

Solving by vector calculation

- The system $\begin{cases} 2x_1 + 4x_2 = 6 \\ x_1 + 2x_2 = 3 \end{cases}$ can be written as $x_1\begin{pmatrix} 2 \\ 1 \end{pmatrix} + x_2\begin{pmatrix} 4 \\ 2 \end{pmatrix} = \begin{pmatrix} 6 \\ 3 \end{pmatrix}$.

Vector $\begin{pmatrix} 6 \\ 3 \end{pmatrix}$ is situated in a space spanned by $\begin{pmatrix} 2 \\ 1 \end{pmatrix}$ and $\begin{pmatrix} 4 \\ 2 \end{pmatrix}$ and can be written in infinitely many ways as a linear combination of: $\begin{pmatrix} 2 \\ 1 \end{pmatrix}$ and $\begin{pmatrix} 4 \\ 2 \end{pmatrix}$, e.g. $3\begin{pmatrix} 2 \\ 1 \end{pmatrix} + 0\begin{pmatrix} 4 \\ 2 \end{pmatrix} = \begin{pmatrix} 6 \\ 3 \end{pmatrix}$ or, $23\begin{pmatrix} 2 \\ 1 \end{pmatrix} - 10\begin{pmatrix} 4 \\ 2 \end{pmatrix} = \begin{pmatrix} 6 \\ 3 \end{pmatrix}$ etc. See Figure A.11.

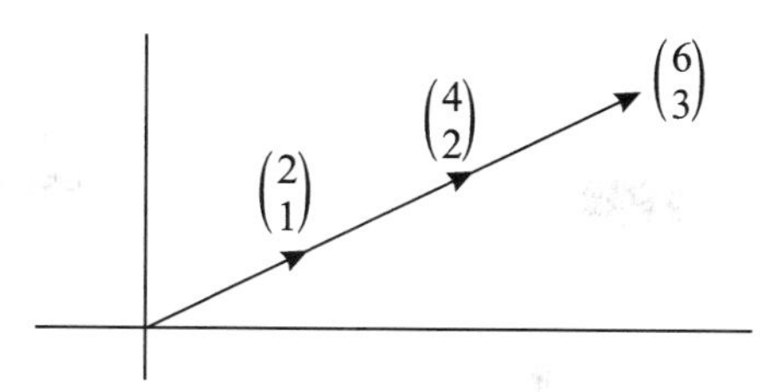

Figure A.11. Solving a system of equations with infinitely many solutions by vector notation.

Inconsistent systems of equations

Consider system $A\underline{x} = \underline{b}$ in which $A = \begin{pmatrix}2 & 4\\1 & 2\end{pmatrix}$; $\underline{b}\begin{pmatrix}8\\2\end{pmatrix}$; $\underline{x} = \begin{pmatrix}x_1\\x_2\end{pmatrix}$

$$A\underline{x} = \underline{b} \quad \Rightarrow \quad \begin{cases}2x_1 + 4x_2 = 8\\ x_1 + 2x_2 = 2\end{cases}$$

Is the system solvable

- r(A) = 1; r(A, $\underline{b}$) = 2; r(A) ≠ r(A, $\underline{b}$) so $A\underline{x} = \underline{b}$ cannot be solved, i.e. $\underline{b}$ is not situated in the space spanned by the column vectors of A.

Solving by pivoting

- Solving the system of equations $\begin{cases}2x_1 + 4x_2 = 8\\ x_1 + 2x_2 = 2\end{cases}$ by pivoting gives:

x_1	x_2	
2	4	8
1	2	2

x_1	x_2	
1	2	4
0	0	–2

The second equality $0x_1 + 0x_2 = -2$ cannot be fulfilled; the system is inconsistent.

Graphical solution

- $\begin{cases}2x_1 + 4x_2 = 8\\ x_1 + 2x_2 = 2\end{cases}$

The two lines are parallel. No point fulfils both equations. The system is inconsistent.

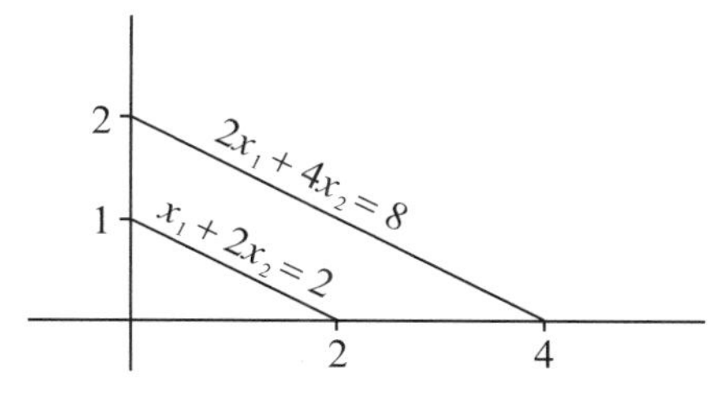

Figure A.12. Solving an inconsistent system of equations, graphically.

Solving by vector calculation

- The system $\begin{cases} 2x_1 + 4x_2 = 8 \\ x_1 + 2x_2 = 2 \end{cases}$ can be written as $x_1\begin{pmatrix}2\\1\end{pmatrix} + x_2\begin{pmatrix}4\\2\end{pmatrix} = \begin{pmatrix}8\\2\end{pmatrix}$. The vectors $\begin{pmatrix}2\\1\end{pmatrix}$ and $\begin{pmatrix}4\\2\end{pmatrix}$ point in the same direction and span a line. As vector $\begin{pmatrix}8\\2\end{pmatrix}$ is not on this line it cannot be written as a linear combination of $\begin{pmatrix}2\\1\end{pmatrix}$ and $\begin{pmatrix}4\\2\end{pmatrix}$.

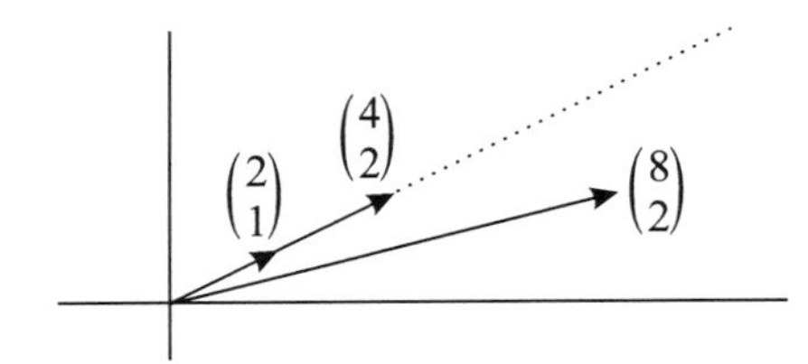

Figure A.13. Solving an inconsistent system of equations by vector notation.

In the three examples (one solution, many solutions, no solutions), the relation between pivoting and the graphical representation in $\mathbb{R}^2$ is given. We extend the examples towards sets of equations in $\mathbb{R}^3$.

Extension towards $\mathbb{R}^3$

In $\mathbb{R}^3$, a plane is described by an equality in three variables (e.g. $2x_1 + 3x_2 + x_3 = 6$)

Consider three equations in three variables in $\mathbb{R}^3$ corresponding to three planes.

- Two arbitrary non parallel planes A and B intersect on a line. The third plane C (not parallel to the intersecting line of A and B) intersects the line in one point. This point is a unique solution of the three equations.
- If the three planes coincide, all points in the plane (infinitely many) are solutions.
- If the planes are parallel, no point fulfils all equations; it is an inconsistent system.

Consider a set of two equations in three variables (two planes in $\mathbb{R}^3$).

- If the two planes A and B are parallel, there is no intersection, so no solution exists; the system is inconsistent.
- If the two planes A and B are not parallel, all points (infinitely many) on the intersecting line of A and B are solutions of the system.
- If the planes A and B coincide, the set also has infinitely many solutions.

Remark

The concept of a line in $\mathbb{R}^2$ and plane in $\mathbb{R}^3$ can be generalised in the space $\mathbb{R}^n$. The notions of infinitely many solutions and inconsistent systems of $\mathbb{R}^2$ and $\mathbb{R}^3$ is also generally valid for arbitrary sets of equations in $\mathbb{R}^n$. In Chapter 3, this general concept is followed.

Concept A.37

A *hyperplane* in $\mathbb{R}^n$ is the set of points for which $a_1x_1 + a_2x_2 + ... + a_nx_n = b$ or $\underline{a}'\underline{x} = b$.

Example of a hyperplane :

$$2x_1 + 7x_2 + ... + 10x_n = 87 \text{ or } (2, 7, ..., 10)\begin{pmatrix} x_1 \\ x_2 \\ . \\ . \\ x_n \end{pmatrix} = 87$$

A.6. The inverse of a matrix

For a given square matrix $A^{n\times n}$, the $n\times n$ matrix A^{-1} is called the *inverse* of A if $AA^{-1} = A^{-1}A = E_n$. The inverse matrix A^{-1} does not always exist. The matrix $A^{n\times n}$ can be inverted if the columns of A are linearly independent; $r(A^{n\times n}) = n$.

Example A.18.

$$A = \begin{pmatrix} 1 & 1 \\ 2 & 3 \end{pmatrix}; \quad A^{-1} = \begin{pmatrix} 3 & -1 \\ -2 & 1 \end{pmatrix}$$

$$AA^{-1} = \begin{pmatrix} 1 & 1 \\ 2 & 3 \end{pmatrix}\begin{pmatrix} 3 & -1 \\ -2 & 1 \end{pmatrix} = \begin{pmatrix} 1 & 0 \\ 0 & 1 \end{pmatrix} = E_2$$

Note that also holds: $A^{-1}A = E_2$.

Several methods exist to determine the inverse of a matrix A. In this book pivoting will be used.

Example A.19.

$A = \begin{pmatrix} 1 & 1 \\ 2 & 3 \end{pmatrix}$, A is a 2×2 matrix and $r(A) = 2$, so the inverse A^{-1} exists. Let us represent the inverse matrix as $A^{-1} = \begin{pmatrix} x_1 & x_3 \\ x_2 & x_4 \end{pmatrix}$. As should hold $AA^{-1} = E$:

$$\begin{pmatrix} 1 & 1 \\ 2 & 3 \end{pmatrix}\begin{pmatrix} x_1 & x_3 \\ x_2 & x_4 \end{pmatrix} = \begin{pmatrix} 1 & 0 \\ 0 & 1 \end{pmatrix} \rightarrow \begin{pmatrix} x_1 + x_2 & x_3 + x_4 \\ 2x_1 + 3x_2 & 2x_3 + 3x_4 \end{pmatrix} = \begin{pmatrix} 1 & 0 \\ 0 & 1 \end{pmatrix}$$

The elements x_1, x_2, x_3 and x_4 of A^{-1} can be determined by solving the two sets of equations:

$$\text{(I)} \quad \begin{cases} x_1 + x_2 = 1 \\ 2x_1 + 3x_2 = 0 \end{cases} \qquad \text{(II)} \quad \begin{cases} x_3 + x_4 = 0 \\ 2x_3 + 3x_4 = 1 \end{cases}$$

The systems (I) and (II) are solved in tableau-representation by pivoting:

x_1	x_2	
1	1	1
2	3	0

x_3	x_4	
1	1	0
2	3	1

x_1	x_2	
1	1	1
0	**1**	–2

x_3	x_4	
1	1	0
0	**1**	1

x_1	x_2	
1	0	3
0	1	–2

x_3	x_4	
1	0	–1
0	1	1

The solution is $x_1 = 3,\ x_2 = -2,\ x_3 = -1,\ x_4 = 1$, such that $A^{-1} = \begin{pmatrix} x_1 & x_3 \\ x_2 & x_4 \end{pmatrix} = \begin{pmatrix} 3 & -1 \\ -2 & 1 \end{pmatrix}$.

It can be seen that the choice of the pivots in both tableau sequences are identical. This means that the sequences can be combined as follows:

$$\left[\begin{array}{cc|cc} 1 & 1 & 1 & 0 \\ 2 & 3 & 0 & 1 \end{array}\right]$$

$$\left[\begin{array}{cc|cc} 1 & 1 & 1 & 0 \\ 0 & 1 & -2 & 1 \end{array}\right]$$

$$\left[\begin{array}{cc|cc} 1 & 0 & 3 & -1 \\ 0 & 1 & -2 & 1 \end{array}\right]$$

From the initial matrix $(A \mid E) = \left(\begin{array}{cc|cc} 1 & 1 & 1 & 0 \\ 2 & 3 & 0 & 1 \end{array}\right)$ the matrix $(E \mid A^{-1}) = \left(\begin{array}{cc|cc} 1 & 0 & 3 & -1 \\ 0 & 1 & -2 & 1 \end{array}\right)$ has been derived by elementary pivot operations. Multiplication of $(A \mid E)$ by A^{-1} leads to:

$$A^{-1}(A \mid E) = (A^{-1}A \mid A^{-1}E) = (E \mid A^{-1}).$$

This implies that performing pivot operations to transform matrix $(A \mid E)$ to $(E \mid A^{-1})$ is the same as multiplication of the matrix by A^{-1}.

Example A.20.

$$A = \begin{pmatrix} 1 & 2 & 1 \\ 0 & 1 & 1 \\ 2 & 4 & 3 \end{pmatrix} \text{ determine } A^{-1}$$

Starting with tableau $(A \mid E)$, pivot operations are used to arrive at $(E \mid A^{-1})$:

$$
\begin{array}{r|ccc|ccc}
(A \mid E) = & \boxed{1} & 2 & 1 & 1 & 0 & 0 \\
 & 0 & 1 & 1 & 0 & 1 & 0 \\
 & 2 & 4 & 3 & 0 & 0 & 1 \\
\hline
 & 1 & 2 & 1 & 1 & 0 & 0 \\
 & 0 & \boxed{1} & 1 & 0 & 1 & 0 \\
 & 0 & 0 & 1 & -2 & 0 & 1 \\
\hline
 & 1 & 0 & -1 & 1 & -2 & 0 \\
 & 0 & 1 & 1 & 0 & 1 & 0 \\
 & 0 & 0 & \boxed{1} & -2 & 0 & 1 \\
\hline
 & 1 & 0 & 0 & -1 & -2 & 1 \\
(E \mid A^{-1}) = & 0 & 1 & 0 & 2 & 1 & -1 \\
 & 0 & 0 & 1 & -2 & 0 & 1
\end{array}
$$

So the inverse A^{-1} is $\begin{pmatrix} -1 & -2 & 1 \\ 2 & 1 & -1 \\ -2 & 0 & 1 \end{pmatrix}$.

Remarks

1. $A = \begin{pmatrix} 0 & 1 & 1 \\ 1 & 2 & 1 \\ 2 & 4 & 3 \end{pmatrix}$, determine A^{-1}.

After constructing tableau $(A \mid E)$, a_{11} cannot be used as a pivot ($a_{11} = 0$). In this case row 1 and 2 can be switched.

$$(A|E) = \begin{pmatrix} 0 & 1 & 1 & | & 1 & 0 & 0 \\ 1 & 2 & 1 & | & 0 & 1 & 0 \\ 2 & 4 & 3 & | & 0 & 0 & 1 \end{pmatrix}; \text{ Exchange row 1 and row 2} \rightarrow \begin{pmatrix} 1 & 2 & 1 & | & 0 & 1 & 0 \\ 0 & 1 & 1 & | & 1 & 0 & 0 \\ 2 & 4 & 3 & | & 0 & 0 & 1 \end{pmatrix}$$

and proceed with pivot operations in the new system. As soon as at the left hand side E has been reached, at the right hand side A^{-1} appears.

In this way $A^{-1} = \begin{pmatrix} -2 & -1 & 1 \\ 1 & 2 & -1 \\ 0 & -2 & 1 \end{pmatrix}$

2. $A = \begin{pmatrix} 1 & 1 & 2 \\ -1 & 0 & -1 \\ 0 & 2 & 2 \end{pmatrix}$

Notice that $\underline{a}_1$, $\underline{a}_2$ and $\underline{a}_3$ are linearly dependent ($\underline{a}_3 = \underline{a}_1 + \underline{a}_2$), so A^{-1} does not exist. This is detected by the pivot process as follows:

$$\begin{bmatrix} \boxed{1} & 1 & 2 & | & 1 & 0 & 0 \\ -1 & 0 & -1 & | & 0 & 1 & 0 \\ 0 & 2 & 2 & | & 0 & 0 & 1 \end{bmatrix} \rightarrow \begin{bmatrix} 1 & 1 & 2 & | & 1 & 0 & 0 \\ 0 & \boxed{1} & 1 & | & 1 & 1 & 0 \\ 0 & 2 & 2 & | & 0 & 0 & 1 \end{bmatrix} \rightarrow \begin{bmatrix} 1 & 0 & 1 & | & 0 & -1 & 0 \\ 0 & 1 & 1 & | & 1 & 1 & 0 \\ 0 & 0 & 0 & | & -2 & -2 & 1 \end{bmatrix}$$

The concepts discussed in this chapter are used in Chapter 3 for solving LP problems.

Appendix B. Random variables and probability distributions

D.L. (Dik) Kettenis and E.M.T. (Eligius) Hendrix

A short introduction is given on the concept of random variables. Useful probability distributions are outlined.

A *random variable* (stochastic variate) is a function that associates a number with each point in an experimental sample space. We denote random variables by letters with tilde: $\tilde{x}$, $\tilde{r}$, $\tilde{d}$ .. A random variable $\tilde{x}$ is discrete if it takes only discrete values x_1, x_2, ...; more precisely, it takes values from a countable set. A *discrete random variable* $\tilde{x}$ is characterized by the fact that we know the probability (mass) that the variable $\tilde{x}$ takes a value x_i. This is written as $P(\tilde{x} = x_i)$ and we call this the *probability mass function.*

The *cumulative distribution function* $F(x)$ (abbreviation: *cdf*) for a random variable $\tilde{x}$ is defined by $F(x) = P(\tilde{x} \leq x)$.

Example B.1.

Let $\tilde{x}$ be the number of dots that show after tossing a die. For i = 1, 2, 3, 4, 5, 6 the probability that $P(\tilde{x} = i) = \frac{1}{6}$ such that $F(5) = P(\tilde{x} \leq 5) = \frac{5}{6}$.

In cases where random variable $\tilde{x}$ can take any value from a certain interval (called support), $\tilde{x}$ is a *continuous random variable*. Continuous random variables $\tilde{x}$ typically have a *probability density function* $f(x)$, abbreviated as *pdf*. The probability density function $f(x)$ for a random variable $\tilde{x}$ can be interpreted as follows. For small Δ:

$$P(x \leq \tilde{x} \leq x + \Delta) \approx \Delta f(x)$$

and $$P(a \leq \tilde{x} \leq b) = \int_a^b f(x)\,\mathrm{d}x \quad \text{for all } a < b. \quad \text{(See Figure B.1)}$$

Moreover, $\int_{-\infty}^{\infty} f(x)\,\mathrm{d}x = 1$.

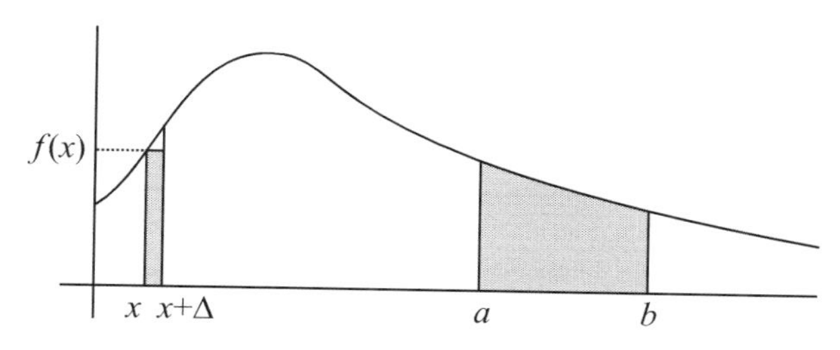

Figure B.1. Probability density function.

The (cumulative) *distribution function* $F(x)$ for a continuous random variable $\tilde{x}$ with density $f(x)$ is given by

$$F(x) = P(\tilde{x} \leq x) = \int_{-\infty}^{x} f(x)\, dx$$

Example B.2.
Consider continuous random variable $\tilde{x}$ with density function $f(x)$ given by

$$f(x) = \begin{cases} 2x & \text{if} \quad 0 \leq x \leq 1 \\ 0 & \text{otherwise} \end{cases}$$

$$F\left(\frac{1}{2}\right) = P\left(\tilde{x} \leq \frac{1}{2}\right) = \int_{0}^{1/2} 2x\, dx = \left[x^2\right]_0^{1/2} = \frac{1}{4} \text{ and } P\left(\tfrac{1}{2} \leq \tilde{x} \leq 1\right) = \int_{1/2}^{1} 2x\, dx = \left[x^2\right]_{1/2}^{1} = \frac{3}{4}$$

The *mean* (the expected value) and the *variance* are two important measures that are often used to summarise information contained in the probability distribution of the random variable. The mean of a random variable $\tilde{x}$ (notation $E(\tilde{x})$) is a measure of central location for the random variable:

Mean of a discrete random variable: $E(\tilde{x}) = \sum_k x_k \, P(\tilde{x} = x_k)$

Mean of a continuous random variable: $E(\tilde{x}) = \int_{-\infty}^{\infty} x \, f(x) dx$

Computing $E(\tilde{x})$ (also written as $E\tilde{x}$), each possible outcome of a random variable is weighted by its probability of occurrence. The mean is the centre of mass of the random variable.

For a function $g(\tilde{x})$ of a random variable $\tilde{x}$, expected value $E(g(\tilde{x}))$ (or $Eg(\tilde{x})$) is defined as:

If $\tilde{x}$ is a discrete random variable: $E(g(\tilde{x})) = \sum_k g(x_k) \, P(\tilde{x} = x_k)$

If $\tilde{x}$ is a continuous random variable: $E(g(\tilde{x})) = \int_{-\infty}^{\infty} g(x) \, f(x) dx$

The variance $V(\tilde{x})$ of a random variable $\tilde{x}$ is a measure for the spread or dispersion of $\tilde{x}$ around $E\tilde{x}$. The variance is defined as $V(\tilde{x}) = E(\tilde{x} - E\tilde{x})^2$.

Variance of a discrete random variable: $V(\tilde{x}) = \sum_k (x_k - E\tilde{x})^2 \, P(\tilde{x} = x_k)$

Variance of a continuous random variable: $V(\tilde{x}) = \int_{-\infty}^{\infty} (x_k - E\tilde{x})^2 \, f(x) dx$

Also $V(\tilde{x})$ may be found from the relation $V(\tilde{x}) = E\tilde{x}^2 - \{E\tilde{x}\}^2$. The *standard deviation* of $\tilde{x}$ is defined as $\sigma_x = \sqrt{V(\tilde{x})}$.

Example B.3.

Consider discrete random variable $\tilde{x}$ of Example B.1.

The expected value $E(\tilde{x}) = \frac{1}{6}(1+2+3+4+5+6) = \frac{21}{6} = 3.5$ and the variance

$$V(\tilde{x}) = \frac{1}{6}(1-3.5)^2 + \frac{1}{6}(2-3.5)^2 + \ldots + \frac{1}{6}(5-3.5)^2 + \frac{1}{6}(6-3.5)^2 = \frac{35}{12}$$

Alternatively, $V(\tilde{x}) = E(\tilde{x}^2) - \{E(\tilde{x})\}^2 = \frac{1}{6}(1+4+9+16+25+36) - (3.5)^2 = \frac{35}{12}$

Example B.4.

Consider random variable $\tilde{x}$ of example B.2.

The expected value $E(\tilde{x}) = \int_{-\infty}^{\infty} x f(x)\,dx = \int_0^1 x\, 2x\,dx = \left[\frac{2}{3}x^3\right]_0^1 = \frac{2}{3}$ and the variance

$$V(\tilde{x}) = \int_{-\infty}^{\infty} (x - E(\tilde{x}))^2 f(x)\,dx = \int_0^1 \left(x - \frac{2}{3}\right)^2 2x\,dx = \int_0^1 \left(2x^3 - \frac{8x^2}{3} + \frac{8x}{9}\right) dx =$$

$$\left[\frac{2x^4}{4} - \frac{8x^3}{9} + \frac{8x^2}{18}\right]_0^1 = \frac{1}{18}$$

Alternatively, $V(\tilde{x}) = E(\tilde{x}^2) - \{E(\tilde{x})\}^2 = \int_0^1 x^2\, 2x\,dx - \left(\frac{2}{3}\right)^2 = \left[\frac{2}{4}x^4\right]_0^1 - \frac{4}{9} = \frac{1}{18}$

In the sequel we present some of the well-known and often used continuous and discrete distributions. In discrete-event simulation, continuous distribution functions are frequently applied for modelling the time an operation takes or inter-arrival times of customers at a facility like a post office. Discrete distribution functions are of value to generate for example the number of ordered copies of a product.

Continuous distribution functions

Among the well-known continuous distribution functions are normal, lognormal, uniform, exponential, triangular, Erlang, Weibull, beta, gamma, and Johnson distributions.

Normal distribution

The usual notation of the normal distribution function is $\tilde{x} \sim N(\mu, \sigma^2)$. The density function of the normal distribution is

$$f(x) = \frac{1}{\sigma\sqrt{2\pi}} e^{-\frac{1}{2}\left(\frac{x-\mu}{\sigma}\right)^2}, -\infty < x < \infty$$

with $-\infty < \mu < \infty$ and $\sigma > 0$. Some graphs of the density function are presented in Figure B.2. The mean and variance are

$$E(\tilde{x}) = \mu$$

$$V(\tilde{x}) = \sigma^2$$

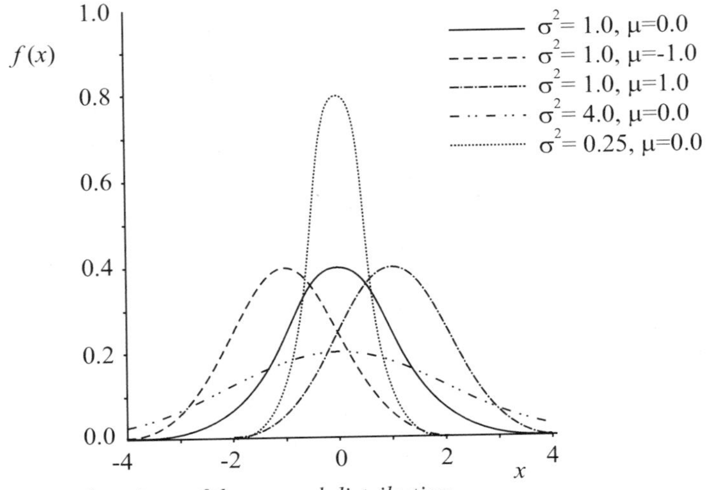

Figure B.2. Density functions of the normal distribution.

If a process consists of a large number of actions and each action takes an interval of approximately the same time, the time needed for the process may be described by a normal distribution. Besides, the normal distribution may be used for processing time of a product when the process is based on actions of a human. Dimensions and weight of a product or something else are other examples where a normal random variable may be applied.
If the normal distribution is used to represent a value larger than, say, a e.g. the minimum time needed to complete a task, then its density should be truncated at $x = a$ (Law and Kelton, 2000). Truncated distribution functions are discussed in Section 13.5.5.

No closed form expression exists for the cumulative distribution. For calculation purposes it is common to transform a variable $\tilde{x} \sim N(\mu, \sigma^2)$ to the standard normal distributed variable $\tilde{y} \sim \mathrm{N}(0,1)$ via $\tilde{y} = \frac{\tilde{x} - \mu}{\sigma}$. For $\tilde{y}$, the cumulative distribution function

$$\Phi(y) = \int_{-\infty}^{y} \frac{1}{\sqrt{2\pi}} e^{-\frac{1}{2}z} \mathrm{d}z$$

is available in tables and $\Phi(y)$ and Φ^{-1} are available in the NORMSDIST and NORMSINV statistical functions in Excel. If one wants to know the probability $F(x) = P(\tilde{x} \leq x)$, one can evaluate $P\left(\frac{\tilde{x}-\mu}{\sigma} \leq \frac{x-\mu}{\sigma}\right) = P\left(\tilde{y} \leq \frac{x-\mu}{\sigma}\right) = \Phi\left(\frac{x-\mu}{\sigma}\right)$. Excel statistical functions also allow to evaluate a general normal cdf $F(x)$ or inverse F^{-1} via NORMDIST and NORMINV.

Lognormal distribution

The usual notation for the lognormal distribution is $\tilde{x} \sim LN(\mu, \sigma^2)$. The density function of the lognormal distribution is

$$f(x) = \begin{cases} \frac{1}{\sigma x\sqrt{2\pi}} e^{-\frac{1}{2}\left(\frac{^e\log x - \mu}{\sigma}\right)^2} & , x > 0 \\ 0 & , x \leq 0 \end{cases}$$

with $-\infty < \mu < \infty$ and $\sigma > 0$. Graphs of the density function are given in Figure B.3. The mean and variance of the lognormal random variable are

$$E(\tilde{x}) = e^{\mu + \frac{\sigma^2}{2}}$$

$$V(\tilde{x}) = e^{2\mu+\sigma^2}\left(e^{\sigma^2} - 1\right)$$

If one has data $X_1, X_2, \ldots, X_n$ that are thought to be lognormal then $\ln X_1, \ln X_2, \ldots, \ln X_n$ can be treated as normally distributed data. The parameters μ and σ used above refer to this normal distribution function.

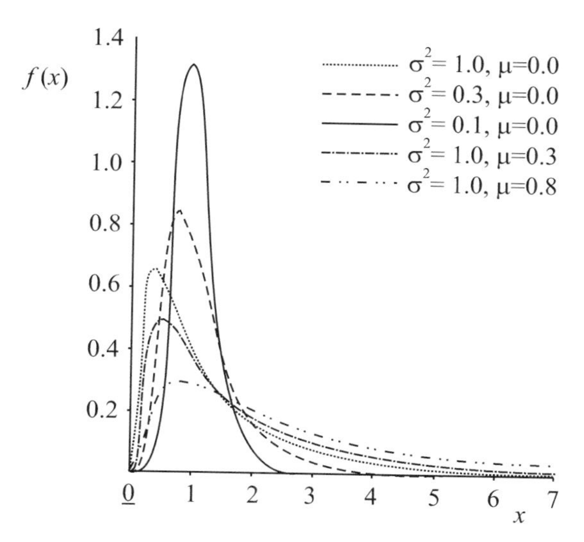

Figure B.3. Density functions of the lognormal distribution.

The density function of the normal distribution is symmetric, whereas the density function of the lognormal distribution is not. In this respect, its distribution is similar to the Erlang distribution we will discuss shortly. The lognormal and Erlang distributions are interchangeable in the sense that the lognormal distribution can be used where the Erlang distribution is applicable. A difference between the two is that the lognormal distribution may have a variance larger than the mean. That means, the lognormal distribution can produce many small numbers and occasionally a large number. Both distributions can be used to represent production time, time to repair a machine and inter-arrival time of clients or requests. A typical application of the lognormal distribution is service time of a clerk where the majority of the tasks take a relative short period and occasionally the period is long. This happens, for instance, if the majority of the clients will be sent to somebody else to be served and only some clients will be served by the server.

Uniform distribution

The usual notation of the uniform distribution function is $\tilde{x} \sim U(a,b)$ with a and b real numbers and $a < b$. The density function of the uniform distribution is

$$f(x) = \begin{cases} \frac{1}{b-a}, & a \leq x \leq b \\ 0 & \text{, otherwise} \end{cases}$$

A graph of the density function is given in Figure B.4. The mean and variance of a uniform random variable are

$$E(\tilde{x}) = \frac{a+b}{2}$$
$$V(\tilde{x}) = \frac{(b-a)^2}{12}$$

One uses the uniform distribution if one lacks information about the distribution. One needs to know (a guess for) a minimum and a maximum value and these boundaries are sharp.

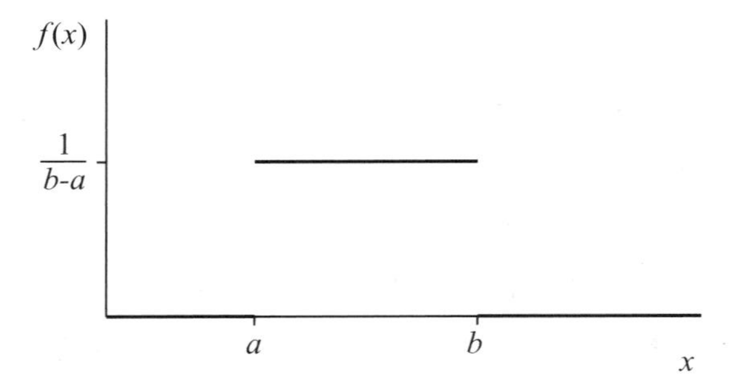

Figure B.4. Density function of the uniform distribution.

The uniform distribution may be used to model events with a certain probability. Suppose an event occurs with a chance of, say 65 percent or, what is equivalent a probability of 0.65. If you want to model this situation and generate the event with a computer, you may do the following. Draw a number R from a uniform distribution function defined on (0, 1]. If $0.0 \leq R < 0.65$, then the event occurs otherwise the event does not occur. For example, suppose $\tilde{x}$ is a random variable that can only take three values 5, 6, and 7, each with a certain probability:

$$P(\tilde{x}=5)=0.1, \quad P(\tilde{x}=6)=0.6, \quad P(\tilde{x}=7)=0.3$$

This means that the probability that $\tilde{x}=5$ is 0.1, that $\tilde{x}=6$ is 0.6, and $\tilde{x}=7$ is 0.3. The sum of all probabilities is one. Again we draw a random variable R from a uniform distribution defined on the interval 0 to 1. If $0.0 \leq R < 0.1$ then $\tilde{x}=5$, if $0.1 \leq R < 0.7$ then $\tilde{x}=6$, and if $0.7 \leq R < 1.0$ then $\tilde{x}=7$.

Exponential distribution

The usual notation for the exponential distribution function is $\tilde{x} \sim Ne(\lambda)$ with $\lambda > 0$. The density function of the exponential distribution is

$$f(x)=\begin{cases}\lambda e^{-\lambda x}, & x>0\\ 0, & x \leq 0\end{cases}$$

Graphs of the density function are given in Figure B.5. The mean and the variance of an exponential random variable are

$$E(\tilde{x})=\frac{1}{\lambda}$$

$$V(\tilde{x})=\frac{1}{\lambda^2}$$

In literature, this distribution is sometimes called negative exponential distribution. The exponential distribution is frequently applied to model inter-arrival times for arrival processes. Examples are arrival of jobs at a job shop, the arrival of aircraft at a runway, the arrival of telephone calls at a switchboard, the breakdown of machines in a factory (this may be seen as arrival of a breakdown at the machine), and arrival of clients at a supermarket. The exponential distribution function may be applied when the number of clients is large. So, for example, arrival of orders at a store with only two clients may not be modelled with an exponential distribution.

Arrival processes may be specified in two ways: specification of the number of arrivals per time unit and specification of the mean of the inter-arrival time. For example, if per hour 30 clients arrive at a shop then the mean of the inter-arrival time is 2 minutes. In other words: on average a client arrives every two minutes. With the exponential distribution function one specifies the mean of the inter-arrival time. With the related Poisson distribution one specifies the number of arrivals per time unit and, therefore, the Poisson distribution is a discrete distribution. In simulation applications, the Poisson distribution is not frequently applied to model arrival processes. This is because in simulation tools, it is easier to work with time intervals than with number of arrivals per unit of time. A description of the Poisson distribution is given further on in this appendix. The exponential distribution is related to the Poisson process. If one speaks about a Poisson process, one means an arrival process where the inter-arrival times are exponentially distributed.

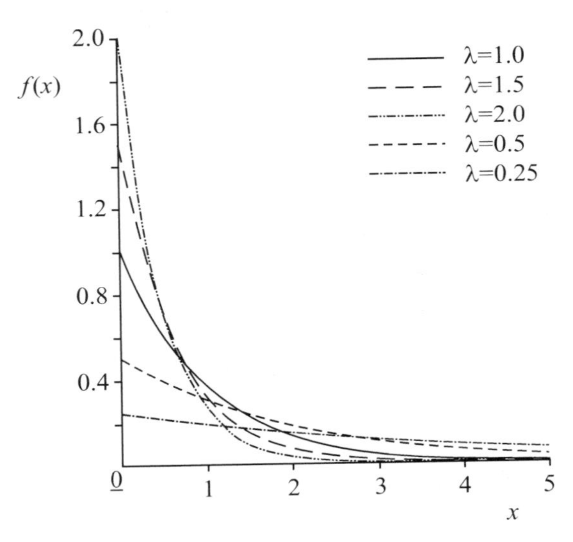

Figure B.5. Density functions of the exponential distribution.

Triangular distribution

The usual notation of the triangular distribution is $\tilde{x} \sim triang(a,b,c)$ with a, b, and c real numbers with $a < c < b$. The density function of the triangular distribution is

$$f(x) = \begin{cases} \dfrac{2(x-a)}{(b-a)(c-a)}, a \leq x \leq c \\ \dfrac{2(b-x)}{(b-a)(b-c)}, c < x \leq b \\ 0 \quad\quad\quad , \text{otherwise} \end{cases}$$

A graph of the density function is given in Figure B.6. The mean and variance of a triangular random variable are

$$E(\tilde{x}) = \frac{a+b+c}{3}$$

$$V(\tilde{x}) = \frac{a^2 + b^2 + c^2 - ab - ac - bc}{18}$$

The triangular distribution is defined on the interval $[a,b]$. The parameter c is the so-called mode. The triangular distribution is frequently applied to situations where it is impossible to collect data on the random variable. This may happen in situations where the system is not available or collection of data costs too much or takes too much time. Subject-matter experts may help in those cases to identify the interval $[a,b]$ by indicating the most optimistic and pessimistic estimates. The mode is identified by indicating the most likely value.

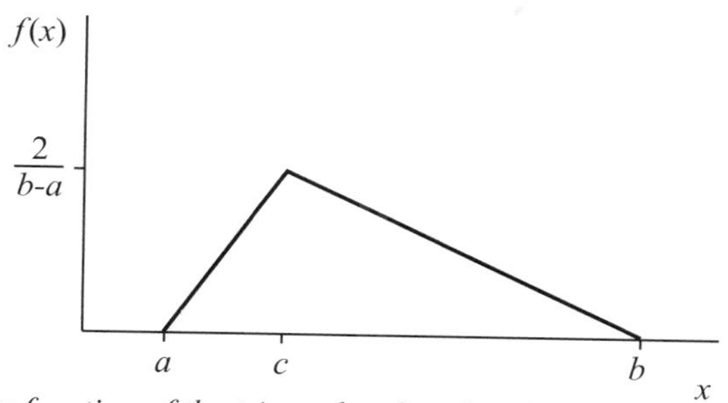

Figure B.6. Density function of the triangular distribution.

Erlang distribution

The usual notation of the Erlang distribution is $\tilde{x} \sim Erl(\alpha,\beta)$. The density function of the Erlang distribution is

$$f(x) = \begin{cases} \dfrac{\alpha^k}{(k-1)!} x^{k-1} e^{-\alpha x}, x > 0, k = 1,2,3,\ldots \\ 0 \qquad , \text{otherwise} \end{cases}$$

with $\alpha > 0$. Graphs of some distributions are given in Figure B.7. The mean and variance of an Erlang random variable are

$$E(\tilde{x}) = \frac{k}{\alpha}$$

$$V(\tilde{x}) = \frac{k}{\alpha^2}$$

The Erlang distribution is used when an activity or service time is considered to occur in phases with each phase being exponentially distributed with parameter α. An additional prerequisite is that the next activity or service cannot start before the previous one has come to an end.

For $k = 1$ the Erlang distribution is the exponential distribution. In that case, the service process contains only one phase. For k large, the Erlang distribution becomes similar to the normal distribution. The Erlang distribution is used to model service times, time to repair a machine, and, occasionally, for inter-arrival times.

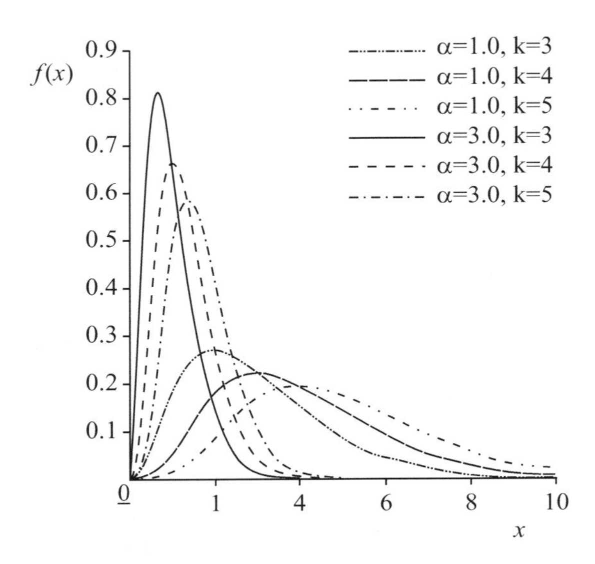

Figure B.7. Density functions of the Erlang distribution.

Weibull distribution

The usual notation of the Weibull distribution is $\tilde{x} \sim Weibull(\alpha, \beta)$, where $\alpha > 0$ and $\beta > 0$. The density function of the Weibull distribution is

$$f(x) = \begin{cases} \frac{\alpha}{\beta}\left(\frac{x}{\beta}\right)^{\alpha-1} e^{-\left(\frac{x}{\beta}\right)^{\alpha}} & , x \geq 0 \\ 0 & , \text{otherwise} \end{cases}$$

Graphs of some density functions are given in Figure B.8. The mean and variance of a Weibull random variable are

$$E(\tilde{x}) = \frac{\beta}{\alpha}\Gamma\left(\frac{1}{\alpha}\right)$$

where $\Gamma(\alpha)$ is the so-called gamma function, defined by $\Gamma(\alpha) = \int_0^{\infty} t^{\alpha-1} e^{-t} dt$ for any real number α,

$$V(\tilde{x}) = \frac{\beta^2}{\alpha}\left\{2\Gamma\left(\frac{2}{\alpha}\right) - \frac{1}{\alpha}\left[\Gamma\left(\frac{1}{\alpha}\right)\right]^2\right\}$$

For $\alpha = 1$, the Weibull distribution is the same as the exponential distribution with $\lambda = \frac{1}{\beta}$.

The Weibull distribution may be used for lifetime of machines or organisms. This is especially true for machines or organisms consisting of a large number of components (parts in a machine and cells or organs in a living creature) where each component may be the reason for breakdown or end of life. Examples are failure of a panel screen and the time it takes for an aircraft to land and clear the runway at a major international airport.

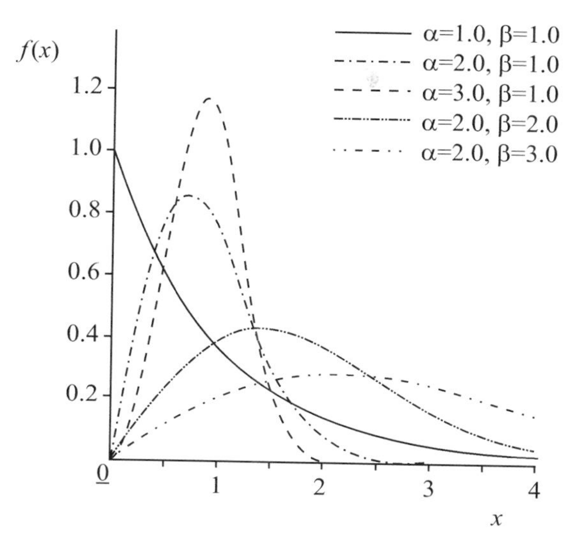

Figure B.8. Density functions of the Weibull distribution.

Beta distribution

The usual notation of the beta distribution is $\tilde{x} \sim beta(\alpha, \beta)$ with $\alpha > 0$, $\beta > 0$. The density function of the beta distribution function is

$$f(x) = \begin{cases} \dfrac{x^{\beta-1}(1-x)^{\alpha-1}}{B(\beta,\alpha)}, & 0 < x < 1 \\ 0 & \text{, otherwise} \end{cases}$$

where B is the complete beta function given by $B(\beta,\alpha) = \int_0^1 t^{\beta-1}(1-t)^{\alpha-1}\,dt = \dfrac{\Gamma(\beta)\Gamma(\alpha)}{\Gamma(\beta+\alpha)}$

Graphs of several density functions are given in Figure B.9. The mean and variance of a beta random variable are

$$E(\tilde{x}) = \frac{\beta}{\beta+\alpha}$$

$$V(\tilde{x}) = \frac{\beta\alpha}{(\beta+\alpha)^2(\beta+\alpha+1)}$$

As shown in Figure B.10, the beta distribution may take a wide variety of shapes. Therefore, it may be applied in stead of an empirical distribution to fit experimental data. The range of the beta distribution is from 0 to 1. By using the equation $\tilde{y} = a + (b - a)\tilde{x}$, one can transform the beta variable $\tilde{x}$ to the scaled beta variable $\tilde{y}$ with a range from a to b.

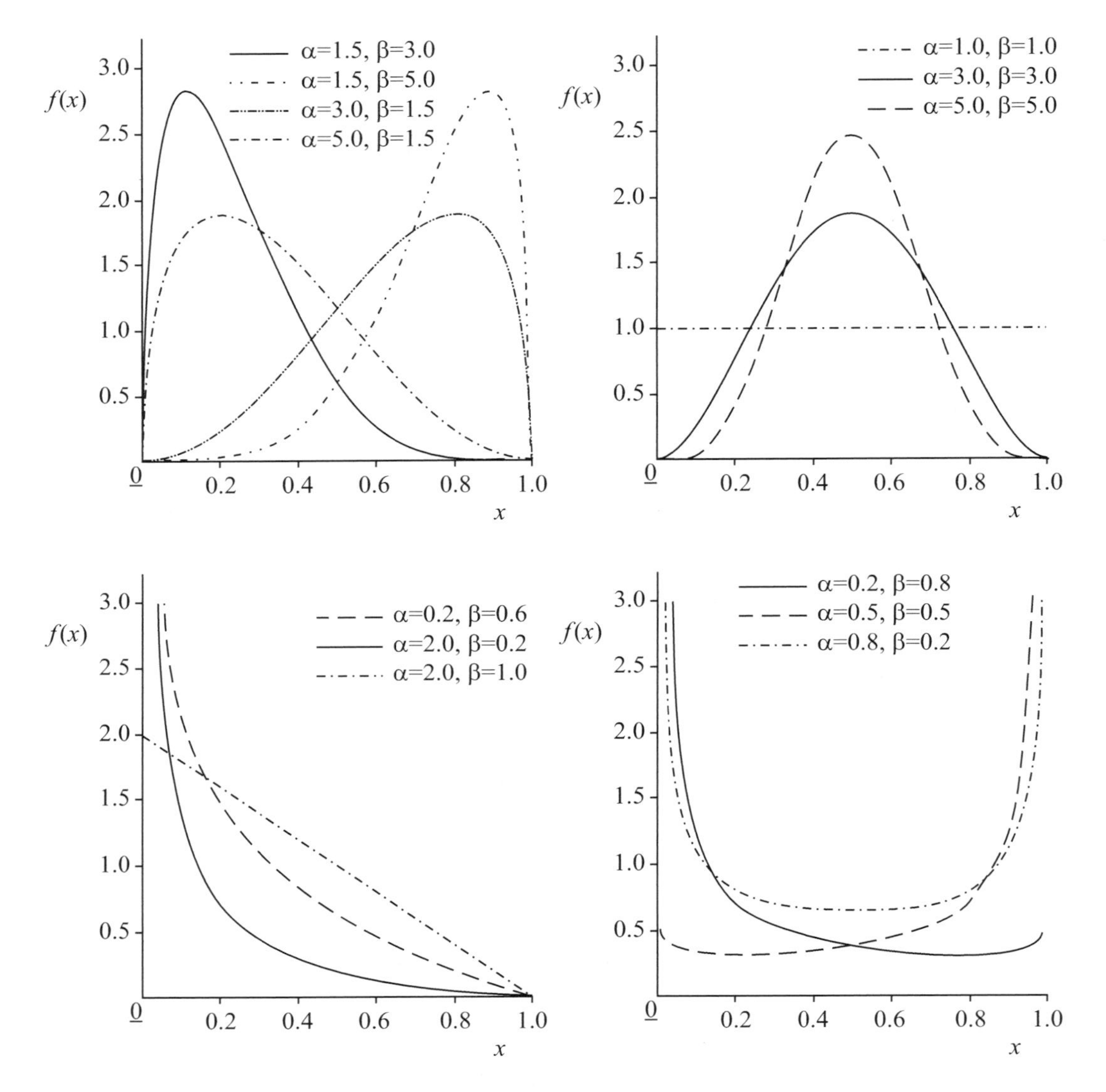

Figure B.9. Density functions of the beta distribution.

Gamma distribution

The usual notation of the gamma distribution is $\tilde{x} \sim gamma(\alpha, \beta)$ with $\alpha > 0$ and $\beta > 0$. The density function of the gamma distribution function is

$$f(x) = \begin{cases} \dfrac{\beta^{-\alpha} x^{\alpha-1} e^{-\frac{x}{\beta}}}{\Gamma(\alpha)}, x > 0 \\ 0 \quad\quad , \text{otherwise} \end{cases}$$

The mean and variance of a gamma random variable are

$$E(\tilde{x}) = \alpha\beta$$

$$V(\tilde{x}) = \alpha\beta^2$$

The gamma distribution is frequently used to represent the time required to complete some task. Graphs of distribution functions of the gamma distribution are given in Figure B.10.

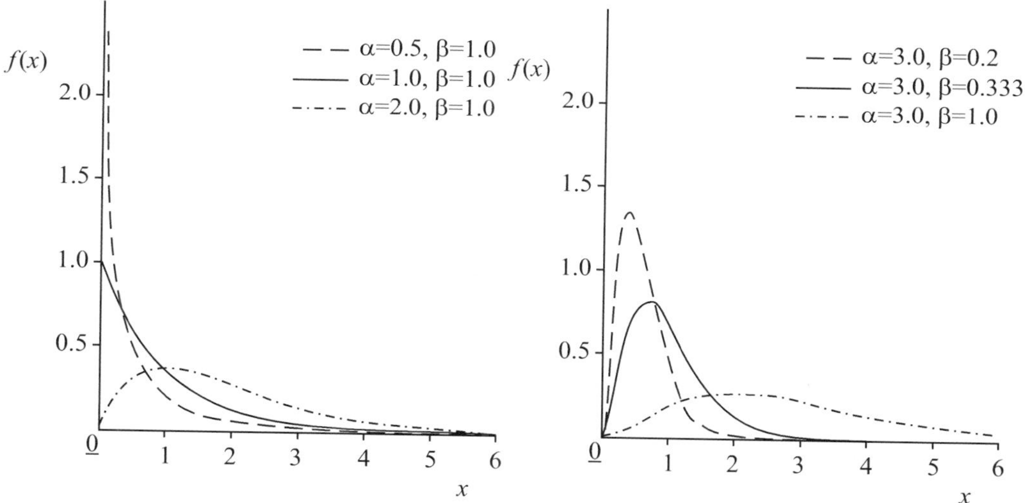

Figure B.10. Density functions of the beta distribution.

Johnson distribution

The Johnson distribution has a bounded version and an unbounded version. For both versions we will provide the density function and some examples to show the flexibility of the Johnson distribution. See (Law and Kelton, 2000) for more information about the mean and variances. Both bounded and unbounded Johnson distributions may have various shapes. Because of its flexibility, it is frequently used to fit datasets.

Bounded Johnson distribution

The usual notation of the bounded Johnson distribution is $\tilde{x} \sim JSB(\alpha_1, \alpha_2, a, b)$, where the parameter a is the location parameter ($-\infty < a < \infty$) and the scale factor is $b - a$ with $b > a$. Shape parameters are α_1 ($-\infty < \alpha_1 < \infty$) and α_2 ($\alpha_2 > 0$).

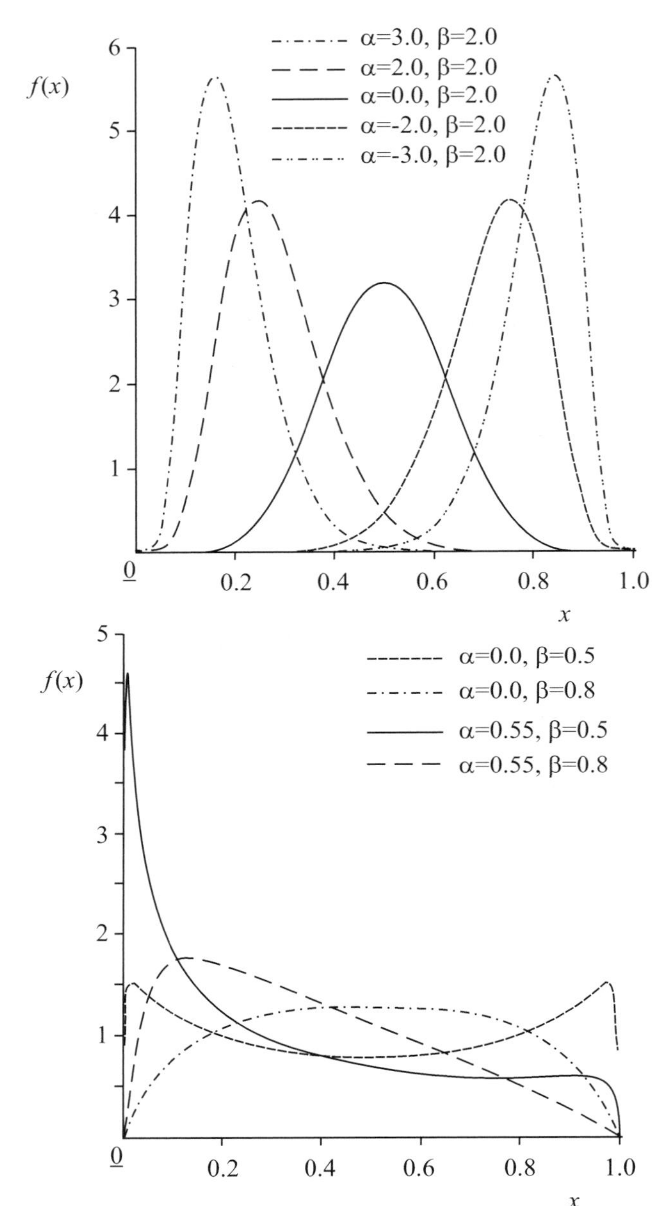

Figure B.11. Graphs of the Johnson bounded distribution (*a = 0 and b = 1*).

The density function of the bounded Johnson distribution is

$$f(x) = \begin{cases} \dfrac{\alpha_2 (b-a)}{(x-a)(b-x)\sqrt{2\pi}} e^{-\frac{1}{2}\left[\alpha_1 + \alpha_2 \ln\left(\frac{x-a}{b-x}\right)\right]^2}, & \text{if } a < x < b \\ 0, & \text{otherwise} \end{cases}$$

Some of the shapes are given in Figure B.11. For both graphs $a = 0$ and $b = 1$.

Johnson unbounded distribution

The usual notation of the Johnson unbounded distribution is $\tilde{x} \sim JSU(\alpha_1, \alpha_2, \gamma, \beta)$, where the location parameter is γ ($-\infty < \gamma < \infty$), the scale parameter is β ($\beta > 0$), and the shape parameters are α_1 ($-\infty < \alpha_1 < \infty$) and α_2 ($\alpha_2 > 0$). The density function of the unbounded Johnson distribution is

$$f(x) = \frac{\alpha_2}{\sqrt{2\pi}\sqrt{(x-\gamma)^2 + \beta^2}} e^{-\frac{1}{2}\left\{\alpha_1 + \alpha_2 \ln\left[\frac{x-\gamma}{\beta} + \sqrt{\left(\frac{x-\gamma}{\beta}\right)^2 + 1}\right]\right\}^2}, -\infty < x < \infty$$

See Figure B.12 for two density functions with $\gamma = 0$ and $\beta = 1.0$.

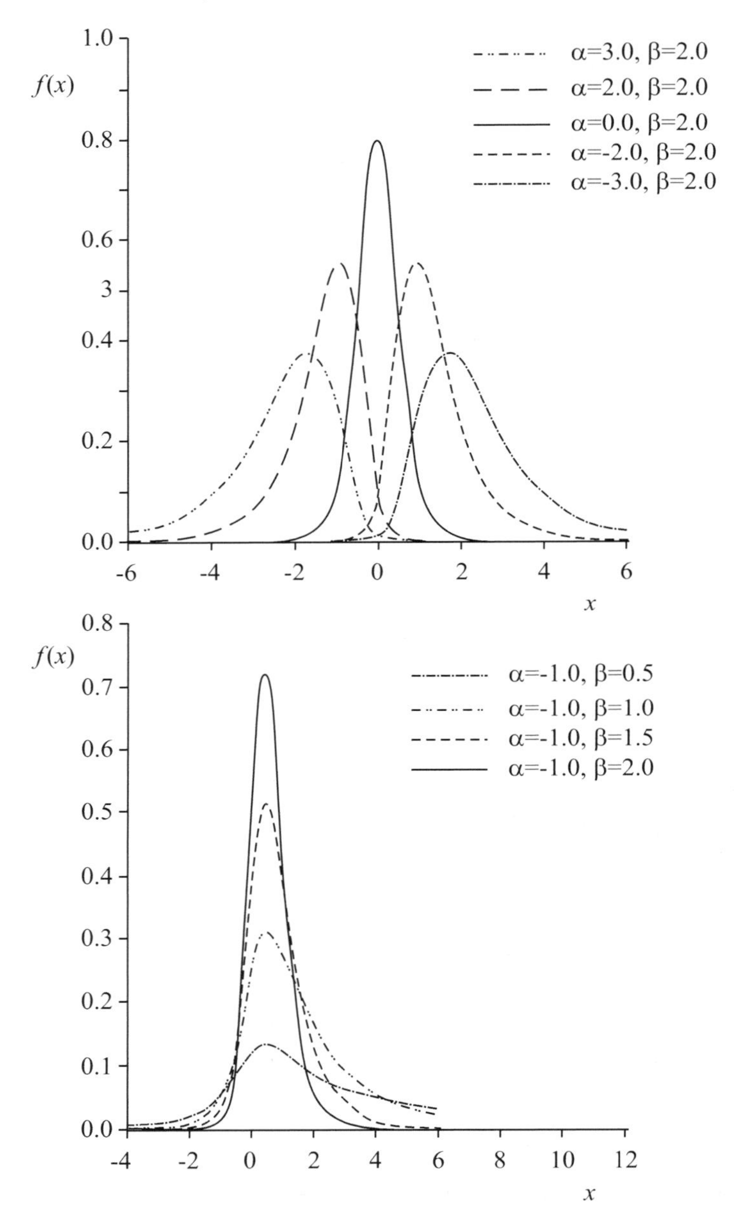

Figure B.12. Graphs of the Johnson unbounded distribution ($\gamma = 0$ and $\beta = 1.0$).

Discrete distribution functions

In this part we will discuss the binomial and Poisson distributions.

Binomial distribution

The usual notation of the binomial distribution is $\tilde{x} \sim bin(t, p)$. The probability mass function $p(x) = P(\tilde{x} = x)$ of the Binomial distribution is

$$p(x) = \binom{t}{x} p^x (1-p)^{t-x} \text{ , for } x = 0, 1, \ldots, t$$

The parameters of the distribution are positive integer t and p ($0 < p < 1$). Several graphs of the binomial distribution are given in Figure B.13.

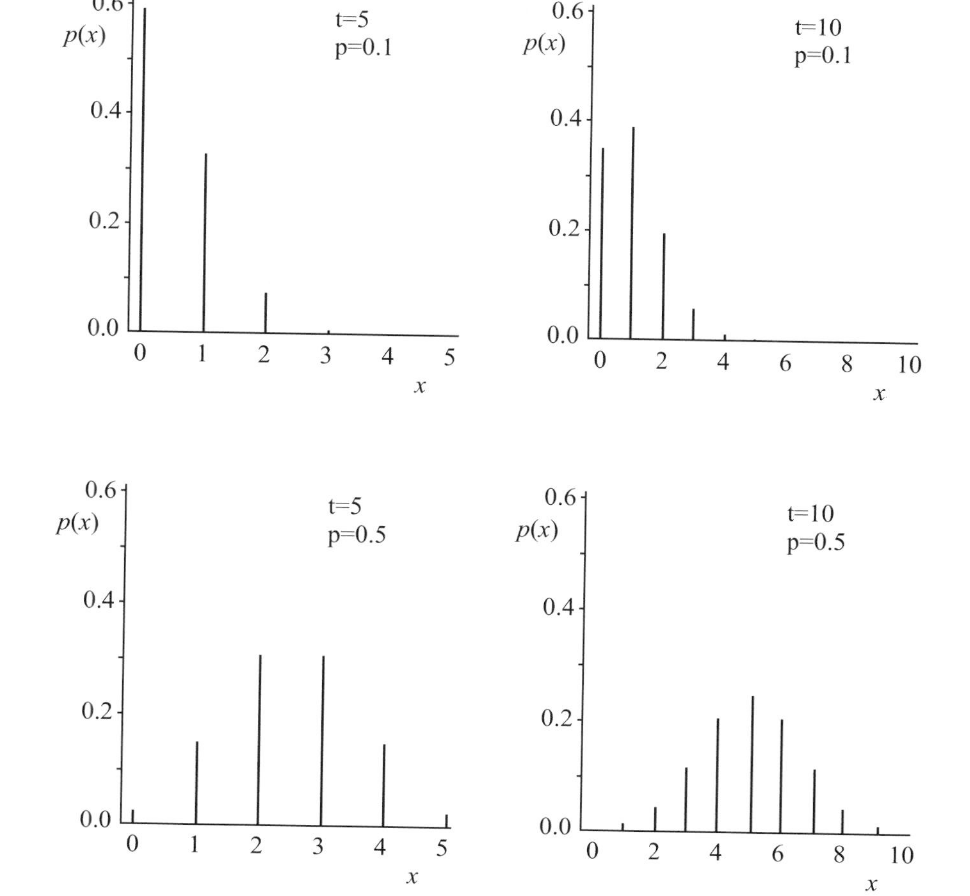

Figure B.13. Graphs of binomial mass functions.

The mean and variance are

$$E(\tilde{x}) = tp$$

$$V(\tilde{x}) = tp(1-p)$$

The binomial distribution function may be applied to model the number of defective items in a batch of size t. Furthermore it may be applied to model the number of items in a batch of random size and number of items demanded from an inventory.

Poisson distribution

A usual notation of the Poisson distribution is $\tilde{x} \sim Poisson(\lambda)$ with $\lambda > 0$. The probability mass function of the Poisson distribution is

$$p(x) = \frac{e^{-\lambda}\lambda^x}{x!} \text{ for } x \in \{0,1,\ldots\}.$$

A graph of the Poisson distribution is given in Figure B.14. The mean and variance of the Poisson distribution are

$$E(\tilde{x}) = \lambda$$

$$V(\tilde{x}) = \lambda$$

The Poisson distribution may be applied to model the number of items in a batch of random size or number of items demanded from an inventory. One may also apply this distribution to generate the number of events, such as number of arrivals within a time period. This distribution is strongly related to the exponential distribution; see the description of the exponential distribution.

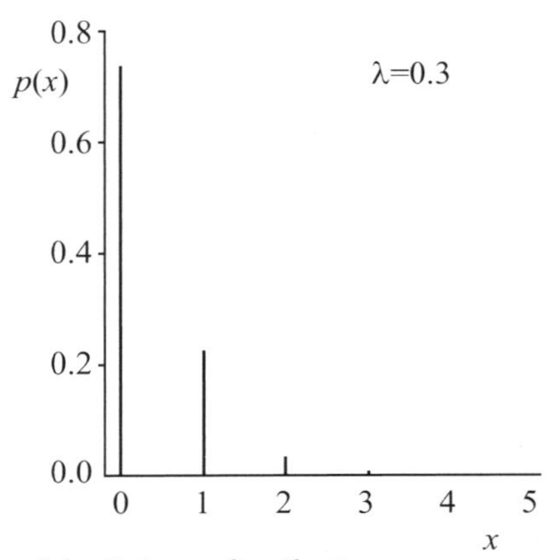

Figure B.14. Mass function of the Poisson distribution.

Appendix C. Output optimisation software

E.M.T. (Eligius) Hendrix

Output for Example 12.2: $$\begin{aligned} &\max\{U(\underline{x}) = x_1 x_2\} \\ &x_1 + 2x_2 \le 6 \\ &x_1 \ge 0, x_2 \ge 0 \end{aligned}$$

Output Excel solver:

Microsoft Excel 8.0e Answer Report
Worksheet: [xlsolver.xls]Sheet1

Target Cell (Max)

Cell	Name	Original Value	Final Value
D9	x1*x2	0	4.5

Adjustable Cells

Cell	Name	Original Value	Final Value
E6	x1	6	3
F6	x2	0	1.5

Constraints

Cell	Name	Cell Value	Formula	Status	Slack
G6	x1+2*x2	6	G6<=6	Binding	0
E6	x1	3	E6>=0	Not Binding	3
F6	x2	1.5	F6>=0	Not Binding	1.5

Microsoft Excel 8.0e Sensitivity Report
Worksheet: [xlsolver.xls]Sheet1

Adjustable Cells

Cell	Name	Final Value	Reduced Gradient
E6	x1	3	0
F6	x2	1.5	0

Constraints

Cell	Name	Final Value	Lagrange Multiplier
G6	x1+2*x2	6	1.5

Output for Example 12.4:

$$\max\{U(\underline{x}) = x_1^2 + x_2^2\}$$
$$x_1 + 2x_2 \le 6$$
$$x_1 \ge 0, x_2 \ge 0$$

Input Gino and output Gino solver:

```
MODEL:
       1) MAX= X1 ^ 2 + X2 ^ 2 ;
       2) X1 + 2 * X2 < 6 ;
       3) X1 > 0 ;
       4) X2 > 0 ;
    END
```

```
 SOLUTION STATUS:  OPTIMAL TO TOLERANCES.  DUAL CONDITIONS:
SATISFIED.

             OBJECTIVE FUNCTION VALUE

        1)        36.000000

    VARIABLE         VALUE          REDUCED COST
          X1        6.000000           .000000
          X2         .000000           .000000

         ROW   SLACK OR SURPLUS          PRICE
          2)         .000000         12.000010
          3)        6.000000           .000000
          4)         .000000        -24.000009Variables
```

Input for Example 12.4 in GAMS:

```
X1
X2
NUT;

POSITIVE VARIABLES X1,X2;

EQUATIONS

BUDGET
NUTD;

BUDGET.. X1+2*X2=L=6;
NUTD.. NUT=E=X1*X1+X2*X2;

MODEL VBNLP /ALL/

SOLVE VBNLP USING NLP MAXIMIZING NUT
```

Output Example 12.4 of GAMS. Somewhere in the output (7 pages) the optimal values of the variables, constraints and shadow prices can be found. Here the solver MINOS finds the global optimum. For another starting value it may find the local optimum. Part of the output:

```
GAMS 2.25.081  386/486 G e n e r a l   A l g e b r a i c   M o d e l i n g
S y s t e m
**** SOLVER STATUS        1 NORMAL COMPLETION
**** MODEL STATUS         2 LOCALLY OPTIMAL
**** OBJECTIVE VALUE                 36.0000

 RESOURCE USAGE, LIMIT            0.220       1000.000
 ITERATION COUNT, LIMIT           0           1000
 EVALUATION ERRORS                0              0

     M I N O S    5.3     (Nov 1990)          Ver: 225-386-02
     = = = = =
     B. A. Murtagh, University of New South Wales
       and
     P. E. Gill,  W. Murray,  M. A. Saunders and M. H. Wright
     Systems Optimization Laboratory, Stanford University.

 EXIT -- OPTIMAL SOLUTION FOUND
 MAJOR ITNS, LIMIT                    1      200
 FUNOBJ, FUNCON CALLS                 4        0
 SUPERBASICS                          0
 INTERPRETER USAGE                 0.00
 NORM RG / NORM PI           0.000E+00

                          LOWER      LEVEL      UPPER     MARGINAL

---- EQU BUDGET           -INF       6.000      6.000      12.000
---- EQU NUTD              .          .          .         -1.000

                          LOWER      LEVEL      UPPER     MARGINAL

---- VAR X1                .         6.000      +INF         .
---- VAR X2                .          .         +INF      -24.000
---- VAR NUT              -INF      36.000      +INF         .
```

References

Aarts, E. H. L. and J.K. Lenstra, 1997. Local Search Algorithms. Wiley, New York.

Ahuja, R.K., T.L. Magnanti and J.B. Orlin, 1993. Network Flows: Theory, Algorithms and Applications. Prentice-Hall, Englewood Cliffs, New Jersey, 846p.

Anthony, R.N., 1965. Planning and control systems: A frame work for analysis. Graduate School of Business Administration, Harvard University.

Banker, R.D., A. Charnes and W.W. Cooper, 1984. Some models for estimating technical and scale inefficiencies in data envelopment analysis. Management Science 30 (9): 1078-1092.

Banks, J., J.S. Carson, B.L. Nelson and D.M. Nicol, 2001. Discrete-Event System Simulation. Prentice-Hall, Upper Saddle River, New Jersey, USA, 594p.

Baritompa, W.P. and E.M.T. Hendrix, 2005. On the investigation of stochastic global optimization algorithms. Journal of Global Optimization 31: 567–578.

Baritompa, W.P., M. Dür, E.M.T. Hendrix, L. Noakes, W. Pullan and G. Wood, 2005. Matching stochastic algorithms to objective function landscapes. Journal of Global Optimization 31: 579–598.

Bazaraa, M., H. Sherali and C. Shetty, 1993. Nonlinear Programming. Wiley, New York.

Beale, E.M.L. and J.A. Tomlin, 1970. Special facilities in a general mathematical programming system for non-convex problems using ordered sets of variables. In: Proceedings of the Fifth International Conference on Operational Research, J. Lawrence (Ed.). Tavistock publications, London, pp 447-554.

Boender, C.G.E. and H.E. Romeijn, 1995. Stochastic methods. In: R. Horst and P.M. Pardalos (Eds.) Handbook of Global Optimization. Kluwer, Dordrecht, pp. 829–871.

Charnes, A. and W.W. Coopers, 1962. Programming with linear fractional functionals. Naval Res. Logistics Quarterly, 9: 181-186

Charnes, A., W.W. Cooper and E. Rhodes, 1978. Measuring the efficiency of decision making units. European Journal of Operations Research, 2: 429-444.

Chopra, S. and P. Meindl, 2007. Supply Chain Management, Strategy, Planning, & Operation, Third Edition. Pearson Education, Upper Saddle River, New Jersey.

Claassen, G.D.H. and Th.H.B. Hendriks, 2007. An application of Special Ordered Sets to a periodic milk collection problem. European Journal of Operational Research 180:754-769.

Dakin, R.J., 1965. A tree-search algorithm for mixed integer programming problems. The Computer Journal, 8: 250-381.

Danilin, Y. and S.A. Piyavskii, 1967. An algorithm for finding the absolute minimum. Theory of Optimal Decisions 2: 25–37. (in Russian).

Dantzig, G.B. and M.N. Thapa, 1997. Linear Programming 1: Introduction. Springer, New York, USA, 435 p.

Dantzig, G.B. and M.N. Thapa, 2003. Linear Programming 2: Theory and Extensions. Springer, New York, USA, 448 p.

Dolan, A. and J. Aldous, 1993. Networks and algorithms, an introductory approach. Wiley, Chichester, West Sussex , 544p.

Eglese, R.W., 1990. Simulated Annealing: A tool for Operational Research. European Journal of Operational Research 46: 271-281.

Erlenkotter, D., 1978. A Dual-based Procedure for Uncapacitated Facility Location. Operations Research, 26: 992-1009.

Ghiani, G., G. Laporte and R. Musmanno, 2004. Introduction to Logistics Systems Planning and Control. Wiley, Chichester, West Sussex.

Gill, P. E., W. Murray and M.H.Wright, 1981. Practical Optimization. Academic Press, New York.

Glover, F. W., 1986. Future paths for integer programming and link to artificial intelligence. Computers and Operations Research 13: 533–554.

Glover, F., J.P. Kelly and M. Laguna, 1999. New advances for wedding optimization and simulation. In: P.A. Farrington, H.B. Nembhard, D.T. Sturrock and G.W. Evans (Eds.) Proceedings of the 1999 Winter Simulation Conference. ACM Press, New York, USA, pp 255-260.

Hax, A.C. and D. Candea, 1984. Production and Inventory Management. Prentice-Hall, Englewood Cliffs, New Jersey.

Hendrix, E.M.T. and J. Roosma, 1996. Global optimization with a limited solution time. Journal of Global Optimization 8: 413–427.

Hendrix, E.M.T. and J.D. Pintér, 1991. An application of lipschitzian global optimization to product design. Journal of Global Optimization 1: 389–401.

Hillen, D.W., 1993. Simulatie in Productie en Logistiek. Academic Service, Schoonhoven, The Netherlands, 252p. (in Dutch)

Hillier, F.S. and G.J. Lieberman, 2005. Introduction to Operations Research. McGraw-Hill, New York, 1061p.

Hoover, S.V. and R.F. Perry, 1989. Simulation; a problem-solving approach. Addison-Wesley, Reading, Mass, USA, 400p.

Keesman, K., 1992. Determination of a minimum-volume orthotopic enclosure of a finite vector set. Technical Report MRS report 92-01, Wageningen Agricultural University.

Kelton, W.D. R.P. Sadowski and D.A. Sadowski, 1998. Simulation with Arena. McGraw-Hill, Boston, MA., USA, 449p.

Khachiyan, L. and M. Todd, 1993. On the complexity of approximating the maximal inscribed ellipsoid for a polytope. Mathematical Programming 61: 137–159.

Kleijnen, J.P.C., 1987. Statistical Tools for Simulation Practitioners. Marcel Dekker, New York, USA, 429p.

Kleijnen, J.P.C., 2001. Ethical issues in modeling: Some reflections. European Journal of Operational Research, 130: 223-230.

Kleijnen, J.P.C. and W. van Groenendaal, 1992. Simulation: a Statistical Perspective. Wiley, Chichester, West Sussex, 241p.

Kuhn, H., 1991. Nonlinear programming: A historical note. In: J.K. Lenstra, A.H.G. Rinnooy Kan and A. Schrijver (Eds.) History of Mathematical Programming. North Holland, Amsterdam, pp. 145–170.

Land, A.H. and A.G. Doig, 1960. An automatic method of solving discrete programming problems. Econometrica, 28: 497-520.

Laporte, G., 1992. The Traveling Salesman Problem: an overview of exact and approximation algorithms. European Journal of Operations Research 59: 231-247.

Law A.M. and W.D. Kelton, 2000. Simulation Modeling and Analysis (3rd Edition). McGraw- Hill, Boston, MA, USA, 760p.

Law, A.M. and S.G. Vincent,1999. ExpertFit User's Guide. Law A.M. & Associates, Inc., Tucson, Arizona, USA.

Lawler, E.L., J.K. Lenstra, A.H.G. Rinnooy Kan and D.B. Schmoys, 1985. The Traveling Salesman Problem, A Guided Tour of Combinatorial Optimization. Wiley, Chichester, West Sussex, 465p.

Markowitz, H., 1959. Portfolio Selection. Wiley, New York.

Naylor, T.H. and J.M. Finger, 1967. Verification of Computer Simulation Models, Management Science 14: 92-101.

Papadimitriou, C.H. and K. Steiglitz, 1982. Combinatorial Optimization, Algorithms and Complexity. Prentice-Hall, Englewood Cliffs, New Jersey, 496p.

Pidd, M., 1999. Just modelling through: A rough guide to modelling. Interfaces 29:118-132.

Pochet, Y. and L.A. Wolsey, 2006. Production Planning by Mixed Integer Programming, Springer, New York, 500 p.

Rayward-Smith, V.J., I.H. Osman, C.R. Reeves and G.D. Smith, 1996. Modern heuristic search methods. Wiley, Chichester, West Sussex, 294p.

Roos, C., T. Terlaky and J.P. Vial, 2006. Interior Point Methods for Linear Optimization. Springer, New York.

Romeijn, H. E., 1992. Global Optimization by Random Walk Sampling Methods. PhD thesis, Erasmus University Rotterdam, Rotterdam.

Scales, L., 1985. Introduction to Non-Linear Opimization. Macmillan, London.

Schrage, L., 1975. Implicit representation of variable upper bounds in linear programming. Mathematical Programming Study, 4: 118-132.

Shannon, R.E., 1975. Systems Simulation: The Art and Science. Prentice-Hall, Englewood Cliffs, New Jersey, 368p.

Shubert, B., 1972. A sequential method seeking the global maximum of a function. SIAM Journal of Numerical Analysis 9: 379–388.

Silver, E.A., D.F. Pyke and R. Peterson, 1998. Inventory Management and Production Planning and Scheduling (3rd Edition). Wiley, New York, 754p.

Slack, N., S. Chambers, R. Johnston and A. Betts, 2006. Operations and Process Management, Principles and practice for strategic impact. Pearson Education, Harlow, Essex.

Taguchi, G., E. Elsayed and T. Hsiang, 1989. Quality Engineering in Production Systems. McGraw-Hill, Boston, MA, USA.

Tijms, H.C., 2002. Operationele Analyse, een inleiding in modellen en methoden. Epsilon Uitgaven, Utrecht. (in Dutch)

Törn, A. and A. Žilinskas, 1989. Global Optimization. Vol. 350 of Lecture Notes in Computer Science, Springer, Berlin.

Törn, A., M.M. Ali and S. Viitanen, 1999. Stochastic global optimization: Problem classes and solution techniques. Journal of Global Optimization 14: 437–447.

Turing, A.M., 1950. Computing machinery and intelligence. Mind 59: 433-460.

Vorst, J.G.A.J. van der, 2000. Effective food supply chains; generating, modelling and evaluating supply chain scenarios. PhD thesis, Wageningen University, 305p.

Williams, H.P., 1990. Model Building in Mathematical Programming. Wiley, Chichester West Sussex, UK, 355 p.

Williams, H.P., 1993. Model Solving in Mathematical Programming. Wiley, Chichester West Sussex, UK, 360 p.

Wilson, B.W., 1990. Systems: Concepts, Methodologies and Applications (2nd Edition). Wiley, Chichester, West Sussex, 410p.

Winston, W.L., 2004., Operations Research: Applications and Algorithms (4th Edition). Thomson, Toronto, 1418p.

Wolsey, L.A., 1998. Integer Programming. Wiley, New York, 264p.

Zee, D.J. van der and J.G.A.J. van der Vorst, 2005. A Modelling Framework for Analyzing Supply Chain Scenarios: Applications in Food Industry. Decision Sciences, 36:65-95.

Zeigler, B.P., H. Praehofer and T.G. Kim, 2000. Theory of Modeling and Simulation (2nd Edition). Academic Press, San Diego, CA, USA, 510p.

Keyword index